Handbook of Optoelectronics
Second Edition

Series in Optics and Optoelectronics

Series Editors:
E. Roy Pike, Kings College, London, UK
Robert G. W. Brown, University of California, Irvine, USA

RECENT TITLES IN THE SERIES

Handbook of Optoelectronics, Second Edition: Concepts, Devices, and Techniques – Volume One
John P. Dakin and Robert G. W. Brown (Eds.)

Handbook of Optoelectronics, Second Edition: Enabling Technologies – Volume Two
John P. Dakin and Robert G. W. Brown (Eds.)

Handbook of Optoelectronics, Second Edition: Applied Optical Electronics – Volume Three
John P. Dakin and Robert G. W. Brown (Eds.)

Handbook of GaN Semiconductor Materials and Devices
Wengang (Wayne) Bi, Hao-chung (Henry) Kuo, Pei-Cheng Ku, and Bo Shen (Eds.)

Handbook of Optoelectronic Device Modeling and Simulation: Fundamentals, Materials, Nanostructures, LEDs, and Amplifiers – Volume One
Joachim Piprek (Ed.)

Handbook of Optoelectronic Device Modeling and Simulation: Lasers, Modulators, Photodetectors, Solar Cells, and Numerical Methods – Volume Two
Joachim Piprek (Ed.)

Nanophotonics and Plasmonics: An Integrated View
Dr. Ching Eng (Jason) Png and Dr. Yuriy Akimov

Handbook of Solid-State Lighting and LEDs
Zhe Chuan Feng (Ed.)

Optical Microring Resonators: Theory, Techniques, and Applications
V. Van

Optical Compressive Imaging
Adrian Stern

Singular Optics
Gregory J. Gbur

The Limits of Resolution
Geoffrey de Villiers and E. Roy Pike

Polarized Light and the Mueller Matrix Approach
José J Gil and Razvigor Ossikovski

Handbook of Optoelectronics

Second Edition

Applications of Optoelectronics

Volume 3

Edited by

John P. Dakin
Robert G. W. Brown

CRC Press
Taylor & Francis Group
Boca Raton London New York

CRC Press is an imprint of the
Taylor & Francis Group, an **informa** business

CRC Press
Taylor & Francis Group
6000 Broken Sound Parkway NW, Suite 300
Boca Raton, FL 33487-2742

Printed on acid-free paper

International Standard Book Number-13: 978-1-1381-0226-2 (Hardback)

Library of Congress Cataloging-in-Publication Data

Names: Dakin, John, 1947- editor. | Brown, Robert G. W., editor.
Title: Handbook of optoelectronics / edited by John P. Dakin, Robert G. W. Brown.
Description: Second edition. | Boca Raton : Taylor & Francis, CRC Press, 2017. | Series: Series in optics and optoelectronics ; volumes 30-32 | Includes bibliographical references and index. Contents: volume 1. Concepts, devices, and techniques -- volume 2. Enabling technologies -- volume 3. Applied optical electronics.
Identifiers: LCCN 2017014570 | ISBN 9781138102262 (hardback : alk. paper)
Subjects: LCSH: Optoelectronic devices--Handbooks, manuals, etc.
Classification: LCC TK8320 .H36 2017 | DDC 621.381/045--dc23
LC record available at https://lccn.loc.gov/2017014570

Visit the Taylor & Francis Web site at
http://www.taylorandfrancis.com

and the CRC Press Web site at
http://www.crcpress.com

Contents

Series Preface

This international series covers all aspects of theoretical and applied optics and optoelectronics. Active since 1986, eminent authors have long been choosing to publish with this series, and it is now established as a premier forum for high-impact monographs and textbooks. The editors are proud of the breadth and depth showcased by published works, with levels ranging from advanced undergraduate and graduate student texts to professional references. Topics addressed are both cutting edge and fundamental, basic science and applications-oriented, on subject matter that includes: lasers, photonic devices, nonlinear optics, interferometry, waves, crystals, optical materials, biomedical optics, optical tweezers, optical metrology, solid-state lighting, nanophotonics, and silicon photonics. Readers of the series are students, scientists, and engineers working in optics, optoelectronics, and related fields in the industry.

Proposals for new volumes in the series may be directed to Lu Han, executive editor at CRC Press, Taylor & Francis Group (lu.han@taylorandfrancis.com).

Preface

This third volume of the *Handbook* is a brand new addition. Its focus on applications is intended to complement the preceding two volumes, which have extensively covered the basic science, key components, and vital enabling technology. The new chapters here have been written by authors presenting their own selected overviews of real-world engineering applications.

The intention of this volume is to concentrate on a number of areas where optoelectronics is either already making, or has great future prospects of making, a major difference to our lives. The objective is not to describe the technology in great physical detail, but rather to give a set of case studies. These cases focus on how the technology can be used, so the scientific and engineering descriptions are much shorter than in earlier chapters, and more examples and photographs are given. We have tried to ensure that, where feasible, descriptions are not related to an obvious commercial product, but, where appropriate, performance data of real systems is occasionally included.

The structure of this volume involves splitting it into major application fields, rather than technology areas. Naturally, this means that inevitably some enabling technologies are applicable to more than one application area. Where this is clearly the case, we have attempted to cross-reference to other relevant chapters, either via a summary table or with a short paragraph or two of how it is used in the other application area.

Due to the huge, and very rapidly growing, number of applications, it is impossible to cover all aspects and applications, even in what is a fairly extensive selection. To try to indicate some of the areas that we may have missed, most of the sections contain a summary table in the introduction, to give a broader picture of the field. Some optoelectronic technologies, such as cameras, light-emitting diodes, and liquid crystal displays, have had extensive applications for many years, whereas others such as solar panels were previously only economically practical for mobile or remote powering, but are now becoming used far more widely as costs reduce.

The short-form treatment of other applications in these tables will inevitably leave some questions unanswered, but we trust that we have presented at least a broad cross-section of case studies, which hopefully succeed in illustrating that optoelectronics is a major force for change in our world.

John P. Dakin

Introduction to the Second Edition

There have been many detailed technological changes since the first edition of the *Handbook* in 2006, with the most dramatic changes seen from the far more widespread applications of the technology. To reflect this, our new revision has a completely new Volume 3 focused on applications and covering many case studies from an ever increasing range of possible topics. Even as recently as 2006, the high cost or poorer performance of many optoelectronics components was still holding back many developments, but now the cost of many high-spec components, particularly ones such as light-emitting diodes (LEDs), lasers, solar cells, and other optical detectors, optoelectronic displays, optical fibers and components, including optical amplifiers, has reduced to such an extent that they are now finding a place in all aspects of our lives. Solid-state optoelectronics now dominates lighting technology and is starting to dominate many other key areas such as power generation. It is revolutionizing our transport by helping to guide fully autonomous vehicles, and CCTV cameras and optoelectronic displays are seen everywhere we go.

In addition to the widespread applications now routinely using optoelectronic components, since 2006, we have witnessed growth of various fundamentally new directions of optoelectronics research and likely new component technologies for the near future. One of the most significant new areas of activity has been in nano-optoelectronics; the use of nanotechnology science, procedures and processes to create ultra-miniature devices across the entire optoelectronics domain: laser and LED sources, optical modulators, photon detectors, and solar cell technology. Two new chapters on silicon photonics and nanophotonics and graphene optoelectronics attempt to cover the wide range of nanotechnology developments in optoelectronics this past decade. It will, however, be a few years before the scale-up to volume manufacturing of nano-based devices becomes an economically feasible reality, but there is much promise for new generations of optoelectronic technologies to come soon.

Original chapters of the first edition have been revised and brought up to date for the second edition, mostly by the original authors, but in some cases by new authors, to whom we are especially grateful.

Robert G. W. Brown and John P. Dakin

Editors

John P. Dakin, PhD, is professor (Emeritus) at the Optoelectronics Research Centre, University of Southampton, UK. He earned a BSc and a PhD at the University of Southampton and remained there as a Research Fellow until 1973, where he supervised research and development of optical fiber sensors and other optical measurement instruments. He then spent 2 years in Germany at AEG Telefunken; 12 years at Plessey, research in Havant and then Romsey, UK; and 2 years with York Limited/York Biodynamics in Chandler's Ford, UK before returning to the University of Southampton.

He has authored more than 150 technical and scientific papers, and more than 120 patent applications. He was previously a visiting professor at the University of Strathclyde, Glasgow.

Dr. Dakin has won a number of awards, including "Inventor of the Year" for Plessey Electronic Systems Limited and the Electronics Divisional Board Premium of the Institute of Electrical and Electronics Engineers, UK. Earlier, he won open scholarships to both Southampton and Manchester Universities.

He has also been responsible for a number of key electro-optic developments. These include the sphere lens optical fiber connector, the first wavelength division multiplexing optical shaft encoder, the Raman optical fiber distributed temperature sensor, the first realization of a fiber optic passive hydrophone array sensor, and the Sagnac location method described here, plus a number of novel optical gas sensing methods. More recently, he was responsible for developing a new distributed acoustic and seismic optical fiber sensing system, which is finding major applications in oil and gas exploration, transport and security systems.

Robert G. W. Brown, PhD, is at the Beckman Laser Institute and Medical Clinic at the University of California, Irvine. He earned a PhD in engineering at the University of Surrey, Surrey, and a BS in physics at Royal Holloway College at the University of London, London. He was previously an applied physicist at Rockwell Collins, Cedar Rapids, IA, where he carried out research in photonic ultrafast computing, optical detectors, and optical materials. Previously, he was an advisor to the UK government, and international and editorial director of the Institute of Physics. He is an elected member of the European Academy of the Sciences and Arts (Academia Europaea) and special professor at the University of Nottingham, Nottingham. He also retains a position as adjunct full professor at the University of California, Irvine, in the Beckman Laser Institute and Medical Clinic, Irvine, California, and as visiting professor in the department of computer science. He has authored more than 120 articles in peer-reviewed journals and holds 34 patents, several of which have been successfully commercialized.

Dr. Brown has been recognized for his entrepreneurship with the UK Ministry of Defence Prize for Outstanding Technology Transfer, a prize from Sharp Corporation (Japan) for his novel laser-diode invention, and, together with his team at the UK Institute of Physics, a Queen's Award for Enterprise, the highest honor bestowed on a UK company. He has guest edited several special issues of *Applied Physics* and was consultant to many companies and government research centers in the United States and the United Kingdom. He is a series editor of the CRC Press "Series in Optics and Optoelectronics."

Contributors

John Arkwright
School of Computer Science, Engineering and Mathematics
Flanders University
Adelaide, South Australia, Australia

Alex Beaton
Ocean Technology and Engineering Group
National Oceanography Centre
Southampton, United Kingdom

Greg Blackman
Imaging and Machine Vision Europe
Europa Science
Cambridge, United Kingdom

Barbara J. Brooks
University of Leeds
Leeds, United Kingdom

Valborg Byfield
National Oceanography Centre
Southampton, United Kingdom

Debbie Clifford
University of Reading
Reading, United Kingdom

John P. Dakin
Optoelectronics Research Centre
University of Southampton
Southampton, United Kingdom

Fernando Araujo de Castro
Materials Division
National Physical Laboratory
Middlesex, United Kingdom

Phil Dinning
Human Physiology and Centre for Neuroscience
Flinders University
Bedford Park, Australia

and

Departments of Gastroenterology and Surgery
Flinders Medical Centre
South Australia, Australia

Gareth J. Edwards
Observant Technology Limited
Hampshire, United Kingdom

Paula R. Fortes
Institute of Analytical and Bioanalytical Chemistry
University of Ulm
Ulm, Germany

André Franzen
Shell International Exploration and Production B.V.
The Hague, the Netherlands

Torsten Frosch
Leibniz Institute of Photonic Technology
and
Institute of Physical Chemistry
Abbe Center of Photonics
Jena, Germany

Go Fujisawa
Schlumberger
Cambridge, United Kingdom

R. Eugene Goodson
Duke University
Durham, North Carolina

Susie E. Goodson
Duke University
Durham, North Carolina

Paul Harrison
SPI Lasers Ltd.
Southampton, United Kingdom

David Hemsley
Oxsensis Ltd.
Oxfordshire, United Kingdom

Jane Hodgkinson
Centre for Engineering Photonics
Cranfield University
Cranfield, United Kingdom

Daniele Inaudi
Roctest Ltd. – SMARTEC SA
Manno, Switzerland

Andreas Knebl
Leibniz Institute of Photonic Technology
and
International Max-Planck Research School
Global Biogeochemical Cycles
Jena, Germany

Vjekoslav Kokoric
Institute of Analytical and Bioanalytical Chemistry
University of Ulm
Ulm, Germany

Gerardo López-Saldaña
Department of Meteorology
University of Reading
Reading, United Kingdom

Candice Majewski
Department of Mechanical Engineering
University of Sheffield
Sheffield, United Kingdom

Michael A. Marcus
Lumetrics Inc.
Rochester, New York

Boris Mizaikoff
Institute of Analytical and Bioanalytical Chemistry
University of Ulm
Ulm, Germany

Matt Mowlem
Ocean Technology and Engineering Group
National Oceanography Centre
Southampton, United Kingdom

Christian M. Müller
Institute of Analytical and Bioanalytical Chemistry
University of Ulm
Ulm, Germany

Oliver C. Mullins
Schlumberger
Cambridge, United Kingdom

Philip Nash
Stingray Geophysical
London, United Kingdom

Dominic O'Brien
Department of Engineering Science
University of Oxford
Oxford, United Kingdom

David Parkyns
Transport for London
London, United Kingdom

Ralf D. Pechstedt
Oxsensis Ltd.
Oxfordshire, United Kingdom

Bobby Pejcic
CSIRO, Energy Flagship
Western Australia, Australia

Constantinos Pitris
Department of Electrical and Computer Engineering
University of Cyprus
Nicosia, Cyprus

Jürgen Popp
Leibniz Institute of Photonic Technology
and
Institute of Physical Chemistry
Abbe Center of Photonics
Jena, Germany

Ivo M. Raimundo
Institute of Analytical and Bioanalytical Chemistry
University of Ulm
Ulm, Germany

Florian Rauh
Institute of Analytical and Bioanalytical Chemistry
University of Ulm
Ulm, Germany

Jessica Rowbury
Imaging and Machine Vision Europe
Europa Science
Cambridge, United Kingdom

Stuart Russell
Optasense, QinetiQ Group plc
Bristol, United Kingdom

Thomas Schädle
Institute of Analytical and Bioanalytical Chemistry
University of Ulm
Ulm, Germany

Matthias Schwenk
Institute of Analytical and Bioanalytical Chemistry
University of Ulm
Ulm, Germany

Felicia Seichter
Institute of Analytical and Bioanalytical Chemistry
University of Ulm
Ulm, Germany

Hillary G. Sillitto
University of Strathclyde
Glasgow, United Kingdom

Gregory Slavik
Ocean Technology and Engineering Group
National Oceanography Centre
Southampton, United Kingdom

Robert Stach
Institute of Analytical and Bioanalytical Chemistry
University of Ulm
Ulm, Germany

Erhan Tütüncü
Institute of Analytical and Bioanalytical Chemistry
University of Ulm
Ulm, Germany

Tuan Vo-Dinh
Department of Biomedical Engineering
Duke University
Durham, North Carolina

Benjamin Watson
3M UK&I
Hampshire, United Kingdom

Andreas Wilk
Institute of Analytical and Bioanalytical Chemistry
University of Ulm
Ulm, Germany

Tsutomu Yamate
Schlumberger
Cambridge, United Kingdom

PART I

Optoelectronics in infrastructure

In this section, we present case studies of the use of optoelectronics in infrastructure. This includes applications on fixed structures, in particular, sensors located next to vital road and rail systems, civil engineering structures, such as bridges and dams, and ones built in or on the structure of buildings or their foundations. (Please note, systems and sensors intended specifically for security and surveillance purposes, for energy, for oil and gas extraction, and ones fitted to moving vehicles will be described in later application sections.)

In order to give a broader introduction to infrastructure applications than we could possibly cover with our selected case studies, we shall first present a summary table. This will give a broader overview than possible in the more specific case study chapters.

As will be the practice in most chapters in this volume, we shall then present a few more detailed case studies. Before commencing, however, it is perhaps appropriate to briefly emphasize where sensors described in this section also have crossover applications to other sections in the volume.

Chapter 1 describes various forms of optical strain sensors for highways, but similar strain sensors, particularly fiber grating types, have numerous other potential application areas, such as structural sensors in wind energy turbines, aircraft and ships, racing yacht masts, and many more.

Chapter 2 describes a distributed optical fiber acoustic/seismic sensor, which has, apart from for infrastructure, obvious applications in security and surveillance. Such sensors are currently employed extensively in the oil and gas industry, and this important application is described more fully by Andre Franzen in Chapter 20.

Chapters 3 and 4 describe camera monitors for roads. These are used for monitoring and identifying vehicles (via number plate readers or still photographs) and determining their speed and position. If the book had been structured in a different way, such sensors would have been considered to be a part of the later sections on "transport" or "security and surveillance," but as they are *fixed* in location, we have chosen to describe them here.

Hopefully, the readers will appreciate that there are inevitable dilemmas of where best to describe these applications and will bear with us in our order of presenting them in the sections that follow.

Table I.1 Summary of applications of optoelectronics in the infrastructure field

Application	Technology	Advantages	Disadvantages	Current situation (at time of writing)	More reading
Sensors for traffic monitoring and control. Vehicle identification. Detection of speeding and other traffic violations.	Visible CCTV cameras IR cameras Time-interval cameras Sophisticated video processing techniques such as vehicle number plate recognition.	Established optoelectronics technology and can use relatively cheap cameras. Camera information can easily be recorded and networked. Individual vehicles can be identified via number plates and acquisition of vehicle flow statistics for high traffic volumes is also possible.	Limited to line-of-sight applications.	Already becoming widely used. Sophistication of networking for traffic monitoring systems is rapidly improving. Already becoming fully integrated with traffic control systems.	See this section. (Part 1)
Inbuilt sensors for road, rail, or airport traffic monitoring, which are buried in highway, or fastened to, or located close to, railway lines, airport runways, etc.	Optical fiber distributed acoustic/seismic sensors (DAS).	Long distance (~30 km) coverage with one sensor. Sensor can be configured along any desired path. Multiple (up to many hundreds) sensing points with one sensing system.	Requires significant in-road or railside works to install. Potential cross talk from seismic signals transmitted through the ground from other areas nearby. Expensive sensors as new sophisticated technology.	A fairly new, but rapidly expanding technology for highway and railway applications, but one that has been used with great success in oil and gas wells.	See also Part VII and Volume II, Chapter 11 (Optical Fiber Sensors).
Sensors for monitoring structural integrity of roads, bridges, dams, buildings, etc.	Optical fiber strain gauges, having short or long gauge length. Sensors are usually embedded in construction materials.	Optical fibers are robust and immune to corrosive liquids such as alkaline cements and road-salt solutions. No conductors, so no problems with electrolytic conduction, galvanic corrosion, and/or lightning strike.	Installers need to be trained in the new technology.	Involves less-established technology than electrical strain gauges, but is rapidly gaining acceptance in many areas.	See this section. (Part 1)

(Continued)

Table I.1 (*Continued*) Summary of applications of optoelectronics in the infrastructure field

Application	Technology	Advantages	Disadvantages	Current situation (at time of writing)	More reading
Very large optical displays for traffic control and providing information to road and rail users (e.g., traffic lights, and overhead gantry or roadside displays).	Originally, these displays used incandescent sources, but most are now replaced by arrays of high-brightness LEDs.	LEDs have had excellent reliability for many years. Most low-cost LEDs packages conveniently emit light in a forward cone, of the type needed to maximize visibility to approaching drivers. Current LEDs are far brighter and more efficient than earlier types and unit costs are falling rapidly.	No real disadvantages.	Unless a revolutionary new technology arrives, these LED displays are here to stay! There is, however, potential competition from large-area organic displays.	See also Volume I, Chapter 10 (LEDs) and Volume II, Chapter 9 (3D Display Systems).
Roadside solar power, for remote powering of LED road signs, traffic lights, speed sensors, emergency telephones, etc.	Photovoltaic optical to electrical energy conversion (often supplemented by small wind generator to generate at night and on cloudy days!).	A very economical way of providing power in locations away from mains electricity supplies.	Needs a storage battery and charge controller to cover periods without sun or wind! Batteries must be replaced regularly, and they could potentially discharge too far in unsuitable weather.	A common feature on cross-country roads and highways. Perhaps the most elegant and artistic ones are those used on French highways, for powering emergency telephones!	See also Volume II, Chapter 16 (Optical to Electrical Energy Conversion: Solar Cells).
Warning strobe lights for high towers, and scanned lights for lighthouses, etc. Dual-pulse flash lamps for roadside speed cameras.	Used to employ arc lamps and gas-filled flash lamps, but most are now LED based.	LED lamps have far greater reliability, easier drive circuitry, and can provide different colors without the need for filters.	Still not as bright as the best arc lamps, but can use arrays of them to help to compensate for this.	As with many areas of lighting, LEDs are gradually taking over.	See also Volume I, Chapter 10 (LEDs).

(*Continued*)

Table I.1 (*Continued*) Summary of applications of optoelectronics in the infrastructure field

Application	Technology	Advantages	Disadvantages	Current situation (at time of writing)	More reading
Road lighting system.	These used to use high- or low-pressure arc lamps (mostly low-pressure sodium lamps), but now changing to white LED-based systems.	Reliability, as very high lifetimes. Present LEDs are far brighter and more efficient than earlier types of lighting. Initial costs are falling rapidly, electricity costs are reduced and maintenance costs are very low.	Somewhat higher initial cost at present, but this is falling very rapidly.	Again, unless a new revolutionary technology arrives, they are here to stay! Expect nearly all lighting systems to soon use LEDs or semiconductor lasers.	See also Volume I, Chapter 10 (LEDs).
Roadside optical pollution sensors.	Optical systems for determination of pollution from vehicles, using spectroscopic analysis.	Can monitor road traffic for identifying vehicles that are emitting noxious fumes or are using illegal fuels.	Expensive systems at present but costs would reduce with greater usage.	A system developed by NASA has been used to monitor automobile emissions in numerous US states.	See Chapter 19 (Raman Gas Spectroscopy) and Part VII chapters on gas sensing for discussion of general concepts.

1

Overview of fiber optic sensing technologies for structural health monitoring

DANIELE INAUDI
SMARTEC/Roctest

1.1 FIBER OPTIC SENSORS

There exist a great variety of fiber optic sensors (FOSs) [1] for structural monitoring in both the academic and industrial areas. In this overview we will concentrate on fiber optic sensing systems for civil health monitoring that have reached an industrial level and have been used in a number of field applications.

Figure 1.1 illustrates the four main types of FOSs:

- Point sensors have a single measurement point at the end of the fiber optic connection cable, similar to most electrical sensors.
- Multiplexed sensors allow the measurement at multiple points along a single fiber line.
- Long-base sensors integrate the measurement over a long measurement base. They are also known as long-gauge sensors.
- Distributed sensors are able to sense at any point along a single fiber line, typically every meter over many kilometers of length

The greatest advantages of FOS are intrinsically linked to the optical fiber itself that is either used as a link between the sensor and the signal conditioner, or becomes the sensor itself in the case of long-gauge and distributed sensors. In almost all FOS applications, the optical fiber is a thin glass fiber that is protected mechanically by a polymer coating (or a metal coating in extreme cases) and further protected by a multilayer cable structure designed to protect the fiber from the environment where it will be installed. Since glass is an inert material very resistant to almost all chemicals, even at extreme temperatures, it is an ideal material for use in harsh environments such as that encountered

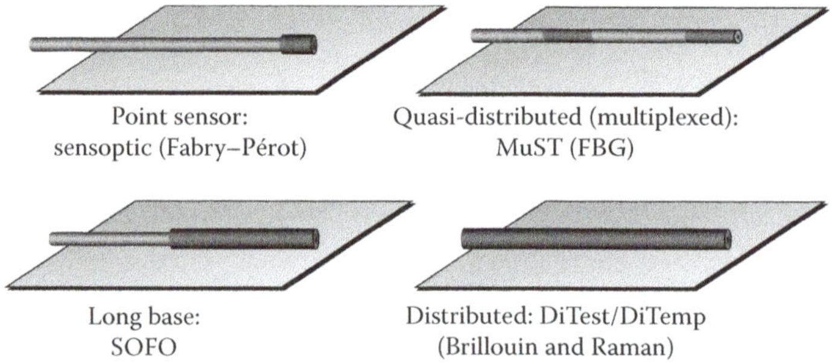

Figure 1.1 Fiber optic sensor types.

in geotechnical applications. Chemical resistance is a great advantage for long-term reliable health monitoring of civil engineering structures, making FOSs particularly durable. Since the light confined to the core of the optical fibers used for sensing purposes does not interact with any surrounding electromagnetic (EM) field, FOSs are intrinsically immune to any EM interferences. With such unique advantage over sensors using electrical signals, FOSs are obviously the ideal sensing solution when the presence of EM, radio frequency, or microwaves cannot be avoided. For instance, FOS will not be affected by EM fields generated by lightning hitting a monitored bridge or dam, nor will they be affected by the interference produced by subway trains running near a monitored zone. FOSs are intrinsically safe and naturally explosion-proof, making them particularly suitable for monitoring applications of risky structures such as gas pipelines or chemical plants. But the greatest and most exclusive advantage of such sensors is their ability to offer long-range distributed sensing capabilities.

Figure 1.2 SOFO system reading unit.

1.1.1 SOFO displacement sensors

The SOFO system (Figure 1.2) is a fiber optic displacement sensor with a resolution in the micrometer range and an excellent long-term stability. It was developed at the Swiss Federal Institute of Technology in Lausanne (EPFL) and is now commercialized by SMARTEC in Switzerland [2].

The measurement setup uses low-coherence interferometry to measure the length difference between two optical fibers installed on the structure to be monitored (Figure 1.3). The measurement fiber is pretensioned and mechanically coupled to the structure at two anchorage points in order to follow its deformations, while the reference fiber is

Figure 1.3 SOFO sensor installed on a rebar.

free and acts as temperature reference. Both fibers are installed inside the same pipe and the measurement basis can be chosen between 200 mm and 10 m. The resolution of the system is of 2 μm independently from the measurement basis and its precision is of 0.2% of the measured deformation even over years of operation.

The SOFO system has been successfully used to monitor more than 150 structures, including bridges, tunnels, piles, anchored walls, dams, historical monuments, nuclear power plants, as well as laboratory models.

1.1.2 Bragg grating strain sensors

Bragg gratings are periodic alterations in the index of refraction of the fiber core that can be produced by adequately exposing the fiber to intense ultraviolet (UV) light. The produced gratings typically have length of the order of 10 mm. If white light is injected in the fiber containing the grating, the wavelength corresponding to the grating pitch will be reflected, while all other wavelengths will pass through the grating undisturbed. Since the grating period is strain and temperature dependent, it becomes possible to measure these two parameters by analyzing the spectrum of the reflected light [3]. This is typically done by using a tunable filter (such as a Fabry–Perot cavity) or a spectrometer. Resolutions of the order of 1 με and 0.1°C can be achieved with the best demodulators. If strain and temperature variations are expected simultaneously, it is necessary to use a free reference grating that measures the temperature alone and uses its reading to correct the strain values. Setups allowing the simultaneous measurement of strain and temperature have been proposed but have yet to prove their reliability in field conditions. The main interest in using Bragg gratings resides in their multiplexing potential. Many gratings can be written in the same fiber at different locations and tuned to reflect at different wavelengths. This allows the measurement of strain at different places along a fiber using a single cable. Typically, 4–16 gratings can be measured on a single fiber line. It has to be noticed that since the gratings have to share the spectrum of the source used to illuminate them, there is a trade-off between the number of gratings and the dynamic range of the measurements on each of them.

Because of their length, fiber Bragg gratings can be used as a replacement for conventional strain gauges and installed by gluing them on metals and other smooth surfaces. With adequate packaging, they can also be used to measure strains in concrete over a basis length of typically 100 mm.

1.1.3 Fabry–Perot strain sensors

An extrinsic Fabry–Perot interferometer (EFPI) consists of a capillary silica tube containing two cleaved optical fibers facing each other, but leaving an air gap of a few microns or tens of microns between them (see Figure 1.4) [4]. When light is launched into one of the fibers, a back-reflected interference signal is obtained. This is due to the reflection of the incoming light on the glass-to-air and on the air-to-glass interfaces. This interference can be demodulated using coherent or low-coherence techniques to reconstruct the changes in the fiber spacing. Since the two fibers are attached to the capillary tube near its two extremities (with a typical spacing of 10 mm), the gap change will correspond to the average strain variation between the two attachment points.

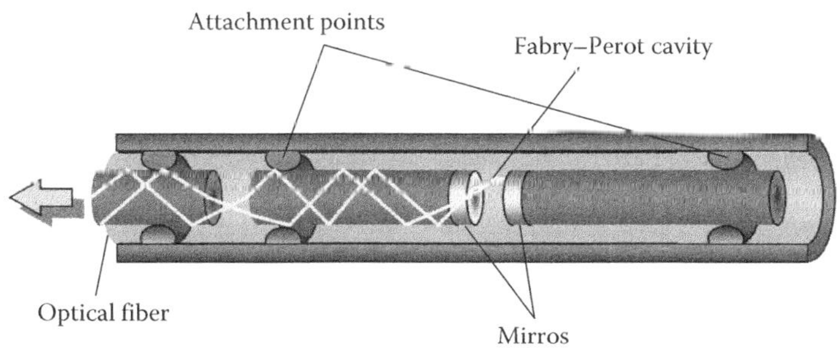

Figure 1.4 Fabry–Perot sensor.

1.1.4 Raman distributed temperature sensors

Raman scattering is the result of a nonlinear interaction between the light traveling in a fiber and silica. When an intense light signal is shined into the fiber, two frequency-shifted components called Raman Stokes and Raman anti-Stokes, respectively, will appear in the back-scattered spectrum. The relative intensity of these two components depends on the local temperature of the fiber. If the light signal is pulsed and the back-scattered intensity is recorded as a function of the round-trip time, it becomes possible to obtain a temperature profile along the fiber [5]. Typically, a temperature resolution of the order of 0.1°C and a spatial resolution of less than 1 m over a measurement range up to 10 km are obtained for multimode fibers. A new system based on the use of single mode fibers should extend the range to about 30 km with a spatial resolution of 8 m and a temperature resolution of 2°C.

1.1.5 Brillouin distributed temperature sensors

Brillouin scattering sensors show an interesting potential for distributed strain and temperature monitoring. Systems able to measure strain or temperature variations of fibers of length up to 50 km with spatial resolution down in the meter range are now demonstrating their potential in field applications. For temperature measurements, the Brillouin sensor is a strong competitor to systems based on Raman scattering, while for strain measurements, it has practically no rivals.

Brillouin scattering is the result of the interaction between optical and sound waves in optical fibers. Thermally excited acoustic waves (phonons) produce a periodic modulation of the refractive index. Brillouin scattering occurs when light propagating in the fiber is diffracted backward by this moving grating, giving rise to a frequency-shifted component by a phenomenon similar to the Doppler shift. This process is called spontaneous Brillouin scattering.

Acoustic waves can also be generated by injecting in the fiber two counterpropagating waves with a frequency difference equal to the Brillouin shift. Through electrostriction, these two waves will give rise to a traveling acoustic wave that reinforces the phonon population. This process is called stimulated Brillouin amplification. If the probe signal consists of a short light pulse and its reflected intensity is plotted against its time of flight and frequency shift, it will be possible to obtain a profile of the Brillouin shift along the fiber length.

The most interesting aspect of Brillouin scattering for sensing applications resides in the temperature and strain dependence of the Brillouin shift [6]. This is the result of the change in the acoustic velocity according to variation in the silica density. The measurement of the Brillouin shift can be approached using spontaneous or stimulated scattering. The main challenge in using spontaneous Brillouin scattering for sensing applications lies in the extremely low level of the detected signal. This requires sophisticated signal processing and relatively long integration times.

Systems based on the stimulated Brillouin amplification have the advantage of working with a relatively stronger signal but face another challenge. To produce a meaningful signal, the two counter-propagating waves must maintain an extremely stable frequency difference. This usually requires the synchronization of two laser sources that must inject the two signals at the opposite ends of the fiber under test. The MET (metrology laboratory) group at Swiss Federal Institute of Technology in Lausanne (EPFL) proposed a more elegant approach [6]. The approach consists in generating both waves from a single laser source using an integrated optics modulator. This arrangement offers the advantage of eliminating the need for two lasers and intrinsically insures that the frequency difference remains stable independently from the laser drift. SMARTEC and Omnisens (Switzerland) commercialized a system based on this setup and named it DiTeSt (Figure 1.5). It features a measurement range of 10 km with a spatial resolution of 1 m or a range of 25 km with a

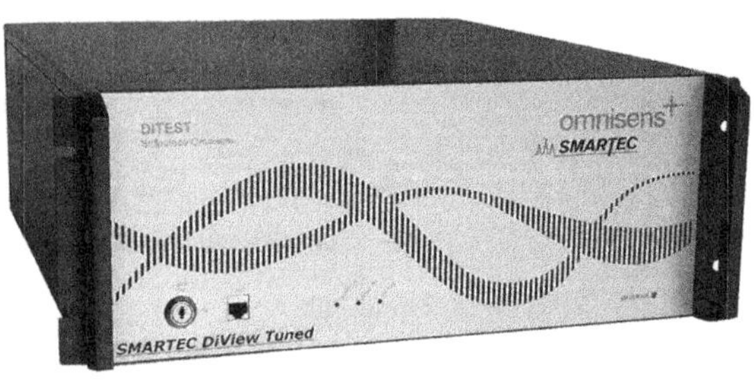

Figure 1.5 DiTeSt reading unit.

resolution of 2 m. The strain resolution is 2 με and the temperature resolution is 0.1°C. The system is portable and can be used for field applications.

Since the Brillouin frequency shift depends on both the local strain and temperature of the fiber, the sensor setup will determine the actual sensitivity of the system. For measuring temperatures, it is sufficient to use a standard telecommunication cable. These cables are designed to shield the optical fibers from elongation of the cable. The fiber will therefore remain in its unstrained state and the frequency shifts can be unambiguously assigned to temperature variations. If the frequency shift of the fiber is known at a reference temperature, it will be possible to calculate the absolute temperature at any point along the fiber. Measuring distributed strains requires a specially designed sensor. A mechanical coupling between the sensor and the host structure along the whole length of the fiber has to be guaranteed. To resolve the cross-sensitivity to temperature variations, it is also necessary to install a reference fiber along the strain sensor. Similar to the temperature case, knowing the frequency shift of the unstrained fiber will allow an absolute strain measurement.

1.2 SELECTED PROJECTS

This section will introduce a few projects showing how fiber optic technology is effectively used for the health monitoring of different types of structures, with different aims and during different phases of the structure's lifetime.

1.2.1 Colle Isarco Bridge

The development of a life extension and/or replacement strategy for highway structures is a crucial point in an effective bridge management system. The monitoring of the Colle d'Isarco viaduct on the Italian Brenner Highway A22 is an example of a global monitoring approach in establishing a bridge management system. The section of the highway that is subject to monitoring activities includes four columns, each of them supporting asymmetrical cantilevers in the north and south direction as can be seen in Figure 1.6 [7].

The overall length of this section is 378 m. The height of the girders near supports 8 and 9 is 11 m; at supports 7 and 10, the height is 4.50 m. The girders have a uniform width of 6 m; the arrangement

Figure 1.6 View of the Colle Isarco Bridge on the Brennero Highway in Italy.

for each road bed is approximately 11 m wide. A wide set of sensors have been installed, including both traditional and SOFO FOSs. Due to the large dimensions of the section, a data acquisition system able to collect widely distributed sensing units was also installed (Figure 1.7). Wireless serial communication was used to transfer the measured data from the almost inaccessible locations on the bridge to the location of the personal computer used to evaluate the measured data.

Data evaluation is performed by a combination of analytical modeling and fine-tuning of the system parameters. The system aims at creating an appropriate match between the nonlinear simulation and the measured data. Since the measurement processes usually introduce a certain amount of variability and uncertainty into the results due to the limited number of measurement points and the partial knowledge on the actions, this randomness can affect the conclusions drawn from measurements. Randomness in measured variables can however be accounted for by their probability density functions. Once a model and its calibration has gained a certain level of completeness, analytical prediction provides a quantitative knowledge and hence, it becomes a useful tool to support structural evaluation, decision-making, and maintenance strategies. This ambitious project aims at a full integration of instrumentation into the decision-support system for structural maintenance.

1.2.2 Pile loading test

A new semiconductor production facility in the Tainan Scientific Park, Taiwan, is to be founded on a soil consisting mainly of clay and sand with poor mechanical properties. To assess the foundation performance, it was decided to perform an axial compression, pullout and flexure test in full-scale

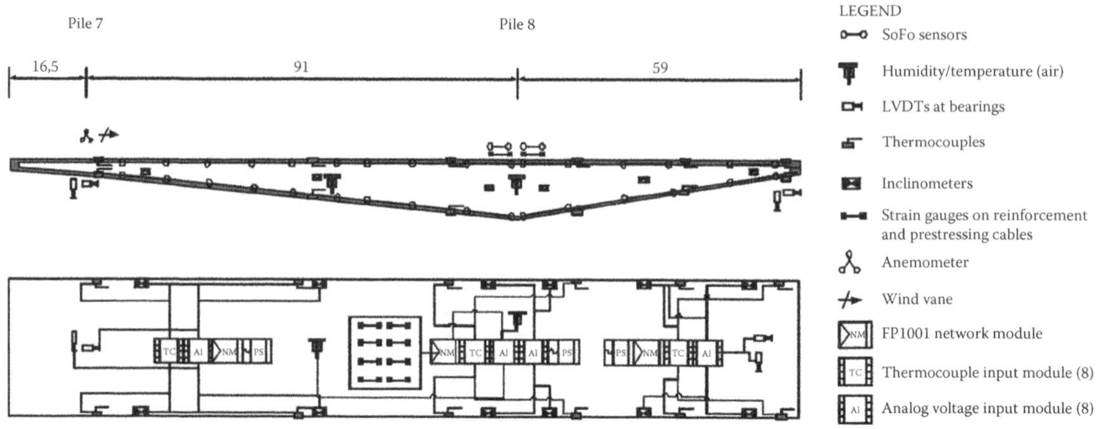

Figure 1.7 Layout of the Colle Isarco Bridge instrumentation. (Courtesy of K. Bergmeister.)

on-site condition. Four meters SOFO sensors were used. The pile was divided into eight zones (called cells). In the case of axial compression and pullout tests, a simple topology was used: the eight sensors were installed in a single chain, placed along one of the main rebars, one sensor in each cell, as shown in Figure 1.8. To detect and compensate for a possible load eccentricity, the top cell was equipped with one more sensor installed on the opposite rebar with respect to the pile axis (see Figure 1.5).

As a result of monitoring, rich information concerning the structural behavior of the piles is collected. Important parameters were determined such as distributions of strain, normal forces (see Figure 1.9), displacement in the pile, distribution of frictional forces between the pile and the soil,

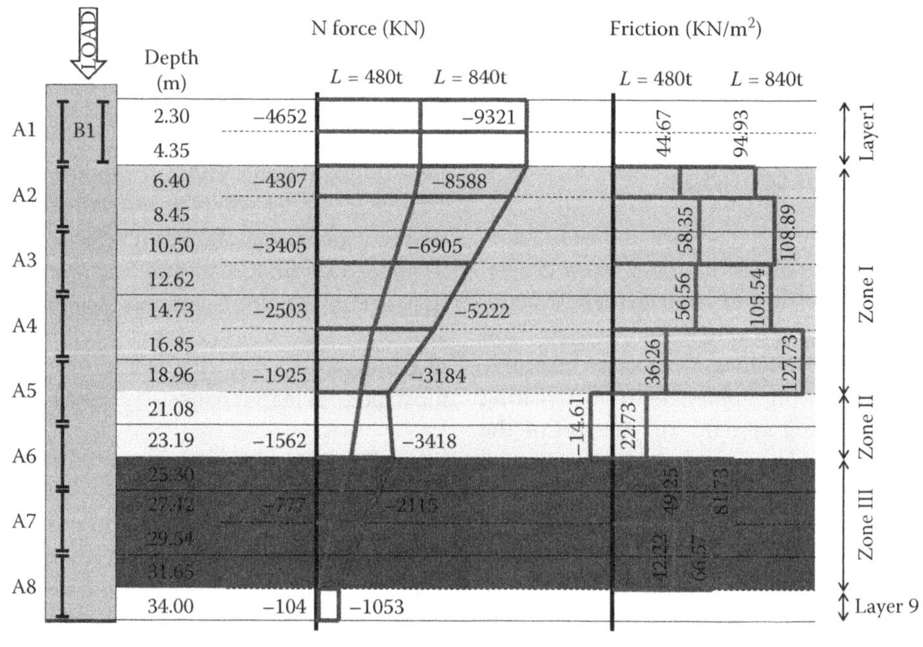

Figure 1.8 Sensor topology and results obtained by monitoring during the axial compression test.

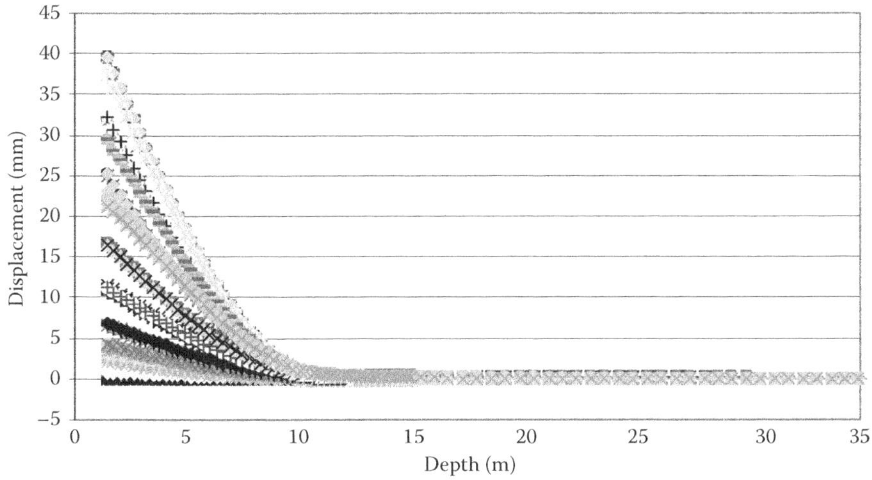

Figure 1.9 Deformed shapes of the pile and identification of failure location.

determination of Young's modulus, ultimate load capacity and failure mode of the piles, as well as qualitative determination of mechanical properties of the soil (three zones are distinguished in Figure 1.8).

In case of flexure test, a parallel topology was used: each cell contained two parallel sensors (as in cell 1 in Figure 1.8) installed on two opposite main rebars, constituting two chains of sensors. This topology allowed determination of average curvature in each cell, calculation of deformed shape, and identification of failure point. The diagram of horizontal displacement for different steps of load as well as failure location on the pile is presented in Figure 1.9 [8].

1.2.3 I35W Bridge, Minneapolis

This application is a good example of a truly integrated structural health monitoring (SHM) system, combining different sensing technologies to achieve the desired level of monitoring.

The collapse of the old I35W Bridge in Minneapolis in 2007 shook the confidence of the public in the safety of the infrastructure that we use every day. As a result, the construction of the replacement bridge (see Figure 1.10) must rebuild this confidence by demonstrating that a high level of safety can not only be attained during construction but also maintained throughout the projected 100-year life span of the bridge.

One of the central factors contributing to this was the design and installation of a comprehensive SHM system, which incorporates many different types of sensors measuring parameters related to the bridge performance and aging behavior. This system continuously gathers data and allows, through appropriate analysis, the acquisition of actionable data on bridge performance and health evolution [9]. The data provided is to be used for operational functions, as well as for the management of ongoing bridge maintenance, complementing and targeting the information gathered through routine inspections.

The monitoring system was designed and implemented through a close cooperation between

Figure 1.10 New I35W Bridge in Minneapolis.

the designer, the owner, the instrumentation supplier, and the University of Minnesota.

The main objectives of the system were to support the construction process, record the structural behavior of the bridge, and contribute to the intelligent transportation system as well as to the bridge security.

The design of the system was an integral part of the overall bridge design process allowing the SHM system to both receive and provide useful information about the bridge's performance, behavior, and expected lifetime evolution.

Monitoring instruments on the new St. Anthony Falls Bridge measure dynamic and static parameter points to enable close behavioral monitoring during the bridge's life span. Hence, this bridge can be considered to be one of the first "smart" bridges of this scale to be built in the United States.

The system includes a range of sensors that are capable of measuring various parameters to enable the behavior of the bridge to be monitored. Strain gauges measure local static strain, local curvature, and concrete creep and shrinkage; thermistors measure temperature, temperature gradient, and thermal strain; while linear potentiometers measure joint movements. At mid-span, accelerometers are incorporated to measure traffic-induced vibrations and modal frequencies (Eigen frequencies). Corrosion sensors are installed to measure the concrete resistivity and corrosion current.

Meanwhile, there are long-gauge SOFO FOSs that measure a wide range of parameters, such as average strains, strain distribution along the main span, average curvature, deformed shape, dynamic strains, dynamic deformed shape, vertical mode shapes, and dynamic damping—they also detect crack formation. Some of the installed sensors are shown in Figure 1.11.

The sensors are located throughout the two bridges, the northbound and southbound lanes, and are in all spans. However, a denser instrumentation is installed in the southbound main span over the Mississippi River, as depicted in Figure 1.12. This span will therefore serve as a sample to observe behaviors that are considered as similar in the other girders and spans.

This project is one of the first to combine very diverse technologies, including vibrating wire sensors, FOSs, corrosion sensors, and concrete humidity sensors into a seamless system using a single database and user interface.

1.2.4 Luzzone Dam

Distributed temperature measurements would be very pertinent for the monitoring of large structures. In the project we will discuss in this section, SMARTEC and the MET-EPFL group used the DiTeSt system to monitor the temperature development of the concrete used to build a dam [10].

The Luzzone Dam was raised by 17 m to increase the capacity of the reservoir (Figure 1.13). The raising was realized by successively concreting 3 m thick blocks. The tests concentrated on the largest block to be poured, the one resting against the rock foundation on one end of the dam. An armored telecom cable installed in a serpentine manner during concrete pouring constituted the Brillouin sensor.

The temperature measurements started immediately after pouring the concrete and extended over 6 months. The measurement system proved reliable even in the demanding environment present at the dam (dust, snow, and temperature excursions). The temperature distributions after 15 days from concrete pouring are shown in Figure 1.14. Comparative measurements obtained locally with conventional thermocouples showed agreement within the error of both systems.

This example shows how it is possible to obtain a large number of measurement points with relatively simple sensors. The distributed nature of Brillouin sensing makes it particularly adapted to the monitoring of large structures where the use of more conventional sensors would require extensive cabling.

1.2.5 Bridge crack detection

Götaälvbron, the bridge over Göta River (Figure 1.15), was built in the 1930s and is now more than 80 years old. The steel girders were cracked, and the steel cracking was caused by two issues: fatigue and mediocre quality of the steel. The bridge authorities repaired the bridge and decided to keep it in service for the next 15 years, but in order to increase the safety and reduce uncertainties related to the bridge performance and integrity, a monitoring system has been mandatory.

The main issue related to the selection of the monitoring system has been the total length of the girders, which is for all the nine girders more than 9 km. It was therefore decided to monitor

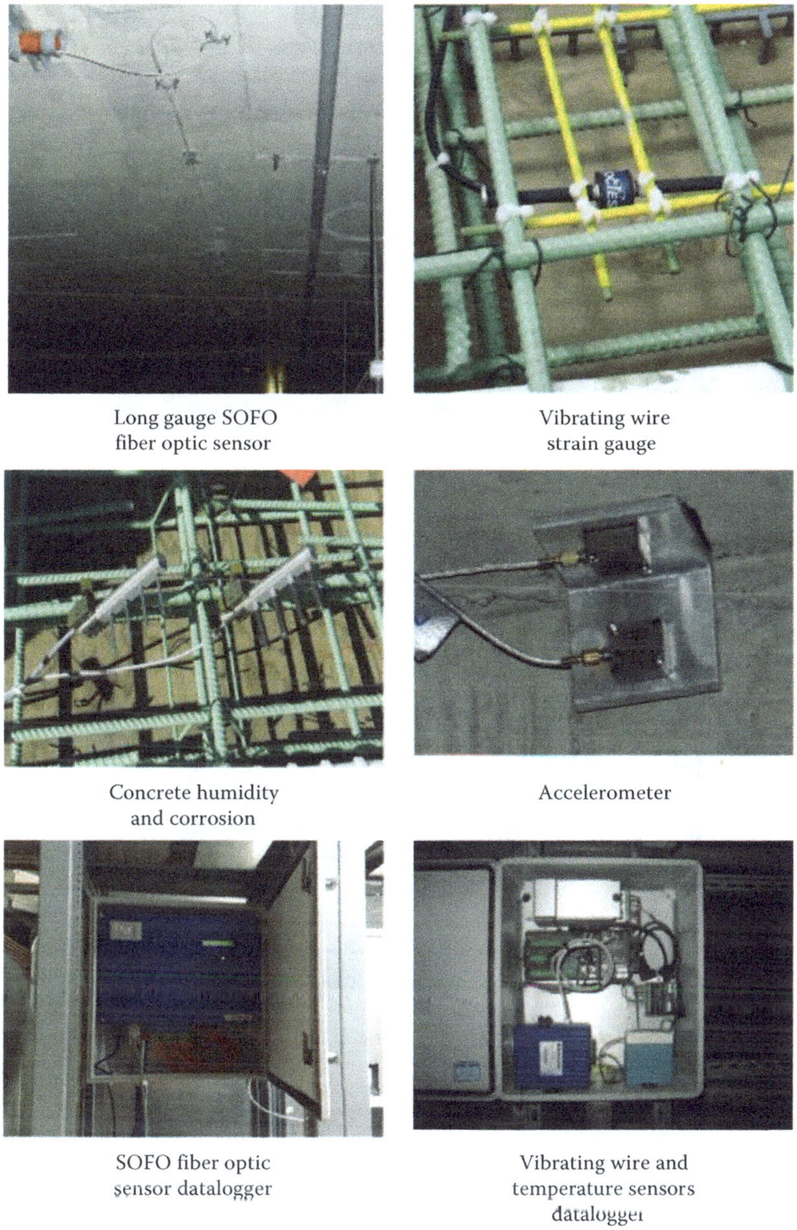

Figure 1.11 Sensing components.

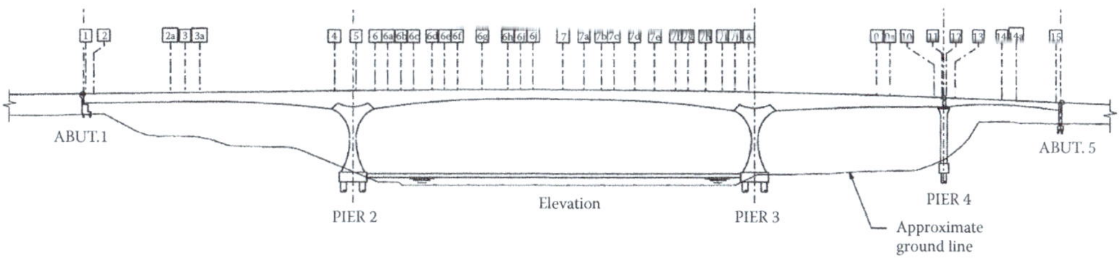

Figure 1.12 Sensor locations.

Figure 1.13 Luzzone Dam raising works.

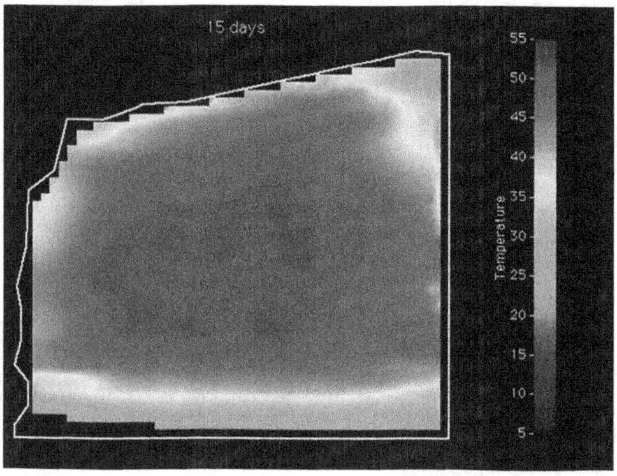

Figure 1.14 Temperature measurements in the Luzzone Dam 15 days after concrete pouring.

Figure 1.15 View of the Götaälvbron Bridge (showing 1 kilometer of it).

Figure 1.16 On-site test of SMARTape gluing procedure (left) and installed SMARTapes.

the most loaded five girders (total length of 5 km approximately). Considering all the alternatives carefully, a fiber optic-distributed sensing system was selected. For the first time, a truly distributed fiber optic sensing system, based on the Brillouin scattering effect, was employed on a large scale to monitor new crack occurrence and unusual strain development [11].

In order for the system to be able to detect the cracks in every point, it was decided to glue SMARTape to the steel girder. The sensor should be placed such that cracks should not damage the sensor, but create its delaminating from the bridge (otherwise the sensor would be damaged and would need to be repaired). The gluing procedure was therefore established and rigorously tested in laboratory and on-site. Photograph of on-site gluing operation is presented in Figure 1.16. The full performance was also tested in laboratory and on-site, and a photograph of the tested SMARTapes installed on the bridge is presented in the same figure.

The installation of SMARTape sensors was a challenge in itself. Good treatment of surfaces was necessary and number of transversal girders had to be crossed. Limited access and working space in the form of lift baskets, often combined with a cold and windy environment and sometimes with the night work, made the installation particularly challenging. The measurements of SMARTape are compensated for temperature using the temperature sensing cable that has also the function of bringing back optical signals to the DiTeSt reading unit.

1.2.6 Bitumen joint monitoring

Plavinu hes is a dam that belongs to a complex of the three most important hydropower stations on

Figure 1.17 Plavinu Dam in Latvia.

Daugava River in Latvia (see Figure 1.17). In terms of capacity, this is the largest hydropower plant (HPP) in Latvia and is considered to be the third level of the Daugavas hydroelectric cascade. It was constructed at a distance of 107 km from the firth of Daugava and is unique in terms of its construction—for the first time in the history of hydro-construction practice; a HPP was built on clay–sand and sand–clay foundations with a maximum pressure limit of 40 m. The HPP building was merged with a water spillway. The entire building complex is extremely compact. There are 10 hydro-aggregates installed at the HPP and its current capacity is 870,000 kW.

One of the dam inspection galleries coincides with a system of three bitumen joints that connects two separate blocks of the dam. Due to

Figure 1.18 SMARTape installation in the inspection gallery.

abrasion of water, the joints lose bitumen and the redistribution of loads in the concrete arms appears. Since the structure is nearly 40 years old, the structural condition of the concrete can be compromised due to aging. Thus, the redistribution of loads can provoke damage of the concrete arm and, as a consequence, the inundation of the gallery. In order to increase the safety and enhance the management activities, it was decided to monitor the average strain in the concrete arm next to the joints [12]. The DiTeSt system with SMARTape deformation (see Figure 1.18) sensors and a temperature sensing cable was used for this. The sensors were installed by VND2 with SMARTEC support and configured remotely from the SMARTEC office. Threshold detection software was installed in order to send pre-warnings and warnings from the DiTeSt instrument to the control office.

1.2.7 Gas pipeline monitoring

About 500 m of a buried, 35-year-old gas pipeline, located near Rimini, Italy, lie in an unstable area. Distributed strain monitoring could be useful in order to improve the vibrating wire strain gauges monitoring system, actually used in the site. Landslides progress with time and could damage pipelines up to the point of being put out of service. Three symmetrically disposed vibrating wires were installed in several sections at a distance typically of 50/100 m chosen as the most stressed ones according to a preliminary engineering evaluation. These sensors were very helpful but could not fully cover the length of the pipeline and only provide local measurements.

Different types of distributed sensors were used: SMARTape and a temperature sensing cable [13]. Three parallel lines constituting five segments of SMARTape sensors were installed over the whole concerned length of the pipeline (see Figure 1.19). The lengths of segments were ranged from 71 m to 132 m, and the position of the sensors with respect to the pipeline axis were at 0°, 120°, and –120° approximately. The strain resolution of the SMARTape was 20 microstrains, with a spatial resolution of 1.5 m (and an acquisition range of 0.25 m). The SMARTape provides monitoring of average strains, average curvatures, and deformed shape of the pipeline. The temperature sensing

Figure 1.19 SMARTape on the gas pipeline.

cable was installed onto the upper line (0°) of the pipeline in order to compensate the strain measurements for temperature. The temperature resolution of the sensor was 1°C with the same resolution and acquisition of the SMARTape. All the sensors were connected to a central measurement point by means of extension optical cables and connection boxes. They were read from this point using a single DiTeSt reading unit. Since the landslide process is slow, the measurements sessions are performed manually once a month. In case of an earthquake, a session of measurements is performed immediately after the event. All the measurements obtained with the DiTeSt system are correlated with the measurements obtained with vibrating wires. The sensors have been measured for a period of 2 years, providing interesting information on the deformation induced by burying and by landslide progression. A gas leakage simulation was also performed with success using the temperature sensing cable.

1.3 CONCLUSIONS

SHM is not a new technology or trend. Since ancient times, engineers, architects, and artisans have been keen on observing the behavior of built structures to discover any sign of degradation and to extend their knowledge and improve the design of future structures. Ancient builders would observe and record crack patterns in stone and masonry bridges. Longer spans and more slender arches were constructed and sometimes failed during construction or after a short time [14]. Those failures and their analysis have led to new insight and improved design of future structures. This continued struggle for improving our structures is not only driven by engineering curiosity but also by economic considerations.

As for any engineering problem, obtaining reliable data is always the first and fundamental step toward finding a solution. Monitoring structures is our way of obtaining quantitative data about our bridges; they help us in taking informed decisions about their health and destiny. This chapter has presented the advantages and challenges related to the implementation of an integrated SHM system, guiding the reader through the process of analyzing the risks, uncertainties, and opportunities associated with the construction and operation of specific bridges and the design of matching monitoring systems and data analysis strategies.

ACKNOWLEDGMENT

The author would like to acknowledge the contribution of all project partners and owners of the structures presented in the case study section. In particular our thanks go to the Brenner Highway Authority, US Federal Highway Administration and Minnesota Department of Transportation, the University of Minnesota, the Goteborg Road Administration, NGI, SNAM Rete Gas, Prof. Luc Thevenaz, Prof. Branko Glisic and all employees at SMARTEC.

REFERENCES

1. B. Glisic and D. Inaudi, *Fibre Optic Methods for Structural Health Monitoring*, John Wiley & Sons, Ltd., Chichester, 2007.
2. D. Inaudi, A. Elemari, S. Vurpillot, Low-coherence interferometry for the monitoring of civil engineering structures, *SPIE, Second European Conference on Smart Structures and Materials*, Glasgow, UK, Vol. 2361, pp. 216–219, 1994.
3. D. Inaudi, Application of civil structural monitoring in Europe using fiber optic sensors, *Progress in Structural Engineering and Materials*, Vol. 2, No. 3, pp. 351–358, 2000.

4. É. Pinet, C. Hamel, B. Glišic, D. Inaudi, N. Miron, Health monitoring with optical fiber sensors: from human body to civil structures, *14th SPIE Annual Symposium on Smart Structures and Materials and Nondestructive Evaluation and Health Monitoring*, San Diego, CA, 653219, 2007.
5. J. P. Dakin, D. J. Pratt, G. W. Bibby, J. N. Ross, Distributed optical fibre Raman temperature sensor using a semiconductor light source and detector, *Electronics Letters*, Vol. 21, No. 13, 20 June 1985.
6. M. Niklès et al., Simple distributed temperature sensor based on Brillouin gain spectrum analysis, *Tenth International Conference on Optical Fiber Sensors OFS 10*, Glasgow, UK, SPIE Vol. 2360, pp. 138–141, 1994.
7. A. Del Grosso, K. Bergmeister, D. Inaudi, U. Santa, Monitoring of bridges and concrete structures with fibre optic sensors in Europe, *IABSE Conference*, Seoul, Korea, August 2001.
8. B. Glisic, D. Inaudi, C. Nan, Piles monitoring during the axial compression, pullout and flexure test using fiber optic sensors, *81st Annual Meeting of the Transportation Research Board (TRB)*, Washington, DC, 02-2701, 2002.
9. D. Inaudi, M. Bolster, R. Deblois, C. French, A. Phipps, J. Sebasky, K. Western, Structural health monitoring system for the new I-35W St Anthony Falls Bridge, *4th International Conference on Structural Health Monitoring on Intelligent Infrastructure (SHMII-4) 2009*, 22–24 July 2009, Zurich, Switzerland, 2009.
10. D. Inaudi, B. Glisic, Distributed Fiber optic strain and temperature sensing for structural health monitoring, *IABMAS'06 The Third International Conference on Bridge Maintenance, Safety and Management*, 16–19 July 2006, Porto, Portugal, 2006.
11. B. Glisic, D. Posenato, D. Inaudi, Integrity monitoring of old steel bridge using fiber optic distributed sensors based on Brillouin scattering, *14th SPIE Annual Symposium on Smart Structures and Materials and Nondestructive Evaluation and Health Monitoring*, San Diego, CA, 6531-25, 2007.
12. D. Inaudi, B. Glisic, Distributed fiber optic sensors: Novel tools for the monitoring of large structures, *Geotechnical News*, Vol. 25, No. 3, pp. 8–12, 2007.
13. D. Inaudi, B. Glisic, Long-range pipeline monitoring by distributed fiber optic sensing, *6th International Pipeline Conference* September 25–29, 2006, Calgary, AB, Canada, 2006.
14. M. Levi, and M. Salvadori, *Why Buildings Fall Down*, W.W. Norton & Company, New York, 1992.

2

Distributed acoustic sensing for infrastructure

STUART RUSSELL
Optasense

2.1 INTRODUCTION

This chapter examines the application of distributed acoustic sensing (DAS) in infrastructure, using a single fiber in an optical fiber sensing cable. Usually, the sensor operates using regular monomode optical fiber cables of the same type as those used for telecommunications, but, in some cases, the cable may occasionally be custom designed to withstand severe environments or to improve sensing performance. A DAS system converts a standard optical fiber into an array of acoustic sensors each with the ability to essentially simultaneously determine the time-varying strain at any given position along the long length of the optical fiber. The operational range of systems is usually limited to approximately 50 km without the inclusion of additional amplification or similar measures. However, unlike the technology that preceded it, DAS is able to measure this strain dynamically, with a frequency response of typically several kilohertz. It is usually the case with DAS that it is not possible for the system to recover the absolute or DC strain as the processing method responds only to the AC content of the strain. However, for many applications, DAS systems are superior to more traditional strain sensing optical technologies, as they are able to respond to relatively fast and low-amplitude *variations* in strain (e.g., AC sensor output signals). In addition, traditional methods are usually based on one or more point sensors, whereas the DAS system can operate in a fully continuous manner over many kilometers of fiber, recovering signals, essentially simultaneously from each and every resolution cell (of length ~1–10 m, depending on the design) along the length.

Before describing the applications of the technology in infrastructure, it is appropriate to describe how the technology used in recent DAS systems differs from the earlier distributed sensing methods described in Chapter 11, Volume II.

2.2 THEORETICAL BACKGROUND OF DAS

We shall now describe the theory of DAS, starting with the basic concepts of optical time domain reflectometry, first introduced in Chapter 11, Volume II.

2.2.1 Conventional optical time domain reflectometry

DAS technology typically employs a modified form of the optical time domain reflectometry (OTDR) principle (see Chapter F2). In conventional OTDR, a short broadband pulse or "probe" is launched into the waveguide under inspection. As this probe propagates along the fiber, a tiny fraction of the transmitted light is scattered due to the linear optical process of Rayleigh scattering. A small fraction of this scattered light is recaptured by the fiber and guided back towards the launch where it can be detected.

In the lowest loss transmission band of optical fibers, usually this Rayleigh scattering process sets the fundamental limit on the transmission; in most application areas, deployed fibers now achieve losses approaching this fundamental limit, which can be as low as 0.18 dB/km for standard single-mode fiber types.

Assuming a constant propagation velocity of the probe pulse, the variation of the measured backscatter intensity, as a function of time, can infer information about the waveguide as a function of position. The interrogation pulse width sets the fundamental resolution of the OTDR method, allowing a typical spatial resolution on the order of half the pulse width. For incoherent illumination, the detected backscatter power as a function of distance can be given as follows:

$$P_{BS} = -b \cdot P \cdot l \cdot \alpha_s \cdot e^{2 \cdot \alpha_a \cdot z}$$

b is the capture coefficient $\sim 1 \times 10^{-3}$ (For SMF 28e)

P is the launch pulse power (W)

l is the launch pulse length (m)

α_s is the scatter coefficient $\sim -4.45 \times 10^{-5}$ (m^{-1})

α_a is the loss coefficient $\sim -4.85 \times 10^{-5}$ (m^{-1})

z is the distance along the fiber (m)

When plotted on a logarithmic scale (Figure 2.1), this relation becomes a straight line representing the loss of the waveguide. The figure also illustrates some typical features that may be observed such as reflections from band splices and unterminated fiber ends.

By inspecting this simple equation, it can be noted that the backscattered intensity is proportional to the product of the backscatter coefficient and capture the coefficient of the fiber, which for a

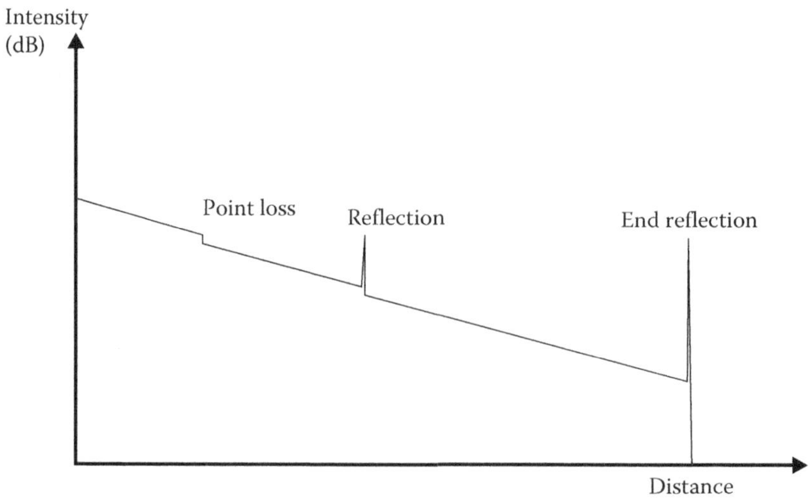

Figure 2.1 A typical standard OTDR trace.

specific fiber type is usually constant. However, the scattered intensity is also proportional to the input pulse power and the pulse length.

The range over which an OTDR can effectively operate is determined by the fundamental noise in the detection system and the detected signal power, which determine the signal-to-noise ratio. Assuming the detection system is optimally designed, the only way to increase the operating range is to increase the pulse launch power or to increase the launch pulse length. However, an upper limit is set on the maximum pulse power that can be launched before the onset of nonlinear optical effects that can degrade the performance of the sensor. Therefore, in order to increase the operational range of the OTDR beyond this, it is necessary to increase the probe pulse width, which in turn necessitates a sacrifice in the achievable spatial resolution.

This approach has been employed for decades to examine the transmission characteristics of installed optical fiber links, allowing the loss budget of the link to be evaluated and bad splices and reflections to be identified, located, and if necessary repaired.

2.2.2 DAS technology

Unlike normal OTDR systems that utilize an incoherent light source, DAS typically uses a coherent illumination pulse, as the technique relies on optical interferometry.

The scattering properties of an optical fiber are determined by microscopic density fluctuations within the material that are frozen in as the glass is quenched from a very high temperature to room temperature during the pulling of the fiber from a preform. These density fluctuations manifest themselves as tiny variations in the refractive index of the material that cause a fraction of the incident light to be reflected or scattered. These variations can be termed "scatter sites." For the purpose of this discussion, the spatial distribution and amplitude variation of these scatter sites can be considered uniformly random.

A typical OTDR optical pulse would, if frozen in time, have a length in the fiber on the order of 1–10 m, which will overlap many millions of these scatter sites. In order to determine the intensity of the scattered field as a function of time, the coherent sum of the electric fields of the light scattered from each one of the millions of scatter sites must be considered. It is key to understanding OTDR that the detector "sees" light that has had to travel to a point in a fiber and back again toward the detector. The apparent illuminated section of fiber as "seen" by the detector is therefore not the same section as observed by somebody watching the pulse propagate from an external point of view but is in fact a section occupying half the (instantaneous) distance of the probe pulse and half its width. Another way of bringing out the key point is to consider a fixed length of fiber. An external observer would see a pulse propagate the length of the fiber and exit the end. However, the detector would not see the light scattered from the distal end of the fiber until the scatter had travelled the entire length back toward the detector. The repetition rate of the pulses is therefore set at a maximum rate of T_{rep}=2Ln/c, where L is the fiber length, n the refractive index, and c the speed of light (Figure 2.2).

It can be shown that when using coherent light, instead of the previously predicted exponential intensity versus distance observed with incoherent OTDR, the coherent trace shows a somewhat similar shape except that it has a highly variable noise intensity, which appears to be randomly modulated with distance. This is indeed the case, and

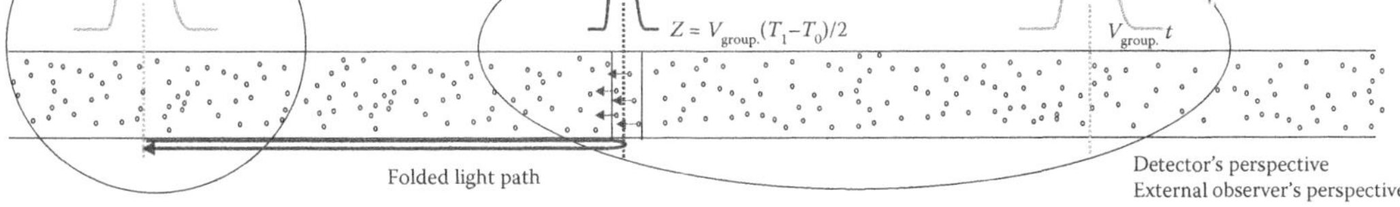

Figure 2.2 Schematic representation of backscatter in an optical fiber.

the probability distribution of this intensity does follow the otherwise expected "Rayleigh" distribution. It is perhaps important to note that the incoherent OTDR result can be obtained via "ensemble averaging" of multiple intensity distributions from coherent illumination of a fiber, provided this is taken at several optical frequencies over a sufficiently wide optical bandwidth (a process that occurs naturally at the detector when using broadband incoherent light sources).

Given a highly stable environment and launch pulse condition, the scattered intensity pattern, which varies randomly as a function of position, will remain unchanged over time. However, a varying strain or temperature acting on the fiber will modulate via the strain optic $n(\varepsilon)$ and thermo-optic $n(T)$ effects the relative positions of the scatter sites, and hence the phase of the scatter from each. As we now have to consider the coherent sum of these waves, the intensity of the detected scatter is in turn modulated by the varying interference of these scattered waves.

This observation and the concept of the environmental sensitivity of this effect were made by Healy [1]. However, only recently, enabling technology has matured to the point where the system design limitations and the intrinsic noise sources present in commercially available optical components have been reduced to a level which enables this sensing method to be useful for real-world applications.

This basic technique with minor modifications has formed the basis of numerous DAS technologies. For this reason, DAS methods are often referred to as coherent OTDR, phase-sensitive OTDR, or ϕ OTDR.

2.2.2.1 LIMITATIONS OF SIMPLE DAS

Up to this point, we have talked about the scatter intensity and how it is modulated by an external environmental change. However, we have not considered in detail the nature of this transduction. This is because, as stated previously, the position of the scatter sites is a random variable, and hence, every position along the fiber will behave differently from its neighbor. This inevitably leads to a transduction coefficient that is also apparently random, nonlinear, and hence unquantifiable without further development. Consideration will now be given to how this problem may be resolved.

The scatter from a single pulse can be considered to be similar to the output from an interferometer. The only difference is that we are considering the interference between millions of waves instead of just two, so higher order effects modify this simple model. The net result from each scatter-site pair is, however, qualitatively similar. Let us assume that the output intensity is a raised cosine function biased around a zero modulation bias phase. This is shown in Figure 2.3.

Figure 2.3 shows the intensity ***response*** of the interferometer as a function of the phase difference of the interfering waves, in the form of a horizontal sinusoidal waveform. The phase ***modulation*** caused by environmental perturbations is, as an example, shown in the form of a vertical sinusoidal waveform. The resulting interferometer output is shown in the right hand diagrams. As stated previously, however, the initial phase bias condition is different for each section of fibre and so the output varies accordingly. The upper and lower parts of figure 2.3 show two different phase conditions and the effect on the interferometer output.

Consider the situation shown in Figure 2.3a. The initial bias is near the "quadrature," that is, at $\pi/2$; for small modulations, the output is approximately linear, and hence, the harmonic distortion is low and the output amplitude is relatively large.

However, in the situation shown in Figure 2.3b, the initial phase bias is near zero. The output is highly nonlinear, and it shows very high harmonic distortion. (In this extreme case, there is frequency doubling of any sinusoidal variation signal.) The output amplitude is very small when compared to the quadrature response. An additional issue is that occasionally, the virtual interferometer will be biased near π, which means that the scattered intensity may be near zero, leading to what is termed as "Rayleigh fading."

Despite these issues, even this simple approach allows for a very powerful sensing system able to determine the location and nature of acoustic events. It has formed the basis of sensing technologies which now find applications in perimeter and border security; infrastructure, that is, rail, road, oil, and gas pipeline monitoring; and telecommunications protection.

2.2.2.2 IMPROVED DAS ARCHITECTURES

When considering use in other more critical application areas, such as seismic surveys, where the quality of the acoustic data yielded is of paramount importance, the limitations of the simple DAS technique

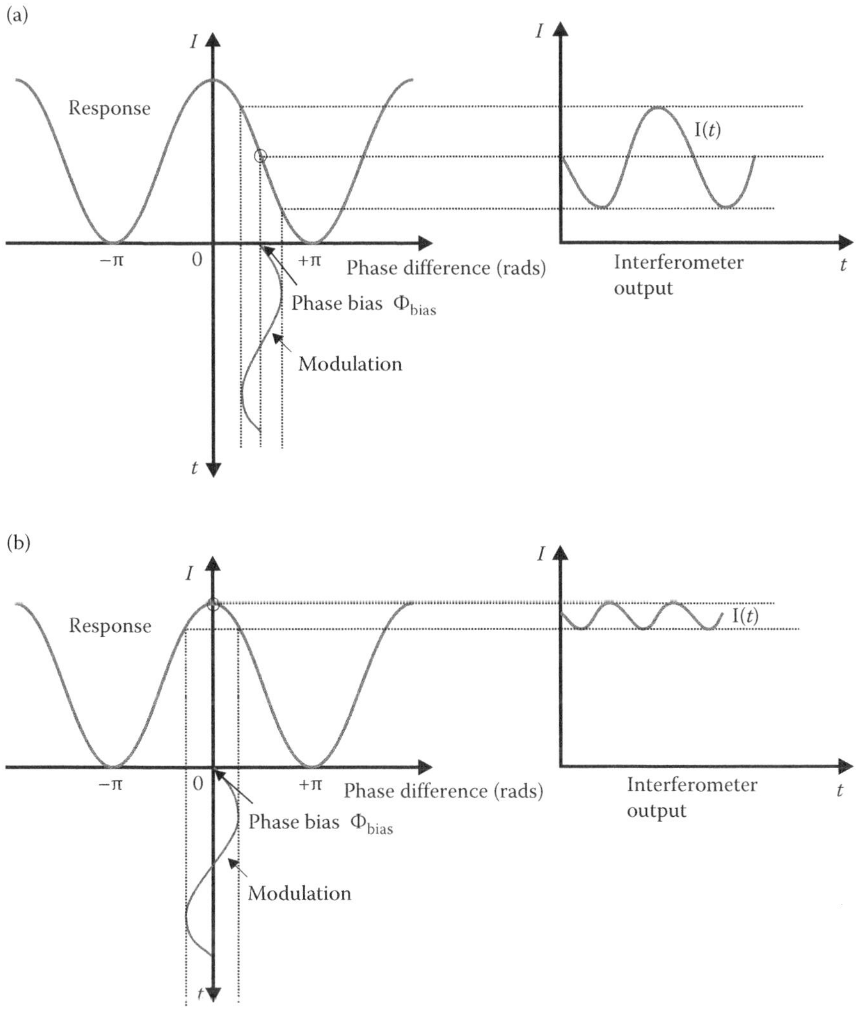

Figure 2.3 Representation of interferometric bias and modulation response in an OTDR single-pulse DAS system. (a) Interferometer with phase bias near quadrature and (b) interferometer with phase bias near zero.

become a limiting factor. Several technologies have been proposed to improve the quality of the acoustic data yielded by the DAS OTDR approach. However, they are all based on similar principles. Instead of yielding a result that is related to the intra-pulse interference terms, they aim to isolate and recover the response of the interference between the probe pulse and a phase reference signal.

1. *Phase-sensitive detection*: One way in which this may be accomplished is by using phase-sensitive detection (Figure 2.4). The scattered light is mixed with a reference local oscillator prior to detection. Consider the interference of two waves: the signal wave with phase ϕ_s and frequency ω_s, and the local oscillator with phase ϕ_{LO} and frequency ω_{LO}.

$$E_s = \frac{1}{2}\left(e^{i(\phi_s(t)-\omega_s t)} + e^{-i(\phi_s(t)-\omega_s t)}\right)$$

$$E_{LO} = \frac{1}{2}\left(e^{i(\phi_{LO}(t)-\omega_{LO} t)} + e^{-i(\phi_{LO}(t)-\omega_{LO} t)}\right)$$

When observed by a square law detector, this yields an intensity

$$I = (E_s + E_{LO})(E_s + E_{LO})^*$$

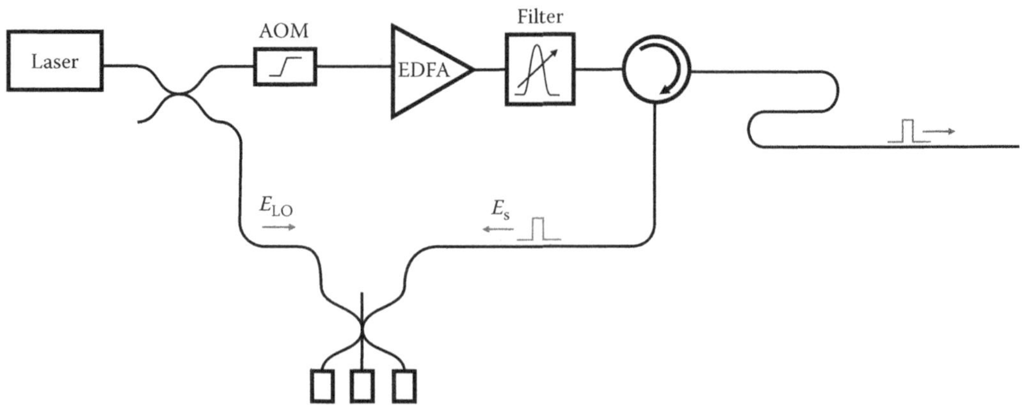

Figure 2.4 Arrangement of phase-sensitive detection DAS. AOM, acousto optic modulator; EDFA, erbium doped amplifier.

When expanded and simplified, this becomes

$$I=\frac{1}{4}|E_s|^2+\frac{1}{4}|E_{LO}|^2$$

$$+\frac{1}{4}|E_s||E_{LO}|\begin{cases} e^{i(\phi_s(t)-\phi_{LO}(t)-(\omega_s-\omega_{LO})t)} \\ +e^{-i(\phi_s(t)-\phi_{LO}(t)-(\omega_s-\omega_{LO})t)} \\ +e^{i(\phi_s(t)+\phi_{LO}(t)-(\omega_s+\omega_{LO})t)} \\ +e^{-i(\phi_s(t)+\phi_{LO}(t)-(\omega_s+\omega_{LO})t)} \end{cases}$$

When interpreted, we see that this generates a DC term; a term at a frequency equal to the sum of the two waves, that is, $(\omega_s+\omega_{LO})$, which is beyond the frequency response of the detection system; and a term at the difference frequency $(\omega_s-\omega_{LO})$, which carries information about the absolute phase difference between the waves $(\phi_s-\phi_{LO})$. Typically, the difference frequency is arranged such that the phase-modulated carrier occurs at a frequency outside of the $1/f$ noise of the detection system. The carrier can then be demodulated to recover the phase difference information.

This method has the benefit of recovering the absolute phase of the scattered light relative to the local oscillator and does not rely on the intra-pulse interference in order to recover the DAS information. However, intra-pulse interference can and still does occur, leading to signal fading as the amount of light scattered from a specific position can reduce to near zero.

The polarization state of the scattered and local fields must also be taken into account. If the fields are cross polarized, no interference occurs and the signal fades again. Polarization diverse detection systems can be employed, but they increase the system complexity. Moreover, since the local oscillator acts as the phase reference, the main interrogator unit can now also be potentially acoustically sensitive. A final potential drawback of this approach is the fact that scattered light is being coherently mixed with light from the local oscillator, which was emitted for a short time after the probe pulse. This places an order-of-magnitude increase on the coherence requirements of the source in order for the phase noise to not be a dominant factor.

2. *Delayed homodyne/heterodyne*: Another method to achieve an improved DAS system is by using an unbalanced interferometer prior to detection (Figure 2.5). In this method, the scatter from one section of the fiber is effectively mixed with the scatter from a section of fiber offset by an amount defined by the imbalance in the detection interferometer. By including a frequency shifting device, such as an acousto optic modulator (AOM) in one arm of the interferometer, this technique can employ either the homodyne or the heterodyne method. The mathematics is virtually identical to the phase-sensitive detection model presented previously; however, this technique offers a few advantages. The response of the system is no longer an absolute phase shift along the fiber but a direct measure of the differential strain over the gauge

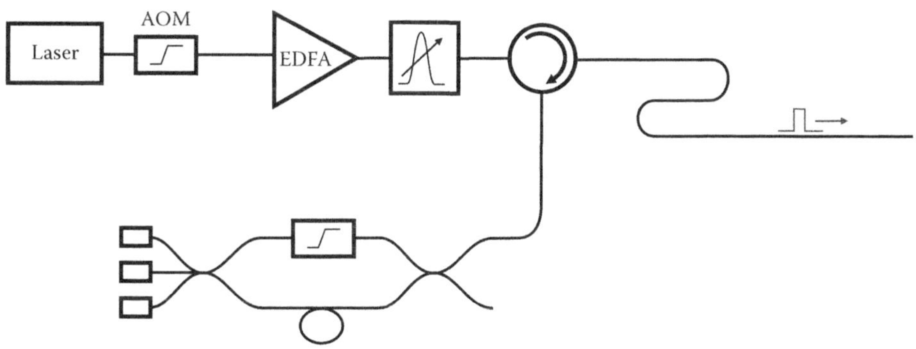

Figure 2.5 Delayed homodyne/heterodyne detection DAS. AMO, acousto optic modulator; EDFA, erbium doped amplifier.

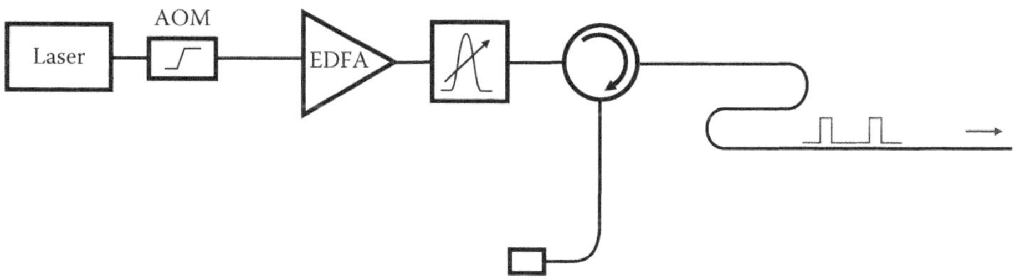

Figure 2.6 Dual pulse interrogation DAS. AMO, acousto optic modulator; EDFA, erbium doped amplifier.

length, as defined by the detection interferometer. The phase noise requirements of the source are no longer so stringent, as the scattered light is always referenced to light with a fixed temporal offset, and, if the gauge length is short relative to the beat length of the fiber, then the polarization states of the two interfering waves are approximately equal.

3. *Dual pulse*: The final technique we will present here is the dual pulse technique (Figure 2.6). This technique varies from the delayed homodyne/heterodyne detection scheme in that instead of a single pulse being launched, two pulses are launched with a defined spatial and frequency offset. This has the advantage of reducing the acoustic sensitivity of the interrogator itself and of simplifying the detection optics. We are also no longer restricted to a gauge length defined by the physical optics and can vary this at will. Phase biasing the virtual interferometer created by the two propagating pulses however requires a little more attention as, unlike in the previous two methods, only a single detector is required and we can no longer take the advantage of the 120° phase shift relationship of the three-port coupler. It is however possible with the proper choice of carrier frequency to have an analytic form of the carrier allowing phase demodulation.

Having presented the technology developments in distributed sensing, the real-world applications of this powerful system will now be described. As the chapter title suggests, the applications listed here are all in the infrastructure area, whereas the subsequent chapters will cover other applications, including security and surveillance and then oil and gas exploration. The use of DAS in oil and gas downhole applications is covered in Chapter 20, Volume III.

2.3 PROTECTION OF COMMUNICATIONS LINKS

World telecommunications and Internet data traffic now rely almost exclusively on fiber optic links. This technology has evolved over time to the point now where a single optical link may carry several million telephone conversations simultaneously, or the equivalent data rate of several terabits per second. These data networks are typically designed with redundancy, but the loss of a link due to accident or malicious act can still cause significant disruption. DAS systems offer a first line of defense, allowing potential threats to a fiber optic link to be monitored in real time over its entire length of many tens of kilometers. Any digging or disturbance near the fiber can then be detected, located, and classified, allowing an appropriate action to be taken. Depending on the threat identified, this may simply involve switching the traffic to a redundant route prior to failure, but the primary aim is to intervene and prevent damage to the network.

DAS can also be deployed to further improve the security of secure data links, such as used, for example, to carry financial data. These links are designed specifically to carry secure data, which are itself deeply encrypted. Additional layers of security are offered by deploying the optical cable within a hardened conduit. Intrusion into these conduits is strictly monitored, and DAS can be used as an additional layer of security. It would be extremely difficult to physically tap a data link monitored by a DAS system in a covert manner.

2.4 MONITORING OF ROADS AND HIGHWAYS

In many countries, congestion and traffic delays have long been a regular feature of commuting life. However, as the number of cars on the road system increases, the need for more carefully managed roads and highways becomes more and more important. Over the last decade, numerous discrete monitoring systems have been trialled, including video cameras, inductive loops, car counting cables, and the global positioning system. None of these systems have provided a total solution, and many are too complex, unreliable, or costly to deploy in volume. All so far are only point sensors, which can miss crucial information occurring in the intervals between them. With increased pressure to monitor and manage traffic flow, especially with the use of driverless vehicles becoming more probable, a cost-effective solution that can be retrofitted to the existing road network is required.

DAS technology potentially offers many of these benefits and could revolutionize the situational awareness of our roads. In most cases, the optical fiber cable required for sensing may have been already installed, buried, or ducted along the highway. Although it may have been originally intended for providing data and telemetry service to traffic signs and control systems already in place, retrofitting DAS to use these cables or fibers will often be a simple and attractive prospect.

2.5 MONITORING OF RAILROADS

Similar to roads, monitoring of vehicles within a rail network is extremely important, not only to ensure an efficient and on-time service but also to operate the network safely. As one might imagine, this is not a simple task, particularly as high electrical interference levels are usually present near electrified lines. Current systems rely on a suite of varied sensing capabilities, but in many areas of current networks, the only measure of the position of a train may be by means of entry and exit gate sensors, usually situated at infrequent intervals along the track. More sophisticated systems with a large number of sensors are not always suitable due to the cost of installation or poor wireless data connection coverage in areas where the existing infrastructure does not support it.

DAS technology is particularly well suited for this application. A single fiber in a cable (as with road highways, now already installed in trackside cables), once connected to a DAS system, can easily be transformed into a series of distributed sensors, thereby delivering a staggering amount of real-time information. With correct handling and processing, this will provide a true multifunction sensing capability suite that is available for every meter of track over distances of many tens of kilometers, a huge improvement on the existing monitoring infrastructure found on railways. This offers great advantages, not only ensuring the safety of the passengers and public, but also allowing the condition of the fleet of vehicles to be monitored and also the network to be run as efficiently as possible. A few of the advantages are listed as follows:

- 100% Coverage of rail track monitored at an affordable cost by exploiting spare, unused, existing trackside fiber optic cabling
- Immediate event detection of trackside incursions that helps protect assets but more importantly helps save lives

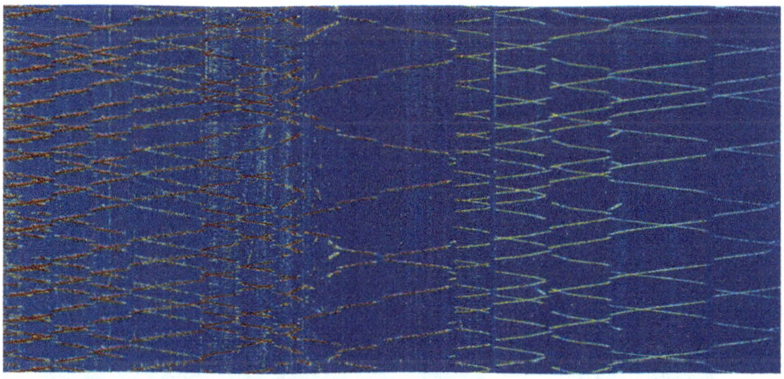

Figure 2.7 Acoustic amplitude output from a DAS system in operation for rail monitoring. (Image Courtesy of Optasense.)

- Detection of rockfall or other disturbances on or near the line, including landslip, mudslide, or tree fall
- Identification of locomotive or carriage "wheel-flat" issues to avoid damaging the track and the vehicles themselves
- Copper cable theft detection that alerts to potential intrusion and other associated activity prior to the occurrence of damage
- Delivery of a safer working infrastructure to protect work parties and the public
- Provision of a critical safety system to alert trackside operators of approaching trains and, with suitable communications links, to warn drivers of other on-track vehicles

Figure 2.7 shows the raw acoustic amplitude data gathered by a DAS system operating alongside a rail network. The vertical axis shows the time, and the horizontal axis shows the distance along the track. Each train is easily distinguishable and this figure shows that the position and velocity of every train are easily traceable.

2.6 PIPELINE SECURITY AND MONITORING

Oil and gas pipelines are valuable assets that carry products with a significant monetary value. In dry countries, even water pipelines are of critical importance. These pipelines traverse desolate, remote, and often disputed regions of the globe. These assets are vulnerable not only to damage caused by natural disasters such as landslip or earthquake but also to attempted theft of contents ("hot tapping") and malicious damage or sabotage by third parties.

DAS technology offers a cost-effective way to monitor the entire length of a pipeline and shield it from potential harm. A single optical fiber cable, either attached to the pipe or buried close to it, can serve as an effective sensor. Complete commercial DAS systems are now available that can automatically monitor and classify potential threats to the health of the pipeline and raise an alarm in time to prevent either damage from occurring or serious leaks from developing.

The sensing features include detection of digging or drilling on or near the pipeline that may be indicative of a third-party intrusion for hot tapping or attempted malicious damage. Obviously, this type of intrusion is unwanted not only due to loss of revenue but also because of the potential environmental damage that can result from the product escaping into the environment. DAS systems are now preventing such hot tapping intrusions all over the globe and have revolutionized this industry. In addition to the malicious activity, the DAS system can be used to detect landslip, avalanche, and earthquake that also threaten the pipeline. Monitoring of routine tasks involved in pipeline maintenance can also be augmented by the use of DAS. For example, Figure 2.8 shows the transit of a cleaning "pig" (a large piston-like object that is pumped along to clean or monitor the inside of a pipeline) along a section of the pipeline. The DAS system is able to clearly detect and locate the pig as it transits the pipeline; it can also detect general noise and more significant pressure pulses generated as the pig transits joints in the pipe. This is of great benefit in assessing the pipeline but is particularly valuable on the occasions when a pig becomes stuck within the pipeline—DAS allows rapid location of the pig, which would otherwise be a potentially difficult and time-consuming task.

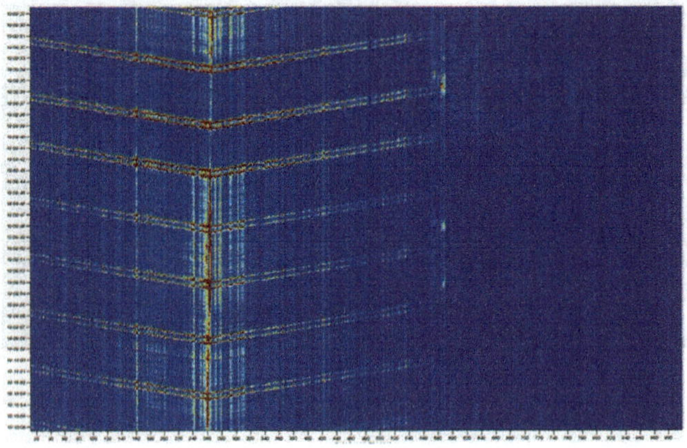

Figure 2.8 A "pig" transiting a pipeline. The pig's location is automatically tracked by the DAS system. Pressure pulse waves can also be detected as the pig crosses joints in the pipeline. (Image Courtesy of Optasense.)

At the time of writing, DAS systems were already being used to monitor more than 11,000 km of pipeline around the world, and their use is still growing rapidly.

2.7 CONCLUSIONS

A few of the major applications of distributed optical fiber acoustic sensors in infrastructure have been described. Many of these were still developing at the time of writing, and, as costs reduce, it is expected that the extent of use and the range of applications will inevitable increase dramatically.

As stated in the introduction, a more detailed account of the technology used will be given in Chapter 7, Volume III. Other applications will also be presented in Chapter 20, Volume III.

REFERENCE

1. Healey, P., Fading in heterodyne OTDR, *Electron. Lett.*, 1984, 20: 30–2.

3

Intelligent infrastructure: Automatic number plate recognition for smart cities

BENJAMIN WATSON
3M UK&I

3.1 INTRODUCTION

This chapter describes how optoelectronics can be used for the purpose of vehicle identification, contributing to the success of large intelligent infrastructure projects. It is illustrated with a series of case study examples for road user charging, access control, and critical infrastructure.

With the investment in smart city transportation set to yield an economic opportunity of USD 800 billion globally (Busch, 2014), it makes sense to first consider what it is that actually makes these cities "smart."

The concept of "smart" intelligent infrastructure is not new; the basic building blocks of the technology have been available for some time, but as with all new developments, full acceptance and use involves evolution, will be described in this chapter. Our present treatment of intelligent infrastructure places a strong focus on electronic vehicle identification (EVI) through automatic number plate recognition (ANPR) and utilizes the characteristics of intelligent infrastructure required to monitor, learn, adapt, predict, protect, and self repair (Paxman 2014).

These core traits closely align to the characteristics of intelligent infrastructure as a framework for the classification of technologies that self-monitor, while controlling their own settings. This is done according to the system input instructions, or is adaptive, responding to what the system has detected and learnt from the aggregation of large data sets. The adoption of learning algorithms can predict the required capacity, optimize the cost and performance, and mitigate security risks.

The widespread realization of this grand vision has been slow, partly due to the inherent fragmentation across stakeholder groups, the diversity of commercial interests, and the physical size of these systems. Fortunately, however, the landscape is now changing and we are set to see an accelerated progression towards all things being connected and interoperable. This is becoming particularly important with the introduction of "autonomous" vehicles set to disrupt the current mobility platforms. In reality, these vehicles are likely, in future, to be not only fully connected to road and highway infrastructure, but also to each other, ensuring far better safety at high speeds and traffic densities, but then rendering the "autonomous" label less applicable

These systems increasingly rely on advanced computational intelligence, open communications protocols, and shared standards that underpin the future of these technologies. Even current infrastructure to control traffic has proven to reduce journey times, ease congestion while improving safety, security, and reliability of the transport-related services shown in Figure 3.2, a process likely to become far more effective and efficient when vehicles are fully controlled by onboard sensors and driving-control software.

3.2 AUTOMATIC LICENSE PLATE RECOGNITION/ELECTRONIC VEHICLE IDENTIFICATION

Infrared cameras are now a widely adopted and rapidly growing application of optoelectronics used to monitor the movement of people and goods across the road network. Such cameras are frequently located on gantries or bridges above highways. These same systems can also be deployed on post structures and buildings at the side of highways or other suitable roadside locations.

This camera technology is particularly useful for law enforcement, especially if the vehicles can be classified according to type; it is even more useful if they can be individually recognized, for example, by their vehicle type, or, more exactly, from their unique number (registration) plate.

Electronic vehicle identification (EVI) through automatic license plate recognition (ALPR) includes the process of detecting a vehicle as it enters the camera's field of view, recognizing the license plate at high speed, and uniquely identifying the vehicle. This can be applied to many applications such as road user charging, law enforcement, and the tracking of vehicles across the transport network (see Figure 3.2) for earlier examples.

The captured image is processed in real time and passed for optical character recognition, analysis, and post processing (see Figure 3.3). The initial number plate image can be detected

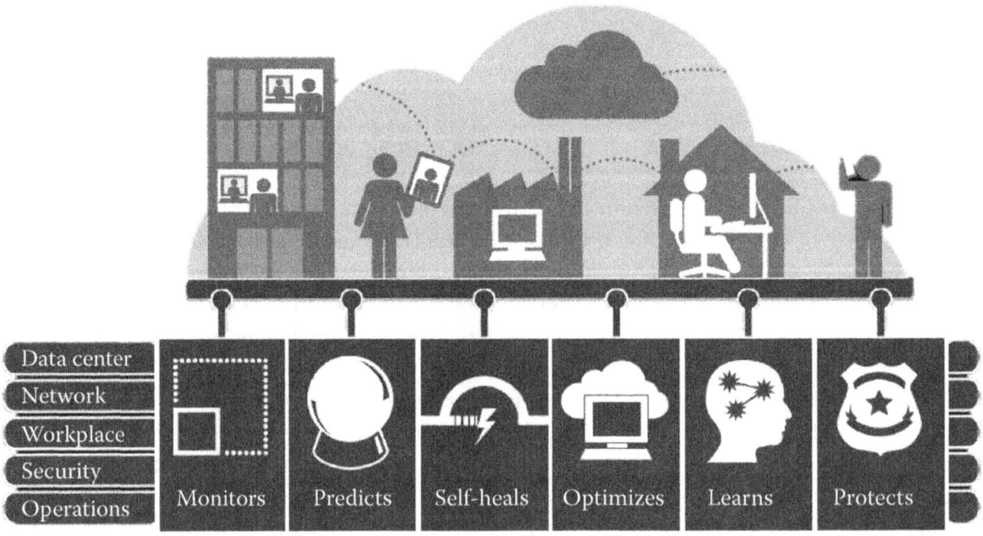

Figure 3.1 The six capabilities of an intelligent infrastructure. (Courtesy of https://www.accenture.com/us-en/insight-intelligent-infrastructure.)

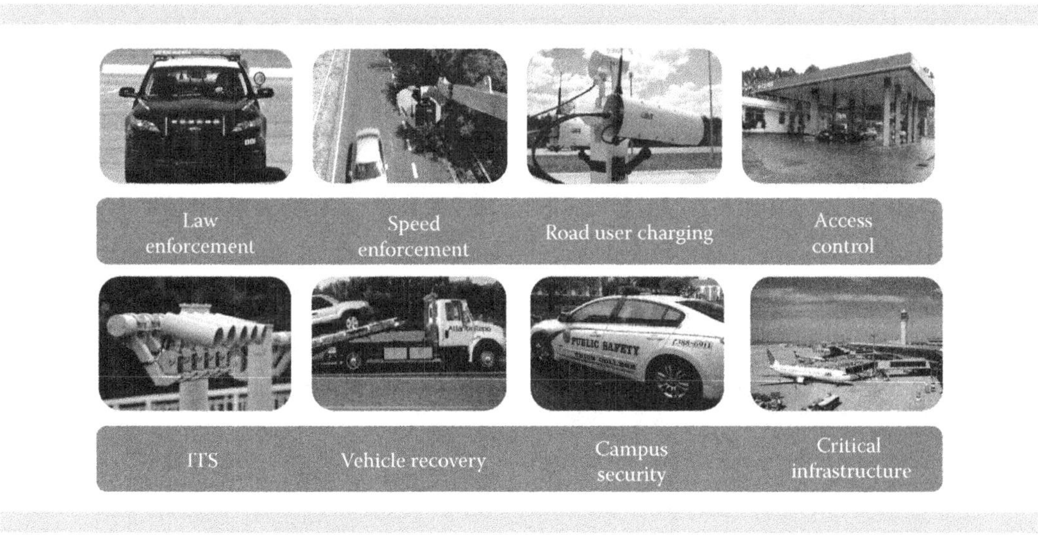

Figure 3.2 Examples of market applications.

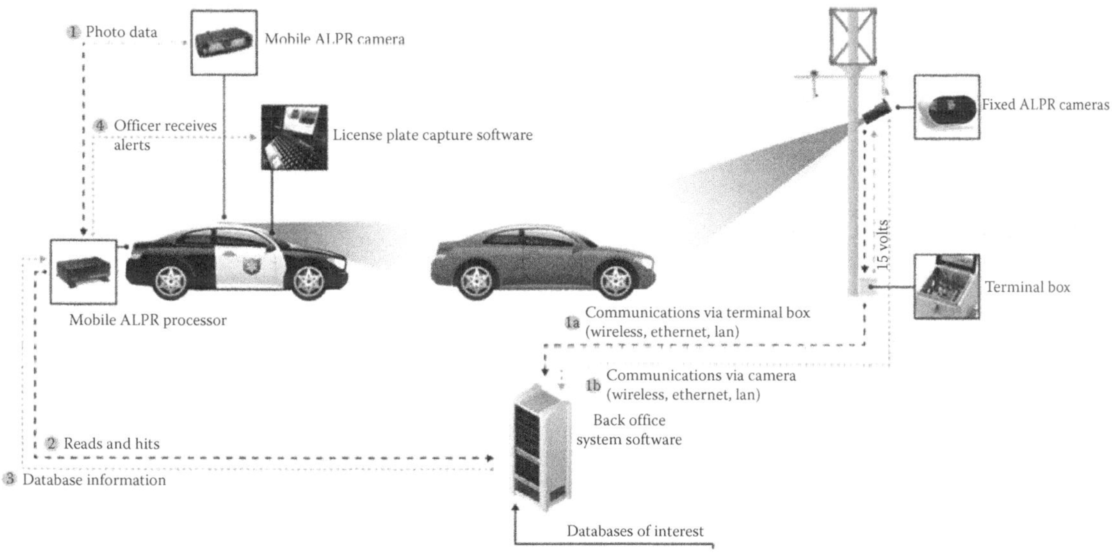

Figure 3.3 Example of ALPR deployment.

and normalized with various techniques, such as sharpening and improved dynamic range, to extract more detail. These normalized images can then be further subjected to noise removal and other image processing techniques, to obtain a "clean" image suitable for character segmentation, syntax checking, and character recognition. All of this happens within a fraction of a second (approximately 20 ms), taking advantage of either the powerful embedded internal processing capabilities of modern day ALPR camera technologies or using post-processing centralized systems.

For each detected vehicle, a "read event" is typically generated, which can then be compiled into a data package suitable for feeding to a software aggregator. This aggregator, equipped with analytical software, turns the data into meaningful intelligence, customized reports, or graphical representations (see Figure 3.4). This intelligence can be used to inform highways support and law

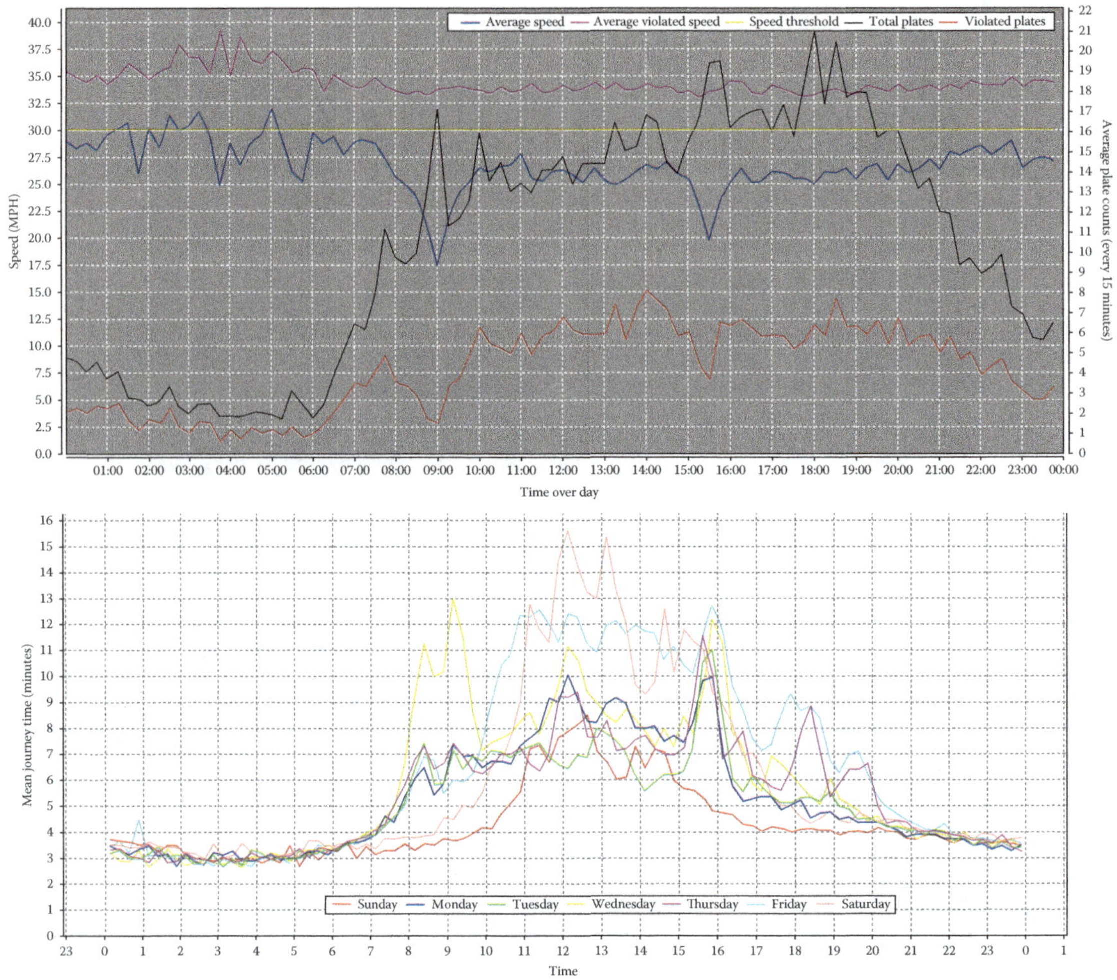

Figure 3.4 Journey time graphs.

enforcement agencies. Alternatively, the read event may be discarded if it is not deemed to be of interest, the decision being made (according to a series of predefined rules) within the license plate reading camera, thanks to onboard processing capabilities.

Partnering EVI technology with powerful enterprise software solutions turns the ANPR data into valuable intelligence for decision-makers. These "enterprise" software platforms enable sophisticated data aggregation, data mining, and analytics.

The vehicle read event can be included as part of a time-stamped data package, with time-stamped consecutive images. These images can be taken either at one point or at geographically separated locations, which allows for measurement of velocity of an individual vehicle and/or the average traffic flow rate for precise temporal information (see Figure 3.4). Furthermore, embedded ALPR cameras can be independently time-locked using GPS time as a primary reference, with high-stability, crystal-oscillator-driven real-time clock as the secondary reference. These systems may be able to detect vehicles traveling up to 220 km/h.

3.3 DATA PROTECTION

Whenever number plate data are routinely captured, it may raise cause, for invasion of privacy concerns. For this reason, suppliers of this technology must adopt robust data privacy approaches and should be encouraged to choose the most stringent criteria for data privacy, data capture, processing, and retention.

Stored data are often encrypted such that only authorized parties are able to decrypt the data. Thus, if an unauthorized person gains access to the data, they will be unable to decipher it.

Communication between cameras and a typical back office may be encrypted to protect sensitive communications from eavesdroppers. Cryptographic techniques can be used to provide strong authentication, thus lending a high degree of mutual certainty that the communicating peers are who they say they are and have the necessary permission for the operation they are attempting.

In the event of unauthorized attempts to access the data, the system can rely on integrated firewalls with default rules that block off any port not associated with authorized and needed services.

With the appropriate handling of data, the "feed" from these systems can also be split to provide key benefits that enable ANPR technology to be used simultaneously for both law enforcement and useful civilian applications, to ease congestion while improving security and safety for the general public.

These combined applications offer a convenient way of spreading the cost of a system between several funding bodies, for example, a local authority and other organizations that may wish to share the ANPR data. The technology is designed to be extensible, easing its evolution to provide additional services if required as part of a smart cities infrastructure. Again, data privacy laws and best practice will apply.

3.4 CAMERA ARCHITECTURE

The embedded camera technology (see Figure 3.5) can include two cameras within a single enclosure. One provides contextual images for color overview, while the other is dedicated to ANPR for optimum performance. Images and video can be streamed via motion JPEG over hypertext transfer protocol, from either camera or the more commonly used H264 advanced video coding. These video compression formats enable the streaming of video content that is increasingly requested as part of these systems. Although not optimized for closed-circuit television streaming, the color overview camera may provide some aspects of this functionality for viewing congestion, accidents, and other incidents.

3.5 EMBEDDED ALPR CAMERA

In terms of communications, the latest generation of ALPR equipment can support Ethernet 3/3.5G/4G, GPRS, cloud-based technology, where no fixed optical fiber or wire links are available or accessible.

Remote communications capability encourages the wider use of mobile applications and may reduce infrastructure and installation costs. Future providers of intelligent machine vision and associated hardware platforms should also be encouraged to consider their role within wider mesh networks, connecting smart cities with self-healing, self-configuring, and non-line-of-sight communication.

In the past, ANPR systems have been sensitive to environmental conditions; however, the use of improved camera housings, often nitrogen purged and usually sealed to IP67, has overcome many of these problems and extended the operating temperature range between −40°C and +60°C. It is therefore only severe weather conditions, for example, where there is severe loss of atmospheric visibility, poorly optimized licence plate materials, or plate damage that will normally cause loss or deterioration of service.

Figure 3.5 3M P392+ Embedded ALPR Camera.

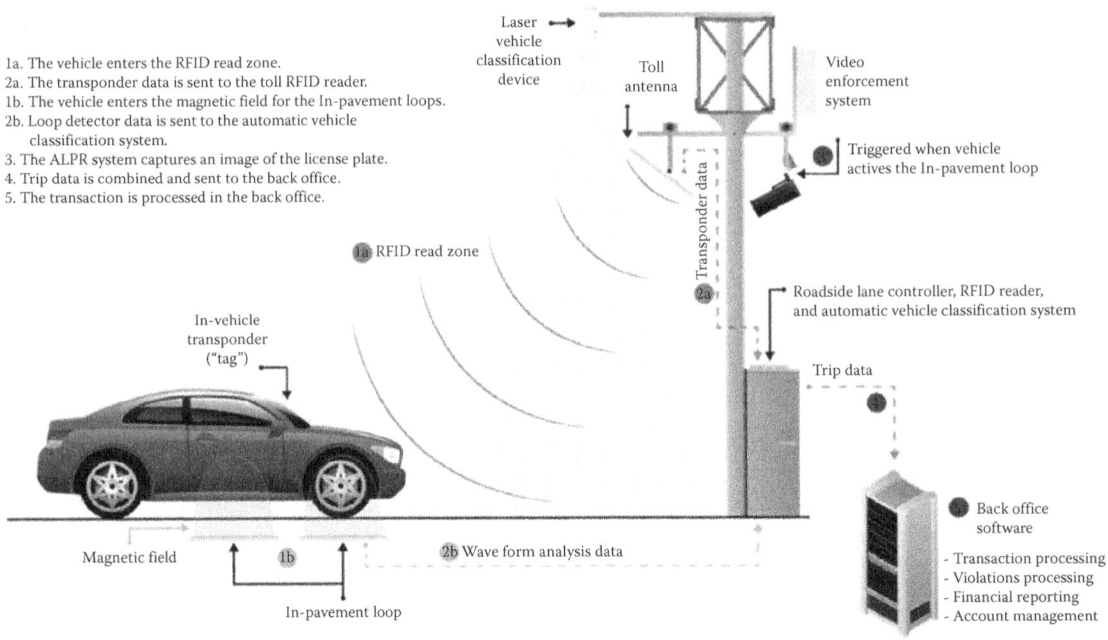

Figure 3.6 Example of ALPR deployment as part of a "total lane solution."

Figure 3.6 shows how ANPR systems can be integrated as part of a "total lane" solution with various third party devices such as dedicated short range communication/radio frequency identification (DSRC/RFID) readers, weigh-in motion (WiM) monitors, variable message signs (VMS), and inductive loop detectors. In the figure, the ANPR system is incorporated in the tolling and parking or ticketing system.

The ANPR system can track vehicles through complicated road networks and over long distances, with a combination of both fixed locations and mobile camera technology. This can be particularly useful as the system provides interoperability across command and control centers. This is important to flag vehicles of interest and apply appropriate rules for the vehicle passage.

3.6 INTELLIGENT TRANSPORT MATERIALS

While the remainder of this chapter focuses on optoelectronics case studies, it is also important for us to recognize the rapidly emerging demand for vehicles to also identify infrastructure by themselves.

This shift in focus is set to significantly accelerate through the introduction of autonomous and connected vehicles. It is important to transfer information from the large variety of other optical devices such as light detection and ranging, light curtains, parking sensors, infrared, and time of flight imaging techniques. These have been described in earlier sections; on-vehicle uses will also be discussed in Chapter G 2.1 on "Optoelectronics for Automobiles, Vans, and Trucks."

As of now, there has been very little focus on improving infrastructure for the interaction between vehicles and roadside materials. This is particularly the case when compared to the significant investment made by automotive manufacturers for the development of these connected vehicles—the autonomous vehicle market is currently estimated to reach USD 42 billion by 2025 and forecasted to reach USD 77 billion by 2035.

Many companies, including 3M, are also working to reinvent the road surface and roadside materials, and their possible modes of interaction with future autonomous vehicles. These new intelligent transport materials, referred to as ITM, will support wayfinding, vehicle security, and safety to optimize journeys with greater situational awareness. Many of these will be passive optical materials, but the use of active materials may also be viable in future. It is reasonable to trust that infrastructure will be more situationally aware, sharing information, before the

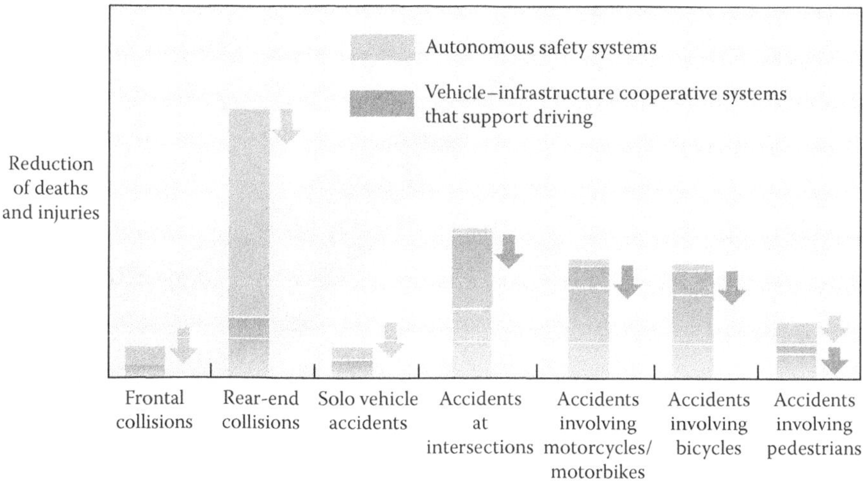

Figure 3.7 Impact of vehicle–infrastructure cooperative systems: Toyota Cooperative ITS. (Courtesy of http://www.toyota-global.com/innovation/intelligent_transport_systems/infrastructure/.)

vehicle approaches, on potential hazards such as level crossings and hidden junctions.

These materials will be part of our future connected infrastructure. The optoelectronics domain must not only consider how to develop the sensing technologies but also the materials being sensed that might include, for example, road markings and road signs with active handshaking between vehicles and transport infrastructure.

Toyota, for example, highlights the importance of cooperative ITS to prevent traffic accidents through supported driving. The infrastructure will bring situational awareness to road users that cannot detect them even using their vehicles' own sensors. Figures 3.7 and 3.8 highlight the important role that infrastructure will play on road safety (see Figure 3.7 and 3.8).

According to the World Health Organization, the total number of road traffic accidents in 2013 resulted in 1.25 million deaths, globally. This can be calculated as 1 death every 25 s. EVI technology is one part of the solution and ITM will follow.

3.7 INTELLIGENT INFRASTRUCTURE: ANPR FOR SMART CITIES

3.7.1 Case study examples

3.7.1.1 PUBLIC SAFETY AND SECURITY

We shall now discuss the use of vehicle identification for safety and security, possibly overlapping a little on the earlier discussions on this in the chapter that concentrates on this aspect.

Camera-based EVI-JTMS (Journey Time Measurement Systems) solutions directly contribute to safer cities through improved security, reduced road traffic accidents, and vehicle-related criminal activity. These EVI solutions provide invaluable data that may also protect the law enforcement agencies and municipal support teams with intelligence from licence plate reading cameras, prior to any intervention into potentially dangerous situations.

The intelligence generated by these optoelectronic EVI systems may also be distributed across wide networks of geographically dispersed teams to drive frontline intelligence from what may be separate data sources, such as known stolen vehicles, road closures, damages to a road network, or a public event that needs to be considered for operatives to be situationally aware.

Towards this goal of interoperable systems for public safety and security, EVI technology, such as smart ALPR cameras can distribute their data to multiple partners, to provide multiple services from the same camera network. This multimodal approach has been seen to both monitor vehicles that may have been involved in criminality in and around a city, and monitor traffic flow to help local authorities manage and optimize the flow of road users in and around the city, reducing the upfront costs of the EVI system. Furthermore, this approach significantly

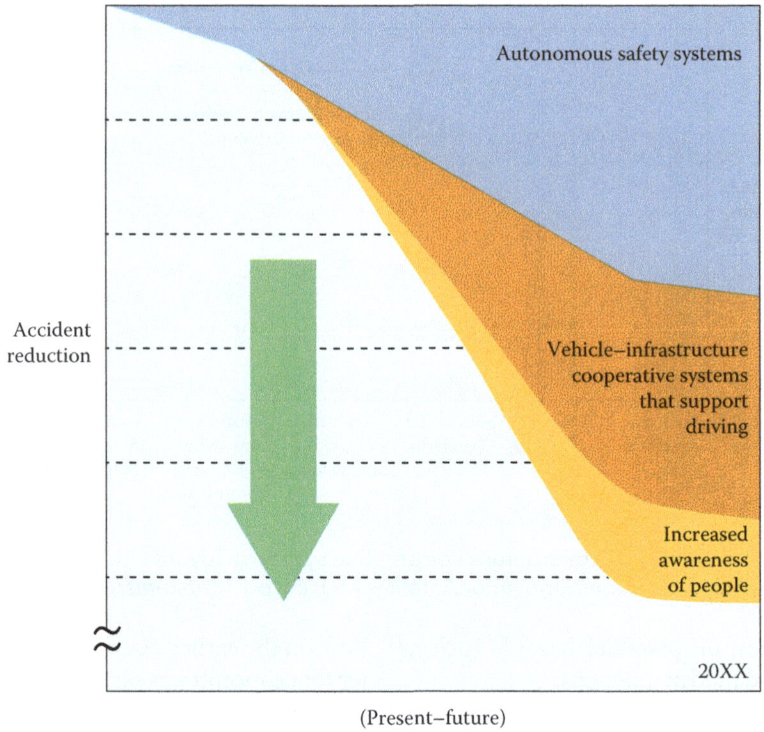

Figure 3.8 Impact of vehicle–infrastructure cooperative systems: Toyota Cooperative ITS. (Courtesy of http://www.toyota-global.com/innovation/intelligent_transport_systems/infrastructure/.)

reduces the environmental impact with a reduced number of installations and disruption to current infrastructure.

ANPR technology is widely used for security purposes to monitor vehicles of interest and for the early detection of vehicle-related crime or more serious security threats. These ANPR cameras will not only capture and read the license plates of regular cars and commercial vehicles, but they will also capture and read the license plates of motorbikes and mopeds, where the system has been configured to do so.

License plate information can be processed locally or distributed to a central application, where comparisons are made against known vehicles of special interest. If a positive comparison is made, the system owner will be notified and the system may be used to track the vehicle. These systems provide law enforcement agencies, with time and date synchronization for each of the detected vehicles that can be compared with a police national database, for example.

Integrated embedded camera systems are particularly useful to meet these challenges, often selected to limit the impact on the cityscape and neighboring infrastructure, with lower maintenance costs and "onboard" processing, with wireless capabilities. These embedded smart ANPR solutions offer remote configuration, real-time processing, and greater redundancy support when compared to distributed cameras driven and controlled by a separate processor.

With a network of embedded cameras, the downtime lost due to hardware, software failure, or even vandalism is limited to the individual camera, without impacting the entire network. Data buffering during times of communication outage can be supported to overcome poor connectivity, for a limited period of time, between the outpost ALPR camera and the in-station aggregator.

The UK National Traffic Information Service is one example of a JTMS system, with a deployment of over 4000 ANPR cameras in their national passive target flow measurement system, for travel-time monitoring.

A typical ANPR-based JTMS system is triggered when any vehicle passes the field of view. The camera then reads the number plate and sends the data via an Ethernet connection to the back office JTMS software. The data that are received by the JTMS software can be compared against all of the other data received and a journey time can be created for each vehicle. These applications enable traffic management teams to monitor the congestion levels and vehicle flow within a city to enact early traffic control measures if congestion starts to build.

In some parts of the world, ANPR technology has been shown to offer in excess of 98% detection accuracy, with a 95% read rate for detected vehicles, The technology can function in most weather conditions unless the license plate itself is obscured. It should be made clear that these performance indicators can vary significantly from country to country due to the differences in license plate design, their composition, and maintenance.

3.7.1.2 RING OF STEEL

ALPR has been globally adopted, by many law enforcement agencies, to deal with vehicle-related crimes. These integrated systems are often referred to as a "ring of steel" to protect borders, counties, and cities monitoring movements of the vehicles as they enter in and out of the EVI perimeter. The most significant and widely known "ring of steel" is the EVI perimeter surrounding the city of London, comprising over 1500 surveillance cameras and ANPR systems that reportedly scanned 75 million vehicles in the first 3 years, as referenced in Haines (2009): "The role of automatic number plate recognition surveillance within policing and public reassurance."

Law enforcement agencies rely on these systems, where a typical deployment may include 30+ cameras located on major routes in and out of the network. The system works by reading the number plates of all the passing vehicles and comparing them with a site-specific, regional, or police national computer database. Vehicles of interest are monitored as they pass the cameras, enabling their whereabouts to be checked and monitored for vehicles of interest. The system can be configured to discard vehicles that do not match predefined criteria.

The primary purpose of these systems is to provide ANPR capability in specific locations to obtain real-time data and information about vehicle movements throughout the city at strategic locations on arterial vehicle routes. These rings of steel can be used to identify matches to police databases to support a pro-active policing response to suspicious vehicles and to view and monitor events as they occur. (Figure 3.3 shows a typical mobile and fixed system.) EVI systems are highly scalable and open to integration with other sensing technologies due to the advances in the onboard processing capability of these "smart" embedded cameras. For example, ANPR/ALPR data can be combined with other data such as vehicle weight, to ensure safe loading as vehicles are weighed in motion, WiM.

The Dutch Ministry of Transport, the Rijkswaterstaat, was concerned about excessive damage to main roads caused by overweight trucks. In view of this, the Rijkswaterstaat awarded the PAT Company in Germany a contract to install seven weight-enforcement systems on the motorway network.

These sites capture images of overweight vehicles detected by WiM sensors. Each monitoring point has four lanes monitored by ANPR cameras and color overview cameras.

When a vehicle is detected by the WiM system and is calculated to be overweight, a set of text and image data on that vehicle can be sent from the processor to the WiM processor and forwarded to a main control office.

3.7.1.3 ROAD USER CHARGING AND JTMS

The adoption of ALPR for traffic flow optimization remains an important application for this technology, utilized by local authorities in large city centers to devise methods of reducing peak-time traffic density in the most critical areas. To help ease this problem, ALPR systems can be installed to monitor and control traffic flow, monitor the overall use of the road network, and improve safety.

Again, JTMS systems automatically read vehicle licence plates at two or more points on the road network. The licence plate data captured at the roadside can then be transmitted to a central computer system where the data from different sites are matched to calculate the exact travel times between pairs of sites.

Calculations of journey times across large networks can be performed at specific intervals, ranging upward from a fraction of a minute and generating accurate and detailed data of traffic conditions across the road network in near real time. This information enables traffic managers to study performance and trends over a long period of time and support the decision to spend budget

Table 3.1 Some of the benefits achieved from the London congestion charging programme, from the consultation impact assessment (November 2012)

Vehicle kilometers fell by almost 19% between 2000 and 2009.
TfL reported £136.8 m net income from congestion charging in the financial year 2011/12.
30% reduction in traffic delays within the first 12 months, within the charging zone.
Traffic levels showed a reduction of 18% in traffic entering the zone during the first year.
29,000 additional bus passengers entered the zone during morning peak periods.
Congestion charging contributes £50 m to London's economy, mainly through quicker and more reliable journeys for road and bus users.
There remains no evidence of any significant adverse traffic impacts from the charge.
The number of penalty charges issued average 165,000 per month (110,000 charge-zone payments per day).
65,000 fewer car trips into or through the charging zone each day.
Taxi, bus, and coach movements have increased by 20%.

on traffic-related initiatives to both councillors and the public.

London was the world's first major city to deploy a congestion charging scheme using ALPR in this way, thereby providing an enforcement infrastructure for the Central London Congestion Charging Scheme. Over 500 cameras were deployed without obstructing the flow of transport in and around the city and without the need for physical tollbooths, barriers, intrusive loop-based technology or supplementary RFID tags.

Embedded ALPR can provide video-based free flow tolling solutions as part of a flexible toll system without the need for physical tickets or passes. The embedded ALPR cameras read the vehicle registration as vehicles enter, drive within or exit the congestion charging zone to be checked against the Transport for London database. The network of camera sites monitor every entrance and exit to the congestion charging zone along the boundary road, and monitor journeys made within the charging zone.

Each camera site consists of both a color camera and a monochrome camera for each lane of traffic being monitored. The cameras provide digital images of the whole vehicle to the ANPR software, which then reads and records each number plate.

Before the congestion charging zone was introduced, London suffered the worst traffic congestion in the UK and among the worst in Europe, with drivers in central London spending 50% of their travel time in queues. It was estimated that the economic loss caused by congestion in London was in the order of USD 6 million every week (see Royal Geographic Society white paper). Table 3.1 shows some of the benefits achieved by the London Congestion Charging Programme.

Alternative examples of road user charging include the London Low Emission Zone (LEZ):

The UK's National Traffic Control Centre (NTCC) and the Alpine and Cross City Tunnels (CCT).

While these serve as good examples to discuss road user charging, there are of course many more deployments worldwide. These include the E470 toll highway in Denver, Colorado and the Elizabeth River Crossing in Virginia, USA (see Figure 3.9).

3.7.1.4 LOW EMISSION ZONE

The LEZ is an extension of the London congestion charging scheme that leverages ALPR as a sustainable technology, reducing emissions through optimized traffic flow, reduced congestion, and reduced journey times. Embedded ANPR technology has a low impact on the streetscape, due to their independence from external hardware and processors that may require separate roadside cabinets and associated infrastructure. These centralized systems can result in greater reconstruction of existing road surfaces and pedestrianized areas during their installation.

The aim of the Transport for London (TfL), LEZ scheme, was to improve air quality in the city by deterring the most polluting vehicles

Figure 3.9 ERC—Elizabeth River Crossings, Virginia, USA.

from driving in the area. The vehicles affected by the LEZ are older diesel-engine lorries, buses, coaches, large vans, minibuses, and other heavy vehicles, derived from lorries and vans. Cars and motorcycles are not affected by the scheme.

The LEZ scheme has over 300 cameras and is applied to all roads and some motorways across the majority of the Greater London area, which is substantially larger than the congestion charging zone, which, although it has been expanded, is still confined to a relatively small area in the city center.

As with the congestion scheme, no barriers or tollbooths are required for the LEZ system; fixed ANPR cameras read the vehicle registration number plate as vehicles drive within the zone. Vehicle numbers are then checked against a database of vehicles (a) which have been tested against the LEZ emissions standards, or (b) exempt or registered for a 100% discount, or (c) for which LEZ charge has been paid.

3.7.1.5 NATIONAL TRAFFIC CONTROL CENTRE

The NTCC project is one of the largest known JTMS systems to reduce the effects of congestion on England's motorways and major trunk roads by informing motorists about incidents and congestion.

The prime function of the NTCC project is to collect, process, and distribute strategic (wide area) traffic information, including the setting of roadside VMS and other dissemination media, using pre-agreed protocols, to assist travelers in planning their journeys.

These systems also support the UK Highways Agency and its operational partners in optimizing the use, management, and operation of the road network.

3.7.1.6 ALPINE AND CROSS CITY TUNNELS

From its opening around April 2005, Sydney's CCT project provided a number of significant benefits to the city, including improved traffic flow, enhanced public transport, dedicated cycle ways, and improved pedestrian amenity. The CCT is a fully operational "electronic" road with no cash booths or barriers, so vehicles can travel through the tunnel without slowing down to pay.

Tolls will be collected either by detecting electronic tags on the front windscreens of vehicles or by automatic identification of the vehicles' number plates. Windscreen tags transmit a signal to the tolling equipment as the vehicle passes one of the tolling points in the tunnel. This registers the vehicle's use of the tunnel so that the appropriate toll can be deducted from the user's prepaid account. Motorists without tags will pay tolls by registering the number plates of their vehicles. Images of vehicle number plates are also captured, read, and matched against registered plates. Motorists without tags will be charged the toll plus an administration fee.

A video enforcement system provides proof of passage of a vehicle at a specific tolling site. The system relies on embedded ALPR recognition processors and infrared illumination, with rear scene images in full color, when deployed with ambient light sources or directional floods for 24h color

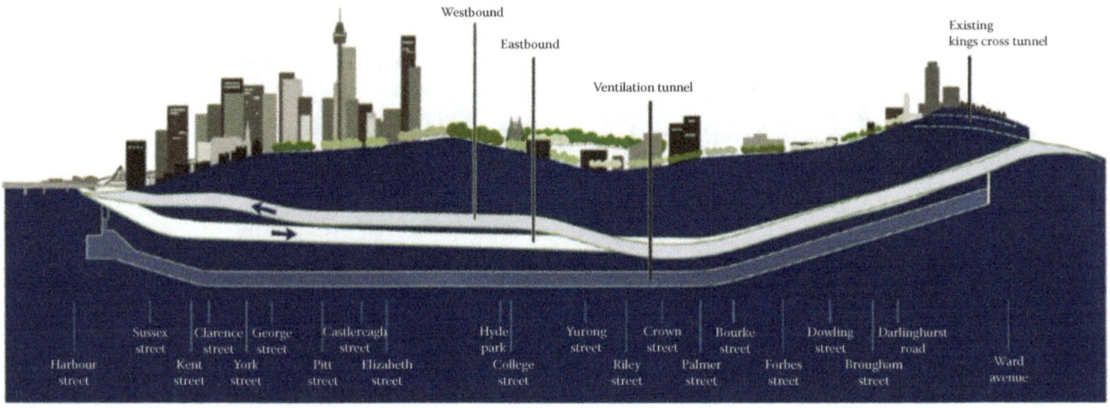

Figure 3.10 Sydney's CCT project. (Courtesy of http://www.rms.nsw.gov.au/projects/sydney-inner/cross-city-tunnel/index.html.)

imaging. This system was designed to achieve high-grade ALPR results and images without distracting drivers (Figure 3.10).

3.7.1.7 CRITICAL INFRASTRUCTURE AND ACCESS CONTROL

ANPR technology is widely used to monitor and safeguard airports, gas stations, supermarket car parks, and motorway service stations. Supermarket car parks are notoriously difficult to monitor, as well as being a target for criminal activity, particularly when the supermarket is closed.

It is widely recognized that barrier-controlled systems can lead to long queues and delays on entrances and exits of car parks, which in turn deters customers from visiting. Parking attendants are not always able to effectively monitor very large-scale car parks and sometimes genuine customers cannot find a parking space. Petrol (gas) station "drive-offs" (leaving without paying for fuel) cost supermarkets and motorway service stations millions of Euros per year, with theft, fraud, and fuel smuggling, costing European governments up to €1.3 billion every year (Kennedy 2013).

The theft of fuel from supermarkets and highway service stations is an increasing problem. Many believe that the most effective way to manage this problem is through the use of ANPR systems that capture the registration details of vehicles that drive off without paying, which can then be passed to the police.

These smart cameras are also very useful for car parks of supermarkets and shopping malls, where free parking is usually allowed for genuine users of the facility. Because ANPR cameras are able to read the number plates of every vehicle as they enter and exit car parks, adding a date and a time stamp. These systems can calculate how long each vehicle has spent in the car park (see Figure 3.11). The owners of any vehicles which have not honored the "free length of stay" policy can then be sent a parking charge notice. ANPR cameras can eliminate the need for any barrier entry systems or ticket validation systems. This, in turn, vastly reduces queuing times at the entry and exit of the car park.

In this context, ANPR systems allow genuine customers to find parking spaces more easily by stopping people from using the car park for other purposes, such as while they are at work for long periods of time.

Meadowhall in Sheffield, UK is an example of a large commercial shopping center, which covers 1.5 million square ft of floor space, contains over 280 shops and caters for 24 million visitors a year, with 6 car parks holding 12,000 free parking spaces.

The Meadowhall installation has the world's first fully integrated digital ANPR cameras that incorporate both the camera and the recognizer/processor in a single sealed enclosure. This ANPR system supplies data on traffic flow into the center's existing security and surveillance system, giving information on vehicles entering and exiting car parks.

The enterprise software, or in-station, is an aggregator for ANPR data that also allows car park operators and management to view the times at which most customers visit the center, which of their car parks are most used, and how long customers typically spend in the center, as well as controlling access to vehicles entering and exiting their service entrances.

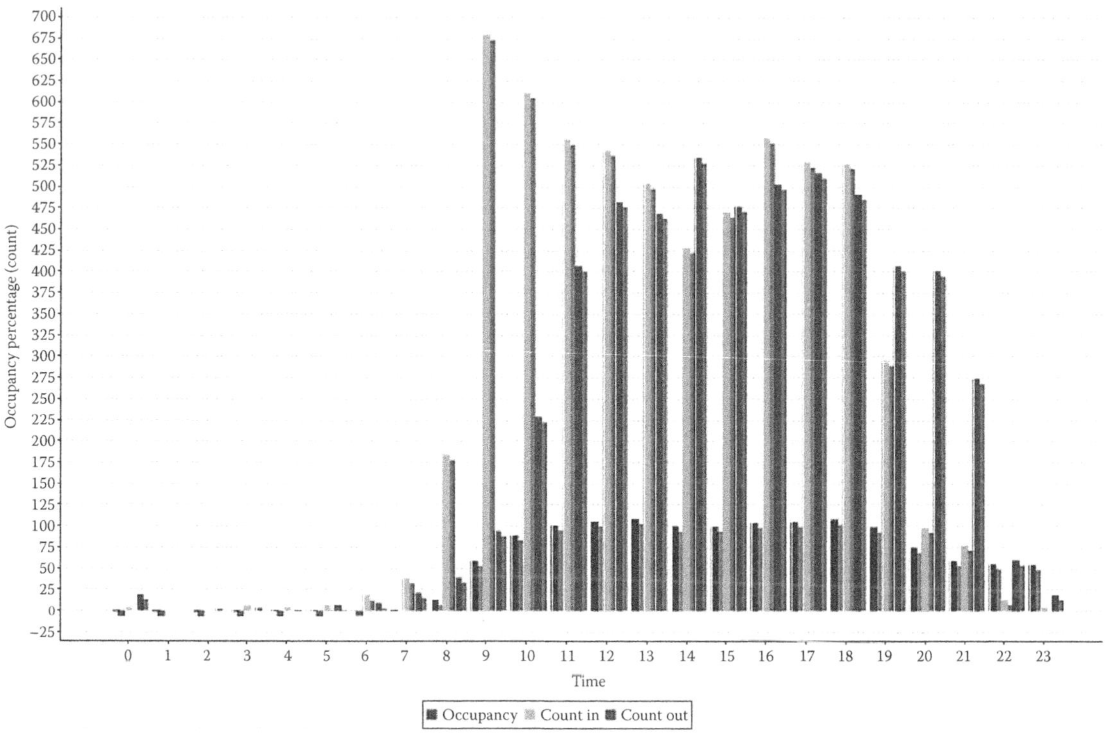

Figure 3.11 Typical car park occupancy results.

3.8 CONCLUSIONS

In this chapter, we have shown the growing importance of vision systems for law enforcement and the control of vehicular traffic in a wide variety of scenarios. The economic value is already evident, and the use of such systems is expanding rapidly to drive greater peace, safety, and security as part of the smart city intelligent infrastructure.

We can see that the increased focus on autonomous vehicles will also become a more central consideration for these smart cities that will drive the development of both ITS cooperative systems and ITM.

There are many more applications that could be discussed, and many more to come, particularly as these intelligent optical systems are well placed to drive the future Internet of Things megatrends, intelligent infrastructure, machine to machine, and more.

While improved traffic flow is the main impact of these solutions, advanced border security and key route surveillance help public and private sector agencies all over the world to manage complex transportation infrastructures more efficiently and to ensure smoother and safer travel experiences.

REFERENCES

Busch, R, Making European Cities Smarter, European Commission: Digital Minds for a New Europe, http://ec.europa.eu/archives/commission_2010-2014/kroes/en/content/making-european-cities-smarter-roland-bush.html, 2014.

Haines, A, The role of automatic number plate recognition surveillance within policing and public reassurance, Doctoral thesis, University of Huddersfield, 2009.

Kennedy, C, Europe loses billions each year to fuel theft and fraud, Oilprice.com, 2013.

Paxman, S, Intelligent infrastructures: unlocking the digital business, Accenture white paper IP Expo Europe, October 9, 2014.

Royal Geographic Society, Congestion Charging Ahead, Royal Geographic Society White Paper, 2003.

Transport for London Consultation and Engagement Portal, Public and stakeholder consultation on a Variation Order to modify the Congestion Charging scheme, Consultation Impact Assessment, 2012.

4

Optoelectronics for control of city transport

DAVID PARKYNS
Transport for London

In this chapter, we discuss the application of optoelectronics for controlling the safe and free passage of road vehicles and pedestrians in London, and for monitoring the possible causes of congestion, such as roadworks. Much of the existing optoelectronic monitoring is done by video cameras linked to intelligent software, but individual sensors such as digital "snapshot" cameras on traffic lights, pyroelectric heat detectors, laser scanners, and optical rangefinders may also have potential application. This may also be a possible application area for all-round (360°) cameras, as discussed in other areas of this volume. Not all sensors, of course, use optoelectronics (e.g., Doppler radar, inductive loops, etc.); a discussion on these sensors is outside the theme of this text, but sensor fusion is important for the overall system.

Optoelectronics is also used extensively in high brightness light-emitting diodes (LEDs), as part of modern traffic lights and active roadside information signs.

4.1 BACKGROUND

Of the 13,800 km road network covered within the Greater London boundary, 580 km of key arterial and orbital roads (the Transport for London Road Network [TLRN]) is managed by Transport for London (TfL). Figure 4.1 shows a map of the overall TLRN area. The road system is vast and complex, with the following journeys taking place daily: 9.8 million by car and motorcycle, 6.3 million by bus, 6.2 million walking trips, 0.5 million cycling, and 0.3 million by taxi. In addition to these journeys, nearly all freight movement in and out of the city is by road (see Figure 4.1).

London's roads are approximately 40% more densely trafficked than in any other UK conurbations; as a result, London experiences 20% of the UK's traffic congestion. This is estimated to cost its economy at least £2 billion a year. This is the current situation, and with London's population set to grow to reach 10.5 million by 2041, the demand on the road network is expected to increase dramatically.

Corridor managers supervise the overall performance of each of the TLRN corridors on a day-to-day basis, identifying and prioritizing improvements to maximize journey time reliability. TfL maintains all traffic signals in London and actively reviews the performance of hundreds of traffic signals every year to ensure that they continue to operate efficiently, maximizing benefits for all road users.

Figure 4.1 Street view of London.

Figure 4.2 Traffic light with camera and sign.

A close cooperation between TfL and the Metropolitan Police Service (MPS) has launched dedicated policing in the Blackwall Tunnel in East London, contributing to improved reliability and faster incident resolution at this critical location. Road closure times have further been reduced using laser scanning equipment to facilitate quicker investigations and establishing a joint roads reopening protocol. The continued funding of the MPS roads policing teams has allowed them to tackle congestion through managing collisions, traffic signals failure, and vehicle breakdowns.

During the London 2012 Olympic Games, TfL utilized a sophisticated traffic management system consisting of a considerable number of junction layout changes, dedicated "games lanes," and an innovative traffic signal strategy in order to meet ambitious journey time targets for games-related vehicles. Extensive partnership working across TfL and with other organizations also provided the capability to react to and address the issues on the road network as they arose. The outcomes of this were overwhelmingly positive, with the agreed network performance targets exceeded during the Games. The technology delivered and the lessons learnt from this experience provided an example of what is possible in road space management, given a suitable level of investment and resource.

4.2 THE SPLIT CYCLE OFFSET OPTIMIZATION TECHNIQUE SYSTEM

The split cycle offset optimization technique (SCOOT) system allows traffic signals to detect vehicles passing along a road and work together to amend their signal timings on a second-by-second basis to adjust traffic flows accordingly through an area, making journeys more reliable. It has helped deliver an average 12.7% reduction in delays for vehicles traveling across the network. At some locations, this increases to almost 20%. Figure 4.2 shows a generic camera on a traffic light.

As a result, London's road networks are now kept running smoothly from day to day by these sophisticated traffic control system technologies; they are able to react to the changing conditions of the road network. At the time of writing, SCOOT is installed on over half of London's more than 6000 traffic signals, utilizing a network of sensors, which feeds information to a central control system used by the control center, which then optimizes traffic signal states to adapt to real-time traffic conditions. TfL is now working on expanding this to cover three-quarters of all traffic signals, helping to expand the benefits to a wider area of London. It also utilizes its wide network of technology to collect additional information on its network, ensuring a real-time awareness on how the network is operating. When events and incidents (both planned and unplanned) have an impact on the road network that is too great for the technology to mitigate without intervention, TfL's 24/7 traffic management resource steps in.

To meet the ever-increasing and often competing demands of TfL's wide range of customers and stakeholders, TfL must constantly develop its capability to monitor and provide for all modes, including buses, pedestrians, and cyclists. Some of the technologies and infrastructure being developed for this purpose include bus and cyclist priority points, which address journey times at key locations on the network.

High-quality bus priority is a key deliverable for ensuring an efficient and reliable service for more than two billion passengers using the bus network each year. TfL continues to develop the interaction between SCOOT and selective vehicle detection for buses, using iBus data to investigate how to improve bus reliability and prioritize buses more effectively. Teams within TfL are also working together closely to understand the pinch points for buses on the network and devise solutions to address these.

The introduction of the "pedestrian SCOOT" is the first of its kind in the world and uses the state-of-the-art video camera technology to automatically detect how many pedestrians are waiting at crossings. It enables the adjustment of traffic signal timings automatically to extend the green pedestrian invitation to cross phase when large numbers of people are waiting, allowing more people to cross the road (Figure 4.3).

In addition, TfL has developed a "call cancel" technology, which can detect when a pedestrian has pushed the crossing button, but has then either crossed before the signal went green or walked away, thus allowing cancellation of the pedestrian crossing phase.

This latest initiative follows on from TfL's successful development of pedestrian countdown technology, which, via a LED display, tells pedestrians how long they still have left to cross the road once the green pedestrian invitation to cross indicator has gone out.

More than 900 pedestrian crossings across London have now been equipped with pedestrian countdown technology with TfL committed to install the technology more widely across the capital in the coming years.

Figure 4.3 Pedestrian SCOOT system in London.

The recent trials of pedestrian SCOOT and the call cancel technology are also an example of how TfL will use innovation to change the management of London's road network to better reflect the character of the local area, Subject to the outcome of the trials, TfL is hopeful that it can further develop the technology to use at other high-footfall areas, such as outside sporting venues or along busy high streets.

4.3 TRAFFIC MONITORING AND MANAGEMENT

The development of intelligent video analysis technologies plays a large role in the management of the road network. An image recognition and incident detection system on some closed-circuit television (CCTV) cameras can recognize abnormal levels of congestion and provide an automated warning.

The London Streets Traffic Control Centre (LSTCC)'s real-time traffic management is dependent on the continued improvement and expansion of technology such as SCOOT and better means of monitoring the network and communicating with customers. Access to more than 1300 CCTV cameras through TfL's own infrastructure and formal camera sharing agreements with the MPS, the Highways Agency, and 20 London boroughs ensures the best use of resources across agencies and covers a large portion of London's strategic road network. (There is no overlap between these dynamic cameras and the fixed view Safety Camera estate.) The operational coverage of cameras is rationalized through a CCTV Steering Group, which coordinates requirements across TfL.

A similar technology, used in our Automatic Roadwork Monitoring project, can measure the level of activity on work sites, making it easier to enforce permit regulations and ensuring completion of work in a safe and timely manner.

TfL also introduced the Transport for London Lane Rental Scheme in 2012. This was a key innovation to reduce roadwork delays, where contractors are charged up to £2,500 a day for working in congested areas and at busy times of the day (Figure 4.4).

Figure 4.4 Transport for London Lane Rental Scheme.

The scheme covers over 200 miles (57%) of the TfL road network, covering the areas most susceptible to major roadwork disruption. As an extra factor to reduce delays, TfL workers are not even exempt from their own rules, ensuring that their works are also delivered with minimal disruption! Not surprisingly, since the introduction of the scheme, approximately 95% of works have avoided the high charge by working outside of peak hours, and serious and severe disruption has been reduced by more than a third. By encouraging companies to carry out their work overnight or during off-peak hours, all road users, including drivers, cyclists, and bus passengers, benefit from more reliable journey times and less disruption.

The recently introduced Traffic Information Management System uses data from SCOOT to provide automatic alerts of areas of congestion. The System Activated Strategy Selection acts as a watchdog at key points on the network and automatically invokes signal timing strategies to prevent congestion from occurring.

Another measure, which increases TfL's ability to understand and control its network, involves the development and deployment of technology to deliver business intelligence and analysis in real time, allowing TfL to develop its capability of predicting the impact of both planned and unplanned events on the road network.

Currently, TfL is able to derive an extensive information on the TLRN but is limited in the extent to which information can be gathered on the rest of the national road network. Just as important as the development of these individual city-based systems is their integration on a national scale. Alongside the expansion of new technologies and capabilities, there is a dedicated effort towards ensuring the cooperative working of all intelligent systems used to manage the road network, ensuring the most efficient use of existing and future infrastructure and resources.

4.4 PROVISION OF TRAFFIC INFORMATION TO ROAD USERS

A key factor to reduce congestion is the provision of timely and accurate traffic information to road users to help them plan their journeys without disruption and delay. The public-access TfL website gives traffic information directly, which is also available by following @TfLTrafficNews on Twitter. The LSTCC also provides real-time information to the travel news media and other stakeholders through a dedicated portal. TfL has embraced the UK open data initiative by allowing more than 2000 third-party providers of travel information services (such as application and web developers, and satellite navigation providers) to get direct and free access to live data feeds through the TfL data portal. Examples of these data are live updates on traffic incidents and road closures, live bus countdown data, and traffic CCTV images refreshed every 2 min. The LSTCC updates over 140 roadside variable LED message signs with relevant, real-time, incident notifications and gives advance warnings of scheduled events and works. TfL also works with the UK Highways Agency to set reciprocal messages on each other's networks.

4.5 THE FUTURE

The ongoing evolution of this technology will allow for a more active approach to the road space management of London's road network, thus getting maximum benefit from the available network resources and ensuring optimum balancing of the competing demands for road space. It will also help TfL to improve the real-time network information that is provided to customers and shared with third parties for their integration into commercial services.

At the time of writing, many of these systems are already in use; research is still ongoing, and business plans and overall strategy for ever-more-comprehensive, more-closely-integrated schemes are continually being developed. It is not possible to reduce the size of the city, but safer and faster transport ensures full use of existing space and shorter travel times.

One final aspect that is set to revolutionize road transport is the likely future use of autonomous vehicles. Although some might still think of them as science fiction, they are so rapidly developing that it is no longer inconceivable that they will dominate our transport system. At the time of writing, progress is such that numerous successful real-city trials of "unmanned" (with a real driver on "standby" so far) vehicles are being reported, and it is becoming more and more likely that eventually only these autopiloted vehicles will be allowed on some major UK roads. Clearly, these vehicles will be capable of extremely rapid two-way communication with the fixed transport infrastructure of cities and highways, allowing even better transport planning options than ever before. As with any software-controlled systems, of course, the need for good cyber reliability and security will be even greater, to avoid a software error or deliberately introduced malware bringing the system to a complete halt.

PART II

On-vehicle applications in transport

This section covers the use of optoelectronics for transport, but deals only *with on-vehicle systems*, hence excluding applications on fixed structures. As we have seen, several aspects of the previous section could have been presented in this one, if we had not now restricted it to on-vehicle uses.

Table II.1 Summary of applications of optoelectronics in the transport field

Application	Technology	Advantages	Disadvantages	Current situation (at time of writing)	More reading
Sensors for observation of wheels and other moving parts of vehicles. Also for measuring speed of a vehicle relative to road or rail.	Smart video cameras. Optical (beam interruption) tachometers. Laser-Doppler velocimeters. Correlation velocimeters (these detect moving patterns such as tire tread or road covering).	Noncontact sensors, which can measure speed relative to stationary objects and also rotation rate of revolving parts. Useful for detection of undesirable locking of vehicle wheels during braking.	Limited to line-of-sight applications. Optics can become dirty or be obscured by snow. In many cases, cost issues still favor nonoptical solutions.	Optical sensors are being used in aircraft turbine engines. General vehicle use is likely to expand, as cost of optoelectronic modules reduces. Major applications exist for driving aids and in autonomous vehicles.	See also Volume II, Chapter 11 (Optical Fiber Sensors) and Part VI (Industrial Applications).
Navigational sensors	Ring laser gyroscope and fiber gyroscope, both based on Sagnac effect. Position updates from fixed stations.	Optical gyros are compact and have no rotational parts. Updates on position and orientation of road vehicles can be obtained from fixed stations (e.g., traffic lights or gantries) using optical communication links.	Ring laser gyros can lock when close to zero rotation, so need to have a small mechanical or optical bias. Very costly compared to Satnavs.	Ring laser gyros have been used on many commercial airliners, but costs of these and fiber gyros are too high for most automobiles. (Most automobiles rely on Satnavs, which are radio signal based.) Fiber gyros are well suited to monitor fast-moving objects, where zero-drift is less of a problem.	See also Volume II, Chapter 11 (Optical Fiber Sensors) for fiber gyroscope.

(Continued)

Table II.1 (*Continued*) Summary of applications of optoelectronics in the transport field

Application	Technology	Advantages	Disadvantages	Current situation (at time of writing)	More reading
Sensors for monitoring and locating external objects, for driving aids and autonomous vehicles	Smart video cameras, including 360° viewing types. Infrared cameras are starting to be used as a night vision aid for automobiles. Scanned lasers and light distance and ranging (LIDAR) systems can monitor 3D topology near vehicles. Pyroelectric detectors for proximity warning.	Many different optical and nonoptical (e.g., ultrasonic) technologies may be combined, allowing multisensor fusion and hence better decision making. Sensors can be coupled to displays and alarms to alert driver of hazards and used to provide automatic collision avoidance. A primary technology area for driver assistance and for fully autonomous vehicles.	Some of the more sophisticated optical sensors are still expensive. There is still some driver resistance to delegating control of vehicles to automated systems. Legal and moral situation for accidents involving autonomous vehicles is still a lively debated issue.	An extremely active field, with almost every major vehicle and vehicle component manufacturer involved. Optical driving aids and fully autonomous vehicles are no longer science fiction and will be a major contributor to improved traffic flow with previously impossible levels of safety! Autonomous working vehicles such as fork-lift trucks, farm tractors, mining trucks are likely to find earliest usage.	See this section.
Sensors to determine whether driver is alert	Cameras to detect driver blinking or becoming drowsy	Cameras with smart processing	Too expensive at present for general vehicle use. In future, may no longer be needed with autonomous vehicles!	A useful safety aid, likely to be more economic if manufactured in quantity.	See this section.

(*Continued*)

Table II. 1 (*Continued*) Summary of applications of optoelectronics in the transport field

Application	Technology	Advantages	Disadvantages	Current situation (at time of writing)	More reading
In-vehicle sensors for in-transit monitoring of the condition of roads, bridges, railway lines, overhead lines, etc.	Smart video cameras, including 360° viewing types. Scanned lasers. LIDAR systems.	Very long lengths of road, rail, etc., can be covered in a reasonable time. Particularly, useful for monitoring railway lines, road surfaces, and overhead power lines.	Expensive systems, only justified by the speed of coverage and huge labor saving. Close manual inspection may still be needed at times.	Established technology in many areas, but their use is expanding as technology improves.	Example: http://www.railway-technology.com/contractors/track/fraunhofer/
Vehicle lighting and displays	Originally used incandescent sources, but now nearly all are being replaced by arrays of high-brightness LEDs. Individual elements of LED arrays can easily be addressed selectively, so this allows them to be used in steerable headlights.	LEDs are ultrareliable, and most emit light in a forward cone, to maximize visibility without need for additional lenses. Addressable LED arrays allow compact displays to be produced. Using high-power focal-plane LED arrays can produce steerable headlights, without moving parts. Current LEDs are far brighter and more efficient than earlier types and costs, even for headlights, are now becoming competitive with tungsten lamps. Even ultrabright devices required for narrow-beam headlights are now being produced using LEDs and/or blue lasers with white-re-emitting phosphors.	No real disadvantages	LED lighting is used everywhere! Displays are used in all modes of transport and also for entertainment screens. White LEDs are the best way of illuminating LCD displays. LED vehicle sidelights and headlights are becoming standard, with higher brightness headlights now using intense blue lasers with visible phosphors to produce white light.	See also Volume I, Chapter 6 (Lasers and Optical Amplifiers), Volume I, Chapter 10 (LEDs), and Volume II, Chapter 9 (3D Display Systems).

(*Continued*)

Table II. 1 (*Continued*) Summary of applications of optoelectronics in the transport field

Application	Technology	Advantages	Disadvantages	Current situation (at time of writing)	More reading
In-vehicle optical communications	Mainly optical fiber links using digital binary amplitude-modulation encoding. Polymer fibers are possible for short-range application in small vehicles. A very good option for aircraft and ships.	Light weight of cables, high bandwidth, and freedom from electrical interference and earth-loop problems are compelling reasons for use in aircraft.	More expensive for simple low-bandwidth connections. Possibility of optical "broadcast" links within vehicle cabins or engine compartments	Commonly used for high-speed data on aircraft and ships, but not widely used in automobiles. May become important in future as the cost of links and on-vehicle data bandwidth increases.	See Volume II, Part I (Enabling Technologies for Communications).
Sensors to detect weather conditions.	Water-on-windscreen detectors. Fog, snow, and ice detectors.	Mainly video cameras, but LIDAR systems can be used in aircraft.	Most still have a rather high cost for automobiles, but commonly used in aircraft.	Use likely to expand, as cost of optoelectronics modules reduces. An important factor in the design of autonomous vehicles, which can potentially be affected by bad weather.	See Volume II, Chapter 12 (Remote Optical Sensing by Laser).

5

Optoelectronics for automobiles, vans, and trucks

JOHN P. DAKIN
University of Southampton

5.1 INTRODUCTION

For many decades, automobiles and many other road vehicles had used essentially the same basic technology for their drive systems and passenger cabins, apart from a gradually evolving stream of improvements to performance and passenger comfort. In the past 20 years or so, however, far more advanced (smart) vehicle technology has started to take a major hold.

This evolution commenced initially with onboard computer controls, engine management systems, satellite navigation features, and various new safety features, in particular, braking control, and collision and skid avoidance. Automatic braking systems and traction control systems have now been a safety feature on all but the lowest cost vehicles for many years, and may eventually be a legal requirement for all road vehicles.

In the past few years, however, progress is taking the technology to far higher levels of sophistication, to the extent that a major part of the cost of vehicles is becoming that involved with provision of the "smarter" features. Optoelectronics started to intrude slowly at first, initially for light emitting diode (LED) panel lights, new, then for ultra-bright arc-lamp-based headlights. Now both discrete and inbuilt satnavs are common; various other flat screen displays have been used to replace traditional instruments and high-reliability semiconductor light sources are replacing all incandescent and discharge lamps.

The most recent developments, however, are far more disruptive in nature, causing a major revolution in driver safety levels, to the extent where they are eventually expected to lead to human drivers having no part in the driving process at all—in fact, progress is so fast that this may have already started to happen by the time you read this text!

To enable these advances, the level of digital information processing required in vehicles is increasing to an extent that it is stretching the limits of the technology, with the computational and sensor/actuator costs already becoming a serious fraction of the overall cost of the vehicles. This situation offers tremendous commercial advantages to major computer hardware and software companies in a way that has been the case in the past with manufacturers of typewriters, watches, and cameras. As such, therefore, it poses an existential threat to many established automobile manufacturers. Any that survive will have to adopt quickly, by developing their own technology very rapidly or buying-in (or collaborating on) the necessary high-tech capabilities.

Many of the various ways in which optoelectronics can play a part in vehicles were presented in Table II.1, but it will now be useful to expand on this preliminary discussion by taking a look at the application sectors in the following sections.

5.2 VEHICLE LIGHTING

5.2.1 LED panel lights

Almost all low power panel/instrument illumination lights in vehicles, originally using old-fashioned tungsten filament lamps, are now replaced by LEDs. LEDs are compact, have far greater reliability, are more rugged and consume far less power. As heat dissipation in highly efficient LEDs is less of a problem, they can be used in greater spatial density; hence, far more dashboard designs and locations are feasible. Lamp housing design is simplified and is often not even necessary. Unlike tungsten lamps, LEDs can also be used in arrays to create displays providing pictorial information. They can also be used to backlight liquid crystal displays (LCDs). As will be discussed later, there is now a general tendency to use various forms of 2D display to replace all the individual panel lights, meters, and other dashboard functions.

5.2.1.1 LED SIDE-LIGHTS, INDICATOR LIGHTS, AND BRAKE-LIGHTS

LEDs have also, for several years, been the lamp of choice for medium power vehicle side lights and as intermittent indicators for direction change or braking. Again, they have far greater reliability, are more rugged, and consume less power. Rear vehicle lights, in particular, are becoming a new minor "art" form with small arrays of LEDs allowing vehicle rear lights to have a much larger area and to have any desired shape to reflect the character of the vehicle.

5.2.1.2 MAIN AUTOMOBILE HEAD-LIGHTS

For many years, LEDs were not bright enough to replace the main vehicle headlights, which require intense (high visible radiance) sources to provide their closely collimated light beams. Merely adding more LEDs in the focal plane only permits an increase in total light output power by broadening the divergence of the beam; it was not useful for well-collimated main headlight beams. The development of ultra-bright LED white light sources has now, however, reached the stage where a single LED or a very compact array of LEDs has sufficient power and brightness (radiance) to produce the required beam, which requires an ultra-bright focal-plane light emitter. From such a near-point-source emitter, it is now possible to produce a well-directed headlamp beam of the desired power (see Figure 5.1).

By using a compact focal-planar array of several such very small LED sources, it is possible to "steer" the beam direction by switching on selected elements of the array. This steering is possible without any mechanical moving parts, because each lateral

Figure 5.1 An advanced multi-LED headlight from Hella.

position of a source in the focal plane of a lens (or of a concave reflector mirror) corresponds to a different beam angle in the far field.

This development means that a headlight beam can be automatically directed to left or right of the forward direction of the vehicle, for example, to follow a bend in the road. To facilitate this, the vehicle requires sensors that will either detect the driver turning the wheel or use smart vision systems to detect, in advance, the lie of the road ahead. The facility to have switchable arrays also allows an even more sophisticated antidazzle feature to be incorporated. This involves the detection of oncoming car headlights (again with optoelectronics, using smart camera sensors, or small detector arrays) and the responsive switching off of selected LEDs in the headlight, so as not to dazzle the oncoming driver or drivers. This has great advantages compared to a driver seeing the oncoming vehicle and just dipping the headlights, as it not only reacts instantly, but it can also still provide full illumination of the road and verges outside of the field-of-view region occupied by the other driver's vehicle and eyes (see Figure 5.2).

Recent developments involve the use of lasers in headlights. In this case, however, the laser light is used indirectly, by directing the output from a blue laser onto a focal-plane phosphor to create an intense white spot. The technology can be used to create multiple spots in a focal plane array, either by steering the laser or by constructing multiple light beam units within the same device. At the time of writing, Audi cars have been equipped with

Figure 5.2 Beam pattern from Hella multi-LED headlamp when dimmed in selected areas to avoid dazzle. (Courtesy of Hella AG.)

Figure 5.3 Optical arrangement of laser headlight, with blue laser exciting a white-light phosphor in the focal plane of a headlight via a silicon micro-optics scanning mirror. (Courtesy of Audi AG.)

Figure 5.4 Light pattern from an Audi headlight using main and dipped LED headlights with brighter laser-augmented far beam. (Courtesy of Audi AG.)

one of the earlier forms of such a headlamp, in a development with Bosch and Osram.

In this, the radiance of a multiple LED headlamp is augmented in each headlight by an additional laser unit. Focused spots, from up to four 450 nm Osram blue lasers per headlight, are scanned across the surface of the phosphor using a steerable silicon micro-optics mirror, enabling the direction of the far field headlamp beam to be varied according to the position of the illuminated spots in the focal plane of the headlamp lens (see Figures 5.3 and 5.4).

The re-emitted light from the phosphor has a color temperature around 5500°C (just a little less than that of sunlight), and the light can be automatically dimmed when camera sensors detect other road users in the path of the beam.

5.2.1.3 IN-VEHICLE DISPLAYS

Most modern automobiles now have a flat-panel display to repl§ace most, if not all, of the panel instruments. This includes all the hazard and warning indicators, general vehicle function, mileage indicators, radio controls and channels, satnav information and any video information from reversing cameras, infrared (IR) night-vision cameras, ultrasonic or camera-based parking aids, and so on. Much as the current case with televisions and computer displays, the most common technology is using back-illuminated LCD displays, but LED, OLED (organic light emitting diode), electroluminescent and plasma displays are also contenders. Clearly, this not only allows more information to be presented at an economic cost compared to the prior use of multiple moving-coil instruments, but also allows the driver to program the display as desired, selecting a wide variety of other display functions or options using steering-wheel-mounted switches, etc.

5.3 OPTOELECTRONICS FOR ADVANCED DRIVING AIDS

5.3.1 Sensors to detect driver's state of attention

An early aid to safety for top-of-range vehicles was a dashboard camera system facing the driver, which, with the aid of smart software, could detect signs of drowsiness or momentary sleeping and sound a loud alarm. Even when the eyes are open, there are various signs that show if a driver is drowsy, such as blink rate or characteristic movement patterns.

5.3.2 Lane-following sensors

Another early optoelectronic driving aid was provided by lane-following hardware. This can use cameras to detect if the vehicle is wandering to the side of or across the white driving lanes markers on the highway, and then to sound a warning to the driver. In a more advanced system, it can provide an automatic mode of operation to take over control of the automobile and hold it within lanes. Such systems are also commonly used in simpler "line-guided" vehicles, such as smart fork-lift trucks that can move material around factories or warehouses, and for simple guided people carriers to take tourists around a park, etc. by following a painted line.

5.3.3 Parking/reversing aids and simple collision avoidance systems

As stated earlier, many automobiles are fitted with miniature external solid-state cameras. The most common is a rear-facing camera having a very wide field of view, which acts as a very convenient parking aid. This can, with the aid of the panel display, help the driver to see the distance to any object close to the rear of the vehicle when reversing. This guidance can, of course, also be carried out with, or assisted by, ultrasonic sensors which are then used to depict close-proximity "collision danger" areas on the same optoelectronics display. Such cameras or ultrasonic sensors can, of course, operate in any desired direction and, with suitable software, be used to not only provide warning of imminent collision, but can also be coupled to engine and braking controls to help prevent any possibility of collision.

5.3.4 All-round viewing systems

All-round viewing is a very useful driving aid to help detect the presence of nearby objects, or, if designed with image processing software to identify their nature. There are several ways of providing this facility. Key objects near the vehicle can be classified as passive stationary features, such as lamp posts, road signs, buildings, kerbs, and objects in the road, or as active objects, which are, or might start, moving. This less-predictable category includes other vehicles (autos, trains, bicycles), pedestrians, animals, etc. Ideally, a full-round system providing 3D distance information is needed. Non-optical methods, such as ultrasonic and conventional radars, can provide some of these features, but optical methods will play a key part in providing higher resolution detail.

One of the methods of achieving, viewing, and mapping of the surrounding 3D environment is with a scanned LIDAR (light detection and ranging) system (see Chapter 12 in Volume II). Short-range LIDAR systems operate over a range of 1–30 m and can provide distance information by measuring the two-way time-of-flight of a laser beam reflected back from nearby objects.

For LIDAR intended for vehicle use, where pedestrians, cyclists, and other drivers may be present in the field of view, a low-power laser is essential, preferably one operating at wavelengths such as 1500 nm, where eye safety limits are higher. Generally, because of the complex nature of the surface of the reflecting objects (pedestrians, cyclists, vehicles, etc.), these monitor delays in the amplitude modulation pattern of the laser, such as time-to-return of simple pulses or orthogonally encoded streams of pulses, rather than using optical coherence methods.

By scanning the direction of the laser beam in azimuth and elevation, and observing delays as a function of direction, a valuable 3D map of objects around the vehicle can be constructed. By measuring the situation at high update rates, the LIDAR signals not only provide range and direction, but can be processed to provide relative velocity information and hence determine the possibility of a collision.

In addition to LIDAR information, which has a limited pictorial resolution, a simpler roof-mounted 360° camera can provide higher definition information to back up the raster-scan type image from the LIDAR (see Figure 5.5). Such a camera, provided with suitable optics, can provide all-round vision with a reasonable vertical field-of-view range. (The system, which is also well suited for law enforcement activities, will be described in more detail in Chapter 6, Volume III on security and surveillance.)

Clearly, if two spatially separated cameras are used, they can also be processed to provide stereoscopic vision, much as human drivers can do, but with the potential for enhanced accuracy when the cameras are more widely spaced, and of course, the more rapid response possible using artificial intelligence.

Although all these optical systems can provide excellent information in good visibility, the situation becomes far less "clear" in bad weather conditions, particularly if the sensing optics is dirty or covered with water from mist, condensation, ice, or rain droplets. Surfaces can be cleaned with suitable miniature wipers, but, as we all know with a dirty window, external situations can become very difficult to decipher at times. There are possibilities of purging the volume near the lenses with clean air, but this does not help in fog, mist, or heavy rain. If the atmospheric visibility is bad, then the system is impaired in just the same manner as for humans, except that, in fog or mist, longer wavelength infrared cameras suffer less scattering and obscuration from small particles in the atmospheric path.

It is possible that, as costs reduce, vehicles may be required to be fitted with camera systems to record driving events, so that, in the case of an accident, they can be used to implicate or vindicate a driver or an external driver, cyclist, or pedestrian.

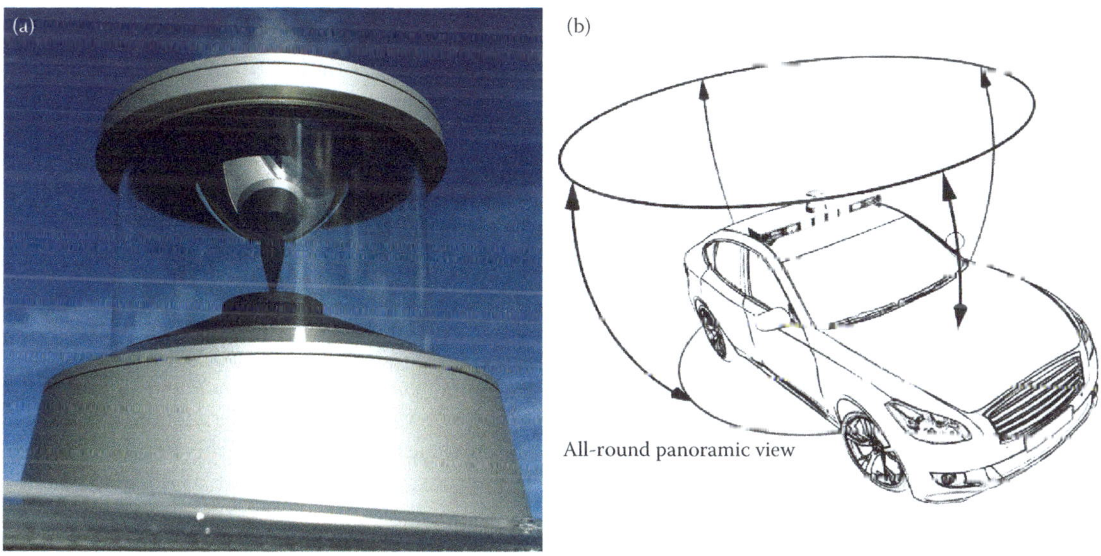

Figure 5.5 (a) A roof-mounted 360° camera to enable all-round viewing from a video camera. (b) Special optics, combined with postdetection video processing is used to produce the all-round view. (Courtesy of Observant Observations plc, UK.)

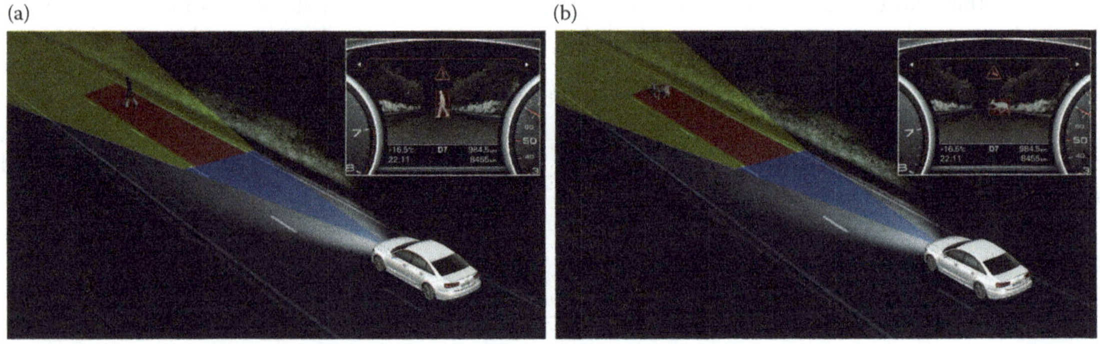

Figure 5.6 Schematic of an infrared night vision system for automobile use, suitable for early detection of unlit vehicles, (a) pedestrians, or (b) animals in the field of view. (Courtesy of Audi AG.)

5.3.5 Infrared night-vision driving aids

A recent development for night driving involves an infrared forward-looking video camera, which can view the road ahead in this useful region of the spectrum, and produce an otherwise conventional video display of the thermal scene (see Figure 5.6).

This facility, which has evolved from many years of military use, is particularly useful for advance warning of hazards, which are warmer than the surroundings such as warm-blooded animals, pedestrians, and even hot parts of vehicles. The camera is set to observe such objects at ranges at or beyond the limit where they might first become easily noticeable to the driver in the visible headlight beam, with the additional advantage that such objects will have a greater visual contrast due to temperature differences from their surroundings, and can also offer better visibility in mist or fog. This contrast can also be used to trigger an audible or visual alarm to alert the driver to the hazard.

5.4 THE NEW AGE OF FULLY AUTONOMOUS VEHICLES

It is the continued development of technologies such as those explained in the previous sections, combined with similar developments in sensors using other possible nonoptical technologies (such as acoustics, tachometers, radar, and millimeter wave technology, GPS satnavs, and inertial and road speed sensors) that is now leading to the greatest single step in motoring for over a century, the possibility of fully intelligent autonomous (self-driving) vehicles!

Human drivers are fallible, but it is expected that faults in autonomous vehicle hardware, and even software equipment can be reduced to near zero, either by careful hardware design or by providing partial redundancy, full duplication, or both. Machines can have multiparameter all-round sensing capability, and can be designed to sense parameters, such as radar echoes and infrared and ultrasonic images, that humans are unable to detect. With the power of modern processors, they can deal with huge quantities of information, yet react in tiny fractions of a second.

It may surprise some to know that similar systems have, in essence, been with us for many years, as most commercial airlines can essentially take off, fly, and land entirely without a pilot, if it was desired, although many of these systems use radar and inertial navigation aids to a greater degree than optics, as distance scales are usually much greater. In addition, many smaller airborne drones and sub-sea remote-piloted vehicles (RPVs) are equipped with many of the necessary sensing functions. In road vehicles, there is, of course, a far greater need to reduce cost, which is a problem.

Except, in the unlikely event of a major fault, an automatic pilot of an aircraft never loses concentration, falls asleep, makes mistakes or misjudgments. He also rarely imbibes intoxicating liquor! The same advantages are clearly attractive for road vehicles, as, even for the best of drivers, reaction times are significant, concentration inevitably varies, and many more are guilty of taking alcohol and drugs than pilots are. This has been borne out by early road tests of autonomous vehicles, which have been proven to have very few accidents despite having already been tested over large distances, albeit,

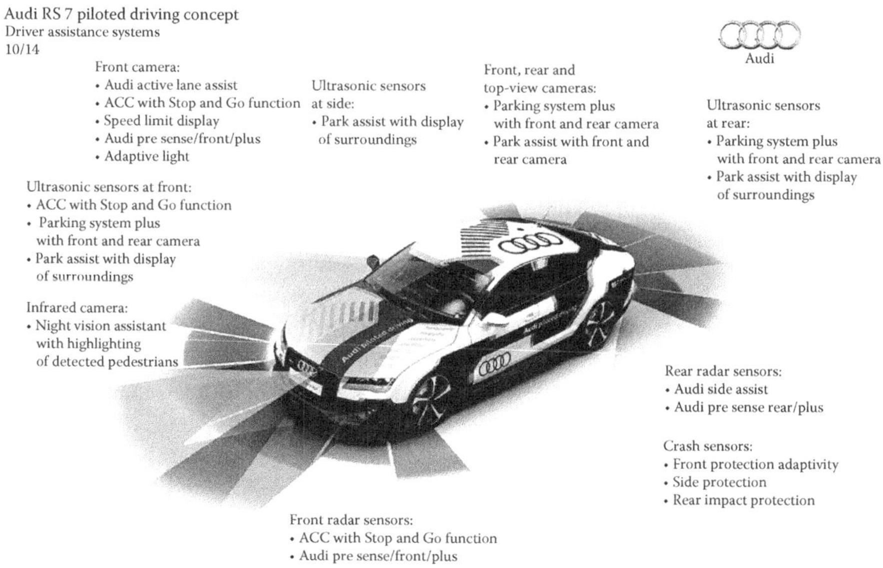

Figure 5.7 Schematic showing a piloted vehicle equipped with multiple sensors to provide driver assistance or permit piloted driving. Many sensing technologies are used, but many of the key ones for 3D viewing use optoelectronics. (Courtesy of Audi AG.)

at the time of writing, only when a human driver was present, in case it was necessary to take over control from the automaton.

The sophisticated systems required for such functions require not only a large number of advanced sensors (see Figure 5.7), but ever-increasing data storage, and computer processing to deal with the huge quantities of data in real time (see Figure 5.8).

The amount of data generated just by using 3D LIDAR scanners and multicamera sensing systems is truly enormous, particularly when rapid updates are required. In addition, highly sophisticated video pattern recognition and decision-making software is required in order to use the data to guide the vehicle in an appropriate manner.

It is probably due to the latter aspects that one of the first long-term road trials of an autonomous vehicle was implemented by Google, primarily, at the time of writing, still a software and computer-related hardware company. Many of the multidisciplinary skill sets required to develop autonomous vehicles and the ever increasing number of electrically powered vehicles are already present in the companies known for developing smartphones and computers. It is therefore of little surprise that, shortly before the time of writing, Apple Inc. announced that they would also be forming a company to develop smart vehicles, presenting a further major threat to existing auto manufacturers. This development is, of course, made all the easier by the advancing trend towards all-electric vehicles, where the design-challenge emphasis is on batteries, electrical power control circuitry for their discharge and recharging, and on peripherals such as motors, actuators, sensors, and displays.

Because of the disruptive nature of the new technology, there is no major motor vehicle manufacturer that does not respond to this threat by

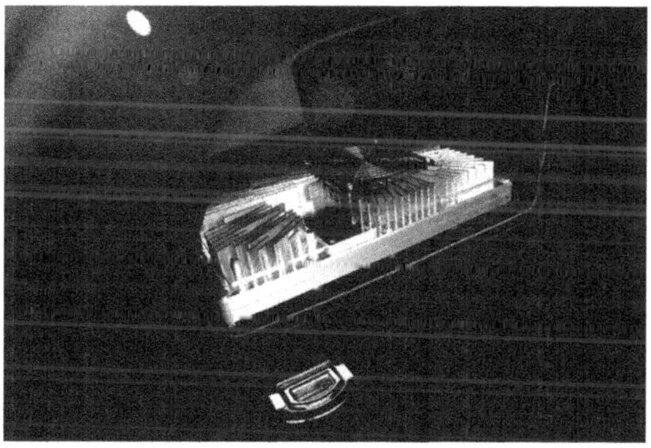

Figure 5.8 Photo of an in-vehicle data processor designed for piloted vehicles. (Courtesy of Audi AG.)

having a significant research activity directed at autonomous vehicles. By the time the reader sees this article, it is not unlikely that fully autonomous vehicles will be permitted on all roads without the requirement for a driver to be there as a backup. Already, low-speed autonomous vehicles are being used in some off-street applications (for example, to provide guided tours for tourist sites) and, as said earlier, it has already been proven that an autonomous car can drive in normal road and highway traffic with a better statistical success rate (lower accident rate) than a human driver.

The impact of autonomous cars will be enormous. First, software companies will have a major position in the vehicle business, even if not directly involved with the manufacture of complete vehicles. It is very likely that the vehicles will also be equipped with "handshaking" communications links to each other and to roadside street architecture. The ability for vehicles to transmit their position, speed, and even their short-term directional intentions to each other will allow road traffic to become far steadier and safer, even with the greater traffic flow rates expected. Complete halting, or even significant slowing down, at junctions will often be unnecessary, as vehicles could interweave seamlessly, even with the narrow gaps between vehicles possible because the autonomous systems have far faster reaction times. Because vehicles can travel in closely spaced trains at constant velocity, fuel economy would be far greater, and vehicle wear greatly reduced. Air resistance would be much lower because of "slipstreaming" and there would be less need for stopping, starting, and unpredictable evasive action.

The availability of autonomous vehicles will also enable many more members of society to be "nominally in charge" of vehicles, as there would be no need for the passenger(s) to have any driving ability. This opens the possibility of travelling freedom to nondrivers such as the infirm and aged and, with pre-programmed travel routes, even the very young (school run without parents?) and the mentally impaired. People can work on the way to and from work, leaving the vehicle to drive itself. Because of the potential safety benefits, it has been suggested that insurance companies specializing only in vehicle accident cover may soon become extinct! This will almost certainly have a major impact on the employment of taxi drivers, as it is highly likely they would be replaced by driverless cars for hire.

Because of the requirement for road travel to be more regimented, to fit in with the travel of closely spaced vehicles, the days of high-performance supercars are likely to soon be over. It may be that there will be a huge increase in vehicle uniformity, with many people having less desire to own a vehicle for sole use. However, for the image conscious, there may still be some residual degree of status remaining in car ownership, particularly if the external visual aspects of car design and the creature comforts of the interior space are important.

Despite all the promise, and some degree of "hype," there are still a few major problems to resolve before autonomous road vehicles are completely accepted. These are mainly based on legal liability issues, and capability in bad weather conditions.

There are particularly difficult problems regarding the decisions a vehicle must make when there is no possibility of avoiding an accident without breaking the traffic law or of causing another separate accident. The legal responsibility if an accident then occurs could potentially be a difficult one to resolve. For example, would blame rest with the manufacturer of the vehicle, the vehicle owner, the "driver" who initially set the vehicle system in operation (and who might initially have an option for taking the vehicle into a hands-on mode, if he chose), or the programmer who wrote the code. Regarding the other issue, if an unexpected event occurs, the vehicle computer may have to decide between several options of evasive action. Some of these options may involve breaking traffic laws (for example, crossing a solid white line, turning into a no-entry lane, or potentially more seriously deciding to turn and deliberately hit another vehicle to avoid a more vulnerable pedestrian). Deciding on the least serious option is a philosophical dilemma, which can often be decided by alert humans, but may require highly intensive programming for a nonhuman driver. It would clearly be particularly difficult for an automaton programmed to obey traffic laws to the letter!

Regarding bad weather, heavy fog, heavy rain, and/or dirt and grime from soiled roads is likely to be a problem. Heavy snow is likely to be particularly troublesome, as it can not only obscure visibility by resting on the optical surfaces, but can present an almost pure-white, featureless background, removing the visual cues of road markings, roadside verges and pavements (sidewalks), etc. on which the optical navigational software depends.

Long before road and highway vehicles are universally allowed to have such systems, there are niche areas where autonomous vehicles are likely to have a much earlier economic impact. This is in working vehicles on mining and construction sites, large-scale farm sites, and for vehicles such as forklift trucks in large factories or storage warehouses. All these work in a more closely controlled working environment, where time-efficient, fuel-efficient, accident-free operation, and labor cost reduction are major issues, and it may, with careful management, be arranged that no humans are normally present. Because of this, legislation may be easier to introduce than for the more complex cases of automobiles on roads and highways. The ability to operate safely, 24/7, regardless of temperature, humidity, and possibly even the need for good lighting, are huge advantages for many industries, so a major technical revolution is expected here, almost certainly over a much shorter timescale.

Once autonomous vehicles are fully accepted for road use, the first most likely area to be dominated by these will be for taxis and buses. Here, there is the additional free seat that would normally be occupied by the driver, the fuel economy, the more significant cost saving of not needing to pay for the salary of a driver and, in the case of the taxis, the avoidance of the potential risk of entering a vehicle with a stranger at the wheel. It appears that, at the time of writing, online taxi services such as Uber are monitoring this activity very closely. By the time of publication, or soon after, this application area may already be starting to be significant in countries that are amenable to having autonomous vehicles on their roads.

5.5 OTHER SENSORS

A fuller description of an all-round (360°) camera system, that may also be applied to vehicles for driver assistance and a possible other sensor for autonomous vehicles, is given in 6, Volume III. This technology allows a single conventional camera to see in all directions around the vehicle by using special optical mirror elements, with subsequent video processing to correct the inevitable distortions in the image.

5.6 CONCLUSIONS

The prospects for optoelectronics in vehicles are clearly exceptional. This is one of the areas where it will make the most disruptive changes to our society. In the future, when autonomous vehicles become the norm, no doubt many people will regret losing the many pleasures of driving, but the savings in time, fuel, human effort, and lives will almost certainly become the deciding features. Eventually, the concept of fallible humans being in charge of high-velocity machines, causing thousands of deaths a year, will no doubt be a cause for incredulity, with historians looking back at yet another aspect of the seemingly primitive behavior of past generations.

ACKNOWLEDGMENTS

The author wishes to thank Audi AG and Hella AG for permission to use photographs and drawings of piloted vehicles and advanced vehicle lighting devices and Observant Innovations for photographs of 360° viewing cameras.

BIBLIOGRAPHY

LED lighting

Haj-Assaad, Sami. LED Lights Go Mainstream for a Reason: From Luxury Rides to Economy Cars. AutoGuide.com (June 5, 2012). http://www.autoguide.com/auto-news/2012/06/led-lights-go-mainstream-for-a-reason-from-luxury-rides-to-economy-cars.html.

Audi MediaCenter. Audi Extends Its Lead with High-Resolution Matrix Laser Technology (April 28, 2015). https://www.audi-mediacenter.com/en/press-releases/audi-extends-its-lead-with-high-resolution-matrix-laser-technology-340.

Autonomous vehicles

The Economist. If autonomous vehicles rule the world: From horseless to driverless. (July 1, 2015). http://worldif.economist.com/article/12123/horseless-driverless.

Audi MediaCenter. Piloted driving (2016). https://www.audiusa.com/newsroom/topics/2016/audi-piloted-driving.

Lin, P. The ethics of autonomous cars. The Atlantic (October 8, 2013). http://www.theatlantic.com/technology/archive/2013/10/the-ethics-of-autonomous-cars/280360/.

6

360° camera systems for surveillance and security

GARETH J. EDWARDS
Observant Technology limited

6.1 INTRODUCTION

This chapter describes how the ubiquitous technology of surveillance cameras can be enhanced to provide a seamless all-round (360°) viewing platform. Standard cameras usually have a very limited field of view and this restricts usage in a variety of vision applications, such as surveillance, robotics, and optoelectronics for vehicle driver assistance. The technology also has potential for fully autonomous vehicles.

360° camera systems (see Table 6.1) have several advantages over standard fixed field of view and motorized pan, tilt, and zoom (PTZ) cameras, and these include the following:

- See everything—"eyes at the back of the head"—all-round visibility.
- Simultaneous imaging of the entire field, rather than sequential shots.
- Increased situational and contextual understanding.
- No motorized or physical camera movement.

There is one principle disadvantage:

- No optical zoom is possible, only post-detection zoom of the digital images.

There are several ways in which a field of view of an imaging system can be increased to cover 360°,

Table 6.1 Various 360° camera systems available in the market

Manufacturer	Name	Type	Description and web link
360FLY	360FLY	Fisheye	HD (high definition) video camera
Bubblescope	Bubblescope	Catadioptric	iPhone 4/4S and 5/5/SE adaptor
Dallmeier	Panomera	Polycamera	Multifocal sensor camera system(www.dallmeier.com)
Homestec	Homestec	Fisheye	HD video camera
Immersive Media	Various	Polycamera	Ultra-high-resolution photography and HD video (immersivemedia.com)
Kodak	PIXPRO SP360	Fisheye	High-resolution photography and HD video (kodakpixpro.com)
Kogeto	Kogeto Dot	Catadioptric	iPhone 4/4S adaptor
NCTech	iSTAR Fusion and iStar Pulsar	Polycamera	Ultra-high-resolution photography and HD video (www.nctechimaging.com)
Observant Innovations	Patrol and Sentry	Catadioptric	High-resolution burst (12 fps) photography (www.observant-innovations.com)
PANONO	PANONO 360°	Polycamera	Polycamera ultra-high-resolution photography
Point Grey	Ladybug (various)	Polycamera	Polycamera ultra-high-resolution photography and HD video (www.ptgrey.com)
Samsung	Gear 360	Polycamera	High-resolution photography and HD video
Sphere Optics	Sphere	Fisheye	Optical adaptor to convert a DSLR (digital single lens reflex) camera into a 360° camera (www.sphereoptics.io)

some of which, including their geometric and mechanical features, are detailed here.

It should be noted that for geometrically accurate 360° viewing or analysis, a full 4π steradian spherical projection must be accurately recreated. To achieve this, the entire captured image must have a single effective viewpoint. Fortunately, there are a variety of different ways in which this can be accomplished. For simplicity and brevity, we shall, in our discussions, consider the production of a single "still images or video". Achieving 360° video acquisition then simply involves the sequential processing of a sequence of such still images.

6.2 360° IMAGE VIEWING

Whichever optical process is used to acquire 360° images of a scene, significant image distortion is inevitable (a common example is seen in 2D atlas maps of the whole world). Fortunately, whichever optical system is used, all of their different heavily distorted formats can be "dewarped." This involves digitally mapping them to go from a distorted to a less distorted (or even undistorted) mapping. They can then be viewed using techniques such as spherical or equirectangular projection.

6.2.1 Spherical projection

To understand the concept of "spherical projection," it is best to consider that you are turning "on the spot" to sweep your view, while looking at the horizon, panning through a complete 360° rotation. By also considering that you could also look up and down through 90° from the horizon during this maneuver, you would then view everything, even to the limits of directly above and below. The term "spherical projection" is derived from imagining that you can view everything that you would see during this exercise, just as if it were "painted" on a sphere, centered on your head. This is the fundamental principle that underlies all imaging that requires the recreation of perspective, and where the entire captured image, or images, must have a single effective viewpoint.

Since a typically sensed image is rectilinear, the problem of representing a spherical panorama is, as mentioned earlier, the same as that of representing the surface of the earth on a flat map (Figure 6.1).

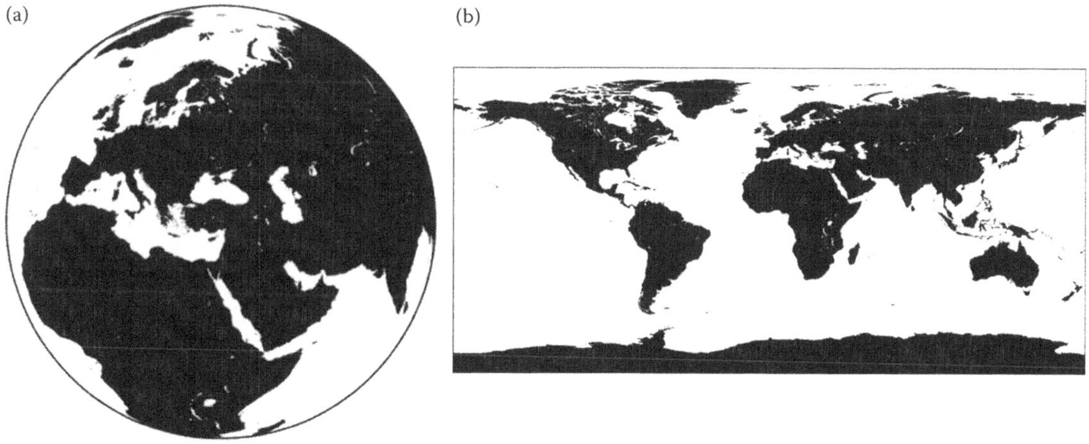

Figure 6.1 (a) Earth projected to (b) a flat surface using Mercator's projection.

Projection methods that have been developed for this include the well-known "Mercator's projection."

6.2.2 Equirectangular projection

Equirectangular projection requires that all lines of latitude and longitude are equally spaced. Because a projection of a complete sphere is typically twice as wide as it is high, and as the points representing the north and south poles are spread along the top and bottom, it is clear that such a projection from a spherical to a flat surface results in extreme distortion (see Figure 6.2). It also follows that digitally transforming and storing such a complete spherical panorama as a flat digital image will inevitably be inefficient as the increasing distortion when close to the poles results in ever greater waste of digital storage space.

6.3 360° IMAGE ACQUISITION

Most 360° cameras, or camera systems, do not image a complete spherical panorama. Typically, they image a field of view that is either a 360° section of a sphere or a portion of a sphere with the areas near the poles removed (like taking the top and bottom off an egg!).

Using these techniques, a complete part-spherical panorama can be digitally acquired and stored as a single image; it can be dewarped and displayed using one or more of the projection techniques described earlier.

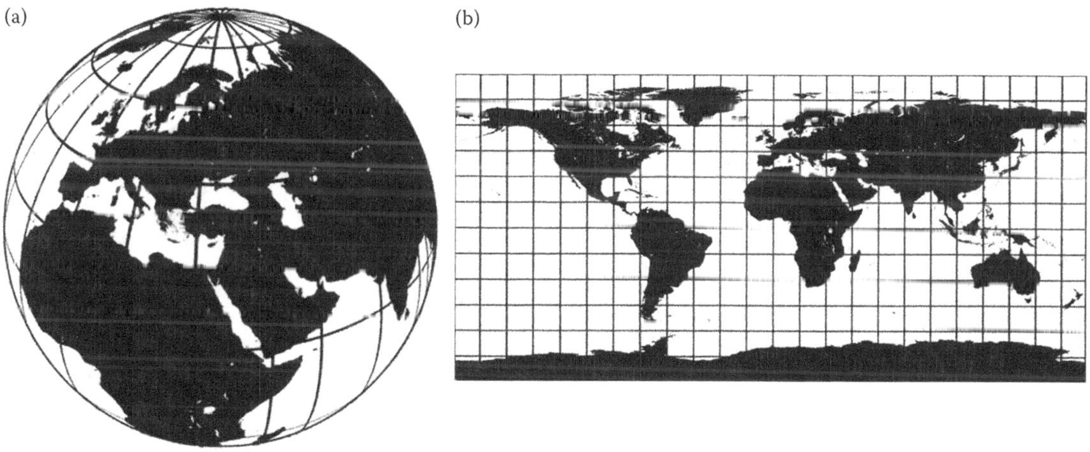

Figure 6.2 (a) Earth projected to (b) a flat surface using equirectangular projection.

There are many different optical arrangements used for 360° camera systems, the most common being categorized as catadioptric, poly-, spherical mosaic, wide-angle or fisheye. These are detailed in the following subsections.

6.3.1 360° Catadioptric camera

6.3.1.1 OVERVIEW OF 360° CATADIOPTRIC CAMERA

A catadioptric camera uses a shaped mirror—such as a parabolic, hyperbolic, or elliptical type to image objects. This is done either directly by reflection of a spherical panoramic scene, or indirectly by arranging a second reflection of part of the reflected spherical panoramic scene. This latter two-mirror arrangement is described as "folded."

To view a relatively undistorted reflection of a spherical panoramic scene, typically, we use a hyperboloidal mirror to yield an image of the scene around its axis. The angles of azimuth and altitude within such an image can easily be computed. For example, it can be clearly seen that all the points with the same azimuth in space appear on any radial line through the center of the image. This is the same for both conical and spherical projection. However, a hyperboloidal mirror has a special feature such that with hyperboloidal projection, the angles are independent of changes in distance and height.

Typically, the choice of whether to employ direct or folded optics is determined by the intended application. For vehicular- and post-mounted catadioptric cameras, where the sensor must look up and down to image a scene both above and below the horizon, a direct arrangement is used for reasons of connectivity. For usage where only looking above the horizon is required, a folded system is preferred.

The principle advantage of 360° catadioptric cameras is that, when geometrically accurate viewing or analysis is required, the recreation of perspective is simple as there is a single effective viewpoint at the focal point of the hyperboloidal mirror. It is also possible to compensate for radial lens aberrations using an image debarreling digital preprocess.

6.3.1.2 ON-VEHICLE USAGE OF A 360° CATADIOPTRIC CAMERA

A very important use of 360° cameras is providing a full 360° field of view around vehicles. The main early applications have been for surveillance and security, for example, on police vehicles, where simultaneous recording of events regardless of direction is particularly useful. There are many potential applications for driver assistance, involving detection of moving driving hazards (such as other vehicles and pedestrians) and stationary objects such as street infrastructure, and the technology may prove to be useful for fully autonomous vehicles (see Chapter 5.4 on transport applications).

We shall now show the types of images captured by a 360° catadioptric camera.

Figure 6.3 shows a spherical panoramic scene "seen" by a catadioptric camera as an annular image. A digital image processing program can "dewarp" the annular 360° image captured from a catadioptric camera and display a part of this as either a 2D flat panorama or as a 3D spherical view. The dewarping process is often referred to as "unwrapping." The 2D flat panorama can be displayed in full or can be split into parts (e.g., two 180° portions, see Figure 6.4). 2D image viewing tools can allow the user to zoom in to magnify a desired part of the 2D view.

A part of the original (typically high-resolution) 3D spherical view can be displayed using virtual 3D pan, tilt, and zoom (VPTZ). To achieve this, a virtual 3D camera, located at the center of the virtual 3D sphere, zooms and pans into portions of the (spherically dewarped and transformed) 3D digital source annular image (Figure 6.5). As this is done with the digital image, no physical camera movement is needed.

Such a VPTZ camera is often referred to as an ePTZ.

The VPTZ field of view can be interactively panned and tilted to simulate what a viewer in the real world might have seen if their point of view had

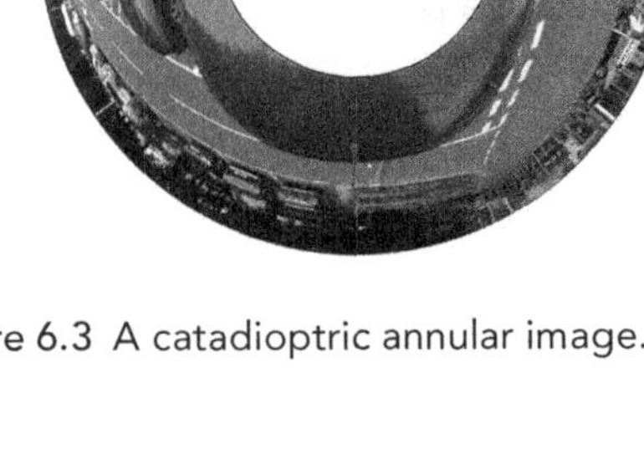

Figure 6.3 A catadioptric annular image.

Figure 6.4 A 2D "unwrapped" annular image.

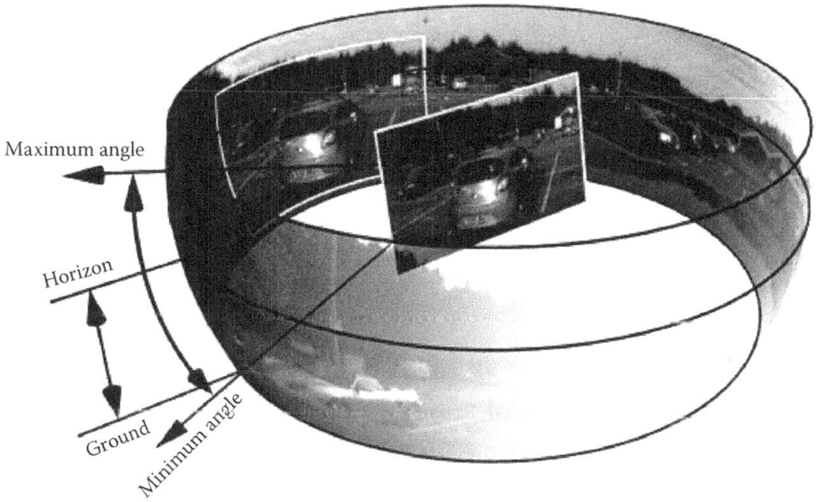

Figure 6.5 Virtual pan, tilt, and zoom.

been collocated and rotated to match that of the 360° camera.

Virtual zoom is simulated by interactively changing the field of view of the virtual camera.

The maximum look-up and look-down angles of the original annular digital image will of course limit the amount of virtual tilt (see Figure 6.5 to the left).

6.3.2 360° Polycamera

6.3.2.1 OVERVIEW OF 360° POLYCAMERA

A polycamera is a tightly packed cluster of cameras, all facing directly outward from a central locus, so that the combination can capture a larger part (or even all) of a 360° field of view.

The tight clustering ensures that the viewpoints of the multiple cameras are in close proximity to each other. This reduces parallax and makes it easier to combine photographic images of multiple views, with overlapping fields of view, into a single seamless large field of view. This process is called image "stitching" by analogy to the process of stitching small pieces of cloth into a larger whole, such as a quilted bedspread.

A polycamera can comprise as few as two back-to-back cameras, each with a matched wide angle lens with an overlapping spherical field of view (e.g., 240°). However, for better resolution and image quality, a polycamera usually comprises four or more cameras (see Figure 6.6, the polycamera here comprises six cameras) arranged such that image overlaps to allow for "stitching" without gaps in coverage (see Figures 6.7 and 6.8). Typically, the greater the number of cameras, the greater the possible resolution, and the lower the aberrations and hence, any visible visual artifacts resulting from "stitching."

The disadvantages, compared to the catadioptric camera described earlier, include the following:

- Lack of a single effective vertically aligned central viewpoint, which results in 3D spatial parallax disparity with respect to objects close to the polycamera in the scene being viewed
- Lens distortion often requiring correction of significant image warping

Figure 6.6 A complete 360° "set" of polycamera images with overlapping horizontal field of view.

Figure 6.7 Polycamera images assembled in sequence before "stitching."

Figure 6.8 Polycamera images combined into a "stitched" panorama.

- Scene motion artifacts that increase significantly when the polycamera is in motion, such as being on a vehicle
- Image to image photographic exposure differences

6.3.2.2 ON VEHICLE USAGE OF A 360° POLYCAMERA

Typically, image acquisition by all cameras within a polycamera is simultaneously triggered, such that an entire scene is acquired with minimal photographic artefacts (e.g., person or car appearing to be in two places at the same time).

The regions of "stitching" are blended compositions of the dewarped and radially undistorted images and the "seams" between the "stitched" source images can often be detected due to poor image processing and the lack of careful control to obtain a satisfactory overall exposure.

Especially problematic, when using a polycamera in which the separate cameras are tilted up or down so as to acquire a non-horizon-aligned vertical field of view, is the increase in spherical distortion and the commensurate increase in image dewarping required and the increase in the number of separate cameras required due to the decrease (proportional to tilt) in overlap.

A further disadvantage is that there is a lack of a simultaneous acquired master image. This leaves the highly unlikely, but also clearly highly undesirable, possibility of tampering with recorded evidence by using time-delayed, or retouched, previous parts of an image.

The complete 360° "set" of polycamera images used in the examples shown in Figures 6.6 and 6.7 are not horizon aligned, each having a tilt down of 20°. The increased image distortion is clearly apparent in the fifth and sixth images.

6.3.3 360° Spherical mosaic

By panning a single camera, it is also possible to assemble a mosaic (including "stitching"), to create a 360° image. As with a 360° polycamera, multiple images are required. These two techniques share many of the same advantages and disadvantages; however, in this case, the mosaics cannot avoid temporal artifacts, as the source images are not simultaneously triggered.

A handheld camera can be used to acquire a full 360° spherical mosaic. These source images can be "stitched" together using advanced image processing techniques such as edge–feature detection. Using this to rescale and align with the other images, a useful single image can be computed. It is not uncommon for enthusiastic photographers to achieve a similar panoramic solution, though this can be tedious and time-consuming and needs great skill to be successful.

A number of applications have been developed for mobile phones (and recently supplied as part of some mobile phones tool set) where a user can take multiple images while "sweeping" their mobile phone around them in a horizontal arc; the images can then be combined using techniques such as those described earlier. However, due to these images being acquired while the mobile camera is held in front of a user, often at arm's length, the resulting lack of a single effective viewpoint for imaging near objects and people, and temporal artifacts, often results in an unsatisfactory 360° spherical mosaic.

By mechanically rotating, instead of smooth panning, a camera in equal incremental steps about a single axis, a very large number of narrow vertical image strips can be acquired, making it easier to stitch them together to form a single image. To do this reliably, a special mechanical mount is required.

Nonmechanical disadvantages of all the panning methods include the difficulty in ensuring complete coverage of a 360° sphere. Whichever 360° spherical mosaic is chosen, the objective is to create a single very-high-resolution image from multiple lower resolution source images.

6.3.4 360° Wide-angle and fisheye

Images taken with wide-angle cameras tend to have severe distortions (aberrations) that pull points towards the optical center causing radially varying distortion. There are many different digital processing techniques available for image correction. Many are available in popular image painting and photographic processing programs on PCs, laptops, and tablets. These techniques are either based on calibrated or noncalibrated cameras. The latter requires that the program must first algorithmically derive the radial and tangential calibration parameters of the imaging system, as it is only with this information that the program can correct the image.

6.4 MAIN APPLICATIONS

6.4.1 Vehicle-mounted 360° surveillance

We referred to the applications of 360° camera systems in the introduction but we shall now go into more detail. The generic advantage is that on-vehicle-mounted 360° panoramic cameras can be considered to be "body worn video for vehicles." As described earlier, with this facility, a vehicle-mounted camera need no longer be looking in the right direction to capture critical footage: these cameras "look" in all directions, all of the time, capturing the entire 360°panoramic all-around view without compromise. For professional use, the cameras must be intrinsically robust and also typically satisfy a myriad of commercial and government electrical, emissions, and commercial standards. Such a process can take many years to achieve. They have to meet the mechanical rigours of vehicle mounting and be rigorously tested and evaluated to exceed the demands of the emergency services community.

Typically, an on-vehicle 360° panoramic camera will be located on top of a light bar or a camera roof mount. This is to provide the panoramic camera with not only a "look out," but also a degree of "look down" capability (see Figure 6.9). The obvious limitation of this is simply the point where the field of view becomes obstructed by the roof of the vehicle (see Figures 6.3 through 6.8). As mentioned earlier, many earlier 360° panoramic cameras were polycameras, comprising multiple individual cameras mounted with the same horizontal plane of view. This is usually an unsatisfactory solution, as the resulting panoramic images often comprise mostly sky, not the local around a vehicle.

For this reason, a catoptric 360° panoramic camera is often used. This allows for the calibration

Figure 6.9 Typically, an on-vehicle 360° panoramic camera will be located on top of a light bar or a camera roof mount.

of imaging such that annular images are acquired between a maximum and a minimum angle providing a balance of viewing between above horizon image information and local viewing (e.g., people near to the vehicle) (see Figures 6.10 and 6.11).

6.4.2 Static 360° surveillance

A 360° surveillance camera is usually a pole- or mast-mounted variant of the on-vehicle camera described earlier (see Figure 6.12).

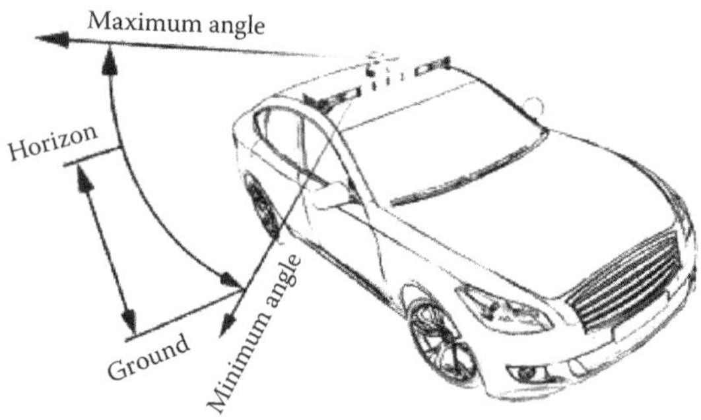

Figure 6.10 Maximum and minimum angle of a catoptric 360° panoramic camera.

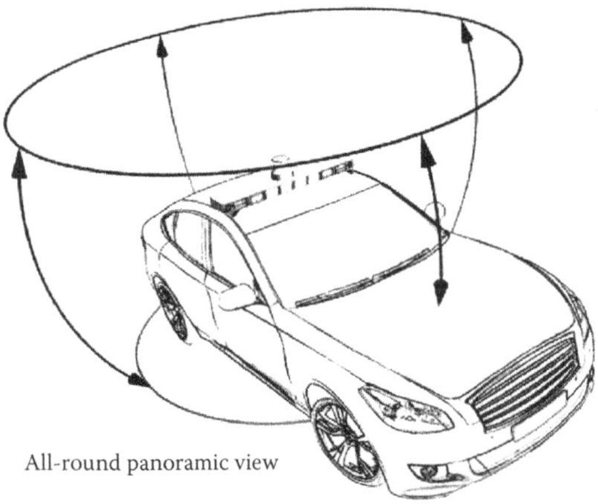

Figure 6.11 Panoramic view offered by a catoptric 360° panoramic camera.

These are designed to be attached to a mast or pole to provide 24 h continuous surveillance in all directions.

Typically, they are IP (Internet protocol) camera systems capable of carrying out complex analysis and image processing. They link to incident management and reporting systems via high bandwidth networks, such that high-quality images can be streamed back to a remote location (e.g., control room, incident room).

They are often designed to be suitable for either fixed sites or sets of temporary ones where rapid redeployment is possible. To allow this, they are built with various mountings for quick and easy installation.

Figure 6.12 360° surveillance camera.

6.4.3 Video conferencing

Various 360° panoramic camera systems are commercially available for video conferencing. Typically, these comprise only a 360° panoramic video camera (often based on fisheye lenses), without any integrated method for simultaneously receiving or viewing 360° panoramic video.

6.4.4 Robotics

Robotics is one of the most disruptive technologies of our period, and reliable sensing and instrumentation have a vital part to play in their ongoing development. As with humans, the most intelligent robots require to "see" what they are doing. Many types of compact 360° camera systems have been mounted on a variety of robots and used to control or drive the robots from remote locations. There is likely to be a huge increase in more intelligent and nearly autonomous or fully autonomous robots in the coming years, and having high-quality visual abilities will be a key factor in their success.

A range of intelligent surveillance systems have been developed that use 360° imaging to simultaneously track multiple objects moving within the large field of view.

6.5 PUBLICATIONS

"Folded Catadioptric Cameras," S.K. Nayar and V.N. Peri, Panoramic Vision, pp. 103–119, R., Springer-Verlag, Apr, 2001. [bib] [©] "Single Viewpoint Catadioptric Cameras," S. Baker and S.K. Nayar, Panoramic Vision, pp. 39–71, R., Springer-Verlag, Apr, 2001. [bib] [©] "A Theory of Single-Viewpoint Catadioptric Image Formation," S. Baker and S.K. Nayar, International Journal on Computer Vision, Vol. 35, No.2, pp. 175–196, Nov, 1999. [PDF] [bib] [©] "Folded Catadioptric Cameras," S.K. Nayar and V. Peri, IEEE Conference on Computer Vision and Pattern Recognition (CVPR), Vol. 2, pp. 217–223, Jun, 1999. [PDF] [bib] [©] "A Theory of Catadioptric Image Formation," S. Baker and S.K. Nayar, IEEE International Conference on Computer Vision (ICCV), pp. 35–42, Jan, 1998. [PDF] [bib] [©] "Catadioptric Omnidirectional Camera," S.K. Nayar, IEEE Conference on Computer Vision and Pattern Recognition (CVPR), pp. 482–488, Jun, 1997. [PDF] [bib] [©] "Generation of Perspective and Panoramic Video from Omnidirectional Video," V.N. Peri and S.K. Nayar, DARPA Image Understanding Workshop (IUW), pp. 243–246, May, 1997. [PDF] [bib] [©].

PART III

Optoelectronics for security and surveillance

This section addresses applications of optoelectronics in security and surveillance. Starting with the ubiquitous CCTV camera, we shall again summarize the field with a table, before presenting chapters on the use of optoelectronics in this area.

Before presenting our selection of case studies, it is again appropriate to do a little cross-referencing of applications for various technologies. It should be noted that several aspects of CCTV camera technology were already discussed in the infrastructure and transport sections. Clearly, the monitoring and identification of vehicles from roadside and highway gantry sites has obvious security and surveillance advantages, and the distributed acoustic and seismic sensor system we have mentioned (which will be covered in more detail under oil and gas applications) has great promise as a security device for the detection of intrusion through perimeters and protection of vital infrastructure, secure buildings, and valuable objects. Many of the surveillance devices discussed here also, quite clearly, have a great potential in the military field.

Table III.1 Summary of applications of optoelectronics in the security and surveillance field

Application	Technology	Advantages	Disadvantages	Current situation (at time of writing)	More reading
General surveillance, using CCTV cameras.	Cameras with simple direct human observer of TV display. "Smart" video cameras with varying levels of software sophistication. Image recognition software can detect and identify vehicles, number-plates, faces, iris patterns, walking gait, and many other vehicular or biometric features.	Enormous value for detection and prevention of crime, terrorism, or potential crowd hazards. Noncontact sensors, which can provide real-time observation. Signals can be recorded for later perusal and use for evidence in criminal prosecution. Solid-state cameras can be robust, very small, and have low cost.	Limited to line-of-sight applications (but, at extra expense, multiple cameras can be used). Public objection because of perceived loss of personal liberties.	Society has never been so closely observed, as cameras are everywhere. Most are publicly accepted as a compromise between the personal and national security advantages and privacy concerns.	See Chapter 6 (360° Camera Systems), Volume II, Part II (Enabling Technologies for Imaging and Displays), Volume II, Chapter 13 (Optical Information Storage and Recovery) and Chapter 14 (Optical Information Processing), and Part I (Infrastructure) in this volume.
Vehicle number-plate recognition.	"Smart" cameras with software to read number-plates from the video signal. Systems for assessing average speed from transit time between cameras.	Automated detection of vehicles and/or their speed for traffic policing. Can be used for automated tollbooths and city areas where access tolls apply.	As above	Being used in more and more situations. They are saving labor and time in many areas, for example at tollbooths for car-ferry or "turnpike" roads. Leading to many congestion-zone toll schemes for busy city centers.	Same as above.

(*Continued*)

Table III.1 (*Continued*) Summary of applications of optoelectronics in the security and surveillance field

Application	Technology	Advantages	Disadvantages	Current situation (at time of writing)	More reading
Biometrics, e.g., facial feature identification, including iris patterns. Finger and palm print recognition. Gait monitoring.	"Smart" video cameras with software to recognize key features.	Automated border control and policing. Can be used to control access to personal computers, financial data, ATMs, and restricted areas.	As above	These are becoming used in more and more situations. They are enhancing security and saving labor and time in many areas.	Same as above
Reading and recognition systems for banknotes and documents.	"Smart" video cameras with software to recognize key features. Fluorescent materials offer additional features such as spectral emission and fluorescent lifetime aspects.	Automated detection of banknotes for financial control and detection of counterfeiting. Automated reading of documents, posted letters, etc.	No real objections.	These are becoming used in more and more situations. They are saving labor and time in many areas.	Same as above.
Distributed acoustic and seismic sensing systems using optical fiber cable sensor.	Optical detection of disturbances to optical fiber sensing cable, which can be range gated using LIDAR-like methods to locate the disturbance.	Fully distributed sensing of acoustic or seismic signals over distances of up to 30 km. Perimeter security, intrusion detection, guarding of valuable objects.	Sophisticated and expensive system, but cost per sensing element is acceptable.	Only recently developed but becoming more widely used to guard perimeters and borders, oil and gas pipelines, etc.	See Volume II, Chapter 11 (Optical Fiber Sensors) and Part I (Infrastructure) in this volume.

(*Continued*)

Table III.1 (*Continued*) Summary of applications of optoelectronics in the security and surveillance field

Application	Technology	Advantages	Disadvantages	Current situation (at time of writing)	More reading
Pyroelectric proximity detectors.	Detection of variation in infrared radiation from vehicle, animal, or person.	Simple and cheap. Acts on a change in signal, so does not respond to steady background.	Limited range. Can have false alarms.	Most commonly used, both in, and outside personal homes to detect intruders and/or switch on illumination of driveways, etc.	Also covered in Chapter 30 (Applications for Home and Mobile Portable Equipment).
Infrared line-of-sight beam systems.	Detection of obstruction of beam of light between a collimated LED source and a detector, to give warning of intrusion in the path between. Can use mirrors to reflect the beam around objects to be guarded.	Simple and relatively low cost.	False alarms due to dirt, rain, fog, snow, or leaves, etc. To guard a large perimeter, many source–detector pairs are needed.	Used indoors to guard valuable objects (e.g., works of art) by surrounding them with multiple beams. Still used as a low-cost outdoor perimeter guard.	
Solar power, for remote powering of road signs, lights, speed sensors, etc.	Photovoltaic optical to electrical energy conversion (often supplemented by small wind generator!)	A very economical way of providing power in locations away from mains electricity supplies.	Needs a storage battery and charge controller to cover periods without sun or wind!	A common feature on cross-country roads and highways.	See Volume II, Chapter 16 (Optical to Electrical Energy Conversion).

7

Applications of distributed acoustic sensing for security and surveillance

STUART RUSSELL
Optasense

7.1 INTRODUCTION

This chapter examines the application of distributed acoustic sensing (DAS) in security and surveillance using optical fiber sensing cables, which contain regular monomode optical fibers of the same type as those used for telecommunications. A DAS system converts a standard optical fiber into an array of acoustic sensors, each with the ability to essentially simultaneously determine the time varying strain at any given position along a long length of optical fiber. The operational range of systems is usually limited to around 50 km without inclusion of additional amplification or similar measures. However, unlike technology that preceded it, DAS is able to measure this strain dynamically with a frequency response of typically several kHz. Generally, with DAS it is not possible for the system to recover the absolute or DC strain as the processing method responds only to the AC content of the strain. However, for many applications, DAS systems are superior to more traditional strain sensing optical technologies, as they are able to respond to relatively fast and small amplitude *variations* in strain (vis., AC sensor output signals). In addition, traditional methods are usually based on one or more point sensors, whereas the DAS system can operate in a fully continuous manner over many kilometers of fiber, recovering signals, essentially simultaneously from each and every resolution cell (of length ~ 1–10 m, depending on design) along the length.

7.2 BORDER SECURITY

International borders are by their very nature difficult to effectively secure. Large sections are often situated in inhospitable areas, far from dense populations. They are therefore often open to infiltration. This offers the potential for illegal intrusion, trafficking, and opportunities for criminal or even terrorist activities. Protecting these borders with fences, cameras, radars, and people is not only time consuming and generally prohibitively expensive, but any precautions taken are relatively easy for intruders to observe.

It is also not always appropriate to fence these borders, not only from a practical or cost point of view but in many cases doing so may also harm wildlife and migration patterns. DAS offers the ability to monitor borders in real time with no overt presence. A simple buried fiber optic cable monitored with DAS capable system provides the enabling means to provide round the clock surveillance.

In some cases, optical cables may already be installed, and the cost of installing DAS is simply the installation of the sensing system itself and not the cable. Natural activities are unaffected and the system itself can be used to inform on the pattern of life and how best to deploy previous active monitoring assets. Although DAS offers a dramatic difference in capability if no further signal processing technology were applied to the raw output of the system, the personnel resource required to continuously monitor the DAS system covering many kilometers of a perimeter or border would be uneconomical and impractical as even trained staff can lose attention over long periods of inactivity. To make a viable commercial system, sophisticated algorithms are employed that can actively monitor the entire length and identify, categorizes, and flag suspicious activity, raising an alarm to a single or limited set of users only when necessary. This allows the system to be almost fully automated with little human interaction required.

Figure 7.1 shows the raw response of the DAS system to a number of varied stimuli. As you can see, a typical DAS installation is sensitive enough to detect actions such as digging in the proximity of a buried cable, or walking laterally over the cable (typical burial depths are of the order of 1 m). As you can see, walking is easily visible in this plot for 20 m on either side of the cable. Vehicles create huge signals, easily distinguished from the signals generated by personnel and more exotic signals such as gunfire and even small aircraft overflying the buried cable are easily detectable.

7.3 DAS FOR PERIMETER SECURITY

This is a similar concept to that of border security, except that the sensing cable will usually surround the facility and often a single system may cover the entire perimeter.

Securing sensitive civilian and military facilities such as airbases, airports, or power stations is of critical importance. The current state of the art of security surveillance has not moved on significantly in the last decade; often, passive fences and security cameras form the only security deterrent to a potential intruder. More sophisticated systems may employ passive infrared sensors, unattended point ground sensors, or leaky feeder sensors to detect movement or intrusion. However, these sensors require remote infrastructure to supply power and data connections. In addition, it becomes difficult to collate the data from several disparate systems to form an overall picture of what is happening to the perimeter. DAS can be used to bolster the security of perimeters, and, in some cases, may be able to utilize an already installed fiber cable to connect and retrieve data from the existing systems. In some cases, DAS may even completely replace the more traditional technologies.

The specific type of installation required for an application of this nature varies depending on the nature of the perimeter or asset to be secured. The simplest installation may be to have fiber buried around a perimeter. This allows the perimeter to be monitored without an overt presence to alert an intruder to the fact that they have been detected.

Secure sites often have multiple layers of security. When multiple layers of fencing are installed,

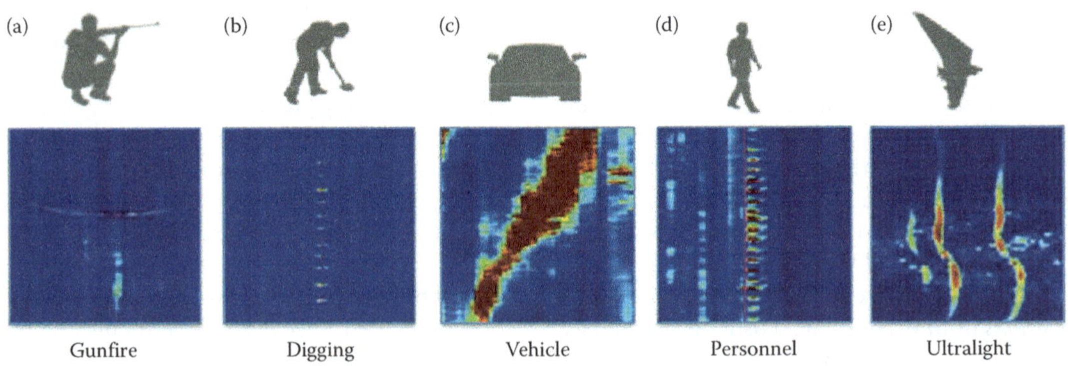

Figure 7.1 Raw DAS response to various events. (a) Rifle shot, (b) manual digging, (c) vehicles, (d) walking, and (e) overflight by a microlight. (Image Courtesy of Optasense.)

the space between the fences is termed a "sterile zone." This zone is typically quiet and the only movement or intrusion into the zone would be a security patrol. Installing a DAS fiber here significantly reduces the potential for false alarms and misclassification of an intrusion event.

Where the asset to be secured is in an urban environment, the acoustic environment is likely to be noisy and cluttered, raising the potential for false alarm. In these situations, it is possible for a fiber to be attached directly to the perimeters fence. It has been shown that DAS systems are able to detect and classify anything from crawling intruders to someone climbing a fence in a 30 mph wind. At the very least, DAS offers an additional layer of information, informing the client clearly when a threat is detected that may have gone undetected with more traditional technology, turning their attention to potential threats, and averting potentially disastrous consequences.

PART IV

Earth resources and environmental monitoring

This section focuses on applications of optoelectronics in the field of earth resources and environmental monitoring. As in industrial areas, spectrometers and camera technologies play a very important part. Again we shall first summarize the field with a table, before presenting more detailed case studies in the chapters.

Table IV.1 Summary of applications of optoelectronics in earth resources and environmental and pollution monitoring

Application	Technology	Advantages	Disadvantages	Current situation (at time of writing)	More reading
Spectroscopic sensors (transmission/ absorption).	Measurement of direct or diffuse transmission of light through air, oil, or water samples. Measurement is usually taken at more than one optical wavelength to compensate for wavelength-independent losses. Applicable to many environments (space, airborne and marine environments are discussed below).	Established technology, with possible real-time information for process control. Strong signal/noise ratio. A wide range of chemicals can be measured. Can have remote sensor heads, if optical fibers are used. Solid-state image detectors enable low-cost spectrometers. Using software, multianalyte detection is sometimes possible.	Optical surfaces can become scratched, etched, dirty, or otherwise contaminated. Not applicable to very strongly absorbing materials, if transmitted signal becomes too weak. Sometimes different compounds or families of compounds can have similar absorption properties, so the method is often not as selective as desired.	Very widely used, either by sampling, for a laboratory-based spectrometer, or using real-time online monitoring. More optical-fiber-remoted online systems are being used, particularly when needed for networking large numbers of sensors (e.g., via optical polling switch) or for examining remote areas, such as deep wells, offshore sites, etc.	See Volume II, Chapter 15 (Spectroscopic Analysis). Also used in oil and gas industry (see Part VII).
Spectroscopic sensors (reflection or frustrated internal reflection).	Measurement of reflection of light directly from a sample or from the internal surface of a glass component (usually part of a prism) in contact with the sample.	Simple to do with camera technology. Most useful for highly absorbing samples, as changes in refection are more substantial. Because of this, the method is complementary to the one above. External examination of contents of glass bottles or plastic packaging is possible.	Optical surfaces can become scratched, etched, or contaminated. Not applicable to weakly absorbing materials.	Used in many process industries (e.g., food industry), because the examination can be done through the transparent packaging.	See Volume II, Chapter 15 (Spectroscopic Analysis). Also used in oil and gas industry (see Part VII).

(Continued)

Table IV.1 (*Continued*) Summary of applications of optoelectronics in earth resources and environmental and pollution monitoring

Application	Technology	Advantages	Disadvantages	Current situation (at time of writing)	More reading
Optical sensors using indicator chemistry.	Use of selective chemical reactions to cause color change in indicators. More selective reactions are possible using molecular binding. Sensing of pH, ions radicals, or complete molecules such as dissolved oxygen is possible.	Can effectively concentrate chemicals to be measured, because of their chemical binding reaction to a sensing layer or film. With highly specific binding reactions, can offer better "fingerprinting" than traditional spectroscopy. Dissolved oxygen sensors using fluorescent lifetime quenching are very attractive ion-insensitive sensors, which, unlike Clark cells, do not consume oxygen.	Nonreversible reactions, in many cases, so not good for long-time online use. Thin-sensing layers are sensitive to contamination. Less easy to get quantitative data than with direct spectroscopy.	A well-used technique, particularly if only needed to sense the presence of a trace pollutant.	See Volume II, Chapter 15 (Spectroscopic Analysis), and Parts VII (Oil and Gas) and Part IX (Medical). Also see: Wolfbeiss, OS. *Fiber Optical Chemical Sensors and Biosensors*, CRC Press (1991).
LIDAR and DIAL spectroscopy.	Use of lasers to probe the atmosphere, usually over very long measurement lengths. Two-wavelength differential measurement (DIAL) compensates for wavelength-independent losses LIDAR is also used to measure atmospheric properties such as cloud height.	Very useful method for probing the atmosphere near industrial sites, etc. Often used in mobile vans, as large high-power lasers and sophisticated detection systems are often used. Cloud height monitors are another useful application. Can also be used to probe atmosphere ahead of commercial aircraft.	Expensive equipment. Care needed to ensure eye safety.	A commonly used method to look for pollution from factory chimneys, industrial complexes, etc. Useful for weather-related cloud height monitoring and for examining upper atmosphere.	See Volume II, Chapter 12 (Remote Optical Sensing by Laser).

(*Continued*)

Table IV.1 (*Continued*) Summary of applications of optoelectronics in earth resources and environmental and pollution monitoring

Application	Technology	Advantages	Disadvantages	Current situation (at time of writing)	More reading
Space-borne optical measurement, usually with multispectral-imaging cameras.	Observation of the Earth from space, using sensitive imaging cameras. Many wavelengths can be measured, including infrared and ultraviolet. (also mm wavelengths, but this is outside topic of book).	The greatest advantage is that huge areas of the Earth can be scanned by orbiting satellites. Applicable to monitoring of atmosphere, sea, or ground. Repeat images can be taken on subsequent orbits to produce full rastor images.	Very expensive, of course, mainly due to the cost of satellite launching and their space qualification. Affected by cloud cover and adverse weather such as fog. Resolution is lower than aircraft-borne measurement.	A widely used technique for monitoring of weather, Earth resources, crop growth, pollution clouds, and many other areas.	See this section.
Airborne optical measurement.	As above, but from lower altitude. Airborne measurement allows in-flight sampling of air or real-time measurement of spectral transmission or turbidity over short direct optical paths.	Higher spatial resolution of Earth coverage possible, as closer to ground. Ability to fly through large volumes of air, measuring (or sampling) during flight.	Fairly expensive, but again large areas covered.	Again, a widely used technique for monitoring of weather, earth resources, crop growth, pollution clouds and many other areas.	See this section.
In situ marine measurements.	As above, but now in the sea. Ship-borne measurement allows in-passage sampling of water, or real-time spectral measurement of transmission or turbidity, as a ship [or remote-powered vehicle (RPV)] travels through the sea. Can also use optical chemical indicators.	Ability to cruise through large volumes of water, measuring (or sampling) during passage. It is common to use "ships of convenience" to save costs, where the instrumentation "hitches a lift", by being towed behind freight-carrying ships.	Moderately expensive. Equipment can easily be lost or damaged by interaction with floating and underwater debris, thick seaweed, wrecks, etc.	A widely used technique for monitoring of marine conditions.	See this section.

Table IV.1 (*Continued*) Summary of applications of optoelectronics in earth resources and environmental and pollution monitoring

Application	Technology	Advantages	Disadvantages	Current situation (at time of writing)	More reading
Terrestrial measurements, for assessing properties of the lower atmosphere.	Measurement of direct or diffuse transmission of light through air, usually with direct path from light source to detector/spectrometer. The light path can also be folded back, by using a distant retro-reflector or scattering screen.	A useful method to detect fog or mist, and also to detect environmental pollution from gas leaks, factory emissions, etc.	Optical surfaces can become scratched, etched, dirty, or otherwise contaminated.	Commonly used as a fog or mist detector.	
Roadside optical pollution sensors.	Optical systems for determination of pollution from vehicles, using spectroscopic analysis.	Can monitor road traffic for identifying vehicles, emitting noxious fumes or using illegal fuels.	Expensive systems at present, but costs would reduce with greater usage.	A system developed by NASA has been used to monitor automobile emissions in numerous US states.	See Volume II, Chapter 15 (Spectroscopic Analysis) and Volume III, Part VII (Gas Sensing).

8

Overview of earth observation from satellites

GREG BLACKMAN AND JESSICA ROWBURY
Imaging and Machine Vision Europe

8.1 INTRODUCTION

Since the launch of Landsat 1 in 1972, NASA's Landsat satellite has collected spectral information of the Earth, providing the longest continuous global record of its surface. Landsat 8 [1,2], the newest satellite in the joint NASA and US Geological Survey (USGS) programme, was launched in February 2013 and collects more than 400 images per day. Five million new and archived images are available to the public, offering scientists the ability to carry out in-depth analyses of changes to the Earth's landscape.

The images from the Landsat satellites, dating back more than 40 years, are used in research on climate change, alterations in ecosystems, as well as the effects humans have on the Earth's landscape. There are images collected before and after Hurricane Katrina (Figure 8.1), for example, so that the extent of the flooding caused by the storm can be analyzed or a study made of the hurricane's impact on the surrounding wetland environment. There are studies looking at urbanization, agriculture, natural disasters, forestry, predicting how diseases spread—Landsat data has been used to investigate the breeding habitats of the malaria vector in order to generate models that predict the spread of the disease, and therefore give insight into how to control it better [3]—and many other aspects of the planet.

Also starting to come online is the European Space Agency's (ESA) Copernicus Earth observation programme [4], a huge environmental monitoring project combining data from around 30 satellites as well as airborne and ground sensors. The first of the Sentinel satellites, which forms the space component of the programme, was launched in April 2014, with Sentinel-2A launched in June 2015, Sentinel-3A in February 2016, and Sentinel-2B in March 2017. Sentinel-1 offers radar images, while Sentinel-2 will deliver high-resolution optical images similar to those captured by Landsat satellites and the French SPOT (Satellite Pour l'Observation de la Terre) satellites.

Later in this section, Gerardo López-Saldaña and Debbie Clifford at the University of Reading (Chapter 9), and Valborg Byfield at the National Oceanography Centre in Southampton (Chapter 10) will both look at satellite-based monitoring, with López-Saldaña and Clifford concentrating on the land, and Byfield on the oceans. In their

Figure 8.1 New Orleans before (a) and two weeks after Hurricane Katrina made landfall in 2005 (b), both captured by Landsat 5. In these false-color images, vegetation appears red and man-made structures appear whitish blue. (NASA GSFC Landsat/LDCM EPO Team.)

discussion on land monitoring, López-Saldaña and Clifford cover measuring the Earth's energy balance, particularly albedo, with satellite data, as well as observing the biogeochemical cycles of the Earth. Satellite imaging is just one area of monitoring the planet, an area focused on here for this introduction, but the section on Earth resources and environmental sensing also includes some terrestrial sensing. Elsewhere in this section, Matt Mowlem, Alex Beaton, and Gregory Slavik at the National Oceanography Centre in Southampton (Chapter 11) cover optical sensors used to measure aquatic environments, while Barbara Brooks of the University of Leeds (Chapter 12) looks at cloud monitoring, in particular tracking the ash cloud thrown into the atmosphere when Iceland's Eyjafjallajokull volcano erupted in 2010.

8.2 LANDSAT IMAGING CAPABILITIES

The latest Landsat satellite, Landsat 8, generates images via two push-broom sensors: the operational land imager (OLI), and the thermal infrared sensor (TIRS) (Figure 8.2) [2]. These instruments do not use oscillating mirrors to sweep across the field of view as in previous Landsat imaging sensors, but a long array of detectors arranged perpendicular to the flight direction of the satellite. This has the advantage of higher sensitivity.

Landsat 7—which is still in operation—has an enhanced thematic mapper (ETM+) sensor on board, the data from which are digitized to 8 bits, meaning 256 levels of light can be distinguished.

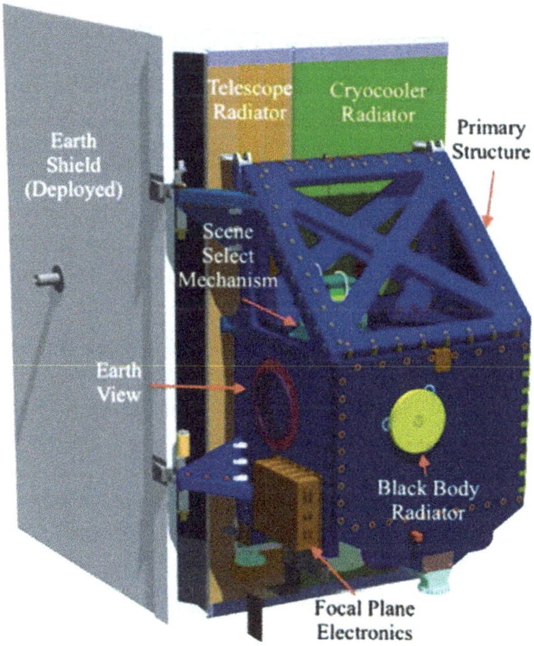

Figure 8.2 Drawing of the TIRS. Credit: NASA.

The data from the Landsat 8 OLI and TIRS sensors, however, are quantized to 12 bits, and are therefore able to distinguish between 4096 levels of light intensity.

The OLI contains more than 7000 detectors in long arrays for each of its nine spectral bands. It covers the visible, near infrared, and shortwave infrared regions of the spectrum (0.43–1.39 µm), with a spatial resolution of 30 m. The TIRS captures thermal infrared data (10.6–12.5 µm) in two bands, and was added to Landsat 8 largely for measuring water consumption and to provide images for managing agricultural water use in the United States. The sensor uses quantum well infrared photodetectors (QWIPs) [5], a lower cost alternative to conventional infrared technology.

The telescope of the OLI and TIRS first directs the field of view across a strip of ground 185 km wide. The detectors collect and read out data across that swath to build up the images as the satellite flies forward. Data collected from the sensors is then interpreted to create an image of the Earth's surface. The distribution of the light intensities across the multiple wavelengths means that these images can be analyzed to identify different features.

One of the main reasons why the TIRS imager was included onboard Landsat 8 in the first place was because of a method to measure evapotranspiration from agricultural land based on remote sensing data in the thermal band. The METRIC (mapping evapotranspiration at high resolution and internalized calibration) computer model [6] was developed by researchers at the University of Idaho in the United States with a grant from NASA in 2000 to investigate ways to estimate evapotranspiration using Landsat data.

The model measures the amount of energy driving evapotranspiration—rather than measuring the evaporation directly—and therefore, how much water is consumed by a particular field or farm. To do this, it requires imagery in the thermal infrared wavelengths to calculate the ground temperature and the energy exchanges associated with evapotranspiration (Figure 8.3).

METRIC is now used across a number of western US states including Idaho, Nevada, Montana, Oregon, California, Arizona, New Mexico, Wyoming, and Nebraska by water resource managers to track how much water farmers are using. Simply monitoring the volume of water pumped onto a field is not an accurate measure of usage, as a lot of water will return to aquifers and other water sources. The usage comes from how much of

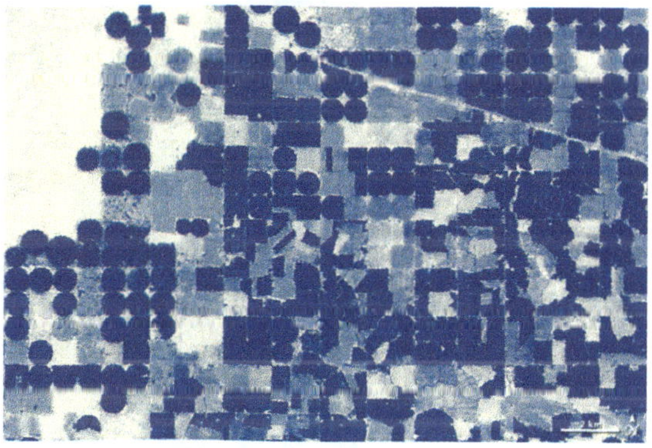

Figure 8.3 This image, based on data collected by the Landsat 5 satellite on August 9, 2006, shows evapotranspiration from vegetation on the Snake River Plain in south-central Idaho. Fields of irrigated crops are dark blue squares or circles, showing that the growing plants are taking up and transpiring water. Fallow and recently harvested fields are lighter blue. Credit: NASA image by Robert Simmon, based on data from the Idaho Department of Water Resources.

the water is drawn by the crops, how much they take up, and how much they transpire. With water being such a valuable commodity in certain parts of the United States, and such a large proportion of some States' use of water attributed to irrigation, water management is hugely important. The information from METRIC has also been used in US Supreme Court cases involving misuse of water.

The Landsat Data Continuity Mission, which became Landsat 8, was not initially designed with a thermal imager onboard; it was only through lobbying from the scientists developing METRIC that the TIRS was included in order to provide thermal data for the METRIC model.

Along with thermal imaging capabilities, Landsat 8 also now carries an extra spectral band to pick up dark blue colors for studies looking at coastal areas and other bodies of water. The higher sensitivity of the 12-bit OLI, plus the extra band in the blue region, allows better analysis of water quality. The Landsat data are being used to study plankton blooms (Figure 8.4) [7], as well as the spread of algae in the Great Lakes—scientists at the Michigan Tech Research Institute have been using Landsat images to map the algae *Cladophora* and monitor its abundance in the lakes of Michigan, Huron, Erie, and Ontario [8].

There have also been studies of natural disasters, Hurricane Katrina being a case in point. The US Geological Survey's National Wetlands Research Centre released a report soon after Katrina made landfall on August 29, 2005, stating that 217 square miles of Louisiana's coastal region was flooded as a result of Hurricanes Katrina and Rita (the latter storm arriving around a month after Katrina on September 24, 2005) [9]. The study used Landsat thermal mapper imagery to make the comparison; the data also provided a baseline for monitoring the recovery of the wetlands. The 217 square miles of land lost as a result of these two hurricanes alone represented 42% of what scientists had predicted—before the storms struck—would be lost over

Figure 8.4 The OLI on Landsat 8 captured this view of a phytoplankton bloom near Alaska's Pribilof Islands on September 22, 2014. The milky green and light blue shading of the water indicates the presence of vast populations of microscopic phytoplankton. Credit: NASA Earth Observatory images by Jesse Allen and Norman Kuring, using Landsat 8 data from the US Geological Survey.

50 years, from 2000 to 2050, even though storms had been factored into the models.

Studying natural disasters and providing emergency response is one of the key uses for satellite imagery in general. Images acquired before and after flooding, for example, offer immediate information on the extent of inundation and support assessments of property and environmental damage.

But there are of course many other uses for Earth observation, and normally, remote sensing satellites will carry various different sensing instruments—imaging and other types including lidar, radar, spectrometers, etc.—depending on what they are designed to measure. The challenge then becomes how to get timely images without the huge costs that goes into building, launching, and retrieving the data from large Earth observation satellites like Landsat.

8.3 SMALL-SATS

In order to continue the long-term Landsat data record, NASA and the US Geological Survey are currently working on a sustained land imaging (SLI) programme, to establish a sustainable satellite system that will provide continuous Landsat-quality measurements while minimizing the risk of gaps in data collection. In the past, Landsat satellites have been developed one or two at a time, and it then becomes a balancing act when building the next satellites between avoiding a data gap and unnecessary repeat measurements.

An SLI Architecture Study Team was set up to study what technology could provide the best balance of measurement capability, data continuity, and cost. The team was tasked to evaluate many different options for sustaining land imaging in the future.

The results of fiscal year 2014 architecture studies [10] indicated that efficiencies can be gained by moving toward smaller satellites, or "small-sats," to take advantage of lower cost and speedier launches.

The team are looking to use technology that is already commercial, to minimize risk, time, and cost; they are looking for something that has high technical readiness, rather than technology that has promise or that is still in development. However, the challenge is to identify new technologies that are not yet, but might, be sufficiently well developed and demonstrated in the future to incorporate into the SLI programme.

The industry is already seeing a trend towards low-cost small satellites for remote sensing. A traditional Earth observation satellite with a resolution of better than 1 m weighs thousands of kilograms—Landsat 8 weighs 2071 kg when fully loaded with fuel—that makes it expensive to launch. There is also greater competition between satellite companies; the emergence of a new breed of space data companies has driven the market towards the use of small satellites. Planet Labs is one such company, offering satellites with optical and near infrared spectral bands for imaging. These sorts of satellites make the whole concept of commercial remote sensing a lot more cost-effective.

Planet Labs has numerous small satellites, called Doves, orbiting Earth, the first of which it released from the International Space Station on February 11, 2014. Each Dove weighs around 5 kg and carries a camera with a resolution of 3–5 m. Planet Labs acquired Terra Bella, formerly Skybox Imaging, from Google in 2017.

Imaging resolutions of the satellites in orbit are currently around 1 m, while the next generation of imagers have 30 cm resolution or better. However, a satellite typically only orbits the Earth once every 100 min, and then, depending on the orbit, it might be a week before it gets back to the same location. This is one of the reasons for building smaller and cheaper satellites, because it allows larger numbers of them to be used to increase the imaging coverage. There are also opportunities to have lots of satellites in a constellation.

Between 2000 and 2750 nano- and micro-satellites are predicted to be launched from 2014 to 2020 [11], according to data from the UK Satellite Applications Catapult, a technology and innovation centre that aims to foster economic growth in the UK space sector. The centre predicts the market for small sats to reach £970 million in 2020, a share of around half of the total commercial Earth observation data market in 2020.

But as satellites become smaller, so must the imaging equipment. The challenge is getting the instruments down to a size that would fit on the smaller satellites. Smaller spectral sensors are becoming available commercially and advances have been made in the sensor technology for remote sensing and hyperspectral analysis.

There is also a trend to make instruments dedicated to a particular purpose. As smaller satellites become more reliable and less expensive, a number of these devices can be built to each carry a different sensor payload and then fly in formation very close to each other—a few kilometers apart—but in the same orbit. This allows many different parameters to be measured almost simultaneously.

For instance, CO_2 levels in the atmosphere and the sea temperature could be measured from two different satellites at the same time. NASA's A-Train (short for Afternoon Train) is one example of a constellation of six imaging satellites operating within the same orbital track. It is made up of: OCO-2, GCOM-W1, Aqua, CloudSat, CALIPSO, and Aura. The constellation is in a polar orbit, moving across the equator at around 1:30 pm local time each afternoon; hence, the name Afternoon Train.

The six satellites contain various instruments operating at multiple wavelengths. The Orbiting Carbon Observatory 2 (OCO-2) is the most recent addition to the train, joining in July 2014. It measures atmospheric CO_2 based on high-resolution grating spectrometer readings, with the aim of improving the understanding of the sinks and sources of CO_2.

The Global Change Observation Mission-Water (GCOM-W1) satellite and the Aqua spacecraft both contain instruments for measuring the water cycle, including precipitation, water vapor, and sea water temperatures. GCOM-W1, launched by the Japan Aerospace Exploration Agency (JAXA) in 2012, follows behind OCO-2 by 11 min, while Aqua runs 4 min behind that.

CloudSat, a joint effort between NASA and the Canadian Space Agency, contains cloud profiling radar to study clouds and the role they play in regulating the Earth's climate. It runs 2 min and 30 s behind Aqua and is followed only 12.5 s later by the Cloud-Aerosol Lidar and Infrared Pathfinder Satellite Observations (CALIPSO) satellite. CALIPSO, developed by the French space agency, Centre National d'Etudes Spatiales (CNES) and NASA, also studies clouds but, in addition, they study the role that aerosols play in regulating climate. It makes measurements via lidar. Because CALIPSO and CloudSat fly so close together, their lidar beam and radar coincide on Earth's surface about 90% of the time.

Finally, Aura brings up the rear of the train, flying at around 15 min behind Aura. It has four instruments that produce vertical maps of greenhouse gases, among other atmospheric chemical constituents.

Constellations of satellites like the A-Train mean observations can be made of various environmental phenomena simultaneously. At the same time, a number of different remote sensing capabilities can be brought together that would otherwise be impossible to install on one large satellite.

The sensors onboard the A-Train have given evidence to suggest why the summer of 2007 experienced large losses of Arctic sea ice, for example [12], 38% less ice during this summer than average. The A-Train captured the environmental conditions during the loss and showed that anomalously high winds and fewer clouds accelerated the ice melt, in addition to high air temperatures and low humidity.

8.4 THE FUTURE OF EARTH OBSERVATION

Constellations of small satellites might be the way commercial companies are going about providing increased coverage of the surface of Earth, but the big space agencies are still designing large remote sensing missions, along with their plans for deploying small-sats—Landsat 9 is planned for launch in 2020.

The European Space Agency (ESA)'s Copernicus [4], Earth Observation programme, is one such giant mission and will add to the multitude of data coming from Earth observation satellites like Landsat, Pléiades, and SPOT. Data will eventually come from around 30 satellites as well as airborne and ground sensors in this huge project.

Sentinel-1A, launched in April 2014, is the first in a fleet of Sentinel satellites as part of the Copernicus programme. Sentinel-1 is a constellation of two satellites carrying radar imaging equipment. Sentinel-2 [13], launched in two parts in 2015 and 2017, carries an optical payload, imaging in the visible, near-infrared, and shortwave infrared regions in 13 spectral bands. It will have a swath width of 290 km and a revisit time of 2–3 days at mid-latitudes. Imagery from Sentinel-2 will provide similar information to the Landsat satellites and be used for areas such as land management, monitoring agriculture and forestry, and for disaster control and humanitarian relief operations.

Sentinel-3 (Figure 8.5), the first part of which was launched in February 2016, monitors the sea and land surface temperature and color for the purposes of supporting ocean forecasting systems

Figure 8.5 Sentinel-3A in the cleanroom at Thales Alenia Space in Cannes, France. Credit: ESA–Stephane Corvaja, 2015.

and for climate research, while Sentinel-4 and -5 will be meteorological missions monitoring the atmosphere. There is also a Sentinel-6 planned, which will carry a radar altimeter to measure sea surface height.

Projects like Copernicus and the continuation of the Landsat programme will provide vital scientific data to advance the understanding of our planet. The addition of small satellites from commercial companies and space agencies will also improve the image coverage of Earth, meaning that almost all natural and man-made events will be imaged from space.

REFERENCES

1. Irons, et al. 2012. The next landsat satellite: The landsat data continuity mission. *Remote Sensing of Environment*, 122, 11–21.
2. Roy, D. P., et al. 2014. Landsat-8: Science and product vision for terrestrial global change research. *Remote Sensing of Environment*, 145, 154–172.
3. Clennon, J. A. et al. 2010. Identifying malaria vector breeding habitats with remote sensing data and terrain-based landscape indices in Zambia. *International Journal of Health Geographics*, 9: 58 http://www.ij-healthgeographics.com/content/9/1/58.
4. Aschbacher and M.Milagro-Pérez. 2012. The European Earth monitoring (GMES) programme: Status and perspectives. *Remote Sensing of Environment*, 120, 3–8.
5. Reuter, D. C. et al. 2015. The Thermal Infrared Sensor (TIRS) on Landsat 8: Design Overview and Pre-Launch Characterization. *Remote Sens*, 7 (1): 1135–1153.
6. Kramber, W. J. et al. 2012. Landsat-based ET data for a Water Delivery Call in Idaho, 2012 Western States Remote Sensing of ET Workshop, http://idwr.idaho.gov/files/gis/RSofET2012-LandsatWaterDelCall-Oct25.pdf.
7. Behrenfeld, M. J. 2014. Climate-mediated dance of the plankton. *Nature Climate Change*, 4, 880–887.
8. Brooks, C. et al. 2015. A satellite-based multi-temporal assessment of the extent of nuisance Cladophora and related submerged aquatic vegetation for the Laurentian Great Lakes. *Remote Sensing of Environment*, 157, 58–71.
9. Barras, J. A. 2007. Land area changes in coastal Louisiana after hurricanes Katrina and Rita, 97–112. Science and the Storms: the USGS Response to the Hurricanes of 2005.
10. Volz, S. 2014. Sustainable Land Imaging (SLI) architecture study interim status briefing
11. Brunskill, C. 2014. Satellite Applications Catapult: Satellite Markets, Missions and Upstream Technologies, Photonex 2014 presentation.
12. Stroeve, J. et al. 2008. Arctic sea ice plummets in 2007. *Eos Trans. Amer. Geophys. Union*, 89 (2): 13–14.
13. M. Drusch et al. 2012. Sentinel-2: ESA's Optical High-Resolution Mission for GMES Operational Services. *Remote sensing of Environment*, 120, 25–36.

9

Satellite-based land monitoring

GERARDO LÓPEZ-SALDAÑA AND DEBBIE CLIFFORD
University of Reading

Environmental monitoring using remote sensing is a top-down approach in which an instrument, usually mounted on a satellite, is used to produce images of Earth's atmosphere and surface. Using such instruments, it is possible to detect features at electromagnetic wavelengths that the human eye cannot sense.

Since the satellites regularly pass over the same area of the planet, it is possible to monitor and assess changes. For instance, at regional scale, we can detect abrupt changes in the land surface such as deforestation and areas affected by wildfires or subtle changes like urban sprawl. At global scale, these images can also provide information about Earth's energy balance, quantifying the amount of radiation that is reflected back into space, which is a key contribution to climate modeling and projections.

9.1 REGIONAL MONITORING

There are several Earth observation (EO) programmes aimed at environmental monitoring. One of the first was the Landsat project, a joint programme between the United States Geological Survey (USGS) and the National Aeronautics and Space Administration (NASA) whose main goal was to collect global land surface images. The first Landsat satellite (Landsat-1) was launched in July 1973 and collected data in a Sun-synchronous, near-polar orbit at an altitude of 917 km, with a repeat cycle of 18 days, meaning that the system was capable of observing the same spot on the land surface every 18 days. The spectral information sensed by Landsat-1 was limited to the visible green (0.5–0.6 μm) and red (0.6–0.7 μm) and two near-infrared (NIR) channels (0.7–0.8 and 0.8–1.1 μm). The ground resolution of each image pixel, also referred to as the spatial resolution, was 80 m.

Several missions followed the first Landsat; the last one, Landsat-8, was launched in February 2013. The Landsat-8 onboard sensor, the operational land imager (OLI) has both improved ground resolution, at 30 m, and a wider range of spectral bands spanning the visible, the NIR, the shortwave infrared (SWIR), a dedicated band for cirrus cloud identification and a panchromatic (PAN) band, which is sensitive to all visible colors of the spectrum at 15 m spatial resolution. Additionally, the platform carries the thermal infrared sensor (TIRS) with two thermal spectral bands (10.6–11.19 and 11.5–12.51 μm) at 100 m spatial resolution.

Figure 9.1 depicts Rondônia, a state in Western Brazil. The first image in the left-hand side is from Landsat-2 acquired in 1975, while on the right-hand side is an image from Landsat-7 taken in 2012. The images are shown in false color, with the bright green highlighting areas of dense vegetation, a color which is absent in the 2012 image. Using

Figure 9.1 Deforestation patterns in Rondônia, Brazil. On the left-hand side is a Landsat false color image from 1975; on the right-hand side is an image from 2012.

Landsat data, Skole and Tucker (1993) found that 6% of the closed-canopy forest had been cleared between 1978 and 1988, and approximately 15% of the forested Amazon was affected by deforestation.

Copernicus is the European Commission programme for monitoring Earth. The observation component is a set of systems collecting data from multiple sensors, covering EO satellites, predominantly the European Space Agency (ESA) Sentinel missions (Fletcher 2012), as well as *in situ* sensors. All these data serve to address specific issues in six thematic areas: land, marine, atmosphere, climate change, emergency management, and security. These are known as the Copernicus services (Copernicus 2015).

Each Sentinel mission is based on a constellation of at least two satellites. The missions use a wide range of technologies including radar and multispectral radiometers. The first satellite of the Sentinel-2 mission was launched in June 2015 carrying the multispectral instrument (MSI) that senses Earth in 13 spectral bands, going from the visible to the SWIR, in different spatial resolutions ranging from 10 to 60 m. At 290 km, MSI has a much wider swath than previous multispectral optical missions such as SPOT and the

Figure 9.2 A false-close-up of an area in the Po Valley—showing Pavia (centre) and the confluence of the Ticino and Po rivers—is a subset from the first image taken by Sentinel-2A acquired on June 27, 2015 at 10:25 UTC (Coordinated Universal Time). (Image obtained from: http://www.esa.int/Our_Activities/Observing_the_Earth/Copernicus/Sentinel-2/Sentinel-2_delivers_first_images.)

aforementioned Landsat. Acquisition of state-of-the-art imagery across the whole wide field of view required a sophisticated optical design that had to be optimized to avoid pixel deformations at the edge of the swath. The design includes a three-mirror anastigmat telescope with a pupil diameter equivalent to 150 mm, mounted on the platform to avoid optical aberrations (Fletcher 2012). Figure 9.2 depicts a false color composite of the first MSI image from the MSI sensor over Italy. The multispectral data provided by Sentinel-2 is now freely available and is helping to monitor the land surface conditions at global and regional scales.

9.2 GLOBAL SCALE MONITORING

For a comprehensive and reliable environmental monitoring system, we require sensors capable of sensing the whole Earth's surface, across a wide range of wavelengths, and on a daily basis. In practice, many compromises have to be considered, such as the power required by the electronics, the cooling, the control systems on board the platform, and the maximum amount of data can be stored by the instrument, or the rate of data transmission back to Earth, which will limit the spatial resolution that can be managed.

One of the first global radiation detection imagers, the Advanced Very High Resolution Radiometer (AVHRR), first carried on TIROS-N and launched in October 1978, was a 4-channel radiometer designed mainly to observe the ocean surface and atmosphere. The AVHRR sensor was improved to a 5-channel instrument (AVHRR/2) covering the visible (0.55–0.68 μm), NIR (0.725–1.10 μm), mid-wave infrared (MWIR) (3.55–3.93 μm), and the TIR (10.3–11.3 and 11.5–12.5 μm) that was initially carried on NOAA-7 and launched in June 1981. This instrument had a spatial resolution of 1.1 km and was flown in a Sun-synchronous polar orbit allowing it to image the whole Earth's surface in one day. The onboard storage devices were limited and only limited data, with a degraded spatial resolution known as global area coverage (GAC) of 5×3 km, could be transmitted to Earth. The local area coverage (LAC) with 1.1 km resolution was only transmitted when the satellite was in view of a ground receiving station, leading to poor coverage at the higher resolution. The latest instrument version is AVHRR/3, with 6 channels, first carried on NOAA-15 launched in May 1998 and still functioning operationally (in 2014) on NOAA-18 (launched in May 2005) and NOAA-19 (launched in October 2009). In addition, the AVHRR sensor is carried on the European Organisation for the Exploitation of Meteorological Satellites (EUMETSAT) satellites Metop-A and Metop-B, launched in October 2006 and September 2012 respectively. The thermal capabilities of the sensor mean it is possible to derive geo-physical variables such as land surface temperature, as shown in Figure 9.3.

One of the strongest controls on Earth system's energy balance is the land surface condition, in particular, the albedo that determines the fraction of the incident radiation that is reflected back into space. It is possible to demonstrate the importance of this environmental variable by calculating the impact of a small change in planetary albedo. Earth absorbs energy at a rate of $(1-A)I_{TS}/4$ (Gray 2010), where A represents Earth's albedo and I_{TS}, the amount of total solar irradiance. If $I_{TS} = 1366\ \mathrm{Wm^{-2}}$ (Gray 2010) and the mean albedo of Earth $A = 0.3$ (Prentice et al. 2012),

Figure 9.3 Average land surface temperature (LST) during July 2014 derived using the AVHRR thermalbands. Temperatures range are from 220 to 340 K.

the solar power available to the whole Earth system is 239.05 Wm^{-2}. A change as small as 0.01 in the albedo leads to a change in the radiation balance of 2.39 Wm^{-2}. By contrast, the global average radiative force exerted by anthropogenic carbon dioxide (CO_2) is only 1.82 ± 0.19 Wm^{-2} (Myhre et al. 2013). Therefore, an accurate quantification of Earth's albedo is critical for climate change monitoring.

The only way to quantify the albedo at global scale is using data acquired from optical instruments on satellites. The main goal of the GlobAlbedo project was to create just such a long-term albedo data record, using images from the European Space Agency (ESA) Medium Resolution Imaging Spectrometer (MERIS) and the Centre National d'Etudes Spatiales (CNES) SPOT Vegetation. The MERIS sensor on board the ESA-ENVISAT platform (now defunct) scanned the Earth in 15 different spectral bands between 0.4125 and 0.9 μm with a spatial resolution of 300 m at nadir. The still-operational SPOT-VGT instrument senses the Earth using 4 spectral bands appropriate to vegetation monitoring in the visible (VIS, 0.43–0.47 and 0.61–0.68 μm), the NIR (0.78–0.89 μm), and the SWIR (1.58–1.75 μm).

Figure 9.4 is a red, green, and blue (RGB) false color composite representing the land surface broadband albedo at a specific wavelength. Each broadband is projected into a different RGB channel to create the composite, the shortwave albedo (SW) is in the red, the NIR in the blue, and the VIS in the green.

9.3 REGIONAL LAND SURFACE CHARACTERIZATION

Characterizing Earth's land surface is a key element to capture the dynamics of the biogeochemical cycles, which in the end influences the climate system. Identifying land cover and land cover changes, therefore, plays a significant role since vegetation can store carbon that could otherwise be released to the atmosphere. Using daily observations of satellite data, it is possible to characterize the land surface dynamics. However, one of the main challenges is actually to observe the surface, first of all because satellites in a polar or near-polar orbit (as is usual for the platforms carrying such instruments) usually pass over each point on the Earth's surface only once per day. This single opportunity to see the surface means that in some regions of the Earth where cloud coverage is persistent or there are heavy aerosols, for instance due to biomass burning, the instrument may not see the underlying surface at all. Figure 9.5 depicts all daily observations from the Moderate Resolution Imaging Spectroradiometer (MODIS) sensor over the United Kingdom during July 4–19, 2007, showing the persistent cloud cover over the UK.

Given the extensive global cloud coverage, especially in tropics, it is necessary to apply a variety of techniques to extract useful information. Techniques employed in practice range from simple temporal compositing that selects the best observations within a time window, to sophisticated data assimilation techniques (based on control theory) where all

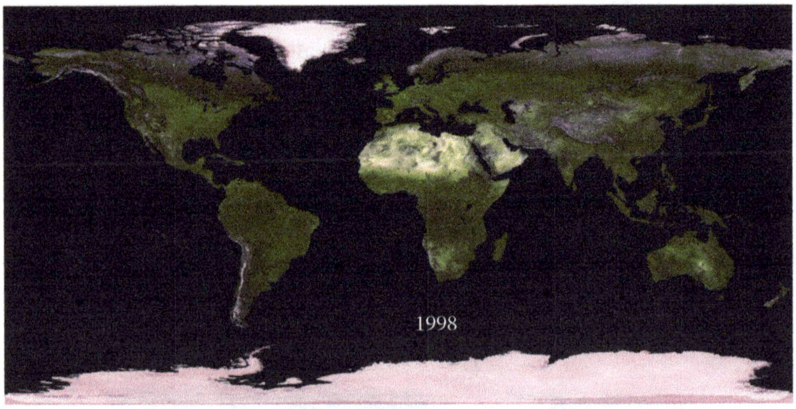

Figure 9.4 False color composite of land surface for June 1998. Bright colors for instance in Greenland indicate a high albedo, hence reflecting the radiation back to the space and dark green colors, e.g. in the Amazon, absorbing the radiation.

Figure 9.5 True color composites of all observations over the UK from the MODIS sensor for July 2007, showing the impact of cloudiness on views of the surface.

Figure 9.6 False (left-hand side) and true (right-hand side) color composites of modeled surface reflectance for July 4–20, 2007. The images are RGB composites where different spectral channels are displayed using three visual primary color bands (red, green, blue), hence RGB image.

observations and their associated uncertainties are ingested into a physical model to determine the most plausible land surface state. Figure 9.6 depicts the result of modeling surface reflectance over the UK using daily observations at optical wavelengths on the left-hand side and NIR and SWIR on the right-hand side. A bidirectional reflectance distribution function (BRDF) model was used to compute the reflectance as a function of the varying combinations of illumination-and viewing-angle.

Optical data going from the VIS to the NIR and the SW are a strong asset in environmental science. Sensors on board satellites have been producing such data since the early 1970s resulting in a long-term data record of the Earth's surface characteristics. Using all these data, it is possible to retrieve geophysical variables to analyze the physical properties of oceans, for instance, sea surface temperature and salinity, and biophysical ones such as chlorophyll-a concentration and fluorescence. Combining this information allows us to derive biogeophysical products such as ocean primary productivity, and ultimately leads to a better understanding of processes like ocean acidification and the monitoring of key marine ecosystems like coral reefs and mangroves.

More recent instruments are now designed to sense with much more spectral detail, taking measurements in very specific slices of the electromagnetic spectrum, which are very sensitive to vegetation health. Once again, using physical models and all the spectral information available, it is possible to make use of very cloudy images and then, in an analogous way to the ocean products described earlier, retrieve geophysical variables such as land surface temperature as the basis for further biogeophysical products such as land cover, land use, and vegetation productivity.

REFERENCES

Copernicus (2015). http://www.copernicus.eu/main/copernicus-brief.

Fletcher, K. (2012). Sentinel-2: ESA's optical high-resolution mission for GMES operational services, SP-1322/2.

Forster, P., Ramaswamy, V., Artaxo, P., Berntsen, T., Betts, R., Fahey, D.W., Haywood, J., Lean, J., Lowe, D.C., Myhre, G., Nganga, J., Prinn, R., Raga, G., Schulz, M., and Van Dorland, R. (2007). Changes in atmospheric constituents and in radiative forcing, in: Climate change 2007: The physical science basis. *Contribution of Working Group I to the Fourth Assessment Report of the Intergovernmental Panel on Climate Change*, edited by: Solomon, S., Qin, D., Manning, M., Chen, Z., Marquis, M., Averyt, K. B., Tignor, M., and Miller, H. L., Cambridge University Press, Cambridge and New York, 129–234.

Gray, L.J., Beer, J., Geller, M., Haigh, J.D., Lockwood, M., Matthes, K., Cubasch, U., Fleitmann, D., Harrison, G., Hood, L., Luterbacher, J., Meehl, G.A., Shindell, D., van Geel, B., and White, W. (2010). Solar influences on climate. *Reviews of Geophysics*, 48, RG4001, doi: 10.1029/2009RG000282.

Myhre, G., Shindell, D., Bréon, F.-M., Collins, W., Fuglestvedt, J., Huang, J., Koch, D., Lamarque, J.-F., Lee, D., Mendoza, B., Nakajima, T., Robock, A., Stephens, G., Takemura, T., and Zhang, H. (2013). Anthropogenic and natural radiative forcing. *Climate Change 2013: The Physical Science Basis. Contribution of Working Group I to the Fifth Assessment Report of the Intergovernmental Panel on Climate Change*, edited by Stocker, T.F., Qin, D., Plattner, G.-K., Tignor, M., Allen, S.K., Boschung, J., Nauels, A., Xia, Y., Bex, V. and Midgley, P.M., Cambridge University Press, Cambridge and New York.

Prentice, I.C., Baines, P.G., Scholze, M., and Wooster, M.J. (2012). Fundamentals of climate change science. In: *Understanding the Earth System: Global Change Science for Application*, edited by Cornell, S.E., Prentice, I.C., House, J.I., and Downy, C.J., Cambridge University Press, Cambridge, 39–71.

Sentinel (2015). http://www.esa.int/Our_Activities/Observing_the_Earth/Copernicus/Overview4.

Skole, D. and Tucker, C. (1993). Tropical deforestation and habitat fragmentation in the Amazon. Satellite data from 1978 to 1988. *Science (Washington)*, 260(5116), 1905–1910.

10

Optical remote sensing of marine, coastal, and inland waters

VALBORG BYFIELD
National Oceanography Centre

"Ocean color" satellite sensors were primarily designed to measure the global variation of chlorophyll concentration present in ocean phytoplankton. This allows the blooming and die-off of these microscopic plants, which form the base of the oceanic food web, to be monitored. The spectral images from the satellites revealed, for the first time, striking evidence of the close link between phytoplankton chlorophyll and global current features such as the oligotrophic ocean gyres, coastal and equatorial upwelling of deep water rich in nutrients (Figure 10.1).

Over the last two decades, orbital sensors addressed a much wider range of ocean color applications. Marine and aquatic applications of optical remote sensing now include the following:

- *Chlorophyll concentration*: Global measurements of chlorophyll concentrations and primary productivity for marine ecology, fisheries, conservation and sustainable management of marine living resources
- *Climate change*: Measurement of ocean uptake and storage of atmospheric carbon dioxide
- *Biological hazards*: Monitoring of harmful algal blooms
- *Pollution*: Monitoring marine pollution and water quality of coastal and inland waters
- *Oil spills*: Oil spill monitoring and response planning

Before going into more detail about these applications, it may be helpful to look at the principles behind optical measurements of water and its constituents.

10.1 PRINCIPLES OF OCEAN COLOR MEASUREMENTS

Figure 10.2 shows the main contribution to water radiance as measured by a sensor from the top of the atmosphere. Broadly speaking, the signal recorded by the sensor consists of the following:

- *Water-leaving radiance*: A proportion of sunlight traveling down through the water is scattered upward by the water in the direction

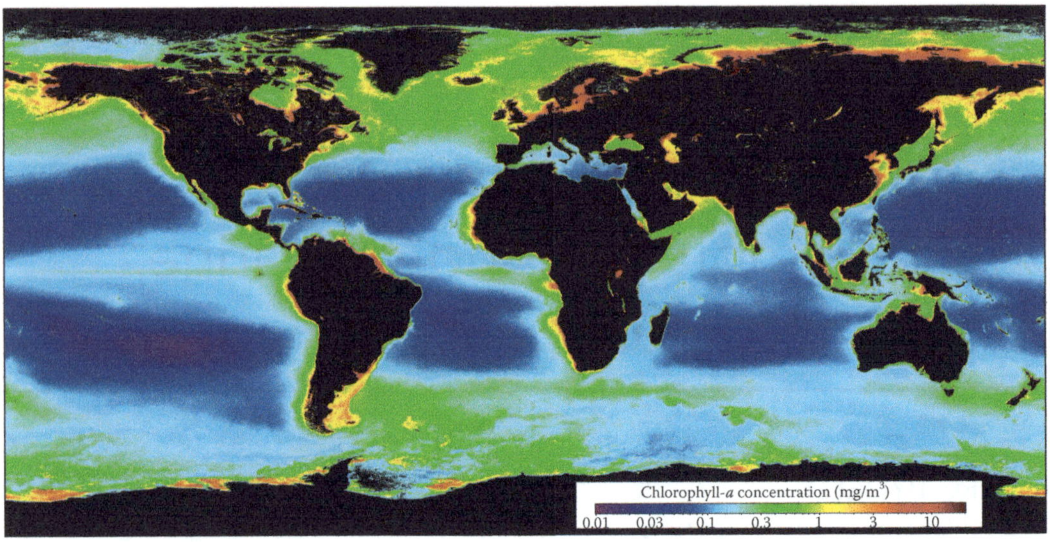

Figure 10.1 Composite chlorophyll concentration map of the Earth created from the whole Envisat MERIS mission (2002–2012). Note the low chlorophyll concentrations in the five subtropical ocean gyres (dark blue) and higher concentrations in the equatorial upwelling zone and in the coastal upwelling regions on the eastern side of Africa and South America, where nutrient-rich deep water brought back to the surface gives rise to higher phytoplankton production. (ESA/NASA.)

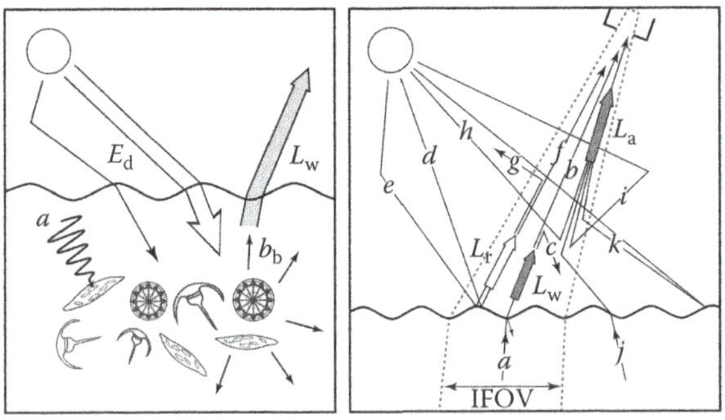

Figure 10.2 Contributions to radiance measured from the top of the atmosphere. Left: Downwelling irradiance, E_d, is the total light entering the water from direct sunlight and sky radiance, light scattered in the atmosphere before reaching the sea surface. The L_w: E_d ratio is known as water-leaving reflectance, and it depends on the absorption coefficient (*a*) and backscattering coefficient (b_b) of the water and its dissolved and suspended constituents. Right: L_w is the water-leaving radiance immediately above the sea surface in the direction corresponding to the sensors instantaneous field of view (IFOV). A proportion of this signal is lost due to atmospheric scattering and absorption (*c*). Light reflected at the sea surface, L_r, includes sunglint (*d*) and reflected sky radiance or sky glint (*e*); some of this is lost before reaching the sensor (*g*). The atmospheric contribution to the top of atmosphere (TOA) radiance, L_a, consists mainly of light scattered by the atmosphere into the sensor field of view (*h* and *i*). However, it is also necessary to consider light from the Earth's surface outside the IFOV (*j* and *k*); this can lead to considerable error in pixels adjacent to land or cloud. (National Oceanography Centre / ESA LearnEO! CC by 3.0 Unported License.)

of the sensor. The spectral properties of this light depends on the absorption and scattering properties of the water itself and its dissolved and suspended constituents.
- *Radiance reflected at the sea surface*: A proportion of sunlight is reflected directly at the sea surface (Fresnel reflection) and by foam from breaking waves.
- *Atmospheric path radiance*: A proportion of light is scattered towards (or away from) the direction of the sensor by air molecules and aerosols such as dust and water droplets.

For most oceanographic applications, the water-leaving radiance is the primary parameter of interest because it contains information about the water, its phytoplankton content and any dissolved organic matter and suspended sediment particles. Suspended particles such as phytoplankton cells or suspended sediments scatter light back up. Under most conditions, this particle backscattering is much stronger than any scattering by pure seawater, so there is usually a clear relationship between particle concentrations and water reflectance.

Two-way absorption by phytoplankton pigments and any colored dissolved organic matter (CDOM) change the spectral properties of the reflected light. The pigment chlorophyll-*a* absorbs blue light, but not green. Thus, there is a close relationship between the blue:green spectral ratio of reflected light and concentrations of chlorophyll-*a* in phytoplankton cells. This was the basis of the first chlorophyll algorithms, and various blue:green ratios are still used to determine open ocean chlorophyll concentrations.

Deep water reflects very little light, unless it contains high concentrations of suspended particles that can scatter the incoming solar radiation back up. Over 90% of the signal received by a satellite sensor comes from the atmosphere, and it is necessary to correct for this contribution before applying algorithms to retrieve phytoplankton chlorophyll concentrations. Atmospheric correction involves several steps. After first masking (ignoring) pixels affected by cloud cover, wind speed information is used to estimate the contributions from surface reflectance and foam from breaking waves. The required correction for scattering and absorption anticipated from a dry, clean atmosphere is relatively constant and depends mainly on atmospheric pressure. Scattering and absorption by aerosols (water droplets and/or dust) are more difficult to estimate, so most algorithms use bands in the near infrared (NIR), where water-leaving radiance is minimal, to predict, the aerosol contribution. Errors in chlorophyll retrieval are most commonly due to problems with the atmospheric correction caused by highly scattering water, when water-leaving radiance in the NIR is higher than usual. Detector noise, particularly in the NIR where the signal is low, can also create problems with the atmospheric correction.

10.2 MAIN OCEAN COLOR SENSORS

To provide reliable information, ocean color sensors need a wide dynamic range and a high signal-to-noise ratio (SNR)—ideally of the order 500:1 or better. Most sensors designed for land remote sensing do not have a sufficiently high SNR to be useful over the open ocean, although their data can be of value in shallow or turbid coastal waters. The usual trade-off between SNR and spatial resolution of camera-based sensors (high resolution involves less optical energy on each pixel) means that most ocean color instruments are made with medium resolution sensors, with pixel sizes of 250 m–1 km (Table 10.1), significantly lower than that used for land monitoring. With higher resolution cameras, of course, the signals from larger groups of pixels can be taken to improve SNR at the expense of effective resolution.

The first true ocean color sensor in orbit was the coastal zone color scanner (CZCS), on NASA's Nimbus 7 satellite, which provided data on the global ocean between 1978 and 1986. A number of short-lived missions followed (Table 10.1). However, ocean color monitoring did not truly come of age until after the launch of NASA's Sea-viewing Wide Field-of-view Sensor (SeaWiFS) in 1997. The color bands chosen for SeaWiFS were similar to those of the earlier CZCS (see Figure 10.3), allowing comparison between the two data sets. Later ocean color sensors have retained these (or similar) bands in order to provide continuity of long-term measurements. SeaWiFS continued to deliver ocean color data to oceanographers and marine ecologists until 2010. In 2002, it was joined by the moderate resolution imaging spectrometer (MODIS) on NASA's Aqua satellite and the medium resolution imaging spectrometer (MERIS) on ESA's Envisat. With the U.S. Visible Infrared Imager Radiometer Suite (VIIRS) and the

Table 10.1 Ocean color sensors

Sensor	Satellite	Agency	Dates	Swath (km)	Resolution (m)	Bands[a]	Wavelength range[a] (nm)	SNR
CZCS	Nimbus7	NASA (USA)	Oct. 78 Jun. 86	1556	825	6 (5)	433–800	>150
OCTS	ADEOS	NASDA (Japan)	Aug. 96 Jul. 97	1400	700	12 (8)	402–885	450–500
MOS-B	IRS P3 (India)	DLR (Germany)	Mar. 96 May 04	200	500	13	408–1010	
SeaWiFS	Orbview-2	NASA (USA)	Aug. 97 Dec. 10	2806	1100	8	402–885	200–>500
MODIS	Aqua	NASA (USA)	May 02 -	2330	1000 (250)	36 (14)	405–965	500–1000
MERIS	Envisat	ESA (Europe)	Mar. 02 Apr. 12	1150	300 or 1200	15	395–1040	500–1000
VIIRS	Suomi-NPP	NOAA NASA	Oct. 11 -	3000	370 or 740	22 (9)	402–885	
OLCI	Sentinel-3	ESA EuMetSat	Feb. 16 -	1270	300 or 1200	21	395–1040	500–>1500

[a] The number of optical (visible to NIR) bands are given in brackets. The range includes only these optical bands.

- Indicates that the satellites are still in orbit.

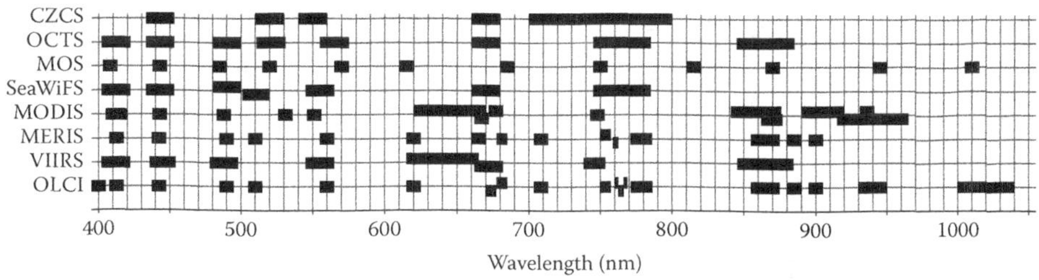

Figure 10.3 Optical bands for the main ocean color sensors described in Table 10.1. (ESA LearnEO! CC by 3.0 Unported License.)

European Ocean Land Colour Instrument (OLCI), optical monitoring of ocean and inland waters has now become truly operational.

10.3 PHYTOPLANKTON PRODUCTIVITY AND MARINE ECOLOGY

The abundance of phytoplankton is a direct indicator of the seas' ability to support life. Phytoplankton is the "grass of the sea," the basis of the marine food web. They also supply much of the world's oxygen. Hence, ocean color measurements provide essential basic information for marine ecological studies, fisheries science, and conservation. Global monitoring of phytoplankton chlorophyll concentrations was the first, and most important of the "ocean color" applications. It dictated the wavebands chosen for CZCS and SeaWiFS, and the need for continuity of measurements for long-term monitoring ensured that broadly similar wavebands were included in all later ocean color sensors.

Starting with SeaWiFS in 1997, we now have a long-time series of chlorophyll data, which makes it possible to understand seasonal and interannual variability in the primary productivity of the global ocean. Ocean color sensors continue to monitor key ecosystems such as the major upwelling regions off the coast of Peru, West Africa, and South Africa/ Namibia, which support some of the world's richest fisheries. In the North Atlantic, numerous studies have tied the spawning of fish and other marine species to the onset of the spring plankton bloom. In high latitude waters, where access is difficult and expensive, satellite measurements of ocean color have shown how massive plankton blooms support a rich abundance of life. The blooms often start near the ice edge and expand into ice-free water. Variations in the recruitment to commercially important fish stocks have been linked to phytoplankton phenology (timing of the seasonal cycle) as observed by ocean color satellites.

Phytoplankton tend to bloom when the water column is sufficiently stable for the plankton cells to remain in the sunlit zone near the surface and where the water is rich in nutrients, which have been mixed up from deeper layers of the ocean. Fronts between different water masses are regions where mixing between surface and deeper water often occurs, so they support higher level of phytoplankton abundance and are clearly visible in satellite chlorophyll data (Figure 10.4). Microscopic zooplankton graze on these blooms, and are in turn food for fish and other marine animals. Hence, the satellite images give an indication of where fish are likely to be found; many commercial fishing vessels use satellite data to their advantage. In a similar way, conservationists may use chlorophyll data to understand the movement and ecology of marine species such as turtles and whales.

Phytoplankton productivity is not measured directly by ocean color satellites, but must be inferred from the knowledge of plankton biology and information about their distribution in the water column. Research in this area is based on optical models with optical *in situ* data from floats and research ships, combined with chlorophyll and photosynthetically available radiation (PAR), both from ocean color satellites.

10.4 CLIMATE CHANGE AND OCEAN CARBON UPTAKE

Phytoplankton productivity varies significantly over both seasonal and interannual time scales, so the time series from 1997 until now is still not sufficient to establish whether there has been a long-term

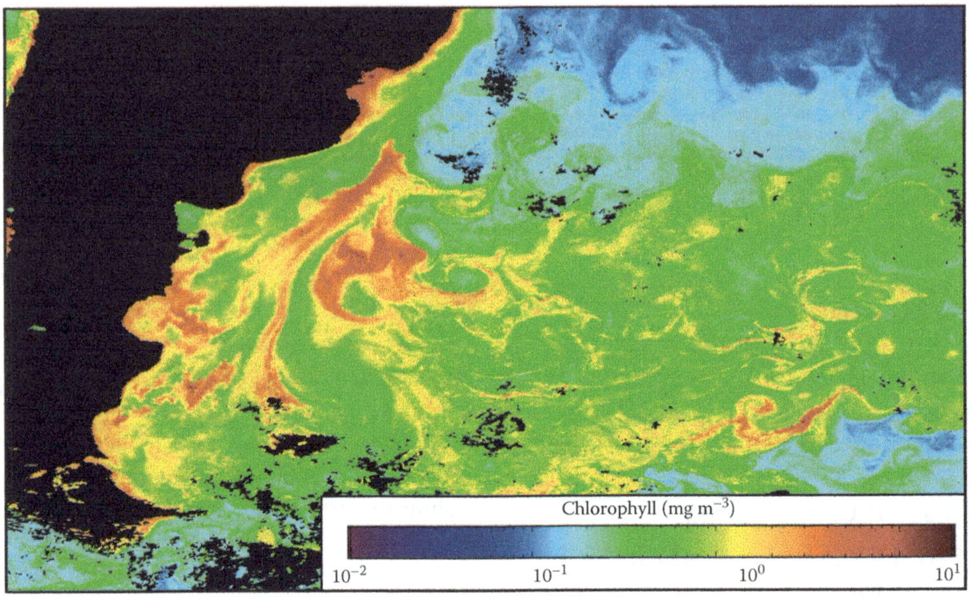

Figure 10.4 January 2007: High concentrations of phytoplankton chlorophyll mark the region east of Argentina where the warm poleward flowing Brazil Current and the cold equatorward flowing Malvinas Current converge. (Plymouth Marine Laboratory/ESA Ocean Colour CCI. Reproduced with permission.)

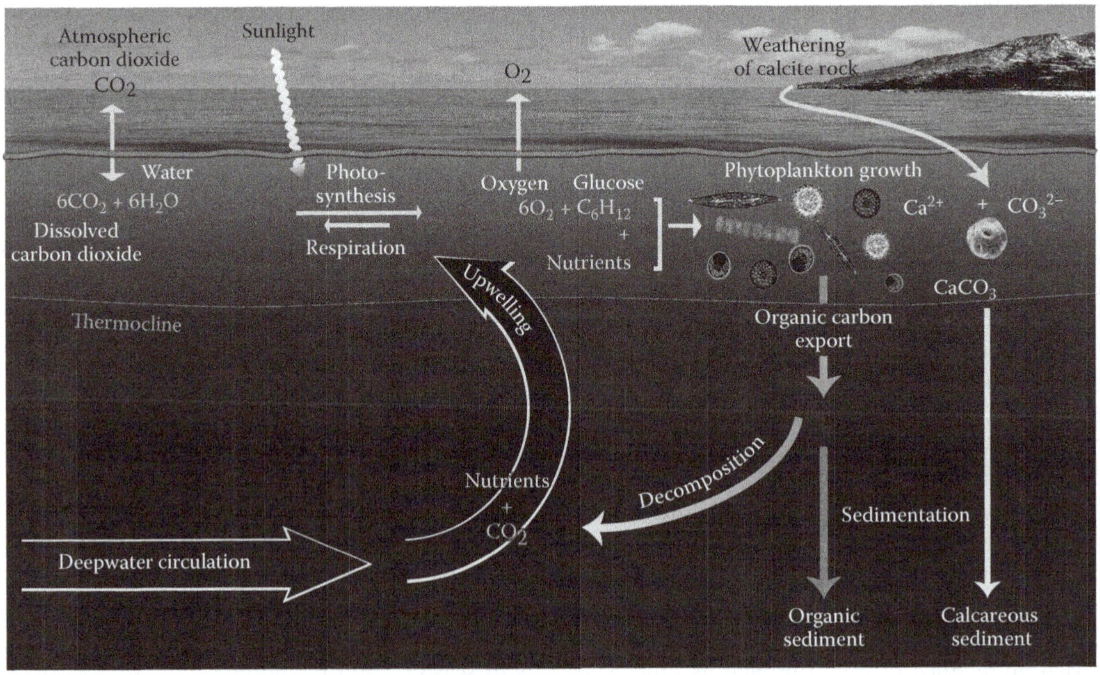

Figure 10.5 The "biological carbon pump." (National Oceanography Centre. CC by 3.0 Unported License.)

change. However, phytoplankton clearly play a central role in the ocean's carbon uptake through photosynthesis. If sufficient nutrients are available, the plankton cells will grow and divide. When the plankton cells die, they sink into deep water, where they decompose, releasing nutrients and CO_2 into the water (remineralization). This process is known as the "biological carbon pump" (Figure 10.5). It contributes to the ocean's uptake and storage of carbon dioxide, and keeps atmospheric CO_2 about 200 ppm lower than it would be if the oceans were without life.

Calculations of biological productivity based on satellite chlorophyll data allow climate scientists to estimate the ocean's biological uptake of CO_2. This is however only part of the story. We also need to know the proportion of the organic carbon that reaches the deep ocean. Most organic matter is remineralized at the surface layer; only a small fraction reaches the deep ocean. Once there, the carbon is stored away from the atmosphere for hundreds of years, only gradually returning to the surface as it slowly decomposes. Some of it remains there, of course, in the form of organic carbon and carbonates in shells of marine creatures buried in seafloor sediments.

Apart from satellite monitoring, above-surface and *in situ* oceanographic measurements of the organic particle flux can also be made from ships, using onboard sensors, as well as sediment traps and sensors on floats and moored buoys. These are clearly more localized, but can make more accurate measurements to determine water turbidity, nutrients, and oxygen concentrations at different depths. When combined with the global satellite data, these provide important information about the marine carbon cycle and its contribution to the marine environment.

10.5 HARMFUL ALGAL BLOOMS

Ocean color monitoring allows real-time observation of red tides and other phytoplankton blooms. When a bloom has negative impacts on other organisms or humans, it is known as a harmful algal bloom. A harmful bloom may be toxic, causing large-scale mortality of fish or shellfish poisoning in humans. Other plankton species, while not containing toxic chemicals, may be present in such number that they smother other organisms, for example, by clogging their gills. Intense blooms can also cause widespread deoxygenation of bottom waters when they decompose; so they can be a major cause of death for bottom-living organisms. Typically, only one or a small number of phytoplankton species are involved.

Harmful algal blooms may occur in marine as well as freshwater environments. An example can be seen in Figure 10.6, which shows a harmful algal bloom (HAB) in Lake Erie in October 2011. It was caused by a record spring rainfall, which washed fertilizer into the lake, promoting the growth of toxic cyanobacteria throughout the summer and autumn.

The monitoring of harmful algal blooms in coastal and inland waters allows responsible authorities to issue warnings of potential shellfish poisoning and other effects of toxins. Such warnings are important, for example, for the aquaculture industry, allowing stocks to be protected or harvested before damage can occur. Unfortunately, satellite data cannot itself determine if a bloom is toxic, but very high chlorophyll concentrations can indicate that a harmful bloom is likely. This allows those responsible for monitoring water quality to investigate further, for example, by taking water samples to identify the species responsible for the bloom and determine whether the bloom is toxic or likely to affect resources of economic importance in other ways. A warning can then be issued to anyone at risk of potential harm.

10.6 POLLUTION MONITORING

High chlorophyll concentrations can indicate eutrophication of coastal and inland waters, often (but not always) arising from sewage pollution, industrial effluents, land-runoff of fertilizer, and even atmospheric deposition of nitrous oxide from car exhausts. In some areas, this can lead to an increased frequency or intensity of algal blooms, some of which may be toxic or harmful in other ways. Satellite chlorophyll data is therefore one of the tools used to carry out environmental impact assessments, often along with other measurements to determine the exact nature of the pollution problem, and its impact on local ecology.

Satellite-derived chlorophyll concentrations are also a good proxy for eutrophication of coastal and inland waters. However, in such waters, the standard global chlorophyll products may not be sufficiently reliable; this is because they do not consider the effects of suspended clay or mud

Figure 10.6 Harmful algal bloom in Lake Erie, October 2011. (NASA Earth Observatory.)

particles and CDOM brought into the sea by rivers. Unfortunately, CDOM absorbs strongly in the blue (much like chlorophyll) range, while suspended mineral particles can scatter light in the same way as plankton cells. It is therefore difficult to determine phytoplankton chlorophyll reliably without simultaneously retrieving CDOM absorption and suspended particle concentrations. In such waters (known to optical oceanographers as Case 2 waters), global open ocean algorithms must be replaced with more complex methods. Most of these algorithms use look-up tables based on radiative transfer modeling and have been developed using neural networks trained with *in situ*—satellite match-up data; an example is the Case 2 algorithm developed for MERIS. A form of this algorithm, implemented in the ESA toolbox for Envisat and Sentinel-3, allows users to adapt the processing to their own region, based on their own optical measurements of backscattering and absorption coefficients.

Early sensors such as CZCS and SeaWiFS did not have the wavebands required to do this, and therefore, tend to overestimate chlorophyll levels in coastal waters. Distinguishing sediments and CDOM from phytoplankton requires a larger number of wavebands in the red region of the reflectance spectrum, 600–700 nm. It was not until data from MERIS and MODIS became available that accurate retrieval of chlorophyll became possible in regions affected by land-runoff.

There are now a number of spectral processing algorithms that can retrieve suspended sediment concentrations, and show how suspended sediments may be transported along the coast. A further useful feature is that many pollutants that are not directly visible (radio nuclides, heavy metals, PCBs) can be adsorbed onto sediment particles. As a result, information about sediment transport derived from optical data may be a useful tool in pollution studies.

10.7 OIL SPILLS

In the last decade, ocean color satellites have increasingly been used to measure and monitor oil spills and guide the oil-spill response to major accidents. A major drawback of optical satellite sensors is that they cannot see through cloud, so synthetic aperture radar (SAR) is the main method of oil detection in regions where cloud is a problem. However, in regions that are relatively cloud free, such as the subtropics, optical data from MERIS and MODIS, are increasingly used to give an overview of larger spills. In combination with SAR data, the full resolution (250–300 m) data from these sensors can help to make the detection of smaller spills more reliable.

Most ocean color applications use atmospherically corrected ("Level 2") data. Oil spill monitoring, however, requires top of atmosphere ("Level 1") data, because specular reflection at the sea surface is an important part of the signal used to distinguish surface oil from clean water. Oil has four properties that affect the signal received by an ocean color sensor at the top of the atmosphere:

1. Oil dampens capillary waves, changing the sunglint distribution from that expected from clean water at a similar wind speed.
2. Oil fluoresces, particularly when excited in the violet and blue ranges, and this can contribute to the signal from surface and dispersed oil.
3. Oil has a higher refractive index than water, so surface reflection from oil is higher than from clean water.
4. Light absorption by crude and heavy refined oil is extremely high at short wavelengths and decreases exponentially with wavelength in the visible and NIR part of the spectrum.

The interaction between these different factors controls the oil–water contrast used to detect surface oil. The appearance of the oil depends on the oil type, layer thickness, and degree of weathering and emulsification of the surface oil, as well as on environmental properties such as wind speed and the position of the sun relative to the sea surface and the satellite. This makes optical oil spill detection relatively complex, and oil–water contrast changes, depending on whether the oil is located in the satellite glint zone or outside it. This is illustrated by two examples from the Gulf of Mexico Deepwater Horizon oil spill shown in Figures 10.7 and 10.8.

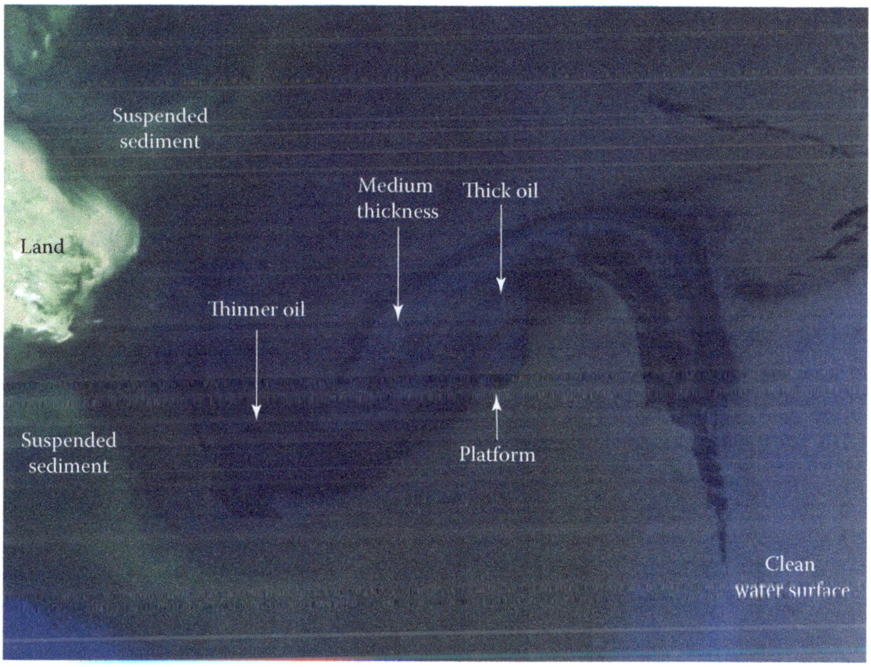

Figure 10.7 True color composite MERIS data from the Gulf of Mexico oil spill on April 29, 2010. From outside the sunglint zone, it is possible to obtain relative oil thickness from the spectral properties of top of atmosphere radiance. This is particularly useful for oil spill response, allowing cleanup efforts to be concentrated where they are most effective. The absorption coefficient of crude and heavy refined oil is very high, so even a thin surface layer (1–5 µm) suppresses the signal from the underlying water. Light absorption decreases exponentially with wavelength, so as oil thickness increases, the peak of the reflectance spectrum shifts increasingly to higher wavelengths. Thick oil will appear red, brown, or black depending on the type of oil and the degree of weathering. In this image, the thickest oil can be seen as red-brown streaks. Credit: ESA LearnEO! (Created by Val Byfield for ESA LearnEO! All LearnEO illustrations are provided under a Creative Commons Attribution 3.0 Unported license. Source https://creativecommons.org/licenses/by/3.0/.)

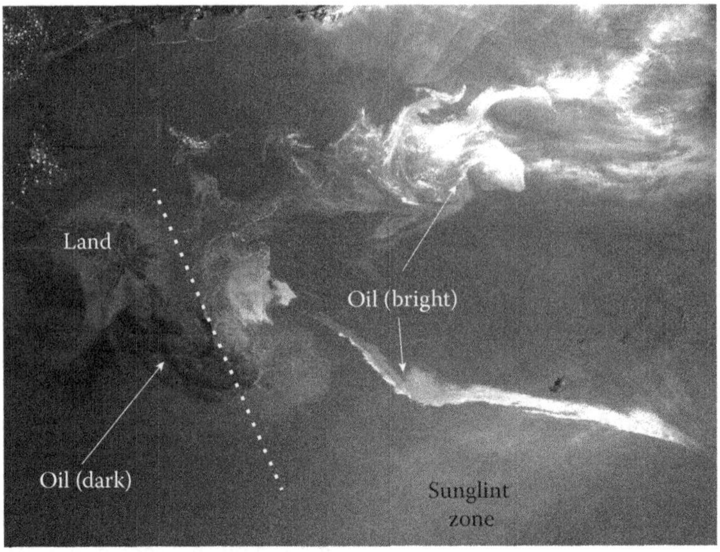

Figure 10.8 MERIS data from the Gulf of Mexico oil spill on May 25, 2010. Most of the surface oil in this image is found in the sunglint zone, so the oil appears brighter than the surrounding water. Oil appears brighter for two reasons. The refractive index of oil is higher than that of water, so the Fresnel reflection from an oil surface is greater than from a clean water surface. Oil also dampens capillary waves, so the proportion of the surface that reflects direct sunlight is higher than for water, where a larger number of wave facets are steep enough to reflect sky radiance, which is orders of magnitude less intense. A transition zone occurs at the edge of the sunglint zone (dotted line in the image) where the oil–water contrast is reversed so that the oil appears darker than the surrounding water because a smaller proportion of the wave facets reflect the direct solar beam. The location of this transition depends on the viewing configuration relative to the sun, the wind speed, the wave-damping capacity of the oil, and the oil refractive index. Within the glint zone, it is more difficult to determine oil thickness than outside it, although, as seen in this image, the thickest surface oil sometimes appears brighter than thinner oil. Credit: ESA LearnEO! (Created by Val Byfield for ESA LearnEO! All LearnEO illustrations are provided under a Creative Commons Attribution 3.0 Unported license. https://creativecommons.org/licenses/by/3.0/.)

10.8 CONCLUSIONS

In this chapter, many different applications of optical remote sensing over water bodies have been described. Usually, the information obtained from optical satellites is most useful when combined with data from other sensors (SAR, thermal sensors of sea surface temperature, satellite altimetry measurements of ocean currents, etc.) In many cases, information from model studies and *in situ* measurements are necessary in order to obtain (or calibrate/verify) the required information. Used sensibly, and with consideration for its strengths and limitations, ocean color remote sensing is an important tool for understanding and monitoring marine and aquatic environments. Its primary advantage is that huge swathes of ocean, large lakes, and long, wide rivers can be monitored in ways that terrestrial measurements could not perform in a timely or economic manner.

REFERENCES AND FURTHER READING

Byfield, V. Monitoring Marine Oil Pollution: Using SAR and optical data to detect and track surface oil. ESA LearnEO! Lesson 2. http://www.learn-eo.org/lessons/l2/.

ESA Ocean Colour Climate Change Initiative (CCI), http://www.esa-oceancolour-cci.org.

International Ocean Colour Coordinating Group (IOCCG) report series (1998–2015) http://ioccg.org/what-we-do/ioccgpublications/ioccg-reports/

Morales, J., V. Stuart, T. Platt and S. Sathyendranath (2011) *Handbook of Satellite Remote Sensing Image Interpretation: Applications for Marine Living Resources Conservation and Management*, EU PRESPO and IOCCG, Dartmouth, Canada, p. 293. http://ioccg.org/what-we-do/ioccgpublications/eupresp oioccghandbook/.

Data access

NASA Ocean Color Web, http://oceancolor.gsfc.nasa.gov/

Ocean Colour CCI data from Plymouth Marine Laboratory, https://www.oceancolour.org.

ESA data access, https://earth.esa.int/web/guest/data-access.

ESA/Eumetsat Sentinel data access, https://scihub.copernicus.eu.

11

Oceanographic and aquatic: Applications of optic sensing technologies

MATT MOWLEM, ALEX BEATON, AND GREGORY SLAVIK
Ocean Technology and Engineering Group

This chapter presents a brief overview of the large range of optical sensing technologies that are used to great effect to measure our oceans and aquatic environments. These environments are important as they supply resources (such as fish and minerals) and services to both society and the natural environment (such as buffering heat and CO_2 concentrations). However, measuring oceans and aquatic environments is a challenge as they are large and difficult to access in their entirety, and a wide range of parameters must be measured in order to understand the complex processes occurring within them. Here we present an overview of the optical sensors that address this challenge and give some example applications of lab-on-chip *in situ* (i.e., submersible) chemical sensor technology that has precise optical absorption measurement at its heart.

11.1 MEASUREMENT REQUIREMENTS

Natural biogeochemical cycles in the oceans alone provide "ecosystem services" valued at US$19 trillion per annum in 1997, equivalent to the global gross national product [1]. Ocean and terrestrial waters provide resources such as food, nutrients, and drinking and irrigation water and play a key role in climate regulation, arguably the most important environmental issue facing humankind [2]. Hence their measurement is a priority for science, industry, and government agencies.

A central problem in oceanography and aquatic applications is the acquisition of data over the vast expanse and depth of the oceans, rivers, and lakes. The required spatial and temporal resolution of measurement varies with the process being studied, but most remain undersampled. For example, phytoplankton blooms may have variability on meter and minute scales, whereas in all but isolated study areas, subsurface biological data are updated in tens of kilometer scales and with sampling intervals at best days, but often years.

Parameters of interest include the following: the temperature and conductivity of the water (often used together to calculate salinity and density, which are primary parameters in physical oceanography and therefore used in studies of climate change); currents and depth of water (including measurement of mass and heat flow, mixing, tides, and changes in mean sea-level); wave-height, period/wavelength, and extent of wave-breaking (used in estimating heat and chemical exchange between water and atmosphere amongst other applications); chemical parameters such as oxygen, the carbonate (dissolved CO_2) system, nutrients, trace metals, and gases (e.g., used to study interplays between biology and chemistry or to determine productivity or net CO_2 uptake); measurements of biological parameters (e.g., populations, species, genes or activity/biogeochemical impact); and inherent and apparent optical properties (such as irradiance, scatter and absorption, often used in conjunction with biological and chemical data to predict biological productivity or response). In addition, other parameters (e.g., sediment or sub sea floor structure) are also routinely measured in specialist applications.

The combination of the size of the aquatic environment, the number of parameters of interest, and the spatial and temporal resolution at which these data are required creates an almost insatiable need for improved measurement systems for which optical technologies offer a number of advantages.

11.2 DIRECT OPTICAL SENSORS

Direct optical sensors make their measurement by quantification of the optical response of the environment that is otherwise unperturbed by alteration or modification. This is in contrast to indirect optical sensors (explained in Section 11.3), where the sample is modified in some way prior to the measurement (e.g., by the addition of a reagent or interaction with an indicator).

11.2.1 Satellite and remote sensing

Direct optical systems are used to great effect in remote sensing/satellite sensors [3,4]. While measurement is limited to surface waters, and accuracy is often inferior to *in situ* (e.g., submersible) sensors, remote sensing enables global coverage with 1–100 km scale pixels with repeat measurements typically every few days.

Imaging spectrophotometers use solar radiation as a light source to measure apparent optical properties of water, which can be used in conjunction with corrections for illumination variation, specular reflection, and scatter and atmospheric modulation to estimate inherent optical properties. These measurements can be combined with additional data or models, or used directly to provide estimates of parameters of interest. For example the outputs of ocean color remote sensing include estimates of the following: chlorophyll *a* concentration, colored dissolved organic matter (CDOM), pigment absorption coefficients, particulate organic carbon (POC), calcite, and phytoplankton carbon, growth rates, and primary production.

Radiometers operating in the windows in the region 4, 8, and 11 μm are routinely used to measure water surface temperature to within ±0.4 K. Light detection and ranging (LIDAR) is well suited to use from aircraft and has been used to measure water depth in coastal waters [5]. With adaptation, LIDAR has also been used to remotely stimulate fluorescence and Raman scattering for estimation of chlorophyll and CDOM [6,7].

11.2.2 *In situ* sensors

In situ sensors can provide high accuracy measurements of subsurface and surface water parameters at high temporal resolutions (seconds to hours) and if deployed in number, or combined with moving platforms (such as autonomous underwater or surface vehicles, profiling floats, or ocean gliders), can make measurements over wide geographic scales. However, aquatic environments are aggressive, particularly for optical sensors. Technologies must resist or mitigate the effects of the following: pressure (typically up to 60 MPa); wide temperature range (typically −2°C to +35°C); corrosive environments; highly variable non-target chemical and biological composition; and biofouling leading to occlusion of optical windows and inlets, modification of the environment, or added mass/drag. In addition, the accuracy and precision demanded in many science and regulatory applications is difficult to achieve; accuracy requirement of one part in 35,000 is not uncommon.

The fundamental measurement of water salinity/density is predominantly made using very mature conductivity detection devices together with accurate temperature measurements (typically to an order better than 0.003 K). However, optical technologies for measuring density more directly via refractive index measured by refractometry are in development [8].

Mature optical sensors are used to measure *inherent optical properties (IOP)* and *apparent optical properties (AOP)* [4]. These include the AOP irradiance (radiant light flux in a given direction and per solid angle, per unit area perpendicular to propagation) that is often separated into directional components, such as upwelling or downwelling irradiance, and with additional spectral restrictions [e.g., 400–700 nm, whence it is termed photosynthetically available (or active) radiation (PAR)]. Scalar (i.e., the integral of irradiance over all solid angles) PAR is typically measured with a diffusing spherical light collector (typically plastic or opal glass) that guides light from all directions to an appropriate detector. Wavelength dependent sensitivity is achieved via the properties of the diffusing sphere, the detector, and if necessary additional filters. Spectroradiometers are used to measure the spectral profile of irradiance. There is a wide variety of detector/collector geometries to investigate the directional variation of irradiance. Irradiance information is typically used to estimate input energy for biological activity or to estimate the contents of the water. For example, phytoplankton contain photosynthetic pigments and their presence gives a distinctive spectral profile. While species to species similarity and individual variability makes precision and quantification difficult, this approach can be used to estimate population makeup and size for different classes of organism.

There is an extensive inventory of sensors to measure IOPs. Principally, these target absorption and scatter (both elastic and inelastic). Transmissometers are widely used to measure the effects of scatter and absorption together and consist of a (typically collimated) light source separated by a fixed distance from a detector that may or may not record spectral information. Spectral absorption attenuation meters measure concurrently spectral absorption and attenuation coefficients by using two tubes filled with the sample: one is black to minimize reflections, whereas the other, which aims to measure only the absorption coefficient that is reflective. Absorption alone is also measured in similar systems using liquid core waveguide capillary cells to confine the light to the sample.

Recent advances are delivering a commercial transmissometer sensor with polarizing filtering [9] that utilizes the extreme birefringence of calcium carbonate relative to other seawater components to make *in situ* estimates of particulate inorganic carbon [10].

There are numerous optical sensors that detect scatter, the most prevalent being nephelometers or turbidity sensors, which are optimized to make measurements of suspended particulate derived scatter. These typically operate using backscatter over a narrow optical window and small angular range and are used, amongst other applications, to investigate sediment load and biological particulate concentration. More complex instruments make measurements at multiple wavelengths and angles and can even enable accurate estimation of the volume scattering function.

Inelastic scatter measurements are principally used to quantify chemical and biological parameters. Fluorescence with pump light in the visible range is widely used to quantify photopigment (usually chlorophyll *a*) concentrations, and time resolved fluorescence, *pump and probe* or fast repetition rate fluorometry (FRRF) can assess the physiological health of photosystem II and hence can assess the health and productivity of photosynthetic organisms in a given sample. However, these techniques require frequent calibration as variability within phytoplankton communities and water matrix results in a drift or time variant calibration.

Fluorescence with excitation light in the UV can also be used to measure organic molecules including tryptophan and CDOM. The current trend is towards multispectral sensors, using this technique to enable more detail in mixed samples/matrices.

UV transmission spectroscopy is routinely used to make measurements of nitrate concentrations in natural waters. This is possible because with limited signal processing, the absorption from nitrate can be separated from other absorbing species such as bromide and CDOM in the same spectral range [11,12]. Instruments typically consist of a UV source (deuterium or xenon lamp) and a grating spectrometer inside a pressure vessel with optical windows or fibers sending and receiving light through the sample. The method is advantageous as it enables high frequency (typically 1 Hz) measurements, but the precision ($\pm 0.5\,\mu M$) and accuracy ($\pm 2\,\mu M$) is limited by interferences and current instrument design and is insufficient in some applications.

Raman spectroscopy can be used to make measurements of a wide range of (inorganic and organic) chemical species in water and in solid substrates in aquatic environments. However, to date, the size, cost, and weight of current instrumentation has limited its widespread application. Moreover, natural waters are typically dilute (nM or pM) with respect to target chemicals such as micronutrients and organic molecules. This presents a challenge to direct Raman spectroscopy and has motivated further research into the application of surface enhanced Raman spectroscopy (SERS) for analysis of water. However, SERS is complicated by nonreversible adsorption effects on the metal substrates typically used as the sensing surface. Active research is addressing this problem.

Flow cytometry, where single cells are measured in a continuous flow at high rate as they pass through a water jet or channel, is a powerful technique for counting and classifying living cells in natural waters. Typical designs use laser light at multiple wavelengths to illuminate the cells, with scattered and fluorescent light measured at multiple angles and wavelengths respectively. To reduce the size and extend the duration of *in situ* cytometers, microflow cytometers [13] are being investigated.

11.3 INDIRECT OPTICAL SENSORS

Indirect optical methods are powerful for creating high specificity and high performance *in situ* sensors because sensitivity to the parameter of interest can be greatly enhanced by modification of the sample. Possible modifications include separations or addition of reagents that complex or react with the target to produce an absorbing, fluorescent, or luminescent product, or the use of a reversible indicator material that changes its optical properties in the presence of the target. These approaches are used to great effect in the analysis of aquatic environments. The addition of reagents is typically more complicated (see Section 11.3.1) as it requires manipulation of the sample (e.g., with pumps and fluidic networks). The use of indicators is simpler to operate as the analyte typically interacts with the indicator as it diffuses through a solid matrix in which the indicator is bound.

Examples of indicator-based *in situ* optical sensors include the optodes for oxygen, pH, and pCO_2 stemming from the seminal work by Wolfbeis (Regensberg) and Klimant (TU Graz) and co-workers to produce *"optodes"* (optical electrodes) for environmental analysis. Typically, these sensors have an indicator immobilized in solid matrix that is permeable to the analyte. The indicator is generally a luminophore with a luminescence lifetime that is quenched in the presence of the analyte. To improve stability and reduce effects from ambient light and bleaching, a second luminophore that is not quenched by the analyte or other matrices in the sample, is included in the same solid matrix. This is termed dual luminophore reference or DLR. The measurement of the indicator and reference luminescence is performed in the time domain or via phase shift between excitation and luminescent light. *In situ* aquatic oxygen sensors based on this technology have been very successful and are widely deployed in a wide range of environments. Optodes for pH, pCO_2, and NH_4 are approaching or in the early stages of commercialization and uptake with technology for further analytes in development. The performance of optodes is excellent with a typical accuracy of 3 μM (O_2) and long-term stability is of the same order over months.

A common separation method partitions the analyte of interest from water typically with a membrane that is permeable to the target but impermeable to water. This either places the target within the membrane matrix, or allows it to pass through to an alternative carrier fluid or gas. Both methods are often used to enable spectrophotometric quantification of the analyte at wavelengths where water is a strong absorber. For example, methane can be detected in the mid-IR (infrared) via the evanescent waves in fibers coated with polydimethylsiloxane (into which methane partitions [14]), or methane and CO_2 can be analyzed in the gas (carrier) phase behind a suitably pressure resistant membrane [15]. The technique requires a suitable permeable membrane, and mitigation of the variable membrane permeability and enrichment factor with pressure, time, and other environmental factors. In addition, finite water (and matrix) permeability/uptake of the membrane must be addressed.

Despite obvious drawbacks of complexity and difficulty in ensuring reliability, considerable research has been directed towards the development of reagent-based indirect sensors. Example devices and some detailed examples of current projects are presented in the following sections.

11.3.1 Reagent-based *in situ* indirect optical sensors

Reagent-based sensors mix the sample with a reagent that reacts or binds with the target to form an absorbing, fluorescent or luminescent product. The advantages include the availability of numerous robust reagent-based analytical methods/assays for numerous chemical and some biological targets and excellent performance (sensitivity, stability, accuracy, precision, and limit of detection) that is afforded—typically orders of magnitude better than direct optical methods. The drawbacks include the complexity of the resultant instrument that typically includes multiple pumps and often valves, a fluidic network, an optical detection cell, reagent stores, power sources, control and data logging electronics, sample input filters, and interconnecting tubes. The reduction of the impact of this complexity on the size, cost, reliability, and ease of use is a key focus of current research.

The first *in situ* reagent-based systems were produced in the 1980s and currently, a number of systems are commercially available (e.g., Refs. [16–20]). However, these have not *overcome* all of the principal disadvantages listed earlier, and therefore, their impact in the market and in aquatic applications has been limited. Recently, a submersible reagent-based phosphate sensor [21] that addresses many of these deficiencies is becoming widely used, demonstrating the large demand for this technology.

Lab-on-chip technologies and microfluidics where the fluidic network is reduced in diameter (typically to <200 µm) and integrated with other systems (such as valves, pumps, or optics) in a manifold or "*chip*" has been investigated by a number of researchers for use in *in situ* sensors. This is because the technology enables the drastic reduction in the scale of the device, the resources it uses. Through miniaturization and simplification of the system components, it can also reduce cost while in some circumstances increasing ease of use. Applications have included the analysis of DNA from microorganisms, measurement of nutrients, the carbonate (dissolved CO_2) system, and trace metals.

One family of lab-on-chip-based *in situ* analyzers developed in Southampton is presented here. This has been applied to the detection of a wide range of nutrient, carbonate, and trace metal parameters using detection of absorbance. Systems monitoring fluorescence for further nutrients and molecular assays (e.g., analysis of nucleic acids, and hydrocarbons) are in development. We focus here on the systems using absorption for chemical detection.

These systems have a microfluidic chip that is formed from layers of poly(methylmethacrylate) (PMMA) that are micromilled (i.e., machined with miniature milling bits, see Figure 11.1) to form channels that are closed by bonding the layers together (explained later, see Figure 11.2). The chip also forms the end cap of a watertight case that contains the pump, valves, electronics, and optics that are mounted on the inside face of the chip. The

Figure 11.1 Micromill facility including multiple computer controlled (CAD-CAM) micromills. Smallest milling bits are 100 µm.

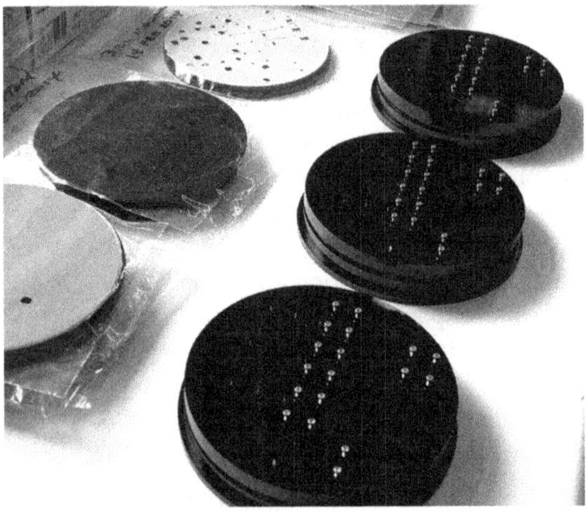

Figure 11.2 Microfluidic chips in various stages of completion during the manufacturing process.

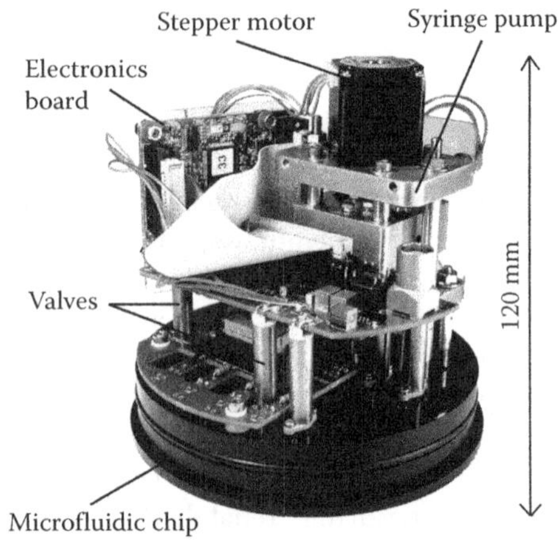

Figure 11.3 Principal internal components of an *in situ* lab-on-chip analyzer. The microfluidic chip forms the end cap to the watertight housing, which minimizes the fluidic distance between the sample inlet and the chip.

outside face of the chip has an inlet connected to the environment via a filter (0.45 μm) and ports for connection of stores of reagent, standards, and blanks. The watertight case is filled with oil that electrically insulates the internal components and includes a bellows or diaphragm to ensure that the external (environment) pressure is communicated to the internal components through the oil. With careful design and selection of pressure resistant components, this pressure balance design enables operation at great depths as there are no internal voids to create pressure differential. The current sensors have proven operation to 6000 m water depth.

The principal internal components of the system are shown in Figure 11.3. The figure depicts the (lab-on-chip) microfluidic chip at the bottom of the image that includes multiple optical cells and three printed circuit boards (PCBs), which from bottom to top form (1) the connections to the LEDs (light emitting diodes) and photodiodes in the optical cells, (2) the main daughter board that controls the pump motor and valves and other actuators, and (3) the main electronics board that includes the microcontroller, data acquisition, LED drive, and logging systems. The daughter board is mounted directly on top of the solenoid valves (Lee, USA, which are used as standoffs,) and is set clear of the syringe pump. The syringe pump is operated with a stepper motor (labeled in Figure 11.3) that turns a threaded rod. The rod engages with a drive nut that moves a drive plate attached to the top of plungers. The plungers move inside static syringe barrels located in and sealed to the chip. Up to four syringe barrels are driven by the single motor and drive plate enabling precise pumping of multiple fluids at very consistent volumetric flow ratios. This is important in achieving consistent reagent to sample ratios and accurate production of the colored product for measurement in the optical cell.

Central to the chip design is the use of a unique optical cell that enables high performance measurement of transmission through the sample/reagent. The cell relies on the use of two innovations. The first is the use of a solvent vapor exposure technique that polishes the marks left from the micromachining fabrication in the optical channel. The other uses a tinted or colored substrate to suppress stray, reflected, and scattered light preventing it from reaching the detector. Figure 11.4 shows optical profilometry data of the base of an optical cell channel of 300 μm width formed in PMMA at increasing durations of exposure to solvent vapor (in this case, chloroform). The vapor appears to preferentially condense and absorb into the rough features causing these and other exposed surfaces to soften and reflow resulting in a polishing effect. This polishing effect is critical in creating high performance optical cells. The smooth surface prevents optical scatter due to the textured substrate, reduces traps for bubbles, and decreases the effect of surface changes (e.g., staining or fouling) on the

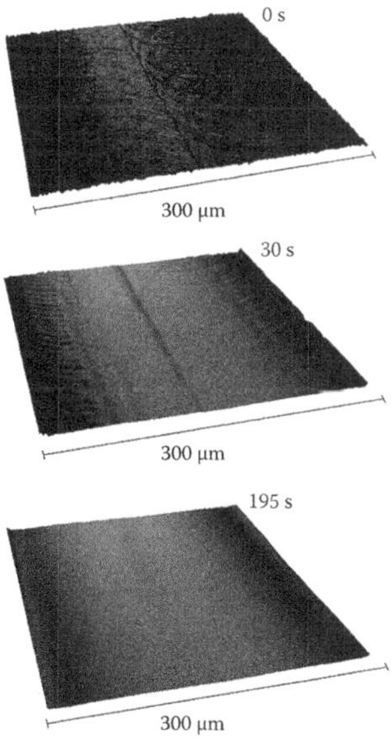

Figure 11.4 Optical profilometry scans of the bottom of the optical/fluidic cell, showing the polishing effect of chloroform vapor after 0, 30 and 195 s of exposure.

measured optical transmission. The softening of the surface also allows for bonding of layers if they are pressed together shortly after exposure to the solvent vapor.

The use of a colored substrate to form the optical cell confers a number of advantages. The principal advantage is that the tint, which is usually chosen to be a broad band absorber in the emission band of the LED, absorbs any light that does not pass directly down the fluid-filled channel. This includes light scattered by the sample, or interfaces or features in the chip, and reflections within the chip, which otherwise dominate the optical power incident on the detector. Using a tint rather than a completely opaque material enables the walls of the optical channel to be thinned in the region of both the LED and the detector. This thinning results in a very small optical loss through the walls and hence, most of the possible optical power is coupled into and out of the optical cell and passes only through the fluid of interest.

Thin tinted walls are used rather than adding clear windows as it enables fabrication of the fluidic/optical channel and the windows in a single material in a single milling/fabrication process, which simplifies optical alignment, fabrication, and therefore reduces cost while increasing robustness.

The effect of the tint can be seen in the photographs shown in Figure 11.5, which compares the scattered and transmitted light in optical cells in clear (top images) and tinted PMMA (bottom images). It can be seen that without the tint source, light from the LED is transmitted and reflected around the chip outside of the analytical channel. In the images on the left-hand side, it is possible to compare the profiles of the light as it exits the cell and impinges on a surface in the cavity made to receive the photodiode (removed for clarity/this demonstration). In the clear chip, light transmitted through the PMMA including that aided by scatter and reflection strongly illuminates a large patch on a piece of paper placed in the detector recess. In the tinted chip, a small less intense spot is observed, which demonstrates that only light traveling directly through the analyte impinges on the detector.

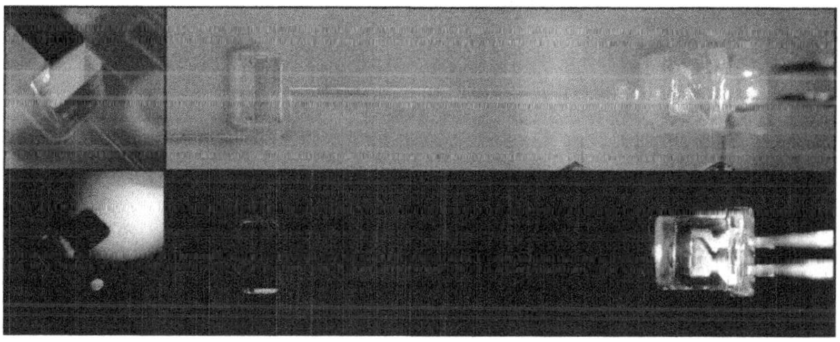

Figure 11.5 Identical optical cells in clear (top) and tinted PMMA (bottom). The panels on the left show the profile of the light as it exits the absorption cell and reaches the cavity for the photodiode. With tinted PMMA, background light has been absorbed and only light that traveled through the absorption cell emerges.

The suppression of stray, scattered, and reflected light is important as it decreases the optical offset, which increases sensitivity, dynamic range, and reduces noise with concomitant impact on precision and hence accuracy.

Because the windows are integral to the optical cell, and optical cells are inexpensive to include in the device, the device can include multiple optical cells. These are principally used to measure the native absorption/scatter in the sample prior to reagent addition, and to provide cells of different length. The latter is done as sensitivity and hence, precision and resolution is maximized when the natural logarithm absorbance in the channel is 1, but above this value nonlinearity may be observed using simple interpretations of the Beer–Lambert law. Using multiple optical cells enables optimization of cell length for a greater range of concentrations and increases the dynamic range.

At very short cell lengths, the tinted PMMA may not be thick enough to absorb all the stray light from the LED, which is close to the detector. To combat this, a light channel (waveguide) is placed between the LED and the channel window to restrict and collimate the light, and to increase the distance between the LED and the detector. By using a high refractive index fill material (e.g., an optical epoxy), it is possible to create a waveguide, though a small numerical aperture is required to achieve the desired stray light suppression. A photograph of an example short cell (prior to bonding) is shown in Figure 11.6. The optofluidic channel is in the center of the image and has length (2.2 mm) determined by two opposed right angled elbows.

Figure 11.6 Optical absorbance cell milled into tinted PMMA. This particular cell is 2.2 mm in length. The LED is further separated from the photodiode by a light channel that is typically filled with optical epoxy.

On the left-hand side, an approximately 1 mm window separates the channel from the photodiode pocket (visible as a rectangle with rounded corners). On the right, the window separates the channel from the light channel, which has a T shape with two circular vias connected to the hanging lobes of the T. Another via is included at the LED end of the base of the T. After bonding and closure of the channel, these vias allow the light channel to be filled, for example, with optical epoxy.

11.4 EXAMPLE DEPLOYMENTS AND APPLICATIONS OF *IN SITU* LAB-ON-CHIP SENSORS

This section summarizes example applications of the lab-on-chip sensor technology when adapted for the measurement of nitrate and nitrite. The sensor has a limit of detection (LOD) of 0.025 μM for nitrate (0.0016 mg/L as NO_3^-) and 0.02 μM for nitrite (0.00092 mg/L as NO_2^-) and a measurement range up to 350 μM (21.7 mg/L as NO_3). This performance is suitable for almost all natural waters, apart from the oligotrophic open ocean where nitrate is depleted and often present at 0.1–10 nM.

The sensor uses the Griess assay, which causes nitrite diazotization with sulphanilamide and subsequent coupling with *N*-(1-naphthyl)-ethylenediamine dihydrochloride (NED) to form an intensely colored azo dye. The dye is pink in color and we observe the intensity via absorption using a 525 nm LED. Nitrate is not detected by the Griess assay, but it may be reduced to nitrite, which can be detected. This we achieve by buffering the sample and passing it through a (catalytic) tube made from cadmium that is weakly alloyed with copper. The tube is connected to the outside of the chip via fluidic connections milled into the chip.

Nitrate is an important macronutrient in natural systems and plays a role in stimulating strong growth and productivity. It is used in agriculture and aquaculture to stimulate plant growth, but in the natural aquatic environment, it may stimulate unwanted growth leading to eutrophication; at high concentrations, it may even be toxic. Therefore, understanding the sources, sinks, and utilization and transport mechanisms is of importance for understanding the environment and the anthropogenic effects upon it.

11.4.1 Macronutrient cycles in the Christchurch estuary

In this project, nitrate sensors are deployed at three sites in the Christchurch estuary. The sites are located in the two source rivers, the Stour and Hampshire Avon, and at the mouth of the estuary at Mudeford.

Estuaries play a vital role in our economy as sites of leisure and commercial activities. They are important nursery grounds for many species of economically important fish yet are also some of the most vulnerable sites for anthropogenic impacts. This can stem from activities in the estuary but also from the accumulation of pollutants gathered by rivers from large catchments. Of particular concern is excess river-borne concentrations of phosphate and nitrate from a variety of sources, such as agricultural (fertilizer, waste) run off and from waste/waste water processing.

Elevated levels can stimulate excess algal growth that can deplete the water of oxygen causing widespread fish kills or cause the growth of poisonous (to both humans and other animals) algal species (red tides) that lead to closure of shell fish fisheries.

While extensively studied and monitored, for example, by the Environment Agency and water companies, there are still major gaps in our data and therefore knowledge about estuaries. In particular, sudden storm events are likely to cause significant spikes in pollution inputs to estuaries, but current measurements performed by sampling (at only a few locations) and later, laboratory analysis are taken infrequently (days or weeks apart) and therefore miss these events.

The use of lab-on-chip sensors seeks to address this gap by providing data on nutrients (initially nitrates, but also nitrites, phosphates and eventually dissolved organic nutrients) at high temporal resolution (6 min to hourly intervals) at three sites in Christchurch Harbour estuary (Dorset). It is used as an example to provide understanding of similar estuaries elsewhere. This high frequency data will enable us to catch transient inputs from storms, and then monitor the fate of these inputs in the estuary or their transport through the estuary mouth to the English Channel.

Our initial results show clearly that there is transient response in nutrient inputs during storms, and observable effects on biological nutrient uptake and productivity.

Figure 11.7 Two lab-on-chip nitrate sensors during changeover for redeployment (Christchurch Harbour Estuary, Dorset) as part of NERC, the Macronutrient Cycles Programme.

Figure 11.7 shows a photograph of two nitrate sensors. One (on the right side) has been retrieved after 3 weeks of continuous operation at an UK Environment Agency/Department for Environment and Rural Affairs gauging station in the Hampshire Avon, while the other has been recently prepared for deployment at the National Oceanography Centre, Southampton. The sensor is the small cylinder at the bottom of a frame used for deployments on mooring. The upper larger cylinders house the liquid stores (reagents, standards, blanks) and waste collecting bag. A photograph of the device deployed in the Hampshire Avon is shown in Figure 11.8.

11.4.2 Development and validation of first generation chemical sensors for icy ecosystems (DELVE)

Icy environments such as glaciers and ice sheets are now known to harbor significant populations of microorganisms despite the challenging environmental conditions of extreme cold, desiccation, freezing, and often high pressure. Many microbes accelerate chemical weathering of sediments and rock producing nutrients for ecosystems downstream. A better knowledge of these processes is important for understanding the following: (1) impacts on global carbon and nutrients cycles, (2) biodiversity and life in extreme environments

Figure 11.8 A lab-on-chip nitrate sensor submerged at the measurement site in an Environment Agency gauging station at the Hampshire Avon, UK.

(e.g., Antarctic subglacial lakes), and (3) hydrological conditions and flow beneath ice sheets.

Currently, data are gathered by infrequent sample collection and analysis in only a few locations. Sensor technology promises to enable more frequent measurements in more locations, but the number of suitable sensors is currently extremely limited.

Building on the developments in oceanographic sensors, this project has developed a first generation of compact chemical sensors for use in glaciers and ice sheets. The sensors include lab-on-chip nutrient and trace metal (Fe and Mn) sensors.

The technology's performance has been evaluated under icy conditions (e.g., at low temperature, under freeze/thaw conditions, at high pressure and with glacial meltwater sample types) in both the laboratory, and in the field (see Figure 11.9, which shows a deployment site in the proglacial stream draining the Greenland Ice Sheet at Leverett Glacier, near Kangerlussuaq, Greenland). This testing has informed a number of enhancements, including the development of freeze/thaw tolerant designs, and assays using reagents adapted to give low freezing temperatures. These adaptations have enabled the sensors to meet the requirements of this application. While providing significant scientific data and understanding of icy environment biogeochemistry in themselves, these developments and deployments lay the foundation for future exploration, measurement, and understanding of subglacial lakes (e.g., Subglacial Lake Ellsworth, Antarctica), marine under-ice environments (e.g., from autonomous vehicles such as Autosub, moving under ice), and across a wide range of icy ecosystems where *in situ* measurements are desirable. The robust technology developed also has strong relevance to water quality monitoring in freshwater environments, which we are exploring through collaboration with the Environment Agency, UK.

Figure 11.9 Deployment of sensors in the proglacial stream draining the Greenland Ice Sheet at Leverett Glacier, near Kangerlussuaq, Greenland.

11.4.3 Application in ocean observatories and on autonomous vehicles

This is the main focus of the sensor development activities in Southampton because of the value of oceans and their importance to society (see Section 11.1), and because of the current paucity of *in situ* data on biogeochemical processes. We are developing a number of sensors (including for the carbonate system, trace metals, cytometry, and nucleic acid analysis (for microbiology)) for this application, but for brevity, here we focus on nutrient sensors and the nitrate/nitrite sensor in particular.

In this application, nitrate is measured, together with other parameters such as salinity, temperature, currents and other biogeochemical parameters. The principal aims are to understand processes that deliver nutrients to the productive sunlit ocean (including cycling and remineralization in the deep sea) and to quantify and model biological processes utilizing nutrients often with the aim of predicting productivity and the extent of CO_2 uptake.

Because of the remoteness and inaccessibility of the majority of the world's oceans, nutrient sensors are integrated on board vehicles or structures termed platforms in the oceanographic community.

The challenge for the technology is to achieve the performance required including robustness in a harsh environment, longevity (many platforms remain at sea without intervention for months or years), and size and cost such that they are small enough and cheap enough to be integrated on all platforms and in large numbers.

The small size and robustness of oxygen optodes have enabled them to be used widely in this application. The lab-on-chip technology is approaching the size, longevity, cost, and performance required in the most widely used, and most demanding platform (in terms of size and longevity particularly)—the profiling float.

There are currently (July 2017) 3833 profiling floats in the world's oceans (www.argo.ucsd.edu). There are a number of manufacturers with floats of a broadly similar design. While deep floats (>4000 m rated) are in production, most of the floats in use are designed to profile the top 2000 m of ocean. They do this by changing their buoyancy through inflating an external bladder with oil (at depth); sometimes, they use air buoyancy (near the surface). Buoyancy is controlled via algorithms and microcontrollers and typically the float will park at depth before slowly surfacing and taking measurements every 10 days. At the surface, the float loiters while it transmits the recorded data via satellite to data processing and storage repositories. Thereafter, it returns to park at depth. The geographical location is determined by ocean currents. The deployment location of new floats is used to control the spacing of the float array. Currently, most floats record temperature and salinity (via electrical conductivity and temperature) as a function of depth (via measurements of pressure). A significant international research effort is developing and integrating a raft of optical and biogeochemical sensors into profiling floats and the array.

Figure 11.10 shows two nitrate lab-on-chip sensors integrated into an NKE PROVOR profiling float and deployed in a Scottish sea loch. To enable

Figure 11.10 Two lab-on-chip nitrate sensors (left and right white cylinders) mounted on an NKE PROVOR profiling float deployed in Loch Etive, Scotland. The orange satellite transmitter arial and black conductivity, temperature, and depth sensors are clear of the water surface, which is also dotted with snow.

widespread uptake, we are further miniaturizing the technology and extending the deployment lifetime. This is currently limited by the (temperature dependent) stability of the reagents and standards which for ocean applications is at worst 3 months. We are hoping to extend this to 6 months in the near term and to over a year in the long term.

A number of other platforms offer advantages over profiling floats in some applications. For example, autonomous underwater vehicles (either propeller driven, or with buoyancy engines and wings to form ocean gliders) have the advantage that they can move against currents, and cover more ground (and depth). They can also enter inaccessible environments such as the waters under ice shelves.

From a sensor's perspective, some also have more available power and space than a profiling float, and can be deployed for shorter durations.

Figure 11.11 shows two lab-on-chip sensors (included in a single cylindrical housing) being integrated with an iRobot/Kongsberg Seaglider for testing as part of the NERC Shelf Seas Biogeochemistry programme where nitrate data are collected with current data across the continental shelf to examine nutrient fluxes and their impact.

Figure 11.11 Lab-on-chip sensors being integrated in the rear cowling of an iRobot/Kongsberg Seaglider. An access hatch is removed revealing the sensor (gray cylinder held in place with cable ties). Flexible bags for storing reagent, standard, blank, and waste that are usually packed in suitable gaps around the sensor are removed and held aloft for clarity.

11.5 CONCLUSIONS

Optical sensors are widely used in oceanographic and aquatic applications. In large part, this is due to their ability to measure important environmental parameters remotely (e.g., from space) and to enable measurement of complex physical, chemical, and biological properties. Coupled with robust analytical techniques such as indicator or reagent-based chemistry, optical sensors are able to measure many complex biogeochemical parameters *in situ* that is, submerged in the environment. The ability to make precise repeatable measurements over a wide dynamic range places optical sensor technologies at the forefront of active technology research to develop systems to meet the demand for ever more data (in both space and time) on a growing list of priority measurements used to address some of the biggest issues faced by mankind.

ACKNOWLEDGMENTS

We would like to acknowledge the following: funding from NERC (particularly the Macronutrient Cycles Programme), EPSRC, and the EU (particularly SenseOCEAN grant agreement no 614141); photography by Alex Beaton, John Walk, Adrian Nightingale and Chris Cardwell; technical input from all at the Centre of Marine Microsystems (past and present) in both NOC and the University of Southampton, particularly Prof Hywel Morgan and his research group; and the support of our many partners in the projects that generated the research and technology presented here.

REFERENCES

1. Costanza, R., R. dArge, R. deGroot, S. Farber, M. Grasso, B. Hannon, K. Limburg, S. Naeem, R.V. Oneill, J. Paruelo, R.G. Raskin, P. Sutton, and M. vandenBelt, The value of the world's ecosystem services and natural capital. *Nature*, 1997, 387(6630): 253–260.
2. King, D., Climate change: the science and the policy. *Journal of Applied Ecology*, 2005, 42(5): 779–783.

3. Martin, S., *An Introduction to Ocean Remote Sensing*. 2004, Cambridge University Press: Cambridge.
4. Dickey, T., M. Lewis, and G. Chang, Optical oceanography: Recent advances and future directions using global remote sensing and in situ observations. *Reviews of Geophysics*, 2006, 44(1).
5. Hoge, F.E., R.N. Swift, and E.B. Frederick, Water depth measurement using an air-borne pulsed neon laser system. *Applied Optics*, 1980, 19(6): 871–883.
6. Reuter, R., D. Diebel, and T. Hengstermann, Oceanographic laser remote-sensing-measurement of hydrographic fronts in the german bight and in the northern adriatic sea. *International Journal of Remote Sensing*, 1993. 14(5): 823–848.
7. Yoder, J.A., J. Aiken, R.N. Swift, F.E. Hoge, and P.M. Stegmann, Spatial variability in near-surface chlorophyll-a fluorescence measured by the Airborne Oceanographic LIDAR (AOL). *Deep-Sea Research Part Ii-Topical Studies in Oceanography*, 1993, 40(1–2): 37–53.
8. Grosso, P., D. Malarde, M. Le Menn, Z.Y. Wu, and J.L.D. de la Tocnaye, Refractometer resolution limits for measuring seawater refractive index. *Optical Engineering*, 2010, 49(10): 103603-1–103603-5.
9. Bishop, J.K.B., Autonomous observations of the ocean biological carbon pump. *Oceanography*, 2009, 22(2): 182–193.
10. Guay, C.K.H., and J.K.B. Bishop, A rapid birefringence method for measuring suspended $CaCO_3$ concentrations in seawater. *Deep-Sea Research Part I-Oceanographic Research Papers*, 2002, 49(1). 197–210.
11. Johnson, K.S., and L.J. Coletti, *In situ* ultraviolet spectrophotometry for high resolution and long-term monitoring of nitrate, bromide and bisulfide in the ocean. *Deep-Sea Research Part I-Oceanographic Research Papers*, 2002, 49(7): 1291–1305.
12. Sakamoto, C.M., K.S. Johnson, and L.J. Coletti, Improved algorithm for the computation of nitrate concentrations in seawater using an in situ ultraviolet spectrophotometer. *Limnology and Oceanography-Methods*, 2009, 7: 132–143.
13. Kim, J.S., and F.S. Ligler, *The Microflow Cytometer*. 2010, Singapore: Pan Stanford Publishing Pte Ltd.
14. Kraft, M., M. Karlowatz, B. Mizaikoff, R. Stuck, M. Steden, M. Ulex, and H. Amann, Sensor head development for mid-infrared fibre-optic underwater sensors. *Measurement Science & Technology*, 2002, 13(8): 1294–1303.
15. Fietzek, P., B. Fiedler, T. Steinhoff, and A. Kortzinger, In situ quality assessment of a novel underwater pCO(2) sensor based on membrane equilibration and NDIR spectrometry. *Journal of Atmospheric and Oceanic Technology*, 2014, 31(1): 181–196.
16. Scholin, C., G. Doucette, S. Jensen, B. Roman, D. Pargett, R. Marin, C. Preston, W. Jones, J. Feldman, C. Everlove, A. Harris, N. Alvarado, E. Massion, J. Birch, D. Greenfield, R. Vrijenhoek, C. Mikulski, and K. Jones, Remote detection of marine microbes, small invertebrates, harmful algae, and biotoxins using the environmental sample processor (ESP). *Oceanography*, 2009, 22(2): 158–167.
17. Hanson, A.K. and IEEE, A new in situ chemical analyzer for mapping coastal nutrient distributions in real time. *Oceans 2000 MTS/IEEE-Where Marine Science and Technology Meet, Vols 1–3, Conference Proceedings*. 2000, New York: IEEE. 1975–1982.
18. Vuillemin, R., L. Sanfilippo, P. Moscetta, L. Zudaire, E. Carbones, E. Maria, C. Tricoire, L. Oriol, S. Blain, N. Le Bris, P. Lebaron, and IEEE, Continuous Nutrient automated monitoring on the Mediterranean Sea using in situ flow analyser, in *Oceans 2009*, Vols 1–3. 2009, IEEE: New York. pp. 517–524.
19. Degrandpre, M.D., T.R. Hammar, S.P. Smith, and F.L. Sayles, In situ measurements of seawater pCO_2. *Limnology and Oceanography*, 1995, 40(5): 969–975.
20. Seidel, M.P., M.D. DeGrandpre, and A.G. Dickson, A sensor for in situ indicator-based measurements of seawater pH. *Marine Chemistry*, 2008, 109(1–2): 18–28.
21. Gilbert, M., J. Needoba, C. Koch, A. Barnard, and A. Baptista, Nutrient loading and transformations in the Columbia RiverEstuary determined by high-resolution *In situ* sensors. *Estuaries and Coasts*, 2013, 36(4): 708–727.

12

Monitoring of volcanic eruptions: An example of the application of optoelectronics instrumentation in atmospheric science

BARBARA J. BROOKS
University of Leeds

12.1 INTRODUCTION

Optoelectronics has been extensively used in the field of atmospheric science, and many of the operational principles and techniques have been covered extensively in this handbook. Presented here is a specific example in which instrumentation utilizing some of these principles was deployed in response to the threat posed by ash emissions from Icelandic volcanoes.

12.2 MONITORING OF ASH FROM ICELANDIC VOLCANOES

After 3 weeks of grumbling seismic activity, Iceland's Eyjafjallajokull volcano erupted on April 14, 2010 for the first time in almost 200 years: Figure 12.1 shows a photograph by Brynjar Gauti taken during the event itself. The eruption ejected a debris plume that reached over 30,000 ft into the atmosphere and was carried by the prevailing wind over the UK and Northern Europe. The danger posed to aviation from volcanic ash is well known and has resulted in loss of power to all engines in the most serious aircraft encounters, most notably those with the ash clouds from Galunggung in 1982 (Hanstrum and Watson 1983) and Mount Redoubt in 1989 (Casadevall 1994).

The aviation industry's standing instructions for dealing safely with volcanic ash are "to avoid all encounters with ash," and this advice has been incorporated into the safety management systems operated by leading air traffic services and airspace management organizations. The "zero tolerance" to ash inherent in this advice substantially reduced the volume air traffic flow through the UK, Irish, and Continental European airspace after the eruption and by April 18, commercial aviation movement through the airspace of 23 European countries had ceased with restrictions in place in further two countries. Over 300 airports, representing 75% of the European network, closed and

Figure 12.1 Eyjafjallajokull eruption. (AP Photo/Brynjar Gauti.)

precautions had to be taken to protect grounded aircraft and airfields from potential ashfall.

The London Volcanic Ash Advisory Centre (London VAAC) based within the UK Met Office is responsible for monitoring and forecasting the movement and dispersion of volcanic ash originating from volcanoes in Iceland. The forecasts comprise maps showing the geographical distribution of ash mass concentration at specific levels in the atmosphere, examples of which are shown in Figure 12.2. To produce these maps, atmospheric dispersion models are required; the NAME (Numerical Atmospheric-dispersion Modeling Environment, Jones et al. 2007) model is used by the London VACC.

In addition to meteorological fields and information concerning the location of the eruption, starting time, release height, and plume height, the accuracy of atmospheric dispersion models depend on the representation of the plume dynamics within the model and the physical properties of the ash and gases in the plume. While in most cases, it is easy to deliver the first three terms (being all static after an eruption has started) to a high degree of accuracy, it can be quite difficult to give an accurate observation of the plume height. Advection and dispersion downstream from the volcano also leads to the downstream ash cloud not necessarily being at the same altitude as the plume top over the volcano and more often than not, this cloud

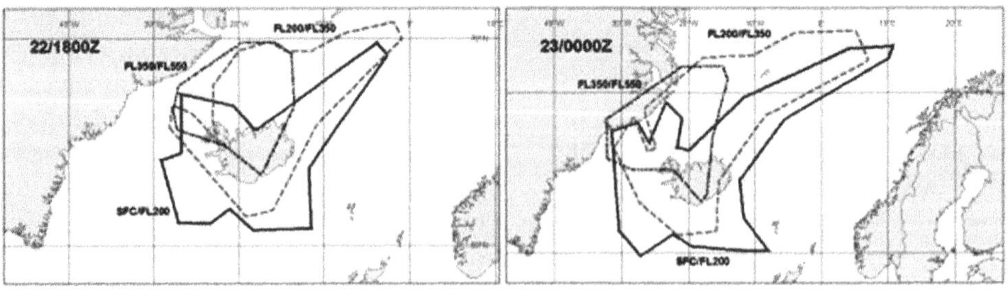

Figure 12.2 Ash forecasts issued by the London VAAC (Volcanic Ash Advisory Centre) at 06:00 UTC (coordinated universal time) on May 22, 2011. The solid lines indicate Flight Level 200 (FL200) and refer to altitudes at and below 20,000 ft (approximately 6 km). The dashed lines indicate FL200–FL350 meaning altitudes between 6and 10 km and the dotted lines indicate FL350–FL550, indicating altitudes between 10and 17 km. These were for the 2011 Grímsvöten eruption.

becomes divided into multiple layers: observations from the Eyjafjallajökull eruption indicated that considerable plume modification process occurred in the first tens to hundred kilometers downstream of the volcano. Better characterization of these processes and the physical properties of the ash are required to improve the performance of the dispersion models and hence, the accuracy of the ash coverage maps issued to the aviation industry.

The range of instrumentation available to make the required observations is large and deciding on the "right tool for the job" requires an appreciation of the phenomena of interest, the measurement principle an instrument employs and the resulting limitations of any measurement made. The physical properties of ash that are of specific interest here include the following:

- Total loading
- The number of particles per unit volume in a given size range: number density
- Single particle shape
- Single particle composition
- Single particle optical properties

Pruppacher and Klett, and Colbeck provide comprehensive discussions on atmospheric aerosol microphysical properties. LIDAR remote sensing techniques have been used to infer unequivocally the presence of ash in an atmospheric layer: achieved by combing a standard aerosol LIDAR (light detection and ranging) with a Raman LIDAR, but they do not provide any information on the microphysical properties of ash that are required to improve model performance. At the present time, only direct, *in situ*, measurement techniques provide these key observations. Table 12.1 provides a summary of both remote sensing and *in situ* instrumentation deployed in the detection of ash clouds.

During the Eyjafjallajökull eruption, the UK Facility for Airborne Atmospheric Measurements (FAAM) BAe-146 aircraft (Figure 12.3, http://www.faam.ac.uk/) flew 12 flights targeting volcanic ash clouds around the UK. The FAAM aircraft is a modified BAe 146-301 carrying a range of instrumentation for the *in situ* observation of atmospheric dynamics, thermodynamics, gas phase composition, and cloud and aerosol properties; a report and discussion of the measurements can be found in Johnson et al. (2012). A wide range of instrumentation, making use of optoelectronic components and concepts, was carried during these sorties and this article cannot discuss them all in detail. The remainder of this commentary will concentrate on two families of instruments: (1) those referred to as optical aerosol spectrometers and (2) condensation particle (nucleus) counters. These are the most widely used family of instruments in aerosol and cloud research and industry.

12.3 OPTICAL AEROSOL SPECTROMETERS

There are a number of instruments including the Passive Cavity Aerosol Spectrometer Probe (PCASP, Droplet Measurement Technologies (DMT), Boulder, USA), Cloud and Aerosol Spectrometer (CAS, DMT), Forward Scattering Aerosol Spectrometer Probe (FSSP, DMT), Dust Monitor (1.108, Grimm Aerosol Technik, Germany) and Compact Lightweight Aerosol Spectrometer Probe (CLASP, University of Leeds, UK), to name but a few that can be termed optical aerosol spectrometers. Although they may vary in sensitivity and measurement range, the basic principle on which they operate is the same: they all assume that the target particle is spherical and of known refractive index, and they all use Mie theory to determine the particle size from the intensity of the scattered light over a given angular sector defined by the instrument's individual physical optical layout.

Consider the FSSP, the optics, which is shown in Figure 12.4. A collimated 633 nm laser beam interacts with the target particles on the particle plane. Light is scattered in all directions, but only that is scattered in the forward direction (relative to the direction of transmission of the incident beam) falls on the right angle prism and is transmitted to the analysis electronics. Note the dump spot: it removes straight through beam that would, if passed to the detection electronics, cause saturation. The size of the dump spot and the collecting optics define an angular range over which the scattered light is collected; for this instrument, the collection area is defined by a region from 3° to 13° in the θ plane (plane of the page) and from 0° to 359° in the Φ plane (the plane perpendicular to the page).

The light collected is initially spit by the 50% beam splitter—one half of the light is projected onto a photodiode, converted to a current, then to a voltage before being amplified and passed through to a pulse height analyzer. The second

Table 12.1 Aerosol and cloud measurement instruments, parameters retrieved and the pros and cons of each system

System	Direct/ remote measurement	Manufacturer	Parameters	Notes
Aerosol LIDAR and ceilometer	Remote	Halo Photonics, Campbell Scientific, Vaisala, Leosphere, MPL, Raymetrics	Aerosol backscatter profile, cloud base and top (if not too attenuating)	• Optically dense clouds cause signal attenuation—it is not always possible to see through a layer so further layers are obscured • Dependent on an retrieval algorithm • Difficult to verify absolute measurement values • If fitted with de-polarization capabilities, particle shape can be inferred • If fitted with Doppler capabilities, particle dynamics can be investigated
Weather and cloud radar	Remote	EWR, Furuno, Metek, ProSensing, Selex ES, Vaisala	Radar reflectivity profiles, cloud droplet and hydrometeor size spectrum, fall velocity	As above
Micro rain radar	Remote	Metek	Range resolved droplet/ particle size spectrum, fall velocity	• Measurements assume a vertical fall path—in high wind or turbulent conditions, this is not always true • Dependent on retrieval algorithm—to use it for ash retrieval, a specific "ash" retrieval parameter has to be employed
Multichannel sun photometer	Remote	Ceimel	Aerosol optical thickness (AOT), aerosol volume distribution, aerosol size distribution, asymmetry parameter, and simple scattering albedo	• Dependent on retrieval algorithm • Column integrated—it does not provide a profile
Cloud imaging probes	Direct	DMT	Droplet/particle optical size spectrum, habit, shape	• Size range restricted • Can get 2 D variant to allow "see" asymmetry in particle shape

(Continued)

Table 12.1 (*Continued*) Aerosol and cloud measurement instruments, parameters retrieved and the pros and cons of each system

System	Direct/ remote measurement	Manufacturer	Parameters	Notes
Aerosol counter	Direct	Grimm, Topas, TSI	Total number concentration	• Some variants require external vacuum • Concentration limited • Can be susceptible to vibration
Aerosol optical spectrometers	Direct	Aerodyne, DMT, Grimm, MetOne	Droplet/particle optical size spectrum	• Assumed spherical particle of known refractive index—specific calibration required for accurate ash particle size density measurements • If fitted with de-polarization capabilities, particle shape can be inferred • Concentration limited
Scanning mobility particle spectrometer	Direct	Grimm, TSI	Particle electrical mobility size spectrum	• Assumed spherical particle of known electrical mobility • Concentration limited
Aerodynamic aerosol spectrometer	Direct	TSI	Particle aerodynamic size spectrum	• Size limited • Assumed spherical particle of known aerodynamic diameter • Concentration limited
Three-wavelength Nephelometer	Direct	TSI	Total scattering, backscatter signal at blue, green, and red wavelengths	• Infer variation in total loading • Infer "rough" separation of scattering signal with particle size

Figure 12.3 The UK Facility for Airborne Atmospheric Measurements (FAAM) BAe-146 aircraft.

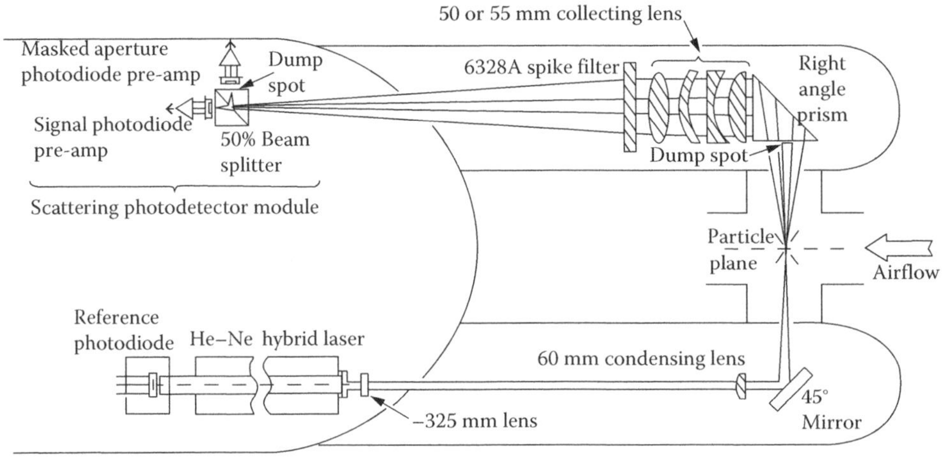

Figure 12.4 Schematic of the FSSP optics.

half is again projected onto a photodiode but in this case, a mask is applied. The resultant pattern is used to determine if the particle in the field of view is being fully illuminated or merely skirting the edges of sample volume. It is this signal that is used to trigger the pulse height analysis of the first half of the signal.

As the instrument name suggests, the data produced are in terms of a spectrum, that is, the number of particles per unit volume in a given size range. Aerosol spectrometers define a series of measurement bins with upper and lower size limits (those for FSSP range 1 are given in Table 12.2) and develop a histogram by incrementing the number within the appropriate bin when a particle is measured in that bin's size range.

In practice, it is voltage levels that have to be compared. Using Mie theory (van de Hulst 1957), the voltage equivalent of the bin boundaries is calculated assuming a spherical particle of known refractive index. The analysis electronics within the instrument compares the measured voltage with those calculated for the bin boundaries in turn until it finds the occurrence where the measured voltage is greater than a bin's lower limit voltage but lesser than the upper limit voltage. When this instance is found, the count in this bin is incremented by one.

The measurement bin voltage levels are calculated assuming totally scattering spherical particles with a refractive index of 1.59 as the physical calibration of the spectrometer is usually performed using polystyrene latex spheres (PSL), particles commercially produced to National Institute of Standards and Technology traceability size standards and users also regularly check the

Table 12.2 Lower and upper diameter limits for FSSP range 1

Bin number	Bin lower limit (μm)	Bin upper limit (μm)
1	2	4
2	4	6
3	6	8
4	8	10
5	10	12
6	12	14
7	14	16
8	16	18
9	18	20
10	20	22
11	22	24
12	24	26
13	26	28
14	28	30
15	30	32

instrument calibration using the same particles with any drift corrected by postprocessing of the data. The composition of atmospheric aerosol is however highly variable (Pruppacher and Klett 1997) and hence so are the refractive indices. For example, the refractive index of sea salt at 633 nm is $1.49-1\times10^{-7}i$, while that of soot at 633 nm is $1.75-0.43i$ (Shettle and Fenn 1979)—the larger complex part of the soot refractive index indicates that soot aerosol is much less efficient at scattering light than sea salt aerosol; in other words, soot is a good absorber. In the situation where there are two particles (one of sea salt and one of soot) of the same geometric size and light is scattered, the voltage of the resultant signal in the spectrometer will be less for the soot particle than for the sea salt particle. This can result in particles being incorrectly sized when the signal is compared to the bin boundary theoretical voltage values. This introduces a source of error into the measurements from optical aerosol spectrometers, but recalibration and data postprocessing, taking into account the approximate aerosol composition, is employed routinely to compensate for this.

Particle sizing and counting relies on the fact that only one particle at a time is in the field of view of the optics and that the transit time of a particle through the sample volume is greater than the time required for the detection electronics to reset itself after analyzing a particle. As supporting electronics advances, this latter consideration has become less of a concern, but particle coincidence (when more than one particle is in the field of view) introduces a concentration limit of the order of 500–1000/cc. Measurements in loadings greater than the concentration limit introduces both counting and sizing errors.

Another source of error is the effect of particle shape—in summary, the greater the deviation of the particle from the spherical, the greater the potential sizing error. In the study of airborne volcanic ash clouds and also in the study of ice clouds, this is a major concern and a combination of aerosol spectrometers with imaging probes is often employed. Imaging probes, as the name suggests, not only size of the particle but also supply the user with images of the particles. They make use of a matrix of photodetectors. An example of the particle imagery produced by the DMT CIP (Cloud Imaging Probe) can be seen in Figure 12.5.

The effect of particle shape on the polarization of the scattered light is now becoming a common place method for analyzing the effects of particle shape. In general, the signal due to the scattered light with the same polarization as the incident light is compared to that from the scattered light with a polarization of 90° with respect to the polarization of the incident beam—the latter being the depolarization signal. The larger the de-polarization signal, the greater the deviation of the particle from the spherical, but it should be noted that a de-polarization signal can also be caused by nonhomogenous particle composition and a particle size large compared to the incident wavelength. Weitkamp (2005) provides an excellent discussion on this topic.

12.4 CONDENSATION PARTICLE COUNTERS

Under the action of an applied vacuum, an aerosol stream (these can be droplets or solids) is drawn into the condensation particle counters (CPCs) at a constant rate through the aerosol inlet (as shown in Figure 12.6). The correct methods of sampling an atmospheric aerosol in order for that measurement to be considered representative of the original will not be covered here, but the reader is recommended to both Hinds (1999) and Colbeck (1998) for further information. Once within the body

Figure 12.5 DMT CIP images of ice particles.

of the instruments, it enters a saturator region, in which the sample air stream is saturated with the vapor of the working fluid—this is achieved by heating and holding this region at a known set temperature. In Figure 12.6, the working fluid is butan-1-ol, but CPCs using water as the active fluid are also available, the fluid used to determine the temperature of the saturator region.

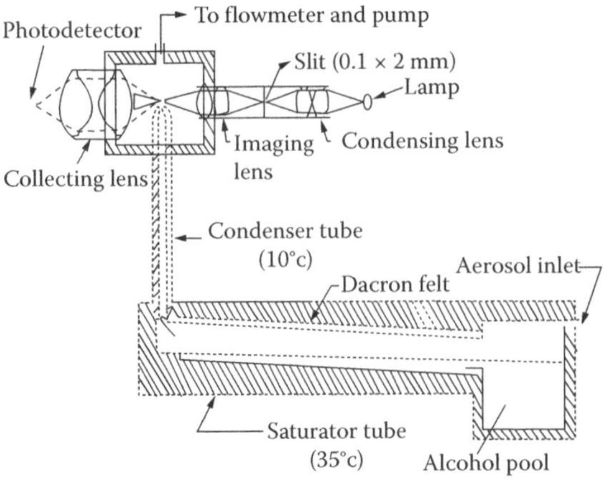

Figure 12.6 Schematic of a CPC.

As the mixture moves through the instrument, it next enters the condenser tube where the mixture is cooled such that the vapor becomes supersaturated (the working fluid used again to determine the temperature of this region) and condenses on the particles/droplets. As a result, the particles grow to a diameter of about 10 µm, allowing for optical detection (Kulkarni et al. 2011; Hinds 1999).

As particle size decreases, the saturation ratio required to ensure detection increases and hence, there is a lower limit to the size of particles that can be detected; for most CPCs, this is in the range of 3–10 nm. The upper limit, however, is dependent on the sample inlet.

Particle counting, for optical aerosol spectrometers, relies on the fact that only one particle at a time (coincidence) is in the field of view of the optics and that the transit time of a particle through the sample volume is greater than the time required for the detection electronics to reset itself after "seeing" the previous particle. For particle concentration typically of the order of 10^4/cc, there is no problem with coincidence; above this level, corrections have to be made and often use is made of the total intensity mode in which concentration is determined from the total intensity of scattered light that is measured. This latter mode is generally subject to a larger error as it requires that all particles grow to the same diameter and that the optical system is frequently calibrated (Wiedensohler 1997).

The application of optoelectronics to the field of *in situ* measurement of atmospheric properties and phenomena is very broad and what has been presented here is just one very specific application.

12.5 SUMMARY

In this chapter, the author has attempted to provide a brief overview of how atmospheric aerosols, in this case, volcanic ash, can be monitored and their microphysical properties determined by measuring the intensity of the light scattered from a particle. Clearly, this is an extensive field and this chapter has only scratched the surface—there is great diversity and range in the instrumentation available.

REFERENCES

Casadevall, T. J. (1994), The 1989–1990 eruption of Redoubt Volcano, Alaska: Impacts on aircraft operations, *J. Volcanol. Geotherm. Res.*, 62, 301–316.

Colbeck, I. (1998), *Physical and Chemical Properties of Aerosols*, Blackie Academic and Professional, London.

Hanstrum, B. N. and A. S. Watson (1983), A case study of two eruptions of Mount Galunggung and an investigation of volcanic eruption cloud characteristics using remote sensing techniques, *Aust. Meteorol. Mag.*, 31, 171–177.

Hinds, W. C. (1999), *Aerosol Technology*, John Willey and Sons, New York.

Johnson, B. et al. (2012), *In situ* observations of volcanic ash clouds from the FAAM aircraft during the eruption of Eyjafjallajökull in 2010, *J. Geophys. Res.*, 117, D00U24, doi:10.1029/2011JD016760.

Jones, A. R., D. J. Thomson, M. Hort, and B. Devenish (2007), The U.K. Met Office's next-generation atmospheric dispersion model, NAME III, in Borrego C. and Norman A.-L. (Eds) *Air Pollution Modeling and its Application XVII* (Proceedings of the 27th NATO/CCMS International Technical Meeting on Air Pollution Modelling and its Application), Springer, New York, pp. 580–589.

Kulkarni, P., P. A. Baron, and K. Willeke (ed.) (2011), *Aerosol Measurement: Principles, Techniques, and Applications*, John Wiley and Sons, Hoboken, NJ.

Pruppacher, H. R. and J. D. Klett (1997), *Microphysics of Clouds and Precipitation*, Kluwer Academic Publishers, London.

Shettle, E. P. and R. W. Fenn (1979), Models for the aerosols of the lower atmosphere and the effect of the humidity variations on their optical properties, *Air Force Geophysics Laboratory*, Report No. TR-79-0214, Bedford, MA.

Van de Hulst, H. C. (1957), *Light Scattering form Small Particles*, Dover Publications, Inc, New York.

Weitkamp, C. (2005), *Lidar. Range-Resolved Optical Remote Sensing of the Atmosphere*, Springer, New York.

Wiedensohler, A. et al. (1997), Intercomparison study of the size dependent counting efficiency of 26 condensation particle counters, *Aerosol Sci. Technol.*, 27, 224–242.

PART V

Military applications

Table V.1 Various optoelectronic technologies and their applications to military needs

Application	Technology	Advantages	Disadvantages	Current situation (at time of writing)
Imaging systems	Visual/near-IR TV	Best angular resolution, easiest to interpret; can now provide near starlight performance at moderate cost	Depends on ambient light, susceptible to visual camouflage, cannot see through smoke	Returning to fashion as the benefits of high-resolution and multiwaveband operation are appreciated
Imaging systems	3–5 μ IR	Works well in most conditions day and night; acceptable angular resolution from moderate apertures	Affected by smoke and fires	Widely used in surveillance and weapon aiming applications
Imaging systems	8–12 μ IR, cooled and uncooled	High sensitivity, can see through smoke	Low scene contrast in hot humid conditions, poorer angular resolution from given aperture	Widely used in surveillance and weapon aiming applications
Imaging systems	Direct vision near-IR night vision goggles	Compatible with starlight and nonvisible illumination	Prone to interference from bright light sources; need fast optics so limited to wide FOV applications	Widely used in infantry and aviators' night vision systems
Stabilized sensor systems	Gyro stabilized turrets or mirrors with one or multiple cameras etc., used on drones, ships, vehicles, aircraft	Can avoid shake due to platform motion, allow smooth tracking of moving target and compensate for own ship motion	Can be difficult to integrate with aerodynamic profile and armor protection	Widely used at various levels of stabilization performance from fractions of a milliradian to a few microradians
Laser rangefinding and designation	Nd:YAG 1.06 μ lasers, glass optics, silicon detectors	Mature technology, compatible with large legacy inventory of laser guided weapons	Not eyesafe	Widely used for laser designation, being superseded by 1.54 μ systems for range-finding
Laser rangefinding	Various at 1.54 μ including erbium glass, diode lasers, and Raman shifted Nd:YAG	Eyesafe, benefit from advances in telecoms technology for laser sources and detectors	Lower power output than Nd:YAG laser for a given package,	Now in widespread use in rangefinders

(Continued)

Table V.1 (*Continued*) Various optoelectronic technologies and their applications to military needs

Application	Technology	Advantages	Disadvantages	Current situation (at time of writing)
Laser rangefinding	CO_2 lasers (9–11 μ), direct and heterodyne detection	Smoke penetration, eyesafe, compatible with thermal imaging optics and sensors	Bigger, more expensive, and more complex for a given level of performance; cryogenic cooling required for detector	Not used much in spite of considerable investment in the 1980s
Laser pointer	Visible and near-IR laser diodes used to show point of aim, and to indicate targets to others	Clear and intuitively obvious hand off	Not passive so may reveal one's position to an enemy	Widely used on infantry weapons for aiming, and to indicate points of interest to night vision systems
Laser weapons	Various	Virtually instantaneous engagement, so aiming is simple; no limitation on number of "rounds" if adequate electrical power available	Expensive, complex, difficult to measure effectiveness, disrupted by atmospheric turbulence and obscurants	Very limited use by a few countries
Missile guidance	Nonimaging infrared, typically cryogenically cooled and mechanically modulated reticule or rosette scan	Relatively simple processing, effective against small bounded targets showing high thermal contrast such as aircraft	Susceptible to clutter and countermeasures; maintenance requirements are significant because of cryogenic cooling and precision mechanics	Widely used in air-to-air and surface-to-air missiles
Missile guidance	Imaging infrared	More options to deal with countermeasures and clutter, more sensitive	More complex processing	Widely used in air-to-ground and anti-tank missiles
Missile guidance	Laser beam riding	Low power guidance beam that is difficult to detect and counter	Require clear line of sight from launch post to target so that operator can track the target throughout the engagement	Used in a variety of short range antiaircraft and anti-tank systems

(*Continued*)

Table V.1 (*Continued*) Various optoelectronic technologies and their applications to military needs

Application	Technology	Advantages	Disadvantages	Current situation (at time of writing)
Missile warning	UV missile warning sensors	Solar blind, low clutter at low levels	Limited range, less effective at high altitude	In widespread use, but starting to be replaced by IR systems
Missile warning	Mid-IR missile warning sensors	Spectrally matched to peak emissions of rocket motors; much longer potential range than UV; useful imagery of surroundings	Need more sophisticated processing to deal with clutter and spurious target-like features	As processing capabilities increase, these are becoming more common
Aerial reconnaissance	Digital visual and IR band sensors on aircraft and drones	Very high-resolution imagery, can fly below cloud; data available in real time if data linked back to base or to forward observers	Need to fly over or near the target, so vulnerable to air defenses; prone to cloud, smoke, and terrain obscuration	Manned aerial reconnaissance systems are still in use, increasingly augmented by drones of all shapes and sizes
Satellite reconnaissance	Digital visual and IR band sensors	High enough to avoid air defenses	Flight paths relatively predictable; takes longer to retask; lower spatial resolution; difficult to make data available in real time to front line users	Significantly used by advanced countries
Missile countermeasures	Modulated and directional jammer using laser or high-intensity flashlamp	Can defeat many missile threats; no significant limitation in number of engagements	Requires understanding of threat characteristics	Increasingly in use to complement or replace flares
Missile countermeasures	Flares	Not directionally critical, can use the same dispenser as chaff	Limited number of rounds can be carried out; some threats require complex flare sequences	Once the only protection means, now used in tandem with directional jammers

(*Continued*)

Table V.1 (*Continued*) Various optoelectronic technologies and their applications to military needs

Application	Technology	Advantages	Disadvantages	Current situation (at time of writing)
Submarine detection	Distributed fiber optic sonar arrays	Sensitive acoustic detection, no electronics required external to pressure hull	Very sophisticated technology	Early production systems deployed in several countries
Perimeter security	Distributed optical fiber sensors providing seismic detection for perimeter security	Critical alignment not necessary, immune to RFI, does not depend on clear sightlines. Not detectable by mental detectors	Need to lay physical cables, subject to deliberate or accidental damage	Initial production systems sold in various countries
High-speed data links	Fiber optic uses ranging from strategic networks to spooling out behind missiles to transfer video to operator station	Immune to RFI; high bandwidth, much lighter and more flexible than equivalent copper, can detect and locate breakages	Fiber vulnerable to damage so need good protective sheathing	Widely used exploiting civil telecoms technologies

13

Military optoelectronics

HILLARY G. SILLITTO
University of Strathclyde

13.1 INTRODUCTION

13.1.1 Purpose

This chapter surveys the military needs for and applications of optoelectronics, and illustrates these with examples. The intention is to give the interested reader a conceptual overview of this very wide subject and provide references for analytical detail.

13.1.2 The military value of electro-optics

Three key benefits of electro-optical technology have led to its widespread adoption by the military. They are as follows:

- High angular resolution through a small aperture, because of the short operating wavelength—making useful electro-optic (EO) systems easy to package on a wide range of platforms.
- Twenty-four-hour operation—passive night vision and thermal imaging systems can "turn night into day."
- The familiarity of the "display metaphor"—images from EO systems look like the images we see with our eyes, which makes them easy to interpret and easy to train operators.

13.1.3 Themes and structure of this chapter

This chapter starts with a brief historical perspective.

The key building blocks of modern military EO systems are imaging and laser subsystems. These and other key enabling technologies are then described.

The uniquely demanding characteristics of the military operating environment are then outlined, since these strongly influence the engineering of optoelectronic technology into military systems.

The roles, functions, operational characteristics, technology, and representative examples of the main classes of military optoelectronic systems are then reviewed.

The chapter concludes with an assessment of the operational impact of optoelectronic technologies, and tempts fate by outlining the characteristics of current trends in the field, which give us insights into how it may evolve in the future.

The author has attempted to provide a neutral overview of the subject and unreservedly apologizes for the "English-speaking/North-west European" bias that probably remains.

13.1.4 The EO environment

Military EO systems are remote sensing systems. Their performance is critically influenced by "the EO environment"—atmospheric transmission and scatter, target and clutter signatures, background emission and reflection, smoke and cloud, fires, and atmospheric turbulence.

Figure 13.1 shows, superimposed on a spectrum from 200 nm to 13 µm, six primary influences on military visual and IR (infrared) system design. They are as follows:

- The transmission of the atmosphere—the main atmospheric "windows" are 0.4–1.6, 1.8–2.5, 3–5 µm (split by the CO_2 absorption band at 4.2), and 8–12 µm; there is much fine structure due to molecular absorption lines within these "windows," and a water vapor absorption continuum in the 8–13 µm band.

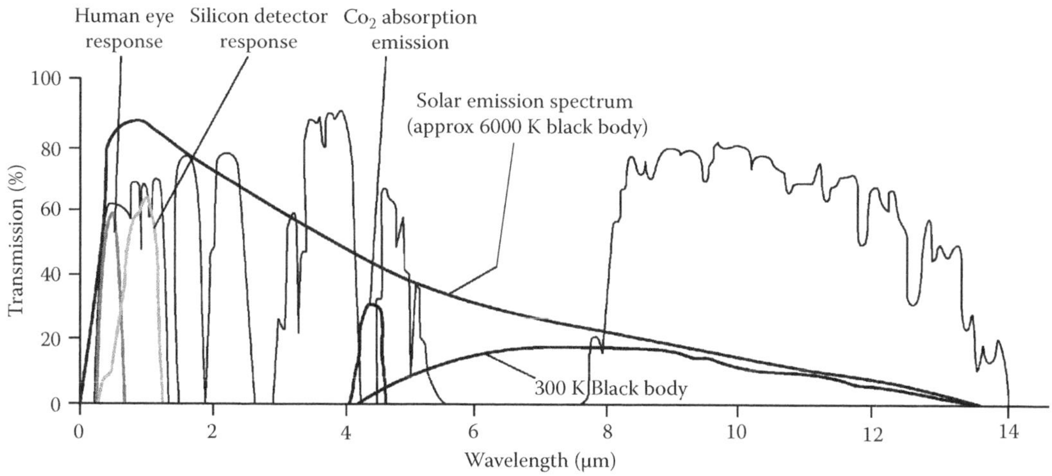

Figure 13.1 The primary spectral influences on military EO system design.

- The solar radiation spectrum, approximating to a 6000 K black body and peaking at about 550 nm when plotted on a wavelength scale (but at 880 nm when plotted on a frequency scale—see Sofer and Lynch [22] for a discussion of the common assumption that the human eye response is matched to the peak of the solar spectrum).
- The human eye response from 400 to 700 nm.
- The silicon detector response from 400 to 1100 nm.
- The hot CO_2 emission band surrounding the absorption band at 4.2 μm, which is the main source of the strong 3–5 μm signature of exhaust plumes and fires.
- The 300 K black body curve representing the emission of natural objects at ambient temperature, showing the dominance of emission in the 8–12 and 4–5 μm regions.

These six factors dominate the design and selection of military EO systems.

Other significant atmospheric factors include the following:

- Atmospheric scatter—Rayleigh scattering has an inverse fourth power relationship with wavelength, explaining why longer wavelength systems are less susceptible to haze and smoke.
- Atmospheric scintillation, which limits the resolution of high-magnification imaging systems and the divergence and beam uniformity of lasers working through atmospheric paths; scintillation is caused by density and humidity fluctuations in the air induced by atmospheric turbulence, and is much stronger near the ground.
- Beam wander due to the same atmospheric mechanisms.
- Clouds, which except when very thin, are essentially opaque to EO systems.
- Screening smoke, which will be discussed later.
- Atmospheric ducting due to refractive index gradients near the ground or sea surface, which can slightly increase the horizon range of low horizontal paths by causing the light to bend and follow the earth's curvature.

Target and clutter phenomena are dominated by reflectance and emission of the surface. These vary with wavelength, sometimes quite markedly; for example, natural vegetation contains chlorophyll that reflects strongly in the near-IR region from just above 700 nm. Shiny metallic surfaces have low IR emissivity and therefore, reduce the apparent temperature of, for example, an aircraft; but of course, they reflect sunlight strongly, making the aircraft easier to detect in the visual band. Hot fires and exhaust gases are extremely prominent in the 3–5 μm region because of the rotational CO_2 emission bands around 4.2 μm, making 3–5 μm imagers less useful for ground-to-ground applications in intense battle conditions than their performance in good conditions would suggest. Measures taken to reduce signature in one band will often increase signatures in another.

13.2 HISTORICAL PERSPECTIVE

13.2.1 Myths, legends, and fantasies: From the Greeks to Star Wars

Man has always fantasized about optical system concepts which would give him an advantage in warfare.

"And I," said Athena (in the legend of Perseus and the Gorgon) in her calm, sweet voice, "will lend you my shield with which I dazzle the eyes of erring mortals who do battle against my wisdom. Any mortal who looks upon the face of Medusa is turned to stone by the terror of it; but if you look only on her reflection in the shield, all will be well." [1]

This use of bright light to dazzle the enemy, the idea of a "death ray" weapon, and the use of indirect viewing optics to mitigate its effect, all seem to anticipate, by several thousand years, a number of the twentieth century's threats, opportunities, and solutions.

In H.G. Wells' science fiction novel *The War of the Worlds* [2], the Martians apparently used some sort of directed energy weapon, which we would now assume was a high-energy laser. Possibly inspired by this, Britain's Tizard commission in the 1930s [3] set a goal of making a death ray, which was not achieved, but led to the successful UK radar program.

In the 1980s, the science fiction film *Robocop* used a helmet-mounted head-up information display and was alleged in a TV documentary to have contributed inspiration to at least one of the Future Soldier Technology programs in the late 1990s.

In the *Star Trek* television series, it takes little imagination to equate phasors with laser weapons; while the *Star Wars* series of films gave its name to Reagan's Strategic Defense Initiative, and anticipated holographic "video-conferencing."

13.2.2 Optical systems and methods—From the Greeks to World War 2

Optical instruments and knowledge were used in warfare since the Ancient Greek times. Around 500 years BC, Thales of Miletus proposed geometry allowing the range of ships to be estimated from a tall tower (Figure 13.2).

The heliograph has been used since ancient times for semi-covert line of sight signaling.

During the age of exploration, from 1400 to 1900 AD, all civilizations placed great importance on navigation by sun and stars. This drove the development of optical instruments and precision navigation techniques—used for surveillance, long range identification, and (in conjunction with signal flags) for signaling.

Around 1900, the Boer War showed the importance and value of camouflage (signature suppression). Boer commandos used camouflage, mobility, and modern rifles to negate British tactics used since the Napoleonic era. The British army introduced khaki uniforms for colonial campaigns in 1880, but for home service only in 1902 [23]. In the days when official war artists rather than CNN formed the public images of overseas warfare, artistic license had perhaps veiled the switch to khaki from the British public; did the army use the Boer War experience as an "excuse" to introduce the change at home?

In the late nineteenth century, Zeiss in Germany and Barr & Stroud in the UK developed

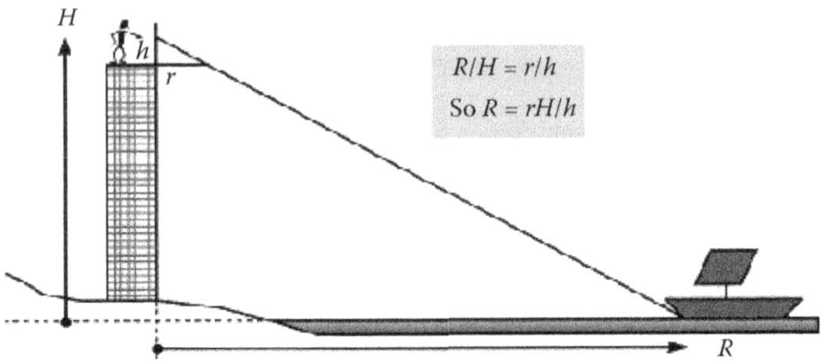

Figure 13.2 Thales of Miletus, 500 BC: How to measure the range to a ship from a tower.

optical rangefinders [10]. By World War 1, Zeiss was recognized as the preeminent optical instrument manufacturer in Europe. Trench periscopes and binoculars were required in vast numbers during the war. The United Kingdom could no longer obtain Zeiss instruments and faced a critical skills shortage. This led to the formation of the Imperial College Optics Group to develop an indigenous capability. Similarly, in the United States, Frankford Arsenal played a key part in the development of military optical instruments [17]. All of these organizations and companies have featured strongly in the recent optoelectronics revolution.

13.2.3 The technology revolution—Lasers, IR, and signal processing, 1955–1980

There has always been a close link between scientific and technological progress and military demands. World War 2 involved the scientific community in the war in an unprecedented manner.

In the United Kingdom from 1936 onward, the Tizard committee spawned many far-reaching technological innovations. These included the following: the development and exploitation of radar; IR aircraft detection [3]; a photoelectric fuze [7]; operational analysis; the Enigma Decoding Organization at Bletchley Park leading to the development of the first electronic computer; and the systematic use of scientific intelligence [3]. The US's crowning technical achievements included the Manhattan Project to develop the nuclear bomb; the efficient industrialization and mass production of all sorts of military technology; and of particular significance for optoelectronics, the development of the communications theory, which underpins all modern signal processing techniques (Shannon, Bell labs). Germany made tremendous strides in many areas, including rocket and jet propulsion and principles of modular design allowing cheap and distributed mass production of aircraft and tanks. During World War 2, the United States and Germany developed and tested the first EO guided missiles, probably using PbS. Britain, Germany, and the United States all developed active near-IR night sights [17].

The subsequent development of military optoelectronics benefited from many aspects of this progress, and notably from the proof that technology managed at a strategic level could win wars.

Night vision was an important technological innovation in the immediate postwar years. But three key breakthroughs after the war transformed optoelectronics from a supporting to a key military technology: in signal processing, IR, and lasers. The transistor was invented in 1948, the first forward-looking infrared (FLIR) was demonstrated in 1956 [9], and the first laser in 1960. These led to a whole new class of integrated EO systems, which developed rapidly in the 1970s and 1980s as the Cold War protagonists sought to achieve and maintain superiority. Government laboratories played a key part in pushing the technology and integrating the first systems. In the United Kingdom, the Royal Radar Establishment was established in Malvern during the war and later expanded its remit to include EO, becoming successively the Royal Signals and Radar Establishment (RSRE), the Defence Research Agency (Electronics Division), and now Qinetiq. In the United States, the Night Vision Lab at Fort Belvoir was founded in 1962 to coordinate night vision development for the US military [17].

13.2.4 Military EOs in use

In Vietnam, IR line scan was deployed on American Grumman Mohawk aircraft in the 1960s. They were credited with being "able to detect the heat where a lorry had been parked hours after it had driven away." Laser designation was used operationally by 1969—a remarkably short deployment time from the laser's invention in 1960—allowing single planes to hit difficult targets such as bridges that had survived months of intense conventional bombing. Night vision scopes helped US soldiers defend themselves against Viet Cong night attacks.

In the British/Argentinean Falklands (Malvinas) conflict in 1982, the American AIM-9L Sidewinder missiles, with a cooled InSb detector and head-on attack capability, allowed Royal Navy Harrier fighters to shoot down nearly 30 Argentinian aircraft without loss. Over 80% of the AIM-9Ls launched hit their targets [15]. Night vision equipment was deployed by the armies of both sides, but does not seem to have been used particularly effectively, nor to have significantly influenced operations. Laser designation allowed effective stand-off bombing attacks without exposing attack aircraft to defending fire.

In Afghanistan in the 1980s, Stinger shoulder launched surface-to-air missiles (SAMs) were used against Soviet attack helicopters, which in turn used TV and IR imaging systems and laser designators for ground attack. Antimissile countermeasures were used to reduce vulnerability of aircraft to SAMs—Soviet aircraft dispensing flares while landing and taking off at Afghan airfields became a familiar sight on TV newsreel shots. Equivalent Soviet missiles (SA-16, etc.) were developed. The Soviet Union maintained a very strong research base in EO technologies throughout the Cold War.

In the Gulf Conflict in 1991, thermal imaging allowed the coalition forces to operate day and night, with an overwhelming advantage at night. The unusually bad weather and thick cloud cover, exacerbated by smoke from burning oil fields, made passive night vision equipment used by both sides almost useless for much of the time, but had little or no effect on thermal imagers. This experience led to an increase in demand for basic low-cost thermal imagers for military roles for which the cheaper passive night vision technology had previously been considered sufficient.

The conflict showed the public, perhaps for the first time, an awesome array of precision weapons, many of them EO guided. The public perception of the conflict was shaped by live TV broadcasts (EO technology again) from the war zone, beamed by satellite and cable TV into homes throughout the world.

The coalition campaign used concepts of "maneuver warfare" developed late in the Cold War. These depend on a high degree of integration of command, control, communication, and information (C3I). EO is a key technology for capturing and displaying this integrated information picture and, in the form of fiber optic communications links, for the secure transmission of high-bandwidth data between fixed command posts. Fiber optic cables were a high-priority target for special forces sabotage missions and precision air strikes during the Gulf War.

Of the many issues for technologists emerging from the conflict, four are important to this discussion.

- Combat identification, to reduce the risk of attacking one's own side, is particularly difficult and important in coalition warfare, where equipment used by friendly armies may not be compatible, and friends and enemies may be using the same types of equipment; and it becomes doubly difficult and important in maneuver warfare, where there is no clear front line.
- Collateral damage—even with precision weapons, not all weapons hit the intended targets (typical figures for laser guided bombs are 70%–80% success rate); and real-time public television from the war zone ensures that the attacked party can exploit such lapses for propaganda purposes.
- The effectiveness of attacks was often overestimated—this is a persistent trend at least since World War 2 and probably since warfare began.
- It took time, often too much time, to get aircraft and satellite reconnaissance imagery to the people on the ground who could make use of it.

13.2.5 Military EOs today

Hence, 40 years after the invention of the laser, EO was a well-established military technology of proven value. The issue for the twenty-first century is not whether, but how best and how widely, to exploit optoelectronics in military applications.

Passive EO sensors, such as thermal imagers, allow operators to detect, recognize, and "identify" objects of interest. With the addition of a laser rangefinder and directional sensing, the object can also be "located" relative to the operator. With the further addition of a navigation capability, it can be located in geographical coordinates. The information can be shared with others via communication links, which in some circumstances may be optoelectronic (free space or fiber); and displayed on situation displays that use optoelectronic technology. When integrated with a weapon system, EO systems provide fire control solutions and help to attack the target. With the advent of high-energy laser "weapons," which in early 2002 were being developed as missile defense systems, EO systems can be used also to attack the target directly.

Hence, EO systems answer the basic questions—where is it? what is it? and (within limits) who is it and what is it doing?—and contribute to a shared information picture that aids military decision making, and in some contexts, closes the loop by attacking targets directly. This can be summarized in a closed loop referred to as the "OODA loop" or "Boyd cycle"[28] (Figure 13.3).

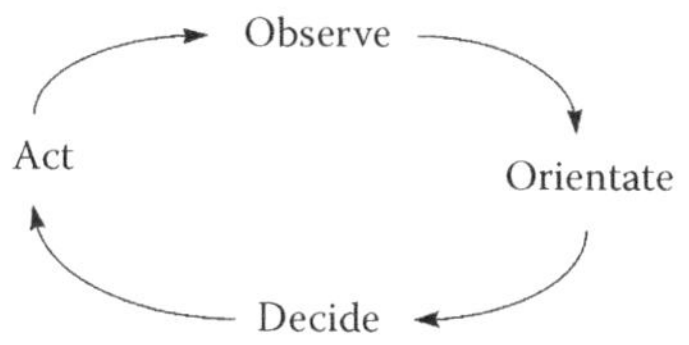

Figure 13.3 "OODA loop" or "Boyd cycle."

13.3 THE KEY BUILDING BLOCKS: IMAGING AND LASER TECHNOLOGY

13.3.1 Imaging

EO imaging systems convert optical radiation from the scene into electrical energy to allow it to be displayed in real time at a more convenient wavelength, brightness level, or location.

They are used for a wide spectrum of military and paramilitary tasks including surveillance, target acquisition, identification, weapon aiming, assessment, situation awareness, threat warning, pilot aid, and covert driving. There is a correspondingly wide range of equipment and technology.

The wavelength range from 400 to about 1700 nm is referred to as the "visible and near-infrared" (VNIR) region. VNIR image sensors depend on reflected light, using TV and image intensifier technology. Early active VNIR imaging systems used IR illuminators to augment natural light. Detection and recognition of objects of interest depends on the contrast between the object of interest and its surroundings, and on the resolution of the sensor. Modern high-resolution TV technology now allows VNIR imaging systems to provide resolution approaching, but not quite matching, that obtained by the human eye through a high-power magnifying sight of similar aperture and field of view (FOV).

Thermal imagers operate in the mid- and far-IR bands (3–5 and 8–12 μm, respectively) by sensing the black body radiation emitted by objects in the scene. Users can passively detect targets that are visually camouflaged or concealed in deep shadow, or when there is no ambient light. In particular, 8–12 μm imagers also see through smoke which blocks shorter wavelengths. In recent conflicts, this has given armies and air forces equipped with thermal imaging an overwhelming operational advantage over those which were not so equipped.

In some applications, the ability to see terrain features is important (flying and driving aids, general surveillance, and orientation). In hot and humid conditions, thermal imagers may experience "thermal wash-out," when everything in the scene is at the same temperature and there is very little thermal contrast. During and after rainfall, terrain contrast is very low in both bands. The 8–12 μm band suffers from water absorption in hot humid conditions; in these conditions, reflected sunlight may provide "TV-like" pictures in the 3–5 μm range. In very cold conditions, by contrast, there is minimal black body emission in the 3–5 μm range, leading to wash-out in this band under some circumstances—tactical targets will still be detected due to their high thermal contrast, but details of surrounding terrain and of man-made structures may be suppressed.

Because of the longer wavelengths, the identification range of thermal imagers is usually limited by the available aperture.

For these reasons, thermal imagers are often complemented by VNIR imaging or direct sighting systems, to allow improved identification range when weather and lighting provide adequate image contrast, and to provide a complementary sensing capability in thermal wash-out conditions.

The following paragraphs discuss in turn the technologies and integration issues involved in the four key imaging methods: image tubes; TV cameras; cooled photon-detecting thermal imagers; and uncooled thermal detecting thermal imagers.

13.3.1.1 IMAGE TUBES

Image tubes were first developed and saw limited use during World War 2. In the British Electric and Musical Industries design, selenium photocathodes released electrons, which were accelerated by an intense electric field through a vacuum onto a photo-emissive zinc sulfide screen. Similar US devices used a phosphor screen [24]. Early devices had low sensitivity and required active illumination. They were referred to as "image convertors" since they converted near-IR radiation into visible light. Other designs used an electron lens. Since the electron lens inverts the image, simple objective and eyepiece lenses could be used with no need for image erecting prisms.

Gibson [17] describes subsequent developments in night vision well, mainly from a US perspective. Key breakthroughs were multistage amplification, which removed the need for active illumination; and the micro-channel plate, which allowed high-amplification image tubes to be made short enough to fit into head- mounted systems, and are much less susceptible to blooming effects from bright lights. "Generation 2+" and "third generation" image tubes are now widely used in night vision goggles (NVG) and light weapon sights. They offer high-resolution imagery that is easy to interpret. They operate in the near-IR (700–1000 nm) region, where grass and vegetation have high reflectivity, giving a good view of most types of natural terrain, and high contrast for most targets of military interest. Modern systems work in "overcast starlight." In very overcast conditions and where terrain contrast is poor (for example, snow or sand), the user's spatial awareness can be augmented by near-IR laser illumination [17]. This, however, runs the risk of being detected by an enemy equipped with similar near-IR imaging systems.

Cockpit and other lighting can interfere with NVG operation. "NVG compatible" lighting is filtered so that it can be detected by NVGs without saturating them.

13.3.1.2 TV CAMERAS

Early TV cameras used vidicon tubes. This technology was in military use in the late 1960s, with TV-guided bombs being used in Vietnam [19]. Solid state image sensors (single-chip silicon photodetector arrays with charge coupled device (CCD) or metal oxide silicon (MOS) on-chip processing) were developed during the 1970s and 1980s and found wide commercial application, for example, for security. These sensors were much smaller and more robust than vidicons, and made a TV camera capability much easier to package into military systems. They are becoming smaller and more sensitive; this reduces the optical aperture required to provide useful performance in a wide range of light levels. There is increasing concern about the harmful effects of laser on the operators' eyes, and an increased desire to mount EO systems in places where it is difficult to provide direct optical channels for the crew. TV sensors are used much more widely in military systems than most observers expected in the early 1990s.

The civil security and broadcast markets have led the demand for lower and lower light level capability in solid state sensors. Image intensifiers have been fitted to the front of TV sensors for many years to give low light TV capability. Image intensified CCD cameras provide similar performance to image intensifiers without the need to get the operator's eye near the sensor. These are also the sensors usually used in laser-gated imaging systems. Emerging technologies such as electron beam CCD (EBCCD), and more recently low-light CMOS, offer the promise of "all light level" TV performance in a package little or no bigger than a daylight TV camera.

13.3.1.3 COOLED PHOTON-DETECTING THERMAL IMAGERS

After the first successful demonstrations in the 1950s, thermal imaging started to shape the whole field of military EO.

TVs and most lasers use glass optics, and could be integrated into existing visual sighting systems. Glass does not transmit thermal IR radiation; hence, thermal imaging was incompatible with existing systems. Besides the detector technology itself, thermal imaging required a completely new family of optomechanical system architectures and parallel developments in optical materials and coatings, lens design, test and calibration techniques, analogue and digital electronics, cryogenics, and servo-mechanical systems.

Early developments, in the West at any rate, were almost exclusively in the 8–12 μm band. Cadmium Mercury Telluride (CMT) became the detector material of choice for first and second generation thermal imagers. CMT is a semi-conductor alloy whose bandgap can be tuned by varying the material ratio. The bandgap is matched to the photon energy of 8–12 μm radiation, of the order of 0.1 eV. When photons of this energy impinge on the detector, electrons are excited from the valence to the conduction bands, and cause current to flow through the detector. This current is amplified and turned into a video signal, which is fed to a display.

For this process to work and produce a usable image, the detector must not be swamped by electrons rising to the conduction band with their own thermal energy. CMT is normally cooled to about liquid nitrogen temperatures, 77 K. Early demonstration systems used liquid nitrogen to cool the detector, which was placed in a "Dewar flask" with an IR window. Joule–Thomson cooling was widely used from the 1970s to the 1990s; nitrogen or air from a high-pressure gas bottle or compressor is forced through a very small orifice at the end of a tube, placed just behind the detector, which is itself mounted within a vacuum enclosure. The expansion of the high-pressure gas cools it to the required temperature. Because of the logistics, safety, and maintenance issues associated with high-pressure gas, this method has now been largely superseded by Stirling cycle cooling engines coupled to the detector. The detectors are mounted within a vacuum Dewar to reduce convection and avoid condensation.

The aim of these systems is to detect small differences in large photon fluxes. First and second generation systems scan the detector across the scene resulting in a modulation appearing on the detector output. This temporal modulation represents the spatial modulation in the scene. The large DC component is subtracted from the signal and the residual difference signal is then displayed as a TV-like image. Many first generation systems sent this modulated output to LEDs (light emitting diodes), which were viewed through the same scanning system, ensuring that scanner errors did not affect the perceived picture geometry. Others produce a Comittee Consultatif International Radiotelecommunique (CCIR) or National Television System Committee (NTSC) video signal; in these systems, scanner timing errors are less acceptable since they result in distortion of the picture.

One method for DC removal is to AC couple the detector output. This is cost-effective and worked well in the US Common Modules, of which many thousands were built. These systems used an oscillating mirror to scan a linear array of 60 or 180 detector elements across the scene. The output is AC coupled, amplified and either sent directly to an array of LEDs that is viewed off the back of the oscillating mirror, or processed through an electronic multiplexer to produce an electronic video signal.

Under some conditions of high scene contrast, AC coupling can generate artifacts from a small hot image that can wipe out a large part of the picture. In the United Kingdom, developments at RSRE to combat this problem led to a very high-performance detector technology called the signal processing in the element (SPRITE) or Tom Elliot's device (TED), named after the inventor. This is an array of eight parallel oblong detectors, each about 10 times as long as it is wide. The drift velocity of the electrons in the CMT is set by electronic biasing to match the image scan rate, providing an improved signal-to-noise ratio equivalent to that obtained by "time delay and integration" (TDI) with discrete detectors. SPRITEs require a more complex "serial/parallel" scanning method, typically with a high-speed spinning polygon to generate the line scan and a small framing mirror oscillating at the video frame rate. The next stage of signal processing is simpler, with only eight amplification stages instead of 60 or 180, and it performs amplification, DC restoration, and channel equalization, resulting in a very stable and uniform picture. This technology is used in the British "Class 2 Thermal Imaging Common Module" (TICM 2) system, and the Thales "IR-18" family, and is operational in many thousand systems worldwide. The technology evolved to 16-element systems with digital processing giving outstanding resolution, sensitivity, and image quality at high-definition TV bandwidths.

Many ingenious optomechanical scanning methods were developed, striving to make the best use of the available detectors. However, the next major innovation was "second generation" detectors that eliminated the need for complex two-axis scanning and for many parallel sets of pre-amplifier electronics.

Second generation thermal imagers use a photovoltaic operating mode in which the incident photons cause a charge buildup that changes the voltage on a capacitor. A focal plane multiplexer carries out time delay and integration for a linear array, and reads the signal off the focal plane serially. Common array sizes are 240, 288, or 480 × 4, or 768 × n in the latest UK high-performance system, STAIRS C. The signal from each pixel is amplified, digitized, and adjusted to compensate for response nonuniformities between detector elements. Modern systems provide a 12- or 14-bit digital output for processing, and 8 bit digital or analog video. Automatic or manual gain/offset correction optimizes the dynamic range of the video output. These systems started to become available around 1990, the French Sofradir 288 × 4 detector being the first to be widely available in series production.

Second generation technology has split into two evolutionary paths. One is "cheaper smaller lighter," as exemplified by the Thales "Sophie" product, which put high-performance thermal imaging into a binocular sized package suitable for use by individual soldiers for the first time. The other is "higher performance in the same package," as exemplified by the US horizontal technology integration (HTI) program and the UK STAIRS C system.

By 2000, the next step in simplification of high-performance thermal imaging systems had appeared in the market. Two-dimensional arrays working in the 8–12 μm band were being developed and entering low-rate production based on two technologies. The first is CMT, an incremental evolution from second generation devices. Efforts are being made to allow operation at higher temperatures to reduce the demands on cooling engine performance. The second, completely new, technology is quantum wells. These use molecular beam epitaxy to build a precisely tailored atomic structure that controls the electron levels by quantum confinement. The spectral band is much narrower than for CMT; this narrower spectral response can be exploited to give a number of system design benefits. Operating temperatures are similar to or colder than the 77 K of first and second generation systems. At the time of writing, the jury is still out as to which, if either, route will dominate the market.

Most cooled 3–5 μm systems use one of the three detector technologies explained earlier. Indium antimonide, a compound rather than an alloy, gives very good intrinsic uniformity and response to about 5.5 μm. CMT, with a different composition and optimized for the 3–5 μm range, allows the long-wave cut-off to be brought in to a shorter wavelength. Platinum silicide exploits silicon TV methods to provide very uniform imagery but with lower quantum efficiency (about 1% compared with about 60% for InSb and CMT). Staring systems of TV resolution are now widely available. There was resistance to 3–5 μm systems for ground-to-ground applications in most NATO armies because of the better performance of 8–12 μm systems in extreme battlefield conditions of fires and smoke. However, the lower cost and better resolution through a given aperture of these systems has led to 3–5 systems being increasingly adopted where the smoke and fire argument does not dominate the choice.

13.3.1.4 UNCOOLED "THERMAL DETECTION" THERMAL IMAGERS

Uncooled thermal imagers use a variety of physical principles to generate electric current or charge as a result of temperature difference. Since they respond to temperature difference, they need not be cooled; this reduced the cost and complexity compared with cooled systems. They, however, need fast optics and large detector elements to achieve good sensitivity; hence, lens sizes for narrow fields of view are large. This limits their applicability to shorter range systems.

Most uncooled technologies respond to high-frequency changes in temperature, and use a chopper to modulate the image of the scene. If the chopper is removed, they can detect moving targets in a static scene by using the motion for modulation—the principle used in passive IR burglar alarms.

The first uncooled imaging technology was the pyroelectric vidicon—essentially, a vidicon tube with the phosphor replaced by a thin pyroelectric membrane. The electron beam was used to "read" the charge built up on the back surface of the membrane. These membranes were not mechanically robust, and would split if irradiated by a focused CO_2 laser!

The first "solid state" thermal arrays used pyroelectric material placed in contact with a readout circuit similar in principle to that used by a solid state TV camera. Early devices suffered from low resolution, nonuniformity, and microphony. These issues have been progressively resolved through a number of generations of process and material technology [18]. The favored technology now seems to be the solid state microbolometer array.

Uncooled thermal imagers are now in service in several western armies as weapon and observation sights. Spin-out technology is now in the market for many applications including automotive night vision, firefighting, and for assessing heat-loss from buildings.

13.3.2 Characterizing imaging system performance

13.3.2.1 VISUAL "IDENTIFICATION" CRITERIA

Target classification with EO systems traditionally depends on the operator's training. The "Johnson criteria" are based on experiments carried out with military operators looking at real or synthetic

Task	Cycles resolved on the target for 50% confidence level of DRI in a single observation
Detection	1–1.5
Recognition	3–4
Identification	6–7

targets using real or simulated imaging systems. The criteria are as follows:

Modeled or measured sensor characteristics are used to calculate the minimum resolved temperature difference, which is plotted against spatial frequency. These criteria are then used along with an assumed target size and temperature difference or contrast to predict "detection, recognition and identification (DRI) ranges" for a particular sensor against a particular target.

This procedure is effective in providing a common benchmark to compare the relative performance of different system designs. It is less effective at predicting actual operational performance—wise suppliers quote minimum figures in their performance specifications, so most systems are better than their specification; and wise customers specify a conservative signature for the calculation, so most targets have higher contrast or temperature than is used in the model. The net effect is that most thermal imager systems are operationally useful at much greater ranges than their calculated DRI figures suggest.

However, it is also important to understand that "identification" specifically refers to the ability of the observer to choose between a limited class of known target types, and refers to "identification of vehicle (or aircraft) type"—for example, F-16 rather than MiG-29 aircraft, T-72 rather than M-1 tank. It bears no relation to the different and vital task of identifying whether the T-72 or F-16 belongs to country A, which is an ally, or country B, which is an enemy.

Typical narrow FOV imaging systems for military applications will have FOV in the range 5–0.5°, and Johnson criteria identification ranges from 1 to 10 km.

Image-based target classification techniques are being developed with a goal of achieving a reasonable level of "classification" confidence—for example, tank rather than cow or car, jet fighter rather than civil airliner—as an aid to clutter discrimination in systems where the operator is automatically alerted to possible targets.

13.3.2.2 NONIMAGING TECHNIQUES FOR CLASSIFICATION AND TRACKING

Image-based automatic classification techniques are augmented by other discriminants such as the trajectory, spectral signature, and temporal signature of the feature. All these methods are used to increase the robustness of infrared search and track (IRST) systems and missile warners. They are also used in automatic video trackers, which are used to keep the optical line of sight of imaging systems pointing at the target as the target and/or the platform maneuvers. These are now able to determine the position of a target in a video image to within one or a few pixels in the presence of clutter, decoys, and target aspect changes.

Identification of a target type is often difficult in automatic systems because methods used by the trained observer are based on combinations of cues, which are difficult to impart to automatic systems. Other techniques used to improve automatic target "identification" or provide images usable by human observers at longer ranges include laser-gated imaging and laser-based frequency domain methods.

13.3.3 Laser systems

13.3.3.1 LASER RANGEFINDERS

Laser rangefinders (LRFs) are used for weapon aiming systems (particularly armored vehicle fire control), for target location in surveillance, reconnaissance, and artillery observation roles, and for navigation, to determine the user's distance from a recognizable navigation feature (Figure 13.4).

Most of these systems use a relatively high-powered short pulse (typically many millijoules in 10–20 ns) and measure the time taken for radiation scattered from the target to return to a sensitive receiver accurately aligned to the transmit beam. They are always associated with direct or indirect sights (Figure 13.5).

The first military laser rangefinders used flashlamp-pumped Q-switched ruby and Nd:YAG lasers operating at 693 and 1064 nm. They presented a significant eye hazard because these wavelengths are transmitted and focused by the human eye, so relatively low power pulses can create high, potentially damaging, power densities on the retina. Therefore, they could not be used freely in training. This in turn made their effective use in combat less certain because operators were not completely familiar with them. Many laser rangefinders now use the "eyesafe" 1.54 μm wavelength—radiation of this wavelength is absorbed within the eye and does not focus down on the retina, so the maximum permitted exposure is much greater than for shorter wavelengths. With careful balancing of transmitted power and receiver sensitivity, modern "eyesafe" laser rangefinders provide militarily useful range performance and can be used with minimal or no operational restrictions.

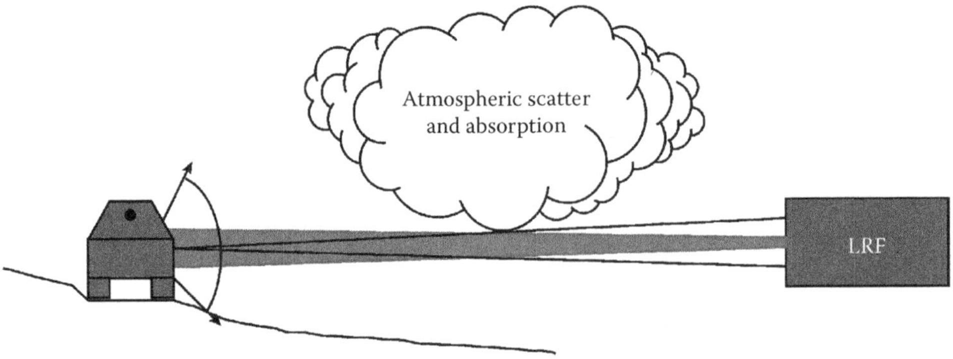

(1) The laser boresight mark is aligned with the target.
(2) The laser pulse is fired at the target (typical pulse lengths are 10–20 ns).
 (i) A small percentage is bled off to the receiver to start the clock.
(3) The laser energy is attenuated by scatter and absorption in the atmospheric path.
(4) With a sufficiently narrow beam divergence, most or all of the remaining energy in the beam hits the target
(5) A percentage of the laser energy is scattered by the target:
 (i) Most targets can be approximated as Lambertian scatterers with a diffuse reflectance of 10%–40%.
(6) The energy collected by the receiver aperture depends on:
 (i) The solid angle subtended by the receiver aperture seen from the target,
 (ii) And atmospheric losses in the return path.
(7) The collected energy is focused onto the detector. Receiver sensitivity depends on a number of factors including detector noise, background noise, quantum efficiency, and optical transmission.
(8) A laser rangefinder's range equation can be derived from this geometry.

Figure 13.4 Principle of operation of a laser rangefinder.

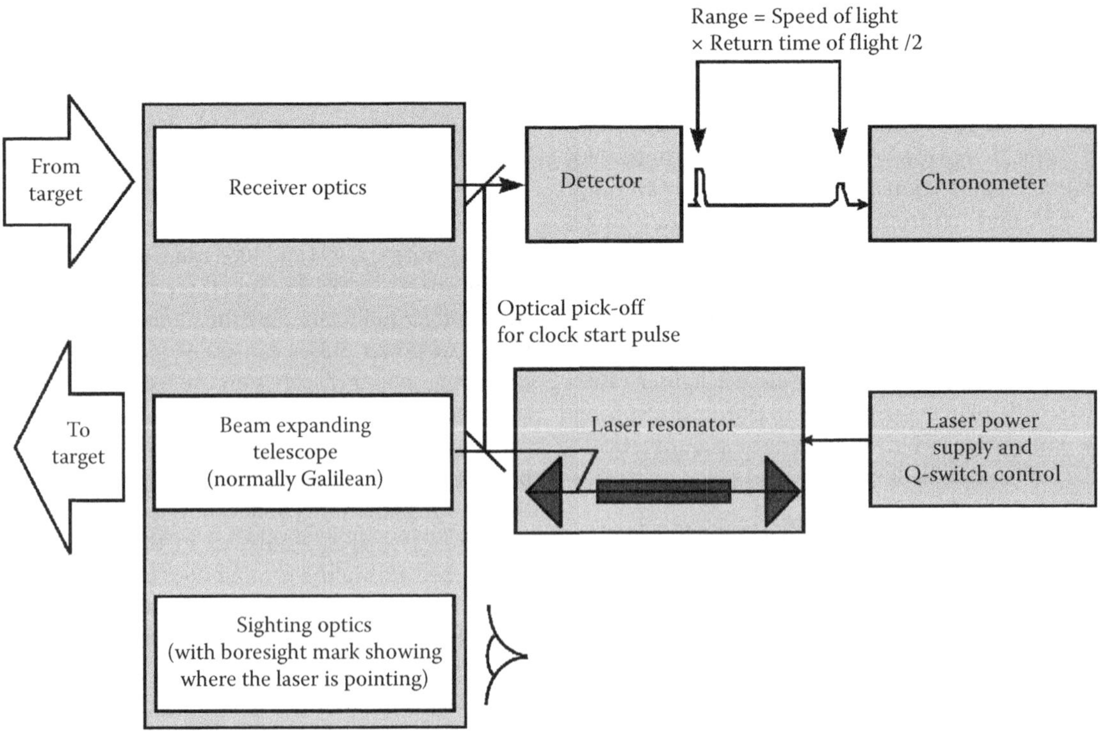

Figure 13.5 Generic architecture of a laser rangefinder.

Most current military LRFs are one of three basic types:

- Nd:Yag or Nd:glass lasers working at 1.06 μm.
- Raman-shifted or optical parametric oscillator (OPO)-shifted Nd:YAG producing 'eyesafe' output at 1.54 or 1.57 μm.
- Erbium glass producing 1.54 μm output directly.

Most are flashlamp-pumped, but diode-pumped systems are starting to enter the market. Q-switching techniques include spinning prisms (used in the first military rangefinders and still common), EO Kerr, or Pockels cells, and saturable absorbers. CO_2 "TEA" lasers are used in some military systems, but have not been widely adopted because they tend to be bigger and more expensive than VNIR systems for a given performance.

Detectors are typically silicon PIN diodes or avalanche photodiodes for the 1.06 μm wavelength. Eyesafe laser receivers benefit from the enormous progress in telecommunications systems at 1.54 μm and use one or other of the GaAs-based compounds. CO_2 systems usually use CMT detectors cooled to liquid nitrogen temperatures. Typical effective ranges are 5–10 km for battlefield systems, more for high-power airborne and ship-borne systems.

Laser rangefinders need to be accurately aligned to the crosswires in the aiming system, otherwise the range obtained will be to the wrong target. They also need to be very robustly engineered to maintain alignment in the severe environment experienced in most military systems.

13.3.3.2 LASER DESIGNATORS

Laser designators are used to provide the guidance illumination for laser-guided ordnance. They are universally Nd:YAG lasers, working at pulse repetition frequencies in the 10–20 Hz region. This is unlikely to change in the foreseeable future because of the massive existing inventory of seekers and designators. Some designators are designed as dual wavelength systems, with a 1.54 μm capability for rangefinding. Most designators have a secondary laser rangefinding capability if only to confirm that the correct target is being illuminated.

Land-based designators are mostly lightweight, tripod mounted, man portable, used by forward air controllers and special forces.

Airborne laser designators are integrated into designator pods for fast jets, or turrets for helicopters or UAVs (unmanned aerial vehicles). These airborne systems typically have a high-performance thermal imager, TV camera, or both; a video tracker; and high-performance stabilization, to allow the system to lock onto a target and maintain a stable line of sight as the aircraft maneuvers.

Designator lasers operate at a high internal power density and are required to have very stable and repeatable boresight, beam divergence, pulse repetition frequency, and pulse energy. They typically use techniques pioneered in the early 1970s in the United Kingdom and the United States—the "crossed Porro resonator" [16], EO Q-switches, and polarization output coupling. Great attention is given to choosing optical architectures and components that are inherently stable. Careful thermal management and rigorous contamination control are also maintained. Rod and slab lasers are used; rod lasers need particular attention to their warm-up characteristics since the optical power of the rod tends to change during laser warm-up. Flashlamps are the most common pump source, but diode lasers are beginning to be used instead. Diode pumping of Nd:YAG lasers is much more efficient than flashlamp pumping, but more expensive and (at least until recently) with significant thermal management problems of its own.

13.3.3.3 HIGH-ENERGY LASERS

After a long period of research and demonstration, high-energy lasers are now being used in developmental antimissile systems. They have demonstrated an ability to cause airframe damage. The first of these systems used chemical lasers, where a chemical reaction creates the energy level population inversion required to allow laser action. This mechanism is very efficient compared to most electrical excitation methods, which makes it feasible to achieve the very high-power levels required in a militarily useful package. Possible use of space-based lasers is discussed by Rogers [20]. Fiber lasers are becoming the technology of choice for delivering large amounts of power, thanks to commercial investment in materials processing equipment.

13.3.3.4 LASER-GATED IMAGING

To achieve longer range in a wider range of ambient lighting and obscuration conditions, a technique known as "range-gated active (or laser) imaging" can be used. A pulsed laser (usually in the visible or near-IR region) is fired at the target, and a time-gated imaging sensor is used to capture a snapshot of the target using the scattered laser light. By matching the time-gating to the round-trip time of flight, the receiver is switched on just before the scattered laser energy arrives back from the target and switched off again just after. Atmospheric backscatter and background clutter can be rejected by the time gating, leaving the target standing out in an otherwise blank image. Such systems can obtain high-resolution imagery at long range with small apertures, independent of ambient light and unaffected by line of sight jitter.

This technique was first developed in the late 1960s. It has been used in a commercially available system for search and rescue in Canada called Albedos [6].

13.3.3.5 LASER RADAR

Laser radar systems have been developed for obstacle avoidance. Typically, they transmit a rapid sequence of laser pulses over a pre-determined scan pattern. A receiver detects the return energy scattered from the ground and from wires or pylons; timing discrimination allows the latter to be distinguished from the former. Warning cues are displayed to the pilot to allow him to take appropriate avoiding action. Imaging laser radar is turning out to be a key technology for unmanned ground vehicles including self-driving cars.

13.3.3.6 LASER BEAMRIDING

Lower power lasers, often near-IR laser diode arrays, are used in laser beamriding systems (see the Section 13.5.3 on weapon guidance).

13.3.4 Important enabling technologies

As well as lasers and imaging, several other technologies are key to the engineering of successful optoelectronic systems for military applications and other similarly harsh environments.

- *Stabilization.* Line of sight stabilization systems typically use gyros to sense disturbance of the line of sight or the optical chassis, and a servo controller and mechanism to adjust the line of sight to compensate for the disturbance. Performance (measured as achieved jitter level under angular disturbance and linear vibration) varies widely, from sub-micro-radians in spacecraft, through 5–20 μrad in high-performance airborne systems, to 50–500 μrad in typical land vehicle systems. Stabilization performance is best understood as an attenuation of the input disturbance; hence, performance figures must be taken in context with the environment. Harsh vibration and shock environments also impose design constraints on stabilization that tend to degrade the achievable performance [11]. Anti-vibration mounts are often used to give the critical stabilization components a softer ride, prevent linear vibration from exciting structural resonances, and match the angular disturbance spectrum to the control bandwidth of the stabilization servo.
- *Processing.* Modern EO systems are largely software driven. The relentless march of Moore's law means that digital processing is becoming the default solution to most control problems, and embedded computing power is now sufficient for real-time image processing functions such as detector channel equalization, contrast optimization, image enhancement, and target detection, tracking, and classification. Many image-processing systems developed during the 1990s used custom silicon devices (application-specific integrated circuits or ASICs). At the time of writing, military production volumes are of little interest to ASIC manufacturers. Field programmable gate arrays (FPGAs) and digital signal processors (DSPs) are now widely used in military systems. Mass-produced devices with high data throughput capability, FPGAs and DSPs are programmed by the individual developer; hence, they are more flexible and economical than ASICs for low-volume applications such as military systems. From around 2010, the advent of the graphical processor unit (GPU) for gaming platforms has shortened development times and provided a powerful, easy-to-use image processing platform.
- *Lens design.* While modular design and reuse are becoming more common, most military systems require custom optical design to satisfy installation constraints and system performance requirements. Particular challenges for the lens designer in military instruments include design for mechanical robustness, high boresight stability and repeatability,

athermalization, and good performance in the presence of bright sources in the FOV (e.g., fires, sunlight). There are numerous references to this subject, notably proceedings of lens design conferences run, for example, by the Society of Photo-electric Instrumentation Engineers (SPIE). IR optical designs are substantially different from visual band ones because of IR materials' higher-refractive index, wider range of dispersions, and larger thermal refractive index coefficients.

- *Optical materials.* Military requirements have driven development and precise characterization of novel optical materials for many uses within EO systems. These include: IR lenses for various sub-regions of the spectrum; mechanically strong broadband windows with good thermal properties; laser and other components with low absorption, low scatter, and high laser damage threshold; and lightweight, thermally, and dynamically stable mirrors and structures.
- *Optical fabrication.* Diamond turning now allows the reliable and repeatable fabrication of aspheric surfaces, which allow simpler and higher-performance IR system designs. "Binary" or diffractive optical components can also be made with diamond turning; this allows color correction (in systems where scatter is not a critical design driver) with even fewer optical components. Diamond fly-cutting is widely used to finish flat and cylindrical mirrors. Great strides have been made in the surface finish, accuracy, repeatability, and flexibility of these techniques, driven mainly by military requirements.
- *Thin film coatings.* Thin film coating products and custom engineering capabilities have been developed to satisfy the many and varied demands of military EO systems. These include: low loss laser coatings with high damage threshold and low scatter; efficient wavelength selective beamsplitters and filters; high-efficiency coatings for a wide range of spectral regions; multispectral coatings that simultaneously offer good performance (usually high transmission) over a number of spectral regions; exceptional environmental durability (sand erosion, rain impact, chemical contamination, and wide temperature range). The products are all compatible with a range of substrate refractive indices and thermal expansion coefficients.
- *Optomechanical design.* Military optoelectronic systems usually need to be compact, lightweight, and robust to a wide range of temperatures and under shock and vibration; they need to be easy to align and maintain, hence modular in construction and achieve high line of sight stability. Often they incorporate mechanisms for switching components in and out of the light path, for scanning or steering the line of sight, and for adjusting focus and magnification. Embedded electronics generate heat, which has to be managed without adverse effects on the rest of the system. Delicate optics have to be housed to maintain critical alignment yet survive severe shocks. These create tremendous challenges for mechanical and optical engineers, leading to the discipline of "optomechanical engineering." Vukobratovich [25] and Yoder [26] address many of the detailed design issues. Godfrey [12] and Jamieson [13] provide insights into the complex issues that occur at the "system architecture" level.
- *Simulation and modeling.* Parametric modeling and physically accurate simulation techniques have been developed to support system requirements definition and validation, system tradeoffs, detailed design, and validation of the final system design. Accurate techniques have been evolved to model all aspects of EO system performance, including scene illumination, target and background signatures, atmospheric absorption and emission, scintillation induced by atmospheric turbulence, image formation, scatter, line of sight stability, and detection and tracking performance. Such models may be integrated to provide realistic end-to-end simulations displaying a representative image from a system yet to be built, or of a scenario too dangerous or expensive to trial—or both. Such models are quite different from those used in training simulators. Simulator systems provide an impression sufficient for their purpose of what an observer would see, using approximations and assumptions to simplify computation and to ensure real-time performance. EO system models provide a physically correct representation of what an observer or image-processing system would see with a real instrument in a specific scenario. Sillitto [29]

provides a perspective on the approach for this type of modeling and how it supports whole system analysis and architectural design.
- *Calibration, test, and alignment.* Numerous test techniques have been developed—both to align, integrate, and characterize EO systems as "black boxes" in their own right; and to assist with their integration with other systems, for example, platforms and weapon systems. Methods of particular note include those associated with characterizing and correlating the subjective and measurable performance of thermal imaging systems; techniques for aligning multi-waveband sensor systems; and "auto-boresight" techniques for dynamically maintaining system alignment during operation.

The various technologies, their applications to military needs, their advantages and limitations are listed in Table 13.1.

13.4 ENVIRONMENTAL FACTORS

We have already alluded to a variety of environmental factors that influence military EO systems. In addition to the "EO environment" already discussed, these can be summarized as follows:

- *Physical threats*: Ballistic threats, rain and sand erosion, air pressure, hydrostatic pressure, own and hostile weapon blast effects
- *Operating environment*: Shock, vibration, temperature, handling, and radio frequency interference
- Line of sight limitations

13.4.1 Physical threats

Physical threats to EO systems include the following.

- *Ballistic threats*: For example, bullets and shrapnel may hit delicate optics causing catastrophic damage. System design is a trade-off between maximizing performance (usually requiring a large aperture) and minimizing vulnerability (for example, by keeping the aperture small). Many systems are designed with shutters that can be lowered to protect the optical aperture, often with slits that allow continued operation with degraded performance.
- *Rain and sand erosion*: Forward facing optics fitted to fast low-flying aircraft are particularly susceptible to damage from sand erosion, which can strip coatings and pit surfaces, and to rain impact, which above a velocity threshold, can cause serious sub-surface damage, leading to rapid loss of transmission. Coatings such as boron phosphide and diamond-like carbon can be applied to protect surfaces while maintaining good optical transmission.
- Chemical erosion may be caused by seawater and by exposure to fuels, cleaning agents, and exhaust efflux. Again, inert coatings can usually be identified to protect while maintaining optical performance.
- Air pressure, hydrostatic pressure, own and hostile weapon blast effects all impose structural loads on external windows, which normally require thick windows to maintain structural integrity and careful mounting to avoid excessive local stresses that can initiate structural failure.

13.4.2 Electromagnetic threats

Enemy systems may use EO or radar sensors to detect and/or degrade EO systems. Notably, submarine periscopes are subject to detection by radar, and reports in the trade press suggest that hostile lasers intended to damage detectors are also seen as a threat. Filters can be fitted to minimize the effect of EO countermeasures but this is often at the expense of the performance of the EO device. Increasingly, attempts are being made to develop detectors that operate outside normal threat wavebands.

13.4.3 Platform characteristics

Different platform types have widely different environmental requirements, mission characteristics, and accepted industry quality standards and interfaces. They also imply different user characteristics—education, training, and tolerance of workload; usability, quality, reliability, and technology complexity; and different production volumes, rates, and support and maintenance philosophy.

Notably, equipment on helicopters have to tolerate very strong low frequency vibrational resonances at harmonics of the rotor frequency. Equipment on main battle tanks (MBTs) and submarines must be hardened to operate after "survivable hits" from

Table 13.1 Various optoelectronic technologies and their applications to military needs

Application	Technology	Advantages	Disadvantages	Current situation (at time of writing)
Imaging systems	Visual/near-IR TV	Best angular resolution, easiest to interpret; can now provide near starlight performance at moderate cost	Depend on ambient light, susceptible to visual camouflage, cannot see through smoke	Returning to fashion as the benefits of high-resolution and multiwaveband operation are appreciated
Imaging systems	3–5 μ IR	Work well in most conditions day and night; acceptable angular resolution from moderate apertures	Affected by smoke and fires	Widely used in surveillance and weapon aiming applications
Imaging systems	8–12 μ IR, cooled and uncooled	High sensitivity, can see through smoke	Low scene contrast in hot humid conditions, poorer angular resolution from given aperture	Widely used in surveillance and weapon aiming applications
Imaging systems	Direct vision near-IR night vision goggles	Compatible with starlight and nonvisible illumination	Prone to interference from bright light sources; need fast optics so limited to wide FOV applications	Widely used in infantry and aviators' night vision systems
Stabilized sensor systems	Gyro stabilized turrets or mirrors with one or multiple cameras etc., used on drones, ships, vehicles, aircraft	Can avoid shake due to platform motion, allow smooth tracking of moving target and compensate for own ship motion	Can be difficult to integrate with aerodynamic profile and armor protection	Widely used at various levels of stabilization performance from fractions of a milliradians to a few microradians
Laser rangefinding and designation	Nd:YAG 1.06 μ lasers, glass optics, silicon detectors	Mature technology, compatible with large legacy inventory of laser guided weapons	Not eyesafe	Widely used for laser designation, being superseded by 1.54 μ systems for range-finding
Laser rangefinding	Various at 1.54 μ including erbium glass, diode lasers, and Raman shifted Nd:YAG	Eyesafe, benefit from advances in telecoms technology for laser sources and detectors	Lower power output than Nd:YAG for a given package,	Now in widespread use as rangefinders

(Continued)

Table 13.1 (*Continued*) Various optoelectronic technologies and their applications to military needs

Application	Technology	Advantages	Disadvantages	Current situation (at time of writing)
Laser rangefinding	CO_2 lasers (9–11 μ), direct and heterodyne detection	Smoke penetration, eyesafe, compatible with thermal imaging optics and sensors	Bigger, more expensive, and more complex for a given level of performance; cryogenic cooling required for detector	Not used much in spite of considerable investment in the 1980s
Laser pointer	Visible and near-IR laser diodes used to show point of aim, and to indicate targets to others	Clear and intuitively obvious hand off	Not passive so may reveal one's position to an enemy	Widely used on infantry weapons for aiming, and to indicate points of interest to night vision systems
Laser weapons	Various	Virtually instantaneous engagement so aiming is simple; no limitation on number of "rounds" if adequate electrical power available	Expensive, complex, difficult to measure effectiveness, disrupted by atmospheric turbulence and obscurants	Very limited use by a few countries
Missile guidance	Nonimaging infrared, typically cryogenically cooled and mechanically modulated reticule or rosette scan	Relatively simple processing, effective against small bounded targets showing high thermal contrast such as aircraft	Susceptible to clutter and countermeasures; maintenance requirements are significant because of cryogenic cooling and precision mechanics	Widely used in air-to-air and surface-to-air missiles
Missile guidance	Imaging infrared	More options to deal with countermeasures and clutter, more sensitive	More complex processing	Widely used in air-to-ground and anti-tank missiles
Missile guidance	Laser beamriding	Low power guidance beam that is difficult to detect and counter	Require clear line of sight from launch post to target so that operator can track the target throughout the engagement	Used in a variety of short range antiaircraft and anti-tank systems

(*Continued*)

Table 13.1 (*Continued*) Various optoelectronic technologies and their applications to military needs

Application	Technology	Advantages	Disadvantages	Current situation (at time of writing)
Missile warning	UV missile warning sensors	Solar blind, low clutter at low levels	Limited range, less effective at high altitude	In widespread use, but starting to be replaced by IR systems
Missile warning	Mid-IR missile warning sensors	Spectrally matched to peak emissions of rocket motors; much longer potential range than UV; useful imagery of surroundings	Need more sophisticated processing to deal with clutter and spurious target-like features	As processing capabilities increase, these are becoming more common
Aerial reconnaissance	Digital visual and IR band sensors on aircraft and drones	Very high-resolution imagery, can fly below cloud; data available in real time if data linked back to base or to forward observers	Need to fly over or near the target, so vulnerable to air defenses; prone to cloud, smoke, and terrain obscuration	Manned aerial reconnaissance systems are still in use, increasingly augmented by drones of all shapes and sizes
Satellite reconnaissance	Digital visual and IR band sensors	High enough to avoid air defenses	Flight paths relatively predictable; takes longer to retask; lower spatial resolution; difficult to make data available in real time to front line users	Significantly used by advanced countries
Missile countermeasures	Modulated and directional jammer using laser or high-intensity flashlamp	Can defeat many missile threats; no significant limitation in number of engagements	Requires understanding of threat characteristics	Increasingly in use to complement or replace flares
Missile countermeasures	Flares	Not directionally critical, can use the same dispenser as chaff	Limited number of rounds can be carried out; some threats require complex flare sequences	Once the only protection means, now used in tandem with directional jammers

(*Continued*)

Table 13.1 (*Continued*) Various optoelectronic technologies and their applications to military needs

Application	Technology	Advantages	Disadvantages	Current situation (at time of writing)
Submarine detection	Distributed fiber optic sonar arrays	Sensitive acoustic detection, no electronics required external to pressure hull	Very sophisticated technology	Early production systems deployed in several countries
Perimeter security	Distributed optical fiber sensors providing seismic detection for perimeter security	Critical alignment not necessary, immune to RFI, does not depend on clear sightlines	Need to lay physical cables, subject to deliberate or accidental damage	Initial production systems sold in various countries
High-speed data links	Fiber optic uses ranging from strategic networks to spooling out behind missiles to transfer video to operator station	Immune to RFI; high bandwidth, much lighter and more flexible than equivalent copper, can detect and locate breakages	Fiber vulnerable to damage so need good protective sheathing	Widely used exploiting civil telecoms technologies

shells or depth charges, sometimes with accelerations specified up to hundreds of *g*. Equipment on aircraft must operate in the presence of high linear vibration caused by engines and aerodynamic buffeting, and high angular disturbance caused by the aircraft's maneuvers.

EO sensors often have to operate near other systems that generate strong radio frequency (RF) fields, for example, radars and communication equipment.

13.4.4 Line of sight limitations

With the rare and uncertain exception of atmospheric ducting, free space EO systems cannot see round corners. Strong signals, for example, from missile plumes, can be detected via atmospheric scatter when there is no direct line of sight. Otherwise, EO systems depend on a clear line of sight from sensor to target.

Hence, terrain geometry strongly influences the use of military EO systems even to the extent that different countries will specify quite different equipment characteristics, depending on whether they expect to be operating with long open lines of sight, for example, desert conditions or short lines of sight in forest, urban, or rolling rural conditions. It is reported that historically most tank engagements occur at less than 1 km, although most modern MBTs are designed to work at up to 3 km. In the 1991 Gulf War, line of sight ranges were often very long, and armor engagements at ranges of up to 5 km were reported. Tactics must be adapted to make best use of the characteristics of available equipment in different terrain.

At sea, the horizon and weather limit lines of sight. In the air, cloud and terrain normally limit the line of sight at low altitude. At medium and high altitude (above 5000 m) and in space, air-to-air lines of sight are very long, limited only by the earth's curvature. At high altitude, the atmosphere is thinner and atmospheric absorption becomes less significant. Air-to-ground visibility is (obviously) affected by cloud cover.

13.5 ROLES AND EXAMPLES OF MILITARY EO

The following sections discuss the main roles in which EO systems are used by the military, under the six principle categories of

- Reconnaissance
- Surveillance and target acquisition (S&TA)
- Target engagement
- Self-defense
- Navigation and piloting
- Training.

The final section deals with some less common systems that do not fit into any of these categories.

Each section discusses the role, gives examples of systems, and discusses some of the underlying design issues.

13.5.1 Reconnaissance

"Time spent in reconnaissance is seldom wasted"—*Military Maxims*

quoted by Mitchell [27]

Reconnaissance is about providing an overall picture of the area of operations, notably where the enemy is and what he or she is doing. It involves observation and orientation and supports decision-making. There is an increased emphasis on using strategic reconnaissance for tracking and targeting as well. This will require improved real-time links between reconnaissance and combat units.

Reconnaissance is performed at all levels, from strategic to tactical. At a strategic level, the area of interest may be worldwide and decision timelines may be in the order of hours, days, or weeks. At lower levels, the area of interest becomes smaller, but the timelines become shorter. At the lowest tactical level, the soldier uses his or her eyes or binoculars, and would dearly love to "see over the hill."

Reconnaissance platforms include satellites, manned aircraft, land vehicles, and foot soldiers. UAVs are increasingly being used or proposed for the full spectrum of reconnaissance. The US "Global Hawk" programme is a strategic reconnaissance UAV. Micro-UAVs, small enough to be carried by an individual soldier, will finally make the "see over the hill" dream a reality for even the lowest level tactical commander.

The human eye has always been used for reconnaissance and always will be. For most of the twentieth century, wet film was the prime recording medium for aircraft.

The advent of satellite reconnaissance forced the development of solid state image sensors, which now deliver astonishing resolution from large aperture satellite systems in low earth orbit. Fixed wing aircraft are fitted with podded or built-in line scan reconnaissance systems using solid state linear array imagers working in the VNIR, and IR line scan systems working in the 8–12 μm band. Two-dimensional staring arrays with stabilized step-stare mirrors are used for long range stand-off reconnaissance in the VNIR and 3–5 μm band. Improvements in sensitivity are allowing VNIR sensors to be used at progressively lower light levels with progressively shorter integration times.

The key performance goals in reconnaissance systems are to maximize area coverage, resolution, and stand-off range of the platform. The ideal reconnaissance EO system would have high resolution, short image capture time, and wide FOV.

13.5.1.1 RECONNAISSANCE SYSTEMS

Airborne fixed wing reconnaissance systems are produced by a number of companies in the United States, the United Kingdom, France, and other countries. A typical fixed wing low level reconnaissance system is the Vinten Vicon product family. Based on a modular architecture and generally podded, these systems are used by about 20 air forces and are being configured for UAV applications.

Most armies maintain ground and helicopter units with a reconnaissance role. These generally operate covertly when possible, behind enemy lines, and offer the benefits of having a human observer able to interpret and prioritize the information, while being relatively immune to cloud cover, which can block satellite and UAV operations. Army reconnaissance units are normally equipped with surveillance systems with varying levels of performance, generally aiming to be able to observe over a reasonable area (2–10 km) from a "hide." Different armies have widely differing philosophies on how to do this difficult task, depending in part on the terrain they expect to operate in, leading to a remarkable diversity of equipment.

Reconnaissance using manned aircraft or land vehicles is not always politically or militarily practical; satellites and UAVs are becoming widely used for reconnaissance.

UAV reconnaissance systems typically use either or both line scan sensors derived from aircraft systems or surveillance turrets derived from helicopter systems. Surveillance turrets typically contain thermal imaging and TV sensors with multiple fields of view, and can be given an area search capability of scanning the whole turret provided the imager integration time is short. The imagery is data-linked to a ground station in real time, and it can also be recorded on board for subsequent recovery and exploitation.

The US "Reaper" UAV has an "armed reconnaissance" capability with a laser designator coupled with Hellfire laser-guided missiles, allowing the UAV to be directed to attack a target observed by the operator at the ground station.

Information about space-based reconnaissance systems is harder to come by. Chaisson [5] refers to KH-11 Keyhole satellites used by the United States. News reports during recent conflicts have emphasized the use of satellite reconnaissance by the United States, and the possible military use of commercially available satellite imagery from many operators throughout the world. US government officials' statements, which can be found on the web, emphasize the priority attached to maintaining a "technological advantage" in the capture and exploitation of information using satellites.

13.5.1.2 RESOLUTION OF RECONNAISSANCE SYSTEMS

Popular fiction suggests that reconnaissance systems can read car number plates from space. To read a number plate, the system would need to resolve about 1–2 cm. We can test this assertion with a simple calculation based on the Rayleigh criterion for optical resolution:

$$\vartheta = 1.22\lambda / d$$

where θ is the minimum angular separation between two resolvable points, λ is the wavelength, and d is the diameter of the aperture. The reciprocal of this angle corresponds to the cut-off frequency (typically quoted in cycles per milliradian) of the modulation transfer function of a diffraction-limited imaging system.

A 600 mm aperture telescope operating at a wavelength of 500 nm would have a diffraction limited optical resolution of 1 μrad. This would allow an imaging resolution of 1m at a height of 1000 km in the absence of image motion and atmospheric degradation. Equivalent resolution for a system working at 10 μm in the thermal IR would be 20 m. This example illustrates the benefit of using the VNIR band where target signatures allow.

Some US reconnaissance satellites are similar in general characteristics to the Hubble Space Telescope [5], with its 8 ft (2.4 m) mirror. The International Space Station is approximately 400 km above the earth, and orbits at 27,600 km h^{-1} (NASA web site). A Hubble-class telescope in this orbit, operating in the visible spectrum at 500 nm, would resolve 0.1 m or 4 in. on the ground, if its line of sight was correctly stabilized on the earth's surface, and if the effects of atmospheric turbulence were overcome.

13.5.2 Surveillance and target acquisition (S&TA)

Reconnaissance imagery is very data-intensive. It tends to be analyzed by interpreters who summarize the information and provide it to senior commanders. The imagery is seldom available in real time to troops on the ground, because the bandwidth required to disseminate raw imagery is very high, but this is changing as broadband network technology becomes pervasive.

Tactical real-time targeting and situation awareness is provided by wide or medium FOV imaging sensors, which can be panned or scanned to cover a wide field of regard. Typical solutions are surveillance turrets, typically carrying a thermal imager, TV camera, and eyesafe laser rangefinder.

Similar sensor suites are integrated into the turrets of many military vehicles for surveillance, target acquisition, and fire control, mounted on pan and tilt heads for border and base security, and on vehicles with elevating masts for mobile patrols. Compact and increasingly integrated systems with similar functionality are used by forward observers, usually on manually operated tripods, and will shortly be small and cheap enough for use by individual soldiers, albeit with scaled down performance. Such systems are sometimes integrated with navigation systems to allow the target grid reference to be calculated.

13.5.2.1 SYSTEM EXAMPLE: SURVEILLANCE TURRETS

Turret systems are used on helicopters, patrol aircraft, and some land vehicle applications.

They are typically stabilized to a level appropriate to the resolution of the sensor payload. They are used for general purpose surveillance, tracking, and targeting, for rescue, and as navigation aids to the crew. This class of product has received widespread publicity on western television because TV programs such as *Police, Camera, Action!* make extensive use of the often spectacular video sequences obtained by police helicopters during surveillance and pursuit operations, mostly car chases.

The turrets usually look like a ball, with an azimuth gimbal, which allows the whole ball to rotate, and an elevation gimbal, which allows the payload within the ball to elevate. Some have a third gimbal, which allows the system to compensate for roll and to keep the image upright while tracking through the nadir. Others have a second pair of gimbals for fine stabilization of the payload. This separates the functions of steering, which is assigned to the outer pair that also has to cope with friction from the environmental seals, from that of stabilization, assigned to the inner pair, which only has to work over a limited angle assisted by the intrinsic inertia of the payload.

Large, high-performance turrets with extremely good stabilization (from 20 µrad down to a few microradians) are used for long range stand-off surveillance, usually with large aperture TV cameras.

Small and agile turrets fitted with thermal imagers and/or intensified CCD cameras are used as visually coupled flying aids. The line of sight is steered to follow signals from a head tracking system, which measures where the pilot is looking; the image is overlaid on the scene by the pilot's helmet-mounted display. Small agile turrets are also used for active countermeasures systems.

Similar systems with appropriately customized or simplified packaging, pointing, and stabilization are used for other surveillance and fire control applications: for example, border surveillance, and ship-borne surveillance and fire control. On ships, they are generally referred to as "directors."

The generic principle of operation of surveillance turrets and "directors" is shown in Figure 13.6.

13.5.2.2 INFRARED SEARCH AND TRACK

Most S&TA sensors are designed primarily to provide images to operators who are responsible for detecting and classifying targets. IRST systems are a special case of S&TA systems designed to detect and track certain classes of targets automatically without operator intervention. Their main application is against aircraft, which are relatively easy to

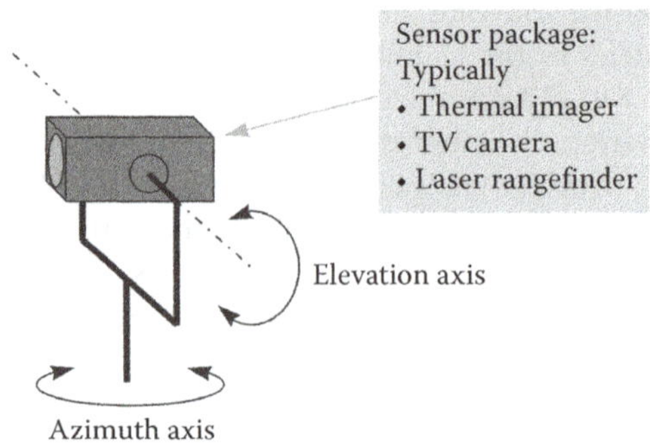

Notes:
Typically, slijp rings are used for electrical connection across the azimuth bearing to allow unrestricted N×360 degree rotation.
3–axis systems have a third rotation axis orthogonal to the other two to provide roll stabilization.
4–axis systems have an outer coarse set of gimbals carrying an environmental cover and a fine inner pair for precise stabilization.
5–axis system have a 2–axis outer set of gimbals for environmental protection and a 3–axis inner set for stabilization and derotation.

Figure 13.6 Principles of surveillance turret operation.

detect because of their high contrast against cold sky backgrounds and their distinctive motion relative to ground clutter. Similar principles can be used from space to detect ballistic missiles.

IRST systems give their users big tactical advantages. They allow them to detect and track aircraft without revealing their presence with radar emissions; they allow more accurate angle tracking of targets than is possible with radar; and most stealth technologies are not as effective against IRST systems as they are against radar. Range to target can be determined, albeit not as accurately as by radar, using kinematic or triangulation techniques. Highly effective as stand-alone systems, IRST systems are even more valuable when integrated into a sensor network including radar, since each sensor, to a considerable extent, complements the other.

The concept of IRST systems dates back to the 1930s, when R.V. Jones demonstrated aircraft detection at operationally useful ranges using cooled detectors, probably PbS. Radar was selected for further development at the time because of its all-weather capability.

First generation airborne IRST systems were fitted to the Mig-29, SU-27 (both 3–5 μm), and F-14 (8–12 μm band). Their design varied; at the simplest, they were conceptually similar to missile seekers with programmed scan patterns. Second generation systems fitted to the Eurofighter Typhoon (Figure 13.7) and the Dassault Rafale use

Figure 13.7 Head-on close-up view of Typhoon aircraft showing FLIR/IRST (left) and head-up display (center)—Photo: SAC Sally Raimondo/MOD© Crown Copyright 2011-Image used under Open Government Licence.

high-performance detector arrays allowing beyond visual range detection, and are in a real sense "passive radars."

Some ships have been fitted with scanned IRST systems to fill the radar gap in detection of sea skimming missiles. Successful production examples include the Dutch Sirius and IRScan systems, using first and second generation long linear array technology respectively, and the French Vampir. The United States, Canada, and the United Kingdom have been developing naval IRST technology for many years; but at the time of writing, they had developed no product.

Of land-based systems, the Thales Air Defence Alerting Device (ADAD) has been notably successful, achieving large orders from the British Army. Working in a static ground clutter environment, it gives fully automatic passive alerts of approaching air threats.

Naval- and land-based systems work in a very complex clutter environment and, for best performance, require good training and a clear doctrine for systems integration and deployment. As technology develops, it will be possible to improve the efficiency and robustness of IRST clutter discrimination, widening the applicability and operational flexibility of these systems.

13.5.2.3 SYSTEM EXAMPLE: SUBMARINE PERISCOPES

Submarine periscopes provide a view of the outside world when submarines are on the surface and at periscope depth. They are used for general navigation and watch-keeping, surveillance, threat and target detection, and to identify potential targets and the range to these targets.

Traditional periscopes are all-glass, relaying a magnified view of the scene to the control room, typically 10 m below the line of sight. This is a demanding optical design problem, with only a small number of companies in the world capable of providing effective systems.

Conventional periscopes now have EO sensors integrated into them. For example, the Thales CK038 product family offers thermal imaging, intensified CCD, color TV, and still camera channels, as well as two or three direct optical fields of view. Indirect sensors can be displayed on a remote console, and "bearing cuts" and "range cuts" provided to the combat system. A variety of antennae for RF sensors can also be carried.

Nonhull-penetrating masts are the next technology step. Thales supplies "optronic masts" for the new Royal Navy Astute class submarines, which are the first in the world to have no direct optical path from inside the hull to the outside world. This innovation reduces the number of holes in the pressure hull and removes constraints on the position of the control room relative to the submarine's "fin." Exceptionally high-quality systems with excellent demonstrated performance, intuitive man-machine interface, and built-in redundancy are required to give users confidence in this approach.

13.5.3 Target Engagement

13.5.3.1 WEAPON AIMING

Weapon aiming involves

- Detecting the target or acquiring it after being cued on by another sensor.
- Determining the information required to launch the weapon—normally, target direction, range and crossing rate, and often also range rate.
- Displaying the results of any fire control computation to the operator to allow him to point the weapon in the right direction for launch.

Typical weapon aiming accuracies for armored vehicle gun systems are in the 50–500 μrad range (10–100 arcs). Shocks in the range 40–400 g may be generated at or near the sensor by the weapon launch. A key part of the skill in weapon aiming system design is to package delicate optical instruments in such a way that they will retain the required accuracy, yet survive the extreme shock and vibration on weapon platforms.

System error contributions come not only from within the EO system, but also from the following sources:

- The weapon system itself—for example, barrel wear and bending under thermal gradients, geometric imperfections and play in linkage mechanisms, and tilt of the weapon platform.
- The environment—cross-winds and air pressure variations.
- Dynamic effects within the weapon platform—vibration, flexing, and servo latencies.
- Target motion—the target may move significantly and unpredictably during the time of flight of the projectile.

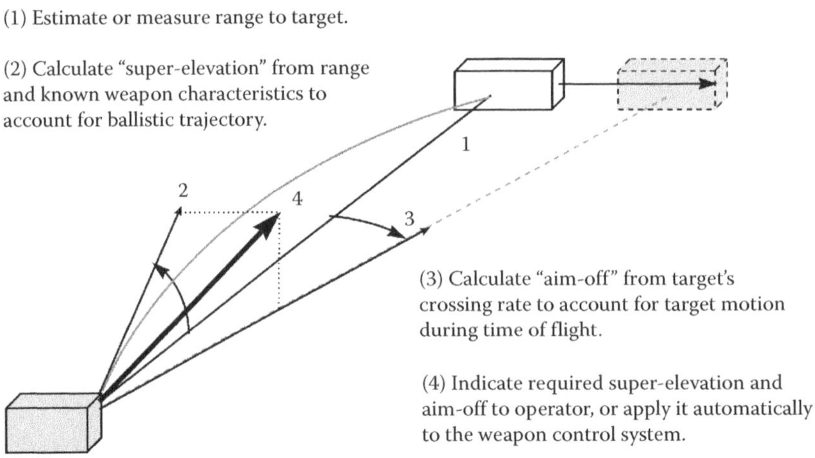

Figure 13.8 Generic weapon aiming system.

Modern high-performance fire control systems measure these contributions and take them into account in the ballistic calculation, minimizing their effect and achieving high overall system accuracy. But conventional gun systems cannot adjust the course of the projectile after it leaves the barrel. Hence, they cannot compensate for target acceleration, and lose accuracy at longer ranges as the projectile slows down and becomes less stable in flight (Figure 13.8).

13.5.3.2 GUIDANCE

Aiming systems for unguided projectiles need to point the weapon accurately at launch and to anticipate the target's motion between launch and impact. These requirements for precision are reduced if the target can be tracked and course corrections made during the flight of the missile. This can be achieved either by tracking the target from the launcher or by homing, either semi-or fully autonomous.

The following common techniques used are as follows:

Command Guidance

- Command to line of sight
- Beamriding

Smart/Terminal Guidance

- Laser designation
- Autonomous homing

13.5.3.2.1 Command Guidance

In command to line of sight (CLOS) systems, both target and missile are tracked. In many systems, the operator keeps a crosswire on the target and a sensor in the sight tracks a flare on the tail of the missile. More modern systems track both target and missile automatically. A guidance system measures the error, and an RF, EO, or wire link transmits control signals to the missile. These systems often use a rangefinder to measure range to the target, and keep the missile off the line of sight to the target until near the expected impact time to prevent the missile or its plume obscuring the target (Figure 13.9).

In a beamriding system, the operator keeps the crosswires in the sight on the target, and a spatially and temporally modulated laser beam is projected along and around this line of sight. A receiver in the tail of the projectile detects the signal; signal processing decodes the modulation and generates

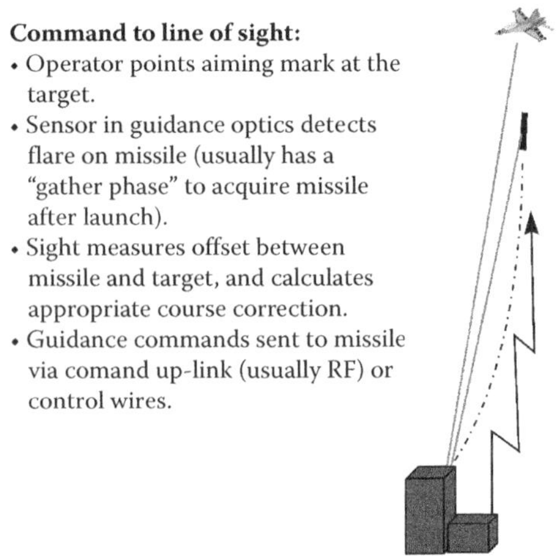

Figure 13.9 Guidance principles: CLOS.

the appropriate guidance correction. This method relies on the operator accurately tracking the target throughout the engagement. It is used in a number of successful antiaircraft missile systems (including the Swedish RBS-70 and the British Starstreak) and some anti-tank missile systems. Most western beam riding systems use diode lasers; a small number of systems use CO_2 lasers. Power levels are much lower than for other military laser systems because it is a one-way co-operative system (Figure 13.10).

13.5.3.2.2 Smart/Terminal Guidance

Laser designation (Figure 13.11) involves illuminating the target with a pulsed laser, which is accurately boresighted to a crosswire in the sighting system. Normally pulsed Nd:YAG lasers are used, associated with a silicon quadrant detector in the seeker head. The seeker, or an optical element within it, is gimballed. A control system moves the seeker line of sight to keep the laser spot on or near the center of the quadrant detector. The missile in turn flies to follow the seeker's line of sight. This system, pioneered by the United States during the Vietnam war, is regarded as the most accurate precision guided system and is in extremely widespread use for anti-tank missiles, air-launched missiles, and a wide range of bombs. Laser-guided artillery shells have been developed but are not in widespread use.

Laser seekers are referred to as "*semi-active* homing" systems—*active* because an active transmitter (the laser) is used to illuminate the target, *semi* because the transmitter is not co-located with the seeker.

Passive missile seekers detect and home onto the contrast between natural radiations emitted or

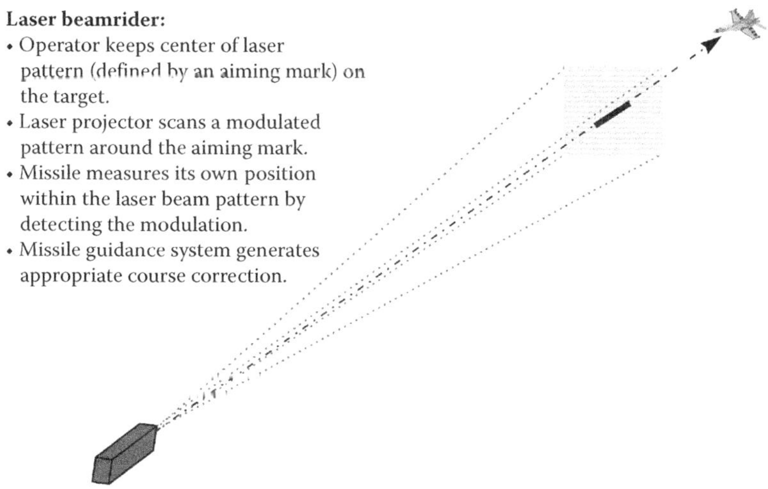

Figure 13.10 Guidance principles: beamriding.

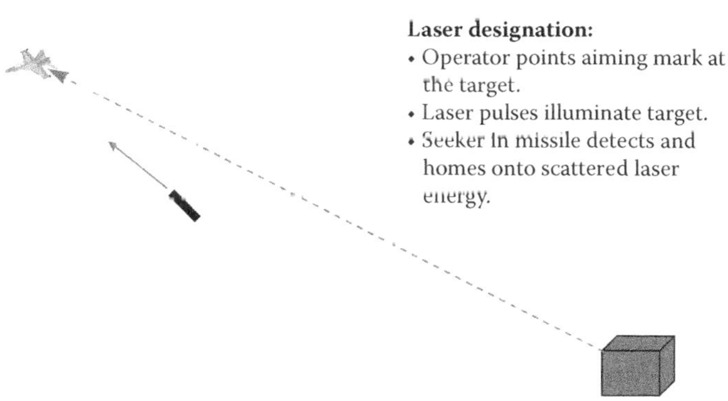

Figure 13.11 Laser designation.

reflected by the target and the background. They can be broadly categorized as follows:

- "Hot spot detectors"
- "Imaging" seekers

Hot spot detectors are simple, easily implemented as all-analog systems, and effective in low clutter environments such as an aircraft against sky background. Antiaircraft systems have evolved through several generations to improve sensitivity and countermeasure resistance. The simplest require a strong signal, for example, from a hot exhaust pipe, and so they can only attack from behind, and are relatively easily decoyed or jammed. The more sophisticated are sensitive enough to home in on black body emission from an airframe, and so are "all-aspect," and employ various anti-decoy techniques, some of them dependent on digital processing.

IR hot spot seeking surface-to-air missiles (SAMs) and air-to-air missiles (AAMs) have accounted for a huge majority of all air combat losses since the 1950s. The best known to western readers are the American Sidewinder family of AAMs, which has evolved incrementally through numerous versions since the 1950s [15], and the Stinger series of SAMs, which evolved from the 1960's Redeye, and has itself progressed through at least three major versions using two different detection principles. Other nations including France, USSR, the United Kingdom, China, and Brazil have developed indigenous IR SAMs and AAMs, some using completely independent seeker principles, others based closely on the successful American designs.

Most antiaircraft seekers work in the 1–2.5 or 3–5 μm bands. Early Sidewinders used uncooled PbS detectors working in the 1 μm region with a glass dome. These systems were easy to engineer using available materials and technologies, but because of the low sensitivity of the uncooled seeker, they needed to look up the hot jet pipe. In the late 1950s and 1960s, sensitivity was improved to allow wider aspects but still tail-on engagements. The PbS detectors were cooled with Peltier coolers, which could be operated continually while the missile was on the launch aircraft, or with high-pressure argon gas, which achieved a lower temperature for a limited period after a very fast cooldown. Later versions were more sensitive still. They used InSb detectors operating in the 3–5 μm band, again cooled with argon, to achieve an all-aspect capability by exploiting the lower temperature emission from skin heating for frontal engagements. These systems took longer to develop because they required a whole range of technology developments in optical and detector materials.

Four different optical principles are commonly used in hot-spot seekers to generate a guidance signal from the target's radiation. They are "amplitude modulated (AM) reticle," "frequency modulated (FM) reticle," "rosette scan," and "cruciform scan." In reticle scanned systems, the detector covers the entire instantaneous FOV of the seeker, and a chopper, or "reticle," modulates the FOV so that small hot areas within the image are preferentially modulated [14]. The phase of the modulation indicates the polar angle, while the amplitude or frequency indicates the off-boresight angle. In AM reticle systems, the reticle is spun to do the chopping, while in FM systems, an optical element (or the entire telescope) is spun to nutate the image on a stationary reticle. In rosette and cruciform scanning, the detector is smaller than and is scanned over the FOV, normally by spinning an optical element.

Hot spot seekers work well in situations where the target is unambiguously differentiated from the background by its intensity and basic dynamic characteristics. Where this is not the case, imaging seekers are necessary to allow more complex processing to distinguish target from background. This more complex situation applies to most antisurface engagements and to more extreme antiair scenarios, for example, those involving ground clutter, complex countermeasures, and low signature targets. Modern imaging seekers can be visualized as any conceivable imaging system with a video tracker processing the output to extract the target position from the image. A well-known and widely used example of an imaging seeker system is the imaging IR variant of the Maverick family [21].

In almost all seeker systems, the optical line of sight is pointed at the target using the tracker output, and the missile follows the seeker according to a control law. This may vary from a simple pursuit trajectory (the missile flies straight toward the target all the time by attempting to null the line of sight angle) through an intercept course (the missile points to the predicted intercept position by attempting to null the line of sight rate) to a more complex pre-programmed trajectory designed to optimize flight characteristics, foil enemy countermeasures or both. Seekers generally have a small

FOV, which can be steered over a large angle. Typical generic seeker parameters are as follows:

Instantaneous detector FOV	0.4–4°
Seeker FOV	2–4°
Seeker field of regard	40°

13.5.3.3 DIRECTED ENERGY

Lasers are used or proposed for a number of directed energy applications:

- Lasers and arc lamps can be used to dazzle or confuse hostile sensors, particularly, missile seekers as part of directed IR countermeasure systems.
- Lasers may be used for sensor damage—there are reports of lasers being used to damage sensors in reconnaissance satellites trying to observe strategic installation.
- Large chemical lasers have been successfully integrated with EO pointing and tracking systems, and these systems have demonstrated an ability to destroy incoming missiles at long ranges—for example, the Airborne Laser Laboratory and THEL programs.

For obvious reasons, there is little open information in this area! Suffice it to say that complex technology integration and detailed analysis, simulation and trials are required to develop effective systems.

13.5.3.4 DAMAGE ASSESSMENT

It is almost as important to the military to know when an engagement has been successful as to carry out the engagement in the first place. Battle damage assessment (BDA) is essential to know whether to continue attacking a target or to switch to another one. There is often a requirement for EO sensors to continue monitoring the target during and after the engagement and to provide imagery or other data, which allows a kill assessment to be made. One aspect of this requirement is for weapon aiming sensors to be able to see through the various EO phenomena produced by the weapon itself.

13.5.3.5 SYSTEM EXAMPLE: VEHICLE SIGHTS

Vehicle sights vary widely in complexity, from simple direct view sights to stabilized multisensor systems. Figure 13.12 illustrates the conceptual layout of a typical high-performance tank fire control sight with thermal imaging, laser, TV, and visual sighting functions.

There is now a trend to lower cost and smaller size, and hence, wider deployment, as imager technology improves and cheapens. A good example of this evolution is shown by a series of products produced by Thales for the British Army.

- The fire control sensor suite for the Challenger 1 tank in the mid-1980s used thermal imager based on the then-new TICM 2 module, mounted in a barbette on the side of the tank's turret, and elevated using a servo system to follow the gun. The Nd:YAG tank laser sight (TLS) originally developed for the Chieftain was mounted on a cradle at the gunner's station and linked mechanically to the gun.
- The Challenger 2 system used essentially the same thermal imager, mounted on the gun barrel to simplify the system and assure best possible boresight to the gun; the TLS was replaced by a high-performance visual gunner's sight with integrated laser rangefinder and two-axis stabilization.
- After the Gulf War, the decision was taken to replace the image intensified sights on the Warrior infantry fighting vehicles with a thermal imager. A low cost second generation thermal imager provides almost as good performance as the TICM 2 system on Challenger in a much smaller and cheaper package; the laser rangefinder is eyesafe; a TV camera is integrated into the commander's sight and displayed on a flat panel display, which also acts as a terminal for the battle management system; a simple low-cost stabilization system provides observation on the move at a fraction of the cost of previous systems; and the whole package is much smaller, lighter, and cheaper than the older high-performance systems.
- For the Scout vehicles that will enter service in the second decade of the twenty-first century, the Warrior system was further developed to provide a panoramic surveillance capability using an azimuth gimbal offering unlimited $n \times 360°$ rotation, a two-axis stabilized head, and a high-performance third-generation thermal imager, with advanced video processing to reduce operator workload and improve orientation and situation awareness (Figure 13.13).

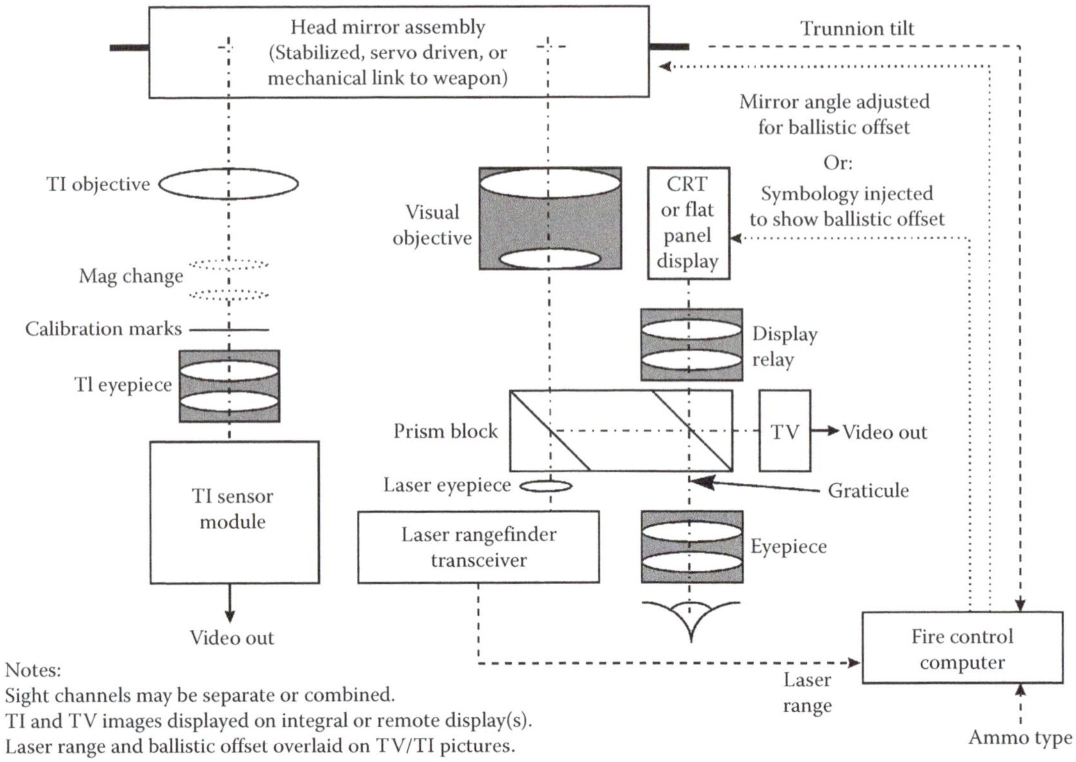

Figure 13.12 Schematic of vehicle weapon sight.

Figure 13.13 "ORION" panoramic vehicle surveillance sight. (Image reproduced with courtesy of Thales UK Ltd.)

13.5.4 Self Defence

13.5.4.1 THREAT WARNING

EO techniques are used for missile and laser threat warning. Missile warning systems depend on detecting the EO emission of the missile motor. Laser warning depends on detecting the direct or scattered energy from the threat laser.

There are two fundamental problems in the design and use of warning systems. The first is to ensure that the detection probability is high and the false alarm rate low, both due to natural phenomena and enemy systems, which may seek to exploit known characteristics of the warning system. A warner that gives too many false alarms will be switched off by the system operators. The second is to link the warning to useful countermeasures against the threat. A warning system is useful only if it increases the probability of surviving the attack and/or successful retaliation.

These issues have exercised technologists since the 1970s at least. Current in-service missile warning systems are usually associated with active IR countermeasure systems on aircraft, using directional energy (lamps or lasers) or flares or both, and work either in the UV or in the mid-IR range. Because of the very short engagement timelines, the linkage between the warning and countermeasure systems is usually automatic. UV warners operating near ground level benefit from a very low background level, since the atmosphere absorbs solar UV before it reaches the earth's surface. By the same token, signal levels are low and the detection range limited, though still operationally useful. In contrast, there is ample signal in the 3–5 μm band, but also high natural and man-made clutter because of sun glint, fires, and other hot sources. A number of spatial, spectral, and temporal techniques are used to classify detections and reject spurious ones before declaring a threat.

Laser warning systems also use a variety of discrimination techniques, including temporal filtering and coherence discrimination, to reject false alarms. Some systems provide accurate direction of arrival information. Deployment of laser warners has been limited because of the "so what?" issue—if the laser warning was going to be followed within a few seconds by a high-velocity shell, what good was the warning? A number of factors are changing this. Improved communication between military vehicles as a result of digitization initiatives will allow threat warning information to be shared quickly between threatened platforms. Until recently, laser rangefinders were normally Nd:YAG, operating at 1.06 μm and therefore a serious eye hazard, and their use was seen as an aggressive act similar to shooting with a gun; but eyesafe lasers will be used much more freely.

13.5.4.2 THREAT SUPPRESSION

Surveillance, target acquisition, and weapon aiming systems aim to help their users by allowing them to detect and engage the enemy. Advantage goes to the side that can complete the OODA loop faster and more consistently. If EO sensors help their user to win, it clearly makes sense to try to reduce their effectiveness, to prevent or delay detection, to make classification more difficult or impossible, to deny the enemy accurate tracking information. The four primary methods for this are camouflage, decoys, jamming, and obscurants.

Camouflage seeks if possible to make the platform look like the background or an innocent natural feature. In World War 2, it was sufficient to break up the visual outline of the platform. With the proliferation of sensors in many different EO bands, camouflage will ideally work simultaneously and consistently in the visible, and near-, mid-, and far-IR bands. This is obviously much more difficult and means that multiband systems are likely to be able to find even a camouflaged target in at least one of their operating bands.

Decoys seek to confuse the enemy sensor by presenting false targets, and either

1. Overloading the system so that the real target is not noticed among the mass of false ones
2. Providing a more attractive target, which is tracked in preference to the real one

Again as sensor systems become more sophisticated, decoys need to be similar to the target in more dimensions simultaneously than used to be the case.

Jamming seeks to overload the sensor so that it is incapable of operating correctly, degrade its sensitivity so that it loses its target, or generate artifacts within the sensor's processing, which have the characteristics of real target signals, effectively producing virtual decoys within the sensor processing. A bizarre early attempt to do this (in WW2) is described in Ref. [8].

Obscurants, most familiarly smoke, seek to block the optical path between the sensor and its target. The principal mechanism by which most smokes works is scatter. Since scatter is wavelength dependent, with shorter wavelengths being scattered more strongly, most smokes which are opaque in the visible range are much less effective, or even completely ineffective, against long-wave IR sensors. This explains the overwhelming preponderance of 8–12 μm systems in NATO (North Atlantic Treaty Organization) armies in spite of the lower volume, aperture, and price of 3–5 μm systems of otherwise similar performance.

13.5.4.3 SYSTEM EXAMPLE: SELF-PROTECTION AND COUNTERMEASURES SYSTEMS

Self-protection and countermeasures systems use EO sensors to detect missile launch or laser illumination, and trigger smoke, decoys, or disruptive modulation using wide angle or directed jamming signals from lasers or arc-lamps. This is a sensitive and rapidly evolving area. For a comprehensive unclassified discussion of the subject, the reader is referred to the *SPIE IR and Electro-Optics Handbook*:

- Laser warning (*IR&EOSH* Vol 7, Chapter 1.7)
- Missile warning (*IR&EOSH* Vol 7, Chapter 1.5)
- Signature suppression (*IR&EOSH* Vol 7, Chapter 3)
- Active infra-red countermeasures (IRCM) (*IR&EOSH* Vol 7, Chapter 3)
- Decoys (IR&EOSH Vol 7, Chapter 4)
- Optical and sensor protection (*IR&EOSH* Vol 7, Chapter 5)
- Smoke (*IR&EOSH* Vol 7, Chapter 6)

13.5.5 Navigation and piloting

There are three main categories of navigation and piloting applications for EO sensors: helping the pilot or driver see where he or she is going, particularly at night; helping to keep track of where the vehicle is relative to the earth by terrain or contour matching; and navigating by the stars.

13.5.5.1 IMAGING: FLYING/DRIVING AIDS

Thermal imagers and image intensified night vision allow military vehicles to move or fly without active emission, such as lights or terrain following radar.

Fixed wide FOV thermal imagers and intensifiers are used as flying aids or drivers' viewers in aircraft and vehicles. In aircraft, the thermal image is often overlaid with accurate registration onto the external scene using a head up display (HUD).

Head-mounted NVG, based on VNIR image intensifiers, are also used, usually as an alternative. But both sensor technologies have advantages and disadvantages. VNIR gives a good and intuitively familiar view of the terrain and horizon under all ambient light conditions down to "overcast starlight." Thermal imaging, on the other hand, is better for detecting most military targets (which are usually hot) and many types of man-made features, and provides a means of seeing the terrain when there is no usable VNIR radiation, as occurred during the Gulf War. The RAF (Royal Air Force) has developed integrated system and procedural solutions for using NVG and head-up thermal imaging together. This gives the benefits of both sensor technologies.

A number of helicopter systems and a few modern fast jets use a head-steered FLIR coupled to a helmet-mounted display to provide all-round thermal imaging slaved to the pilot's line of sight. These systems require low latency in the image, the monitoring of pilot's head position, and control of the thermal imager line of sight. Otherwise, the image will not appear stable relative to the outside world; objectionable swimming effects may develop in the displayed image, which make it difficult for the operators to use the system for any length of time.

There are significant benefits in these applications in sensor fusion, aiming to provide the best of both thermal and VNIR imagery to the pilot. Fully automatic processing is required to reduce pilot workload. Advanced technology demonstrations in this area look promising. The cost of the processing required is coming down rapidly and such systems should soon be affordable for the more high-value applications. This illustrates the way that EO system design is now constrained more by data-handling and bandwidth than by the basic sensor technologies.

13.5.5.2 TERRAIN MATCHING

Cruise missiles use a number of techniques to navigate very precisely over the terrain. Conventional inertial navigation techniques do not have the required accuracy, having a drift of typically 1 nautical mile per hour. While the advent of global positioning system (GPS) may have reduced the need for other methods, video scene matching and laser altimeter techniques have both been proposed to allow missiles to compare the terrain they are flying over with a stored three-dimensional terrain map of the planned route.

Stabilized EO systems such as surveillance turrets and laser designator pods fitted to an aircraft can measure the aircraft's position relative to an identifiable feature on the ground. If the feature's coordinates are known, this information can be used to update the aircraft navigation system, allowing the accumulated drift in the estimated position to be nulled out.

13.5.5.3 STAR TRACKING

The US and Soviet sea-launched ballistic missile programs in the 1950s and 1960s wrestled with the problem of hitting a fixed point on earth several thousand miles away from a moving submarine. Part of the solution was an incredibly accurate gyro system on the submarine itself. Another part was the use of star trackers in the ballistic missiles. Star trackers are commonly used in satellites to provide an error signal to the satellite's attitude control and navigation system to help maintain a precise attitude and position estimate. In the ballistic missile application, the star tracker takes a fix on one or more stars, allowing errors in the launch position estimate to be determined and compensated for. The American Trident system and the Soviet SS-N-8 and SS-N-18 all used different "stellar-inertial guidance" methods. The Soviets deployed this technology before the Americans [4].

13.5.5.4 OBSTACLE AVOIDANCE

Obstacles such as pylons and cables present a serious hazard to low-flying helicopters and aircraft. Obstacle databases can be used to generate warning symbols on head-up displays to indicate known permanent obstacles. Passive sensors will help the pilot to spot temporary and smaller features. Neither method can warn against thin temporary wires and cables.

Obstacle warning laser radar systems have been developed to fill this gap, and have been demonstrated for both helicopters and fixed wing aircraft. Helicopters fly relatively slowly (30–100 ms^{-1}) and fairly short-range systems will give sufficient warning for pilots to take appropriate avoiding action. Practical systems are available using diode lasers with obstacle detection ranges of a few 100 m. Fast jets need longer range warning and faster scanning. Demonstrator systems have been built using CO_2 lasers with coherent detection. These work, but are currently too large to integrate easily into fast jets.

13.5.6 Training

Effective and realistic training is key to the effectiveness of modern military forces. If they cannot train in realistic combat conditions, they will not be effective when they first experience real combat. But if the training is too realistic, for example, with a lot of live ammunition, not all the trainees would survive the experience!

Hence, there is a need to make training as realistic as possible without using live ammunition. Traditionally, "umpires" would move around training areas, making fairly arbitrary judgments about whether troops would have been hit. This was subjective and unrealistic. Similarly, blank ammunition produces noise and smoke, but the people doing the shooting get no feedback as to whether they pointed their weapons in the right direction. And if their equipment includes other hazardous systems, such as non-eyesafe lasers, which cannot be used in a training environment, they are unlikely to use them effectively under the stress of real war.

The need for effective training created a whole new class of EO systems, the "direct fire weapon effect simulator." A low power laser diode transmits a coded pulse train, which if aimed correctly is detected by a detector on the target (or sometimes is returned to the transmitter by a retroreflector on the target and detected by a coaligned receiver). The coding indicates the type of weapon being simulated, the divergence can be adjusted to simulate area effects, and round-trip systems can measure distance to the target and therefore calculate whether the system was within effective range. Modern training systems inject video symbology into the sights of the attacking system to simulate weapon trajectories, tracer, and impact signatures; can electronically disable the weapons of the target if it has been "killed'; and can keep a log of time and position of every engagement so that the engagement can be reconstructed in a virtual reality replay for debriefing and postoperation analysis. Hence, laser training systems can be used to prove the effectiveness (or otherwise) of new doctrine and tactics as well as to train soldiers. The systems contain detailed weapon effect and target information to ensure that engagements are correctly registered (e.g., a rifle cannot kill a tank).

These systems have immense entertainment value, and lightweight short range versions have become popular in the last few years in entertainment arcades!

Another class of training system is the crew training simulator, where very wide FOV video systems are used to generate a synthetic environment as seen from an aircraft cockpit or vehicle, which itself is on a six-axis motion platform. The combination of physical motion and visual cues from the projected video creates a remarkably real istic experience. These systems save users massive amounts of money by avoiding the need to use real aircraft for many training tasks. Again, derivatives of these systems are now well established in the entertainment industry.

Training requirements reflect back on the specification of tactical military systems. For example, laser rangefinders are often required to be "eyesafe," so that they can be operated in accordance with operational drills in a training context; and similarly, modern fire control systems are required to be compatible with laser training systems and sometimes to incorporate various embedded training capabilities.

13.5.7 Other applications

13.5.7.1 COVERT COMMUNICATION

Optical methods have been used for line of sight communication since the discovery of the heliograph, probably in prehistoric times. Signal flags were used by navies in the olden days when sailing ships traveled the world. Once electricity was invented, signaling lamps allowed effective signaling at night with Morse code while maintaining radio silence.

During World War 2, there was a need for covert line of sight signaling, particularly for special operations. IR filters and image converters (the precursors of modern image intensifier tubes) were evolved for this purpose by the United Kingdom, Germany, and United States . Nowadays similar principles can still be used. CO_2 lasers can be used as beacons in the 8–12 μm region, near-IR lasers are often fitted to weapons and surveillance turrets and used as "laser pointers" to indicate targets to soldiers equipped with night vision goggles, and optical point-to-point data links have been developed that allow high-bandwidth data to be sent over any clear line of sight—in the extreme case, between low earth orbit and geostationary satellites.

13.5.7.2 INFANTRY DEVICES

Infantry soldiers use hand-held, weapon-mounted, and more recently, helmet-mounted EO sensors for observation, night vision, and weapon aiming. These have to be simple, compact, low power, and very robust against mishandling.

A number of countries, including the United States, France, the United Kingdom, Canada, Australia, and others, are now running "future soldier" programs that aim to integrate EO sensors, navigation, communications, command and control, and protection functions into a "system" suitable and affordable for use by infantry. The US Land Warrior program, the British Future Infantry System Technology, and the French "Felin" have all demonstrated hardware and at the time of writing, are moving toward full development and production.

13.5.7.3 ALL-ROUND VISION SYSTEMS

All-round vision systems for military vehicles are an emerging market given that low-cost sensitive sensors and flat panel displays are now available. The trend to smaller more compact platforms is creating a demand from designers to remove the vehicle design constraints (such as manned turrets in vehicles, bubble cockpit canopies in aircraft) imposed by the crew's expectation of good all-round vision. The fear of possible future laser threats is creating a demand from operators for better situational awareness from indirect sensors and when vehicle hatches are closed down.

The DAIRS system for the F-35 Joint Strike Fighter (JSF) reportedly uses six 1000 x 100 elements thermal imaging arrays to provide all-round vision for the pilot and to provide data for threat warning and target detection systems. Similar systems for land vehicles entered service in the first decade of the twenty-first century using low-light VNIR and uncooled thermal imaging sensors.

13.5.7.4 IDENTIFYING FRIEND OR FOE

Identifying friend or foe is a complex and difficult problem. Western military organizations in particular are increasingly expected to operate in coalitions with many different nationalities, with different types of equipment used on the same side—and sometimes, the same kind of equipment used on opposing sides. Better visual identification makes a very important contribution to reducing "friendly fire" incidents but cannot prevent them completely.

Many procedural and technical solutions have been tried over the years. In World War 2, distinctive black and white stripes were painted on all allied aircraft just before the D-Day invasion (of Normandy) to allow "instant" recognition of a friend without need to positively identify the aircraft type. On a similar principle, active beacons matched to EO sensors are often used nowadays to give a distinct signature. These methods are very effective in specific operations, but do not provide a permanent solution, since over time they can be copied.

Laser interrogation is used as an operating principle in some modern combat identification systems.

13.6 OPERATIONAL IMPACT

Thermal imaging and night vision can "turn night into day"—allowing forces to operate 24h a day instead of 12, without having to use special tactics for night operation; or indeed to work preferentially at night to exploit the superiority resulting from thermal imaging and night vision technology. The US/allies' EO capabilities provided a huge advantage in the ground battle during the 1991 Gulf War.

Laser designation allows precision targets to be hit (fairly) reliably—about 30% missed in the 1993 Gulf War. There is no evidence that 100% accuracy can be achieved in operational systems; too many environmental and human factors influence the operational outcome. But precision weapons give a massive reduction in circular error probability (CEP), resulting in much lower mission costs and much less collateral damage to achieve a given objective. According to a quote in a web download,

> On May 10, 1972 the first major combat use of these new weapons, along with TV guided Walleye bombs, took place (in Vietnam) resulting in the destruction of the Paul Doumer bridge, which had withstood 3 years of aerial bombardment, totalling 1250 tons of munitions. This single mission was a revolution in air-to-ground warfare.

The "1250 tons of munitions" had all landed somewhere near the bridge, devastating the surrounding area, and probably involving around 1000 operational sorties. We shall not attempt to preempt readers' own judgment about the environmental and human costs and moral issues. But it is clear that there is a very large financial payback from the use of precision weapons in terms of reduced cost of wasted ammunition, reduced cost of aviation fuel, and reduced combat losses. Similarly, laser rangefinders give a direct cost saving to the user calculable from the reduced number of rounds required to achieve a given number of hits, and the consequent savings throughout the logistics chain. Other EO technologies are less widely deployed—because the return on investment for the user is less clear and less easy to calculate.

These "value for money" arguments will assure the place of EO in military inventories for the foreseeable future. Equally, interesting technologies that cannot provide similarly robust financial justification will have a much harder time gaining interest, let alone acceptance into service, from the military.

13.7 TRENDS

Performance of EO systems is reaching operationally useful limits in many areas—hence, emphasis is switching from the "performance at any cost" focus of the Cold War years.

There is a strong drive to lower cost to allow wider deployment—for example, thermal imaging is migrating from tanks (1985–1990 in the United Kingdom) to all combat vehicles (2000–2010 in the United Kingdom) to the individual infantryman and unmanned vehicles (2000 onward). This trend is aided by and is accelerating technology developments that allow simpler and cheaper systems to be developed, using staring and often uncooled IR detectors.

There is a strong drive to integrate with other systems in order to reduce workload (Battle Management Systems, "soldier systems"); to see "over the hill," using UAVs at all levels of command; and to large-scale system integration throughout the "battlespace" to improve "tempo" or speed of operations, under the banner of "C4ISR"—command, communication, control, computing, information, surveillance, and reconnaissance. EO sensor systems (aided by their operators) now have to produce "information" not just "images."

The increasing use of unmanned vehicles has increased the demand for lower cost light-weight EO sensor systems and for intelligent sensor integration and processing on remote vehicles to reduce the dependency on high data link bandwidth between remote vehicles and operators by automating some of the operator's traditional functions.

Some new areas of R&D now receiving much attention are the following:

1. High-power solid state fiber lasers for directed energy
2. Mid-IR lasers for IRCM

3. Three-dimensional imaging laser radar for target identification
4. Hyper-spectral imaging
5. Mine detection using thermal imagers and active imaging

Finally, commercial technologies are moving faster than military ones in the areas of telecoms, image sensor, and processing; the military are now exploiting commercial technology instead of the reverse. This may lead to the widespread use of ultralow cost disposable EO systems on miniature autonomous vehicles and aircraft.

ACKNOWLEDGMENTS

Thanks are due to Steve McGeoch of Thales UK, who contributed his knowledge of the current state of the art for the second edition.

REFERENCES

1. Green, R. L. (1958). *Tales of the Greek Heros*. London: Penguin.
2. Wells, H. G. (1898). *War of the Worlds*.
3. Jones, R. V.(1978). *Most Secret War*. London: Hamish Hamilton.
4. Mackenzie, D. (1990). *Inventing Accuracy*. Cambridge: MIT.
5. Chaisson, E. J. (1994). *The Hubble Wars*. New York: HarperCollins.
6. Albedos. http://www.drev.dnd.ca/poolpdf/e/61_e.pdf, 2002.
7. Burns, R. W. (1993). Early history of the proximity Fuze (1937–1940). *IEE Proc.* A 140: 224–236.
8. Reid, B. H. (1983). The attack by illumination: the strange case of Canal Defence Lights. *RUSIJ*. 128: 44–49.
9. Lloyd, J. M. (1975). *Thermal Imaging Systems*. New York: Plenum.
10. Moss, M. and Russell, I. (1998). *Range and Vision (the first hundred years of Barr & Stroud)*. Edinburgh: Mainstream Publishing.
11. Netzer, Y. (1982). Line-of-sight steering and stabilization. *Opt. Eng.* 21 (1): 210196 (reviews various methods of line of sight stabilization and their suitability—fundamental techniques have not changed though improvements in gyro components and processing availability have changed some of the trade-offs!). doi:10.1117/12.7972866.
12. Godfrey, T. E. and Clark, W. M. (1981). Boresighting of airborne laser designator systems. *Opt. Eng.* 20 (6): 206854 (A useful historical survey well describing many of the engineering issues). doi:10.1117/12.7972826.
13. Jamieson, T. H. (1984). Channel interaction in an optical fire control sight. *Opt. Eng.* 23. (3): 233331. doi: 10.1117/12.7973289.
14. Wolfe, W. L. (1965). *Handbook of Military Infra-red Technology*. 2nd edn. Washington: Office of Naval Research, pp. 645-660.
15. Kopp, C. (1994). The sidewinder story—the evolution of the AIM-9 missile. *Australian Aviation*. http://www.ausairpower.net/TE-Sidewinder-94.html.
16. Ward, R. D. (1973). *Crossed Porro Resonator patent* ca 1974. UK Patent GB 1358023.
17. Gibson, T. Night vision information. http://www.arctic-1.com/works2.htm.
18. McEwan, K. Uncooled IR systems various refs 1990–2002.
19. Pike, J. (2005). Walleye guided bomb description. http://www.intellnet.org/documents/100/060/165.htm.
20. Rogers, M. E. (1997). USAF: Lasers in space—Technological options for enhancing US military capabilities. Occasional Paper No. 2 (Center for Strategy and Technology, Air War College, Maxwell Air Force Base, Alabama).
21. US Navy. Maverick missile description. http://pma205.navair.navy.mil/pma2053 h/products/maverick/index.html.
22. Sofer, B. H. and Lynch, D. K. (1999). Some paradoxes, errors, and resolutions concerning the spectral optimization of human vision. *Am. J. Phys.* 67: 946–953.
23. Bobbitt, P. (2002). *The Shield of Achilles* footnote P. 204: reference to John Lynn, camouflage, *the reader's companion to military history, London*, p. 68.
24. Sillitto, R. M. (2002). Private conversations with author.
25. Vukobratovich, D. (2002). *Optomechanical Design Course Notes* SIRA, various years, CAPT.

26. Yoder, P. (1992). *Optomechanical Systems Design, M Dekker, New York.*
27. Mitchell, C. (1970). *Having been a Soldier.* London: Mayflower Paperbacks.
28. Osinga, F. (2007). *Science, Strategy and War, the strategic theory of John Boyd,* Routledge collates Boyd's various contributions and presents them in an orderly way.
29. Sillitto, H. (2014). *Architecting Systems: Concepts, principles and Practice.* College Publications Systems Series Vol. 6, see particularly Chapter 23.

PART VI

Industrial applications

We shall now present case study applications of optoelectronics in industry, again preceding these more specific items with another summary table. In this area, there are many applications, but cost is usually a critical issue. Yet again, the ubiquitous video camera plays a major part, but there is also far greater potential applications for sensors which measure at discrete points. Spectroscopy is a particularly important area, as are sensors for metrology.

Again, before going on to describe case studies, there is a need to do some cross-referencing of the methods mentioned here, as many are applicable in other areas. Perhaps the most generic sets of techniques are those used in video cameras and in spectroscopic instruments, as these apply to many other sections. Regarding the case studies presented, the other main cross referencing is needed for:

1. The types of fiber sensors used for gas turbine monitoring are also extensively used for electrical power generation.
2. The optical gas sensors are used for the mineral resources industry because flammable (e.g., methane) and toxic (e.g., hydrogen sulfide) gases are also a huge problem in the oil, gas, and mining areas.
3. Solar power generation. Apart from inclusion in the summary table, this is not discussed in this part, as the basics were covered in the chapter "optical to electrical energy conversion: solar cells" in Volume II. The technology for microgeneration for the home will also be discussed in Chapter 30, as this is the most widely used scenario Solar technology is, of course, applicable to nearly all of the sections in Volume III, particularly when there is a need to power remote equipment.

Table VI.1 Summary of applications of optoelectronics in industry

Application	Technology	Advantages	Disadvantages	Current situation (at time of writing)	More reading
Spectroscopic sensors (transmission).	Measurement of direct or diffuse transmission of light through samples. Usually taken at many optical wavelengths. This is one of the main technologies used for industrial gas sensing, mostly with infrared sources and detectors.	Established technology, with possible real-time information for process control. Good signal/noise ratio, unless absorption is very weak. A wide range of solid, liquid, and gaseous chemicals can be measured. Can have remote sensor heads, if optical fibers are used. Solid-state detector array technology enables some low-cost spectrometers for visible region.	In severe environments, optical surfaces can become scratched, etched, or contaminated. Not applicable to strongly absorbing materials, if no light gets through!	Very widely used, particularly in chemical, petrochemical, and pharmaceutical industries. Most measurements are still done on samples in the laboratory, but real-time online monitoring is becoming more common, using fiber-optic links.	See Volume II, Chapter 15 (Spectroscopic Analysis) and Part VII (Oil and Gas).
Spectroscopic sensors (emission, reflection, or frustrated internal reflection [FIR]).	Emission of infrared light is used in industrial pyrometers for surface temperature measurement. Measurement of reflection of light sample or from an internal glass surface (usually part of a prism) in contact with samples.	Pyrometers are valuable noncontact temperature measuring devices. Surface reflection spectroscopy is simple to do with camera technology. FIR is very useful for highly absorbing samples, as the changes in reflective spectrum are higher with these, so the method is complimentary to transmission spectrometry.	Optical surfaces can become scratched, etched, or contaminated. Not applicable to weakly absorbing materials.	Pyrometers are widely used in the iron and steel industry, and many others involving high temperatures (e.g., kilns). Reflection spectroscopy is used in many process industries (e.g., food industry).	See Volume II, Chapter 15 (Spectroscopic Analysis) and Part VII (Oil and Gas).

(*Continued*)

Table VI.1 (*Continued*) Summary of applications of optoelectronics in industry

Application	Technology	Advantages	Disadvantages	Current situation (at time of writing)	More reading
Solar power.	Local electricity microgeneration using photovoltaic solar panels to provide power for factories. Direct heating using thermal solar collectors is also used, but this does not involve optoelectronics directly.	Cost savings are currently possible with the aid of subsidies, but this should soon be the most cost-effective way to provide power for a factory. Energy storage methods are needed to cover dark periods. Already fully viable for providing power for many remote applications.	Current cost is still high in countries with lower sunshine levels. Economic means of storing electrical energy is a key need to make the technology fully viable.	Already cost-effective without subsidies in many hot countries. Already fully viable for providing moderate power levels in remote applications.	See Volume II, Chapter 16 (Optical to Electrical Energy Conversion).
Spectroscopic sensors (Raman scattering).	Measures the weak Raman scattering from samples illuminated with laser light.	Can be used with samples, which scatter light very strongly (e.g., turbid liquids, powders, pharmaceutical pills). The use of visible or near infrared light means that conventional glass optical materials are suitable. The method can, therefore, be combined with microscopy, and optical fiber leads are also possible.	The scattered light generated is very weak, particularly if it has to be collected using optical fiber leads. Powerful semiconductor and solid-state lasers are leading to lower cost, more compact Raman systems.	Now used in many process industries (e.g., pharmaceuticals, oil, and gas, food industry). High-power semiconductor lasers are making fiber-remoted Raman more practical.	See Volume II, Chapter 15 (Spectroscopic Analysis) and Part VII (Oil and Gas).

(*Continued*)

Table VI.1 (*Continued*) Summary of applications of optoelectronics in industry

Application	Technology	Advantages	Disadvantages	Current situation (at time of writing)	More reading
Industrial lighting.	Originally incandescent sources, but now nearly all industrial lighting is being replaced by arrays of high-brightness LEDs.	LEDs have had excellent reliability for many years, current LEDs are far brighter and more efficient than earlier types and unit costs are falling rapidly.	No real disadvantages.	Unless a new revolutionary technology arrives, they are here to stay!	See Volume I, Chapter 10 (LEDs) and Volume II, Part II chapters on display technology.
Camera observation of process lines, conveyor belts, and industrial products, etc.	Usually, using cheap visible charge coupled device cameras. However, I–R cameras can be used.	The capability of such systems can be extended dramatically using sophisticated video processing and pattern-recognition technology.	Mainly limited to line-of-sight.	Very widely used in many industries.	See Volume II, Chapter 4 (Camera Technology).
Industrial robotics.	Monitoring and control of industrial robotics, using cameras, proximity sensors, pyrometers, optical metrology. Flame sensing and material surface sensing systems for industrial welding. Optical fibers are being used in grip sensors.	Intelligent sensing systems are an essential feature of robotics. Optical systems for distance measuring and many other metrology requirements are well developed. Cameras with smart video processing are a powerful tool for object recognition and measurement.	More difficult to monitor interior features, as a line of sight is needed for most measurements.	Robotics is one of the fastest growing areas of the industrial scene, as robots become more capable of complex tasks.	See this section for metrology systems.

(*Continued*)

Table VI.1 (*Continued*) Summary of applications of optoelectronics in industry

Application	Technology	Advantages	Disadvantages	Current situation (at time of writing)	More reading
Laser Doppler velocimetry (LDV) and photon correlation spectroscopy (PCS). Optical particle sizing.	LDV and PCS involve measurement of movement of solid objects or particles, using coherent laser sources. The basic process involves interference between reflections, either from the moving object or particle and a reference reflection from a stationary object (LDV) or from other moving particles (PCS). Other particle sensing systems utilize the angular light scattering distribution (e.g., Mie scattering) or simply monitor "turbidity", the measured attenuation from beam obstruction.	*LDV* is a useful tool for monitoring: Moving engine parts such as pistons, turbine blades, etc. Air or gas flows, for example, in wind tunnels and turbines. *PCS* is a very useful tool for monitoring particle sizes, particularly of monodisperse suspensions such as paints. Other methods are also widely used for larger particles. Fiber optics can also be used with all of these, allowing inspection away from the main optical equipment.	Need line of sight path for most methods.	These methods are well established in industry.	See this section.

(*Continued*)

Table VI.1 (*Continued*) Summary of applications of optoelectronics in industry

Application	Technology	Advantages	Disadvantages	Current situation (at time of writing)	More reading
Other optical sensing technology for industrial monitoring.	Numerous interferometric and time-of-flight sensors (distance) sensing are used for metrology. Widespread use of optical proximity detectors, using fiber cone, focal plane, or crossed-beam methods.	Real-time observation is possible. Interferometric sensors can give remarkable precision of order 10^{-6} of a wavelength. A wide range of laser-based straight-line, plumb-line, level, inclination gauges, and optical time of flight sensors are available. These are used for measurements in the manufacturing, building, and construction industries.	Mainly limited to line-of-sight.	Widely used in many industries. Industrial use is increasing strongly, as cost of these sensors reduces.	See this section.
Optical fiber sensors.	Many different types of sensor have been developed, e.g., temperature, strain, pressure, etc.	Freedom from interference in electrically noisy environments. Can be designed to measure many points over long lengths of fiber. Can be designed to withstand very high temperatures.	The need for low cost and reluctance to use optical technology is still holding back many discrete (point) sensors for industrial applications, despite their widespread use in many other areas such as oil and gas, naval sensing, and medical applications.	Optical fiber distributed temperature sensors are well established, despite being expensive. Optical fiber point sensors for strain temperature and pressure are becoming accepted for niche areas, particularly ones using in-fiber gratings.	See optical fiber sensors chapter.

(*Continued*)

Table VI.1 (*Continued*) Summary of applications of optoelectronics in industry

Application	Technology	Advantages	Disadvantages	Current situation (at time of writing)	More reading
Industrial laser machining and surface treatment.	Cutting, welding, and surface treatments using high-power lasers.	Modern high-power lasers can cut more cleanly and precisely than most other methods. Laser surface treatment, including melting and ablation, can be applied in a very controlled manner.	Higher initial costs, but this is falling very rapidly with time.	Again, unless a new revolutionary technology arrives, laser machining is here to stay!	See this section and Volume I, Chapter 6 (Lasers) and Chapter 11 (Semiconductor Lasers).
3D printing.	Production of 3-D objects using one of following methods: (1) Curing of polymers using a digital projector to give a 2-D pattern of curing light. (2) As above, but using a scanned laser (or fixed laser, with scanned stage holding the object) to produce a similar 2-D light pattern. The higher power allows faster curing or even sintering of powders to produce solid objects.These will also usually involve a programmable position manipulator and material-depositing nozzle. Not all deposition techniques involve optics, but the digital data to produce the object will often be derived by 3D scanning of a real master object.	This is one of the most "disruptive" of manufacturing. technologies! Almost any type of material (metal alloys, ceramics, glasses, and polymers) can be manufactured in this way, to produce items ranging from high-performance aircraft turbine blades to plastic toys. The ability to scan "master" objects (or create them by computer), store dimensions as digital data, and then produce replicas has huge advantages. The data can be transmitted to wherever it is desired, hence enabling objects to be created anyway in the world at the push of a button.	High initial cost of precision equipment, but huge potential labor saving.	This technique is being developed across all industries, including even house building. Optoelectronics can also be used in cameras for monitoring the process as it develops.	http://www.sciencedirect.com/science/article/pii/S1875389212026259

14

Optical gas-sensing methods for industry

JANE HODGKINSON
Cranfield University

JOHN P. DAKIN
University of Southampton

14.1 INTRODUCTION

In this chapter, we discuss the potential of optical gas sensing in industry. The emphasis here will be to summarize the major types of sensors used in industry, their advantages, and their potential applications, with a brief discussion of the basic concepts behind them.

When considering optical methods of detection, it is clear that most simple gases are invisible to the human eye, with the exception of a few, such as chlorine or nitrogen dioxide, and even these gases are only visible at high concentration. Most gases are therefore only detectable optically using other spectral regions or by indirect means.

1. By optical absorption spectroscopy, using the ultraviolet, visible, or infrared (IR) regions of the spectrum—many heteronuclear molecules exhibit absorption at characteristic wavelengths or spectral bands, and this can form the basis of their detection.
2. Using some form of chemical reaction, where a spectral change occurs in a solid or liquid

"indicator" material when the gas interacts with it—the gas is therefore detected indirectly, using changes in the spectral properties of the indicator material.

For industrial applications, a notable exception is nitrogen, which being homonuclear, has only negligibly weak absorption, and being chemically rather inert, exhibits minimal interaction with chemical indicators. However, it is an important process gas that must be determined in many applications, especially in the petrochemical sector. This gas is often achieved by nonoptical mass balance methods, by first measuring everything else likely to be present, and then assuming that nitrogen makes up the remainder of the mass.

14.1.1 Advantages of optical gas detection

It is unfortunate that even the most cost-efficient optical sensors are still often several times more expensive than other sensor types. Industrial markets are highly sensitive to cost. It is therefore important to emphasize the other attractions of optical methods.

Quantitative detection of gases is traditionally dominated by laboratory analytical equipment such as gas chromatographs. These offer low (ppb—parts per billion) sensitivity. They also take time to sample the gas and, being large expensive laboratory instruments, are usually remotely located. A common alternative for *in situ* measurement is to use small ultra-low-cost devices such as pellistors, or semiconductor and electrochemical gas sensors. The latter use electronic transduction of an interaction between a gas and an appropriately chosen material. A pellistor consists of a catalyst bead in a tiny metal-gauze safety cage, which induces combustion of a gas and changes its resistance in response to the heat of combustion [1]. Pellistors are often used in flammable gas detection close to the lower explosive limit, but they cannot distinguish between many types of flammable gas, typically suffer from a zero drift of hundreds of parts per million (ppm), and often require frequent (e.g., weekly) calibration. They can also be chemically poisoned by some airborne impurities. Semiconductor gas sensors are sensitive at low ppm level [2], but they can also drift and cross-respond to other gases or changing humidity. Electrochemical gas sensors can be relatively specific to individual gases and sensitive at ppm or ppb levels [3], but they have limited lifetimes and also suffer from some known cross-response issues, for example, to humidity.

In contrast, gas sensors based on optical absorption offer fast responses (time constants below 1s are possible), minimal drift, and high gas specificity (close to zero cross-response to other gases) as long as their design is carefully considered. Measurements can be made in real time and those based on absorption are essentially noncontact, so they do not perturb the sample, which can be important in process control [4]. Because the transduction method makes a direct measurement of a molecule's physical properties (its absorption at a specific wavelength), drift is reduced and, provided the incident light intensity can be determined, measurements are self-referenced, making them inherently reliable [5]. In this way, optical gas sensing fills an important gap between lower cost sensors with inferior performance and high-end laboratory equipment.

Sensors based on transduction of a signal via an optical indicator have characteristics falling between these examples. The chemistry of the optical indicator is critical to the sensor's success, as it determines the specificity to the target gas, cross-sensitivity to temperature and humidity, and often the response time (since the optical measurement need not limit the response). The sensor recovery time must also be considered, as many indicator reactions are irreversible, or only slowly reversible once the target gas is no longer present.

14.2 DIRECT DETECTION OF GASES

Direct optical detection of gases offers the potential for a rapid, almost instantaneous response, typically determined by the integration time constant of detectors needed to provide a good signal-to-noise ratio. This allows leaks or pollution to be detected more rapidly than with many other sensor types and helps in rapid determination of the dynamics of processes such as combustion. Use of collimated optical beams can also enable measurements to be taken over extended optical paths. In addition, direct optical detection allows quantitative estimates of concentration, which are not always possible using indicator chemistry.

The most common method of direct gas detection is via absorption spectroscopy, where the transmission loss of light during passage through a gas sample is measured.

Optical gas detection using absorption spectroscopy is governed by the well-known Beer–Lambert law [6],

$$I = I_0 \exp(-\alpha\ell), \quad (14.1)$$

where I is the light transmitted through the gas cell when gas is present, I_0 is the light transmitted through the gas cell in the absence of gas, α is the absorption coefficient of the sample (cm^{-1}), and ℓ is the cell's optical path length (cm). The absorption coefficient α is the product of the gas concentration (e.g., in atm of partial pressure) and the specific absorptivity of the gas ε ($cm^{-1}atm^{-1}$).

Note that compared to its common use in liquid phase samples, for gas sensing, the Beer–Lambert law is often described using log base e rather than base 10, and therefore values of α are 2.3 times larger despite having the same apparent units. The consequence is that users of spectral data should always familiarize themselves with the underlying basis of the information that they are using.

Unfortunately, the absorption coefficient of gas molecules is usually quite low compared to that of liquids or solids, primarily owing to a lesser density of molecules occupying the sample. Even with extended optical paths, detectors must often be able to determine very small changes in spectral properties. Thus, sophisticated instrumentation is often required to detect low concentration levels of gas in the parts per million (ppm) range.

Fortunately, for absorption-based sensors, many gases have reasonable vibrational absorption coefficients in the IR region of the spectrum, as shown in Figure 14.1. However, for simple gas molecules, the highest absorption is usually only present in closely spaced sets of narrow rotational–vibrational spectral lines. This fine-line structure means that broadband sources see a relatively low average absorption level.

However, light from narrowband sources, such as lasers, undergoes a far greater absorption, provided the laser emission wavelength is tuned to match (or scan through) one or more of the lines. Hence, lasers tend to be used for the lowest concentration measurements. Lower cost broadband sources, such as incandescent lamps or LEDs (light emitting diodes) can still be used to measure concentrations of gas down to ppm levels if care is take in the design.

The characteristic wavelengths at which the gas absorbs can often be used to identify the gas species. Clearly, provided two gases do not coincidentally absorb on the same narrow line, greater selectivity can be obtained by measuring the absorption of an individual gas absorption line, but measuring on several narrow lines can nearly always overcome this possibility. When using broader band sources, practitioners must be particularly careful to ensure that cross-sensitivity to other potential interferent

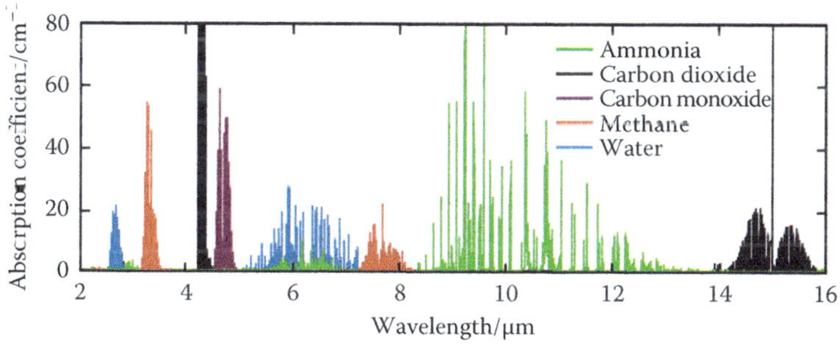

Figure 14.1 Absorption spectra of five gases in the mid-IR region of the spectrum (all at 100% vol.), taken from the PNNL database [7].

species is limited by having no spectral overlap with the target species.

Examples of absorption spectra are shown for some industrially important and common gases across the mid-IR region in Figure 14.1. A more detailed list for the UV-visible–IR regions is given in Table 14.1.

14.2.1 Broadband absorption detectors

Broadband sources such as incandescent lamps and LEDS are low cost, compact, and usually highly reliable light sources for gas detection. Except at high concentrations, broadband detection requires

Table 14.1 Examples of measurement wavelengths for various gases

Gas	Wavelength	Example references
Ammonia (NH_3)	200–230 nm	[8]
	1.51–1.53 μm	[9]
	2.0–2.3 μm	[10]
	9–12 μm	[11]
Benzene (C_6H_6)	230–260 nm	[12]
	5.1 μm	[13]
	14.8 μm	[14]
Carbon dioxide (CO_2)	1.53–1.60 μm	[15]
	2 μm	[10]
	4.2 μm	[16]
Carbon monoxide (CO)	1.56–1.58 μm	[15]
	2.3 μm	[17]
	4.4–4.9 μm	[18]
Ethane (C_2H_6)	1.68 μm	[19]
	3.3 μm	[20]
Formaldehyde (H_2CO)	2.3 μm	[21]
	3.5 μm	[21]
Hydrogen sulfide (H_2S)	190–230 nm	[12]
	1.57–1.59 μm	[22]
Methane (CH_4)	1.65–1.68 μm	[19]
	2.3 μm	[23]
	3.2–3.4 μm	[24]
	7.3–8.0 μm	[25]
Nitric oxide (NO)	190–230 nm	[12]
	1.81 μm	[26]
	5.2–5.4 μm	[27]
Nitrous oxide (N_2O)	4.5 μm	[25]
	5.4 μm	[28]
	7.8–8.0 μm	[25]
Nitrogen dioxide (NO_2)	400–450 nm	[29]
	630–640 nm	[30]
	6.0–6.3 μm	[31]
Sulfur dioxide (SO_2)	190-nm	[12]
	290–310 nm	[32]
	7.5 μm	[33]
Water vapor (H_2O)	1.39–1.40 μm	[34]
	6.7 μm	[35]

relatively large absorption coefficients and is therefore most often used in the mid-IR and ultraviolet regions of the spectrum, where it may be termed as nondispersive IR (NDIR) or ultraviolet (NDUV) detection.

In recent years, commercial LEDs have emerged to cover many regions of the IR spectrum where gases often have their fundamental vibrational absorption region (see Figure 14.1). The most common arrangement is to direct the source emission through a gas detection region, and then detect it with a suitable IR detector. The same, or similar, semiconductor technology used for the LEDs can be adapted to form suitable PN junction devices to detect the transmitted light; broader-band photoconductive, pyroelectric or thermopile detectors can also be used instead. A bandpass filter is often used to provide a wavelength selective element to pass light in a band matching the absorption spectrum. Recent developments of incandescent lamps include the use of micro hotplates with microstructured surfaces, the latter providing a wavelength selective element such that the device emits within a narrow wavelength band.

The simplest detectors for high gas concentrations have just a simple line-of-sight path between source and detector, as shown in Figure 14.2. However, for more sensitive detection, there is an advantage in extending the length of the optical path through the gas. There are many ways of arranging this, including a simple folded path using plane mirrors, more sophisticated folded paths using concave mirrors, and multiple diffusely-scattered paths inside integrating spheres (hollow chambers lined with a highly reflective material, which scatters incident light diffusely in all directions, such that the average path lengths of light, before emerging from a small exit hole onto the detector, is many times the diameter of the chamber). The advantage of the simple line-of-sight arrangement is that gas can enter the path relatively easily compared with some of the other arrangements.

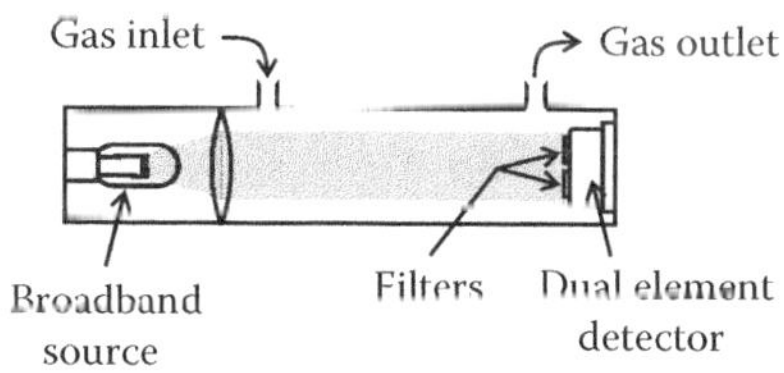

Figure 14.2 Typical arrangement for a single line-of-sight gas sensor using measurement of broadband absorption in two wavelength bands defined by bandpass filters.

There are a number of fundamental noise issues and other practical issues that must be taken care of in real instruments. Two particular issues are discussed as follows:

1. Variations in temperature can cause changes in light output of the source and responsivity of the detectors. Additionally, light sources suffer noise from flicker effects, photodiode detector receivers suffer from fundamental photon noise and additional amplifier noise, and many thermal detectors (e.g., pyroelectric) may respond to background acoustic noise. When measuring small changes in transmitted signals, it may also be necessary to employ some form of intensity referencing by comparing a measurement channel with a reference channel. This referencing can take several forms:
   - A two-path system, whereby part of the light from the source travels through the gas-containing region and another part of the light travels through a gas-free reference path to a matched detector.
   - A two wavelength system, whereby two detectors are used with bandpass filters defining the wavelength regions, the first being absorbed by the target gas, and the second being unabsorbed.
   - Use of reference detectors to compensate for temperature changes and background noise, whereby the reference detector is shielded from the incoming light and matched to a neighboring active detector.

By having a reference path of this nature, changes in light-source intensity and detector responsivity changes with temperature can be effectively cancelled:

2. In some environments, optical paths can be affected by dust, dirt, or spray. As with all types of gas detectors, this is a difficult problem to deal with in severe cases, but it is a particularly serious one for optical detectors. The only real solution is to either pump gas into the optical path via a dust filter, or to use some form of gas-permeable dust filter around the optics. Depending on the relative sizes of

the filter and gas sample cell, this can lead to delays in response due to finite gas pumping time or, when no pump is used, to gas diffusion times. The use of intensity referencing techniques described earlier also helps to prevent problems, so long as optical scattering by the dust does not affect the reference or measurement channels differently. Fortunately, unlike many other types of gas sensors, the presence of serious contamination is detected because of the reduction in detected signal, so safety related sensors can be configured to operate in a fail-safe manner.

14.2.2 Laser-based sensors

Because of the narrow-band nature of gas absorption lines at atmospheric pressure, lasers are inherently suitable for measurement of low concentrations of gases. As mentioned earlier, a laser can be tuned to match the peak absorption of a gas line such that a much larger absorption coefficient is seen than possible with an LED source. Even if the latter is matched to cover a gas absorption manifold of many vibrational/rotational lines, it will only observe the average attenuation of light over this range. By using two laser lines, one matching a peak absorption and one in a spectral region close to the line, a much larger differential absorption measurement can be observed. However, the attenuation is most easily observed using a tunable laser, which is scanned through a strong absorption line. This powerful method is termed *tunable diode laser spectroscopy* (TDLS), and the term is still used generically for unconventional (nonstandard diode) laser sources such as quantum cascade lasers and interband cascade lasers in the mid-IR region. The fractional change in received signal as the laser tunes through the line provides a convenient intensity referencing method. A further sophistication is the use of wavelength modulation spectroscopy (WMS), in which a sinusoidal modulation is applied to the laser current and, in the presence of a narrow absorption feature, the resulting modulation in intensity may be recorded at the detector. For an applied signal at f, the demodulated signal at $2f$ reaches a peak at the line center, is proportional to the gas absorption (and hence, concentration at low values of $\alpha\ell$), and has a zero baseline.

Using a modulated signal offers many advantages, not least that the modulated signal may be recovered in the presence of background light, albeit with loss of self-referencing of the sensor. This can be important, for example, in high temperature applications where there may be a thermal IR emission from the gas mixture or apparatus. For industrial applications, having a background light level presents the disadvantage of requiring calibration, since the recovered signal is affected by the gas concentration and its linewidth, the latter being increased by pressure, and the overall (DC) intensity reaching the detector. However, in recent years, a number of authors have developed so-called calibration-free WMS techniques, using DC and/or $1f$ demodulated signals as normalizing factors, and utilized them in applications such as high temperature monitoring of processes in an industrial scale coal gasifier [36].

The absorption in gases can also be conveniently measured using photoacoustic spectroscopy (PAS). In this method, the gas is passed into an acoustic chamber coupled to a sensitive microphone, and it is excited by a wavelength- or amplitude-modulated laser such that the absorption results in cyclic heating of the gas. The amplitude of the acoustic signal is then proportional to the energy absorbed by the gas. The photoacoustic technique is essentially a sensitive method of detecting this absorbed energy. It is also a method that can be used with broadband light sources. It has the advantage that photoacoustic response is invariant to optical path length, since increases in sensitivity caused by an increase in ℓ are offset by a corresponding decrease in the modulated pressure signal resulting from the increase in cell volume. Photoacoustic sensors may therefore be very small, for example, in the case of the quartz enhanced photoacoustic spectroscopy technique, based on a quadrupole tuning fork [37]. For industrial applications, a disadvantage is that the timescale within which absorbed light is converted to heat can depend critically on the background matrix of gases, as some species are known to quench the energy conversion process for others. This again can lead to calibration issues if the background is unknown, or variable. A further disadvantage is a lack of immunity to background acoustic noise and vibration, which can be mitigated to an extent by clever acoustic design [37]. Despite these disadvantages, PAS based sensing of hydrogen sulfide has been successfully field tested on a petrochemical works to measure the difference between sour (containing H_2S) and sweet (with H_2S removed) gas streams [38].

There are now many commercially available compact semiconductor lasers, such as interband cascade lasers (ICLs) and quantum cascade lasers (QCLs), which have been developed for use in the IR region of the spectrum. Most of these can be wavelength tuned simply by changing their temperature, or, in a more rapid (frequency-agile) manner, by changing their drive current. The availability of such lasers has opened up the possibility of far more sensitive gas detection at sensible cost, still using compact sources and detectors.

Before assuming that narrowband lasers are a perfect panacea for low-level gas detection, however, there is a design problem that must be considered in any practical system using them. This problem is brought about as a result of the apparent advantage of the laser—that of its excellent coherence or narrow spectral linewidth. When directing narrowband sources through optical components and into an optical medium, there are numerous optical surfaces that may reflect or scatter light, and it is possible that mist droplets or dust particles might do the same. If any of the scattered or reflected beams are then coincident on the detector, together with the transmitted signal to be measured, then coherent optical interference can occur. The interference can exhibit a spectral signature, typically as sinusoidal fringes, which can be confused with the gas absorption signature. Interference can be reduced by careful design, such as anti-reflection coatings or windows at the Brewster angle, use of inclined optics, and reducing beam coherence by path length modulation, but it does require careful design compared to the simpler broader band LED-based methods.

Even if measures are taken to avoid such unwanted reflection, the resulting interference fringes are often the performance-limiting factor. Their effect on the spectral baseline is not static, since the relative optical path lengths of the interfering beams will usually vary with temperature of the sensor system or, if it is mechanically strained, the precise dimensions of the optical system. Furthermore, if path lengths are appropriate, the spectral interference fringes can have a linewidth that is similar to the very spectral features being detected such that they could reinforce or reduce the desired frequency modulation spectroscopy. This is clearly a potentially dangerous situation for safety-related sensors, as not only could false alarms be created, but safety warnings could be canceled.

Use of highly collimated laser beams, however, offers long path geometries that would be difficult to implement with other light sources. Directing the laser beam to a retroreflective target and measuring the returned signal permits the use of TDLS in perimeter detection systems. Absorption of light by gas in the space between the laser emission aperture and the reflective target results in a line-integrated measurement in which the extent of the gas cloud defines the value of ℓ in Equation 14.1 and is strictly unknown. Thus, measurements are often expressed in terms of the line-integrated measurement ppm·m. The concept may be extended to standoff measurement of gases, where the retroreflective target is replaced by an ordinary diffusely reflective background. The laser beam is diffusely reflected from background surfaces and a small fraction of this light may be detected with enough signal-to-noise ratio to permit measurement of gas concentration. Stand-off detection has, for example, been used to measure leaks of sulfur hexafluoride (SF_6) from electrical transformers, an application where safety considerations require a remote detection system.

14.2.3 Gas imaging systems

One of the problems with detection of gas leaks is the extended 3-D nature of most industrial sites. In many cases, there may be a possibility of leaks in many small areas of a large industrial site, such as a refinery. This means that, in order to detect leaks, a large number of discrete sensors or a significant number of extended path sensors are needed to cover all the space where gas may first present itself. One means of overcoming this problem has been the development of gas-sensing video surveillance systems, which are capable of "seeing" the small clouds of gas arising from a leak. As most gases have virtually no visible absorption, these systems usually operate in the IR region, and, at the very least, require some form of narrowband optical filter to obtain a good visual contrast. More powerful active filters can also be used, such that the camera forms part of a Fourier transform imaging system or similar systems.

The systems can be divided into two types: passive IR systems and active (usually laser-based) systems. Passive IR imaging makes use of IR imaging cameras developed for other applications, in combination with a narrowband filter centered over the gas absorption band. They have the advantage of excellent resolution and fast potential response at the frame rates of the underlying IR imaging technology. However, for fail-safe systems, there is a disadvantage that passive imaging requires a thermal contrast between the background scene and the gas being imaged. Unfortunately, at equilibrium, with gas at the same temperature as its surroundings, the laws of thermodynamics ensure that the light being emitted thermally by the gas is exactly balanced by the background light being absorbed, resulting in zero contrast. As long as this fact is recognized, in cases where thermal contrast can be guaranteed (for example, in industrial sites where the gas leaking out is either hot or suffers adiabatic cooling due to expansion from a pressure vessel), passive gas imaging still has a part to play. Because of this, such sensors have been successfully deployed in petrochemical works [39], to complement existing sensors that measure at a single point.

Active imaging systems make use of tunable diode laser spectroscopy, as previously discussed in the standoff detection configuration. Combining this with scanning of the beam (and possibly synchronous scanning of the collection optics) results in an image of gas absorption regions which may be overlaid onto a (usually higher resolution) conventional black and white image of the scene [40]. The main advantage is absolute quantification of the scale of the gas leak, again expressed in line-integrated units of ppm·m. A disadvantage is that the interrelated parameters of sensitivity, resolution, and video frame rate are critically limited by the number of photons received within a given time from the diffusely reflecting surface. Performance is only improved by the use of large lenses (e.g., 15 cm diameter lenses have been used in the near-IR), low noise detectors, and high power sources, particularly lasers.

14.3 GASES OFTEN MEASURED USING DIRECT OPTICAL METHODS

Some of the most commonly measured gases are discussed here:

Methane (CH_4): This is one of the most important gases to be measured in industry due to the explosion hazard of this invisible odorless gas, and its particularly potent greenhouse gas effect. The strongest absorption bands of methane, arising from the C–H bonds, occur at 7.8, 3.4, and 2.4 µm, with another somewhat weaker band around 1.6 µm. The spectra consist of many fine lines, which are suitable for more sensitive detection with lasers.

Methane is used in many processes, particularly in the chemical and petrochemical industries, and is the main constituent of natural gas. It is also generated in bioreactors, by the digestive system of many farm animals and by anaerobic digesters and composters, where it may be put to good use as a biogas.

Other alkane gases (e.g., ethane, propane, butane): These absorb in similar spectral regions to methane, but the spectral lines rapidly become less distinct as the molecular size increases due to the molecular vibrations having more degrees of freedom, with their molecular spectra eventually losing any fine-line details if more than four carbon atoms are present.

Alkyne gases (e.g., ethyne): The simplest alkyne gas, acetylene (ethyne) is spectroscopically similar to methane and is a commonly used gas for high-temperature cutting or welding.

Aromatic gases (e.g., benzene, toluene, xylene): These gases have limited fine line structure, but are very important hazards in view of their carcinogenic properties. They show particularly strong absorption signatures in the UV region.

Carbon dioxide (CO_2): Carbon dioxide exhibits strong absorption at 4.2 µm and 10.6 µm, and additionally, it has very weak overtone bands in the near-IR (within the 1520–1570 nm telecoms band) regions, which are more easily addressable with semiconductor sources. This gas is used in many industries, particularly the effervescent drinks industry and as an enhanced recovery agent, when injected into oil wells. It is a greenhouse gas, produced in large quantities by the combustion of fossil materials, volcanic action, and in fermentation processes.

14.4 CASE-STUDY EXAMPLES OF COMMERCIAL OPTICAL SENSORS

We shall now show some specific case-study examples of commercial optical gas sensors using direct optical detection.

14.4.1 NDIR measurement of carbon dioxide

Arguably, the most commercial optical gas detector currently is the NDIR system for measurement of carbon dioxide (CO_2). This has been successfully implemented in a small footprint (20 mm diameter × 16.5 mm high) similar to many other types of sensors based on electrochemical, semiconductor, or pellistor technology [16]. The optical technology has been driven by the fact that competitive technologies for carbon dioxide are largely inadequate to the task. The case for optics has been further strengthened by the use of low-cost components and the small form factor, making the sensor acceptable to system manufacturers more familiar with alternative technologies. Typically, a two wavelength technique is employed, described schematically in Figure 14.3. It helps, of course, that carbon dioxide exhibits strong absorption in a band centered at 4.2 μm and also that a neighboring waveband centered at 3.95 μm, exhibiting minimal absorption from many common gases at the levels normally encountered, is available as a zero CO_2 reference band.

One of the most important developments in this technology has been the integrated packaging into a single, 9 mm diameter TO-5 can, of the two bandpass filters, their two respective pyroelectric or thermopile detectors, and the preamplifier. For the highest performance, two additional matched, light-shielded reference detectors are used plus two ultra-high gain preamplification stages (e.g., employing a 30 GΩ transimpedance amplifier to ensure minimal thermal differences and little electrical crosstalk between the detectors) [41].

The challenge of positioning a 9 mm diameter TO-5 can, an incandescent microbulb source, and an optical absorption path of up to approximately 30 mm into the 20 mm × 16.5 mm format is considerable. This design problem has yielded a number of creative solutions including integrating cavities [42], a reflective spiral optical path [43], ellipsoidal reflectors [44], and the use of injection moulded, high NA (numerical aperture) reflective surfaces [45] (see Figure 14.4). Dust filters are routinely

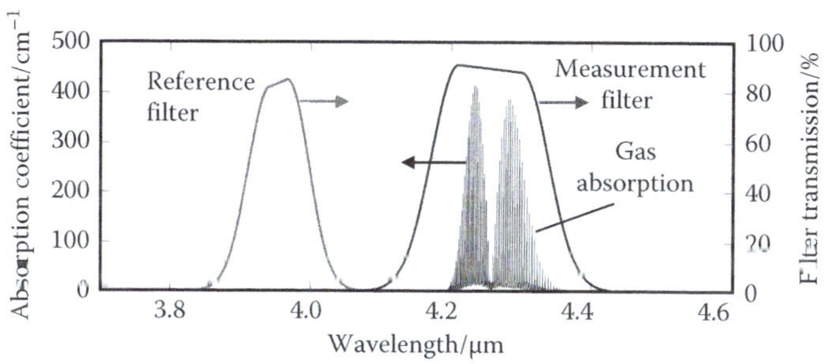

Figure 14.3 Spectral description of NDIR-based measurement of carbon dioxide gas. (Adapted from Hodgkinson, J. et al., *Sens. Actuators B*, 186, 580–588, 2013.)

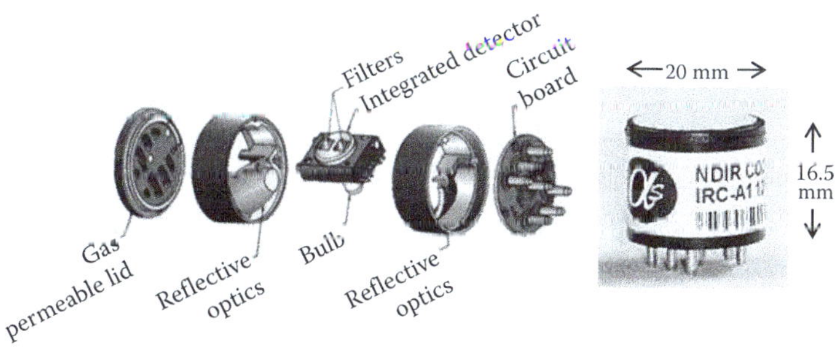

Figure 14.4 Exploded diagram and photograph of an example NDIR sensor for carbon dioxide. (Adapted from Hodgkinson, J. et al., *Sens. Actuators B*, 186, 580–588, 2013.)

incorporated into the top of the sensors without compromising the speed of response, which is largely dictated by the slow (1–2 Hz) modulation frequency of the source. The performance achieved with such apparently simple technology is impressive—detection limits of 20 ppm, in the face of wide temperature or pressure changes, can be demonstrated for stand-alone sensors, without the need for frequent calibration.

Carbon dioxide is a gas that needs to be monitored in many applications. It forms an important element of industrial processes in the brewing and wine industries. Yet, it is an asphyxiant and therefore measured for safety reasons in premises used for both production (fermentation plant) and storage (bar cellars) of beer and wine. As a gas exhaled by people, it can form a measurable marker for indoor air quality and sensors are often used in demand-actuated heating and ventilation systems, enabling building managers to save energy (vis. in limiting the volume of fresh air introduced, which requires heating of cold air). Finally, CO_2 is an important product of combustion.

14.4.2 TDLS-based detection of leaks from pressurized cans

An example of industrial use of gas detection using TDLS is that of measurement of leaks from aerosol cans in the final stages of production. This application is driven by legislation mandating a leak test for each can, which traditionally has been performed by observation of bubbles released when the cans are in a water bath.

This application demanded detection of small leaks of a variety of propellants, depending on what is in use on the production line, including liquid petroleum gas, Freon R134a, CO_2, N_2O, and dimethyl ether. Fortunately, the measurement principles for one measurand can often be applied to others, simply via a change of laser wavelength and, if required, a change of detector. Instruments are now commercially available with several lasers combined into a collinear beam, enabling flexibility in production of multigas sensors (Figure 14.5).

The example shown makes use of the intra-pulse modulation technique for QCLs. Early development of QCLs required that they be pulsed rather than CW (Continuous Wave), and for many wavelengths, they are still only available in the pulsed form. The intra-pulse method activates the laser with a very short (hundreds of ns) current pulse, resulting in a wavelength modulation over the duration of the pulse. Averaging over multiple pulses improves the signal-to-noise ratio. Advantages of this technique include an improvement in the normally observed laser linewidth and high detection speed, since the pulse repetition rate is of the order of tens of kHz. This enables real-time gas leak detection at production speeds of 500 cans/min. Note that the required detection speed is made more challenging by the fact that gas leaks are often only transiently detectable.

The inherent detection speed of this technique makes it particularly suited to challenging

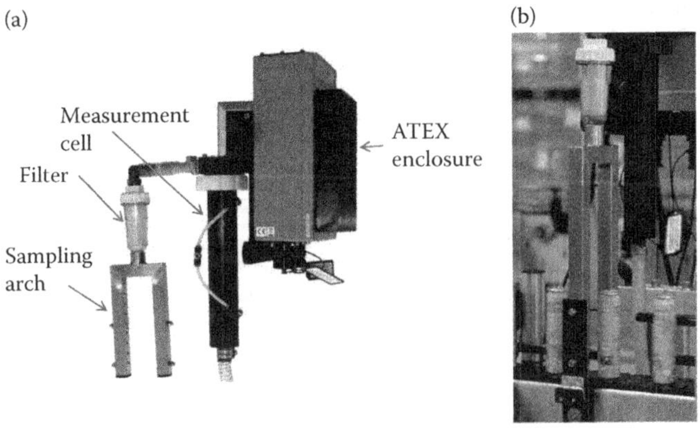

Figure 14.5 Aerosol leak detection on production lines at 500 cans per minute. (a) Schematic diagram of measurement system. (b) Implementation on a production line showing sampling arch. (Images Courtesy of Cascade Technologies Ltd., Stirling, UK.)

industrial environments, where levels of background vibration may be high. Because an entire spectrum is acquired within a few hundreds of nanoseconds, lower frequency vibrations (in the kHz to Hz range) are unable to present any spectral artifacts. This technique has been implemented in a wide variety of industrial applications including engine exhaust monitoring in the automotive and aerospace industries, stack gas analysis, including *in situ* measurement of emissions on board ships, analysis of process stream gases in the petrochemical and semiconductor industries, as well as production line leak detection from pressurized cans and modified atmosphere food packaging.

14.5 INDIRECT DETECTION OF GASES USING INDICATOR CHEMISTRY

We shall now consider indirect detection of gases using indicators. These are chemical compounds that interact with the gas to produce a change in their optical properties (vis., changes in color, fluorescence behavior, Raman scattering coefficients, refractive index, etc.) of the new material.

If this change is permanent, via an irreversible reaction, then the sensor is only suitable for once-only use, for example, to sound an alarm in the case of a dangerous leak. In most cases, it is preferable to have a reversible reaction, where the indicator compound reverses back to its previous state when the gas concentration returns to zero. As the kinetics of the forward and reverse reactions is different, response and recovery times are also different, so sensor recovery times can often be longer.

As examples of indicators, the reader is probably familiar with the Litmus papers used in chemistry classes to detect acid and alkaline materials, and maybe the more sophisticated pH papers capable of measuring a wider range of alkalinity or acidity via a color change. For gas detection, the well-known equivalent of this is the Draeger tube, where gas is drawn into a clear glass or polymer tube containing an indicator, which changes color after reacting with the gas. The same principle is also used in certain law-enforcement "breathalyzers" to detect alcohol vapor. In these cases, optical interrogation of the response is provided by the human eye; therefore, care must be taken to ensure that color changes are distinct, since the perception of color can vary between individuals (who in the most extreme case may be colorblind).

To make a very simple optoelectronic sensor, it is possible to simply view one or more such tubes using a color camera or a simpler optical sensor array, but there are now many more sophisticated and quantitative methods than this, which have been developed to measure gases by virtue of the following types of chemical reactions:

1. pH sensing for acid or alkaline gases using pH indicators similar to those used in Litmus and pH papers: Sensing is via measurement of transmission or scattering of light. This can be used to detect acid gases (e.g., CO_2, HCl, SO_2) or alkaline gases (e.g., NH_3) albeit in a rather unselective manner.
2. Sensing of gases using other color-change indicators, selective to particular compounds or radicals, where the absorption is varied by the reaction with the gas. Sensing is again via measurement of transmission or scattering. This can be used to detect several gases or families of gases, provided suitable indicators are available. The sensor selectivity clearly depends on the indicator used.
3. Sensing of gases using fluorescent compounds, where the fluorescent spectrum or fluorescent lifetime is varied by the reaction of the compound with the gas: A commonly used example of this is sensing of oxygen, which has the property of quenching fluorescence of compounds such as metallo-organic ruthenium complexes. The fluorescent lifetime reduces in a quantitative manner when oxygen is present. This technique is used to detect dissolved oxygen in water treatment processes.
4. Sensing of gases using particularly selective binding interactions in more sophisticated compounds. Again, some change in absorption, fluorescence, or refractive index properties is needed. This can be used for sensing of larger molecules in solution. A few gas-selective compounds are available for reversible sensing of simple gases.
5. Sensing of gases using 2-D arrays of small chemical sensor dots, each specific to particular gases to varying degrees. Such an array

can be conveniently addressed via a miniature video camera to form a simple "optical nose." Even if many of the dots respond to a chemical family or even to several different compounds, if enough are used, the pattern of response may be sufficiently unique to identify a particular single compound. This is a very powerful method to provide good selectivity of gas detection by pattern recognition, even if each individual dot may not be so selective in its response. It has also been developed for many applications such as detection of biomedical compounds, pollution components, and even chemical warfare agents, most of which are however not true gases but fine droplets or mists.
6. Sensing of gases via a refractive index or mechanical change within a material. This is an important technique for the detection of molecular hydrogen H_2, via its effect on palladium Pd and certain alloys, where mechanical changes are measured by changes in the light propagation within optical fibers or integrated optical guides bonded to the material.

It is not always necessary to have a true chemical reaction for sensors of this type as some polymers or other organic layers can selectively adsorb gases from the air. This adsorption process often leads to an effective concentration of the gas, if the partition coefficient causes higher molecular concentration within the polymer. Once in the polymer matrix, all fine-line absorption spectra details are clearly lost due to molecular interactions, but the concentration effects can lead to stronger absorption and easier detection. Fully substituted fluoropolymers and chloro-polymers have low mid-IR absorption (as no C–H bonds are present) if there are no contaminants, so this is a useful area where strong transmission changes can be observed once adsorption has occurred. In addition, the higher concentration of the previously gaseous molecule may lead to the possibility of using Raman scattering, which can be used in a highly selective manner to detect specific gases.

It should be clear that this sensing area has too many variations in chemistry and optical technology to cover in detail in this short applications chapter, so interested readers are directed to the bibliography for further reading. However, it is useful to mention some examples sensors which are finding use in industry.

14.6 CASE-STUDY EXAMPLES OF OPTICAL INDICATOR METHODS

As mentioned earlier, hydrogen is a particularly difficult gas to detect selectively, but Pd has a selective affinity for the gas, forming a Pd hydride where the H_2 molecules dissociate and migrate to interstices in the Pd lattice. The effect is to expand the lattice, leading to changes in the refractive index or mechanical properties of the metal. Thin films of Pd may be deposited, for example, onto optical fiber. Numerous methods exist to probe the material's refractive index, reflectivity, or induced mechanical strain.

Measurement of hydrogen is a growing need with development of the hydrogen economy. As with natural gas, techniques are required across a wide range of concentrations from ppm to %volume for leak detection, leak quantification, and quality measurement. Alternative methods for detecting hydrogen are limited, with semiconductor sensors a possibility (with inherent cross-response and drift issues). In environments where a specific, reliable measurement of hydrogen is required, mass spectrometry may be the only option.

Detection of dissolved oxygen via fluorescence quenching is a technique finding use in the water processing industry, where dissolved oxygen is an important water quality parameter, particularly for treated wastewater. Suitable fluorescent materials are available commercially, as are optical fiber probes onto which these materials are deposited. Measurement of the fluorescence decay lifetime has the advantage of being a self-referenced technique, in that measurement is unaffected by the overall intensity of the signal. Different indicator coatings have been developed for different applications. The indicator may also be a fluorescence patch, which is interrogated remotely from the optical system. Using line-of sight optics, this allows monitoring of transparent closed systems such as food packages or glass bioreactor vessels without the engineering difficulty of organizing leak-tight feedthroughs. Using optical fibers to transmit light to and from the patch allows more remote sensing, albeit at the expense of reduced optical signals.

14.7 CONCLUSIONS

It is clear that there are many types of gas sensors that can be constructed using optoelectronics, but in industry, cost is usually at a premium. Optical sensors must therefore find a niche where there is no other suitable method to detect the gas, or where optics offers particular advantages. Even the lowest cost technologies such as NDIR are not cost-competitive with their nonoptical counterparts. Regarding direct-sensing methods, sensors using broadband sources tend to be used most, but laser-based systems are important for detection over extended paths in petrochemical plants, for checking for noxious emissions from factories, and for ultra-low-level detection of suitable gases. The chief advantage of laser-based detection in this market is its fast response and high degree of specificity to the target gas in a potentially complex background.

REFERENCES

1. Jones, E. 1987. The pellistor catalytic gas sensor. Chapter 2 in *Solid State Gas Sensors*, ed. P. T. Moseley and B. C. Tofield, Adam Hilger (Bristol), pp. 17–31.
2. Jones, T. A. 1987. Characterisation of semiconductor gas sensors. Chapter 4 in *Solid State Gas Sensors*, ed. P. T. Moseley and B. C. Tofield, Adam Hilger (Bristol), pp. 51–70.
3. Bakker, E. and Telting-Diaz, M. 2002. Electrochemical sensors. *Anal. Chem.* 74, 2781–2800.
4. Lackner, M. 2007. Tunable diode laser absorption spectroscopy (TDLAS) in the process industries—A review. *Rev. Chem. Eng.* 23, 65–147.
5. Hodgkinson, J. and Tatam, R. P. 2013. Optical gas sensing: A review. *Meas. Sci. Technol.* 24, 012004 (59 pp).
6. Ingle, J. D. and Crouch, S. R. 1988. *Spectrochemical Analysis*. Prentice Hall (London).
7. Sharpe, S. W., Johnson, T. J., Sams, R. L., Chu, P. M., Rhoderick, G. C. and Johnson, P. A. 2004. Gas-phase databases for quantitative infrared spectroscopy. *Appl. Spectrosc.* 58, 1452–1461.
8. Mount, G. H., Rumburg, B., Havig, J., Lamb, B., Westberg, H., Yonge, D., Johnson, K. and Kincaid, R. 2002. Measurement of atmospheric ammonia at a dairy using differential optical absorption spectroscopy in the mid-ultraviolet. *Atmos. Environ.* 36, 1799–1810.
9. Claps, R., Englich, F. V., Leleux, D. P., Richter, D., Tittel, F. K. and Curl, R. F. 2001. Ammonia detection by use of near-infrared diode-laser-based overtone spectroscopy. *Appl. Opt.* 40(24), 4387–4394.
10. Lewicki, R., Wysocki, G., Kosterev, A. A. and Tittel, F. K. 2007. Carbon dioxide and ammonia detection using 2 μm diode laser based quartz-enhanced photoacoustic spectroscopy. *Appl. Phys. B* 87, 157–162.
11. Pushkarsky, M. B., Webber, M. E., Baghdassarian, O., Narasimhan, L. R. and Patel, C. K. N. 2002. Laser-based photoacoustic ammonia sensors for industrial applications. *Appl. Phys. B*75(2 3), 391–396.
12. Patterson, B. A., Lenney, J. P., Sibbett, W., Hirst, B., Hedges, N. K. and Padgett, M. J. 1998. Detection of benzene and other gases with an open-path, static Fourier-transform UV spectrometer. *Appl. Opt.* 37(15), 3172–3175.
13. Jeffers, J. D., Roller, C. B., Namjou, K., Evans, M. A., McSpadden, L., Grego, J. and McCann, P. J. 2004. Real-time diode laser measurements of vapor-phase benzene. *Anal. Chem.* 76(2), 424–432.
14. Chen, W., Cazier, F., Tittel, F. and Boucher, D. 2000. Measurements of benzene concentration by difference-frequency laser absorption spectroscopy. *Appl. Opt.* 39(33), 6238–6242.
15. Engelbrecht, R. 2004. A compact NIR fiber-optic diode laser spectrometer for CO and CO_2: Analysis of observed 2f wavelength modulation spectroscopy line shapes. *Spectrochim. Acta A*60, 3291–3298.
16. Hodgkinson, J., Smith, R., Ho, W. O., Saffell, J. R. and Tatam, R. P. 2013. Non-dispersive infra-red (NDIR) measurement of carbon dioxide at 4.2 μm in a compact and optically efficient sensor. *Sens. Actuators B*186, 580–588.

17. Hangauer, A., Chen, J., Strzoda, R., Ortsiefer, M. and Amann, M.-C. 2008. Wavelength modulation spectroscopy with a widely tunable InP-based 2.3 μm vertical-cavity surface-emitting laser. *Opt. Lett.* 33(14), 1566–1568.
18. Vargas-Rodríguez, E. and Rutt, H. N. 2009. Design of CO, CO_2 and CH_4 gas sensors based on correlation spectroscopy using a Fabry–Perot interferometer. *Sens. Actuators B*137, 410–419.
19. Hennig, O., Strzoda, R., Magori, E., Chemisky, E., Trump, C., Fleischer, M., Meixner, H. and Eisele, I.2003. Hand-held unit for simultaneous detection of methane and ethane based on NIR-absorption spectroscopy. *Sens. Actuators B*95, 151–156.
20. Parameswaran, K. R., Rosen, D. I., Allen, M. G., Ganz, A. M. and Risby, T. H. 2009. Off-axis integrated cavity output spectroscopy with a mid-infrared interband cascade laser for real-time breath ethane measurements. *Appl. Opt.* 48(4), B73–B79.
21. Cihelka, J., Matulková, I. and Civiš, S. 2009. Laser diode photoacoustic and FTIR laser spectroscopy of formaldehyde in the 2.3 μm and 3.5 μm spectral range. *J. Mol. Spectrosc.* 256(1), 68–74.
22. Varga, A., Bozóki, Z., Szakall, M. and Szabo, G. 2006. Photoacoustic system for on-line process monitoring of hydrogen sulfide (H_2S) concentration in natural gas streams. *Appl. Phys. B*85, 315–321.
23. Kassi, S., Chenevier, M., Gianfrani, L., Salhi, A., Rouillard, Y., Ouvrard, A. and Romanini, D. 2006. Looking into the volcano with a Mid-IR DFB diode laser and cavity enhanced absorption spectroscopy. *Opt. Express* 14(23), 11442–11452.
24. Sonnenfroh, D. M., Wainner, R. T., Allen, M. G. and Varner, R. K. 2010. Interband cascade laser–based sensor for ambient CH_4. *Opt. Eng.* 49(11), 111118 (10 pp).
25. Mcmanus, J. B., Shorter, H., Nelson, D. D., Zahniser, M. S., Glenn, D. E. and McGovern, R. M. 2008. Pulsed quantum cascade laser instrument with compact design for rapid, high sensitivity measurements of trace gases in air. *Appl. Phys. B*92, 387–392.
26. Sonnenfroh, D. M. and Allen, M. G. 1997. Absorption measurements of the second overtone band of NO in ambient and combustion gases with a 1.8-mm room-temperature diode laser. *Appl. Opt.* 36(30), 7970–7977.
27. Fetzer, G. J., Pittner, A. S. and Silkoff, P. E. 2003. Mid-infrared laser absorption spectroscopy in coiled hollow optical waveguides. *Proc. SPIE* 4957, 124–133.
28. Sonnenfroh, D. M., Rawlins, W. T., Allen, M. G., Gmachl, C., Capasso, F., Hutchinson, A. L., Sivco, D. L., Baillargeon, J. N. and Cho, A. Y. 2001. Application of balanced detection to absorption measurements of trace gases with room-temperature, quasi-CW quantum-cascade lasers. *Appl. Opt.* 40(6), 812–820.
29. Kebabian, P. L., Wood, E. C., Herndon, S. C. and Freedman, A. 2008. A practical alternative to chemiluminescence-based detection of nitrogen dioxide: Cavity attenuated phase shift spectroscopy. *Environ. Sci. Technol.* 42, 6040–6045.
30. Sonnenfroh, D. M. and Allen, M. G. 1996. Ultrasensitive, visible tunable diode laser detection of NO_2. *Appl. Opt.* 35, 4053–4058.
31. Rao, G. N. and Karpf, A. 2010. High sensitivity detection of NO_2 employing cavity ringdown spectroscopy and an external cavity continuously tunable quantum cascade laser. *Appl. Opt.* 49(26), 4906–4914.
32. Xu, F., Lv, Z., Zhang, Y. G., Somesfalean, G. and Zhang, Z. G. 2006. Concentration evaluation method using broadband absorption spectroscopy for sulfur dioxide monitoring. *Appl. Phys. Lett.* 88, 231109.
33. Rawlins, W. T., Hensley, J. M., Sonnenfroh, D. M., Oakes, D. B. and Allen, M. G. 2005. Quantum cascade laser sensor for SO_2 and SO_3 for application to combustor exhaust streams. *Appl. Opt.* 44, 6635–6643.
34. Hovde, D. C., Hodges, J. T., Scace, G. E. and Silver, J. A. 2001. Wavelength-modulation laser hygrometer for ultrasensitive detection of water vapor in semiconductor gases. *Appl. Opt.* 40(6), 829–839.
35. Moyer, E. J., Sayres, D. S., Engel, G. S., St Clair, J. M., Keutsch, F. N., Allen, N. T., Kroll, J. H. and Anderson, J. G. 2008. Design

considerations in high-sensitivity off-axis integrated cavity output spectroscopy. *Appl. Phys. B*92, 467–474.
36. Sur, R., Suna, K., Jeffries, J. B., Sochab, J. G. and Hanson, R. K. 2015. Scanned-wavelength-modulation-spectroscopy sensor for CO, CO_2, CH_4 and H_2O in a high-pressure engineering-scale transport-reactor coal gasifier. *Fuel* 150, 102–111.
37. Kosterev, A. A., Tittel, F. K., Serebryakov, D. V., Malinovsky, A. L. and Morozov, I. V. 2005. Applications of quartz tuning forks in spectroscopic gas sensing. *Rev. Sci. Instrum.* 76, 043105 (9 pp).
38. Varga, A., Bozóki, Z., Szakall, M. and Szabo, G. 2006. Photoacoustic system for on-line process monitoring of hydrogen sulfide (H_2S) concentration in natural gas streams. *Appl. Phys. B*85, 315–321.
39. Sandsten, J., Edner, H. and Svanberg, S. 2004. Gas visualization of industrial hydrocarbon Emissions. *Opt. Express* 12(7), 1443–1451.
40. Gibson, G., van Well, B., Hodgkinson, J., Pride, R., Strzoda, R., Murray, S., Bishton, S. and Padgett, M. 2006. Imaging of methane gas using a scanning, open-path laser system. *New J. Phys.* 8, 26 (8 pp).
41. Infratec GmbH, 2005. *Pyroelectric and Multispectral Detectors*, Product Brochure, Infratec GmbH, Dresden, Germany.
42. Cutler, S. C. and Vass, A. 2005. Gas sensor. Patent WO2005054827 (A1).
43. Stuttard, D. M. 2001. Gas sensor. Patent WO 02/063283.
44. Hayward, A. S.and Hopkins, G. P. 2005. Gas sensors. UK Patent GB 2,395,260.
45. Hodgkinson, J., Saffell, J. R. and Smith, R. 2008. Optical absorption gas sensor. European Patent EP1972923 (A3).

Web resources

Gas Sensing Solutions (GSS): http://www.gassensing.co.uk/products/
Pranalytica: http://www.pranalytica.com/
FLIR Gas Detection Systems: http://flir.com/ogi/content/?id=66693
MSA Safety Company: http://www.msasafety.com/
Environmental XPRT: https://www.environmental-expert.com/products
Crowcon Laser Methane Detectors: http://www.crowcon.com/uk/products/portables/LMm-Gen-2.html
Crowcon Gas Pro Detectors: http://www.crowcon.com/uk/products/portables/gas-pro.html

15

Laser applications in industry

PAUL HARRISON
SPI Lasers Ltd.

15.1 LASER MACHINING

The laser machining marketplace is immense, and for the last few years, it has been averaging a growth rate of over 8% per annum [1]. The scope of this sector includes all types of metal processing (such as welding, cutting, and drilling), semiconductor and micro-electronics manufacturing, marking of all materials, rapid prototyping, and micromachining. Also worthy of inclusion are heat treatment, cladding, bending, texturing, cleaning, and engraving. The range of laser materials processing (LMP) applications is continually expanding and a decade ago, it was estimated that less than 50% of potential applications had been achieved so far [2].

Most types of laser machining are based upon thermally modifying the workpiece. In this case, the objective is to either change the surface properties with the heat treatment or actually remove material by fusion or ablation, or even cut right through the material. However, two classes of laser are capable of avoiding deep-seated thermal effects. Ultraviolet lasers, which have an output wavelength less than 400 nm (either an excimer laser or diode pumped solid state (DPSS) laser with a frequency tripled or quadrupled output), are capable of generating high energy photons that can exceed the bond energy for many organic materials and therefore can remove material cleanly without edge damage. Pico- and femto-second lasers can generate ultrashort pulses with extremely high intensity, which can remove material from the surface faster than heat can flow into the part. Industrial LMP lasers can broadly be classified into five main groups, with each group having many subcategories, as outlined in Table 15.1.

The most recent (last decade) laser source development for use in the laser machining area is the use of high power fiber lasers, pumped by arrays of high-power semiconductor lasers. This energy conversion into the small diameter monomode fiber core (the active part of the laser) improves the beam quality and coherence of the source, and this, combined with the reasonably short wavelength (usually of order 1 μm) of fiber lasers can give very precise focused spots.

To give an example of the advantages of thermal laser processing, consider a laser with an optical output of 1 kW. When focused to a spot of 100 μm diameter, it provides an average power density over the focused spot of order 13 MW/cm^2, and if focused to a smaller 10 μm diameter spot, it has a power density of 1300 MW/cm^2. This is more than enough to melt, or vaporize, and hence cut, the most stubborn of materials. Industrial lasers with this level of *continuous* power output or even at least an order of magnitude higher are commercially available.

Table 15.1 Five general classifications of industrial LMP laser

	Typical wavelength (μm)	Output power range (kW)	Beam quality	CW or pulsed operation	Required maintenance level
Fiber laser	1.0 μm+harmonics	Up to 100 kW	Very good	CW ns pulsing, ps and fs pulsing	Low
Solid state	1.0 μm+harmonics	Up to 16 kW	Very good	All types, including Q-switching	Some, including regular servicing
Diode laser	0.8–0.9 μm	Up to 6 kW	Fair	CW or modulated only	Low
CO_2 laser	10.6 μm	20 kW+	Very good	Mainly CW operation	Medium, including regular servicing
Excimer	157–351 nm	Up to 2 kW	Can generate uniform beams	ns pulsing	High, including regular servicing

There are three main industrial groups of machining applications, comprising marking, micromachining, and kW-level materials processing, which are outlined later. Generally, the initial cost of a laser processing workstation is usually higher than that using alternative technologies, but the overall cost of ownership over the working lifetime is almost always significantly lower—this is due to the lack of tool wear (since the focused laser beam does not deteriorate) and the minimal need for consumables, such as cutting fluids. An example of this would be a marking station for industrial products (for example, date coding for the food industry), where a laser-based system has no need for any additives to create a mark, whereas an inkjet printing station needs a continual supply of ink and a mechanical marking station (such as a solenoid marker) will gradually wear out, resulting in a varying quality of mark.

15.2 LASER MARKING

This is part of a wider group of laser surface treatment (LST) applications that includes processes such as heat treatment, laser cladding, laser cleaning, paint stripping, and laser shock hardening. By far the most commercially successful is laser marking, which is applied to a wide range of materials and industries. Three main types of lasers are used, with the choice usually driven by the workpiece material (Table 15.2), although the boundaries are not absolute.

There are several different ways in which a laser can modify the material to create a mark, including (for metals) engraving, layer removal and oxide formation, and (for plastics) thermal degradation, carbonization, bleaching, and foaming. Figure 15.1 shows two applications, (left) engraving of mild steel to create an indelible mark, and

Table 15.2 Lasers used for marking applications

Laser Type	Material type
Low power CO_2 (<100 W)	Wood, labels, glass, rubber, plastics
Pulsed fiber laser	Metals (stainless steel, copper, aluminum, etc)
Fiber-coupled diode laser	Plastics

Figure 15.1 (left) Laser engraved mild steel, (right) laser marked PCB substrate. (Both Images Courtesy of SPI Lasers UK Ltd.)

(right) marking of data-matrix identity tracking codes onto PCB (printed circuit board) substrates. Laser marking applications can be found in almost every industrial market, from marking traceability 2D data matrices on aerospace components, date stamping of food packaging, and "day/night" marking of vehicle dashboards to name a few.

15.3 LASER MICROMACHINING

Laser micromaching is a group of processes that involves fine machining at the sub-millimetre scale, sometimes involving micron level features. Such applications include metal engraving, silicon machining (for example, producing cantilevers for accelerometers and other MEMS (Micro Electro-Mechanical Systems) sensors), glass engraving (for example, glass planar waveguides), and creating identity coding marks within diamonds.

Lasers used for micromachining applications are almost always operated in pulsed mode, typically using pulses shorter than a microsecond. Since the thermal penetration depth is related to the square-root of the pulse duration, shorter pulses can produce finer features. Thus, the emergence of pico- and femto-second lasers has lead to a wide range of new applications. The optical absorption depth is also important for producing fine features. In most materials, shorter optical wavelengths are absorbed in a smaller depth so collateral damage caused by additional heating is reduced. As the feature size reduces, the work station stability becomes more critical in terms of motion system accuracy and the need to suppress vibration and thermal expansion. Many fine-feature workstations use granite as the chassis and are located in temperature-controlled clean-rooms.

For more mainstream applications such as metal engraving of coin die stamps, pulsed fiber lasers are commonly used. These have good beam quality and are capable of reasonably high average power (typically up to 200 W), even with ns pulses, a combination which allows commercially acceptable engraving rates on many types of metal. One such example, engraving of aluminum, is shown in Figure 15.2, where the same laser can be used for engraving, cutting, and marking, despite the high thermal conductivity of this material.

Having small focused spot diameters enables very fine cuts and precise machining, particularly of thin materials, even allowing complex shapes,

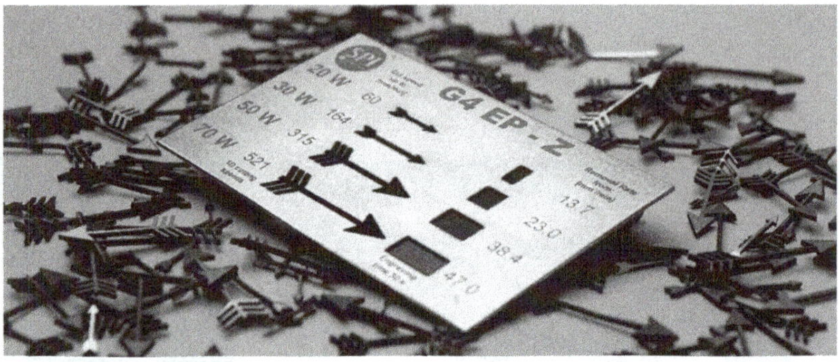

Figure 15.2 An example of cutting, engraving, and laser marking using the same laser workstation. (Image Courtesy of SPI Lasers UK Ltd.)

such as medical stents (springy tubes for holding open arteries and other bodily vessels), to be cut from cylindrical tubes. It also allows surfaces to be marked or engraved with the laser, again scanning the beam or moving the target to write the desired patterns on the surface. Clearly, using this method, it is possible to mark very hard or even brittle materials with very fine patterns, without risking any serious damage or distortion to the articles.

15.4 HIGH POWER (KW) LASER PROCESSING

This processing group comprises applications such as cutting, welding, cladding, and rapid 3D prototyping. Such applications are typically dominated by the CO_2 laser, high power fiber laser, solid state disk laser, and the fiber-coupled diode laser. These lasers are usually operated in CW (continuous wave) mode, but for speed-dependant applications (such as for laser cutting at variable speeds to allow cutting of high quality corners), the lasers are modulated, so the ability to pulse-modulate up to 100 kHz can be important.

When laser cutting metals, the laser beam is focused through a copper nozzle to allow a jet of high-pressure gas to be co-axially aligned with the beam, creating a "melt and blow" effect. This is typically achieved through the use of a laser cutting head. When cutting steels (usually mild steel, sometimes stainless steel), oxygen is often used as an assist gas, which exothermically reacts with the iron, generating more heat and increasing processing speeds, a process known as reactive cutting. Nitrogen assist gas is often used for cutting other metals such as stainless steel and in this case the process is known as inert (or fusion) cutting.

In certain applications where the workpiece is soft or thin (such as gasket material or rubber sheet), laser cutting has a distinct advantage over mechanical cutting (such as a knife cutting or guillotining) as there is no mechanical stress to cause distortion, so the dimensional accuracy of the finished component is much better.

Fiber and disk lasers are used for cutting many types of metal, including highly reflective metals such as copper and aluminum. An example of cutting a range of metals using a 2 kW fiber laser is shown in Figure 15.3. In many instances, the cost of ownership, high efficiency, reduced maintenance, and the lack of need for regular laser and beam delivery alignment make these laser a very attractive option. CO_2 lasers are also used for metal cutting and have the advantage of being able to cut other materials such as wood, plastic and rubber, making them an appealing option for laser job shop companies (see Figure 15.3).

Figure 15.3 Laser cutting of a range of metals including stainless and mild steel, copper, brass, and aluminum. (Image Courtesy of SPI Lasers UK Ltd.)

Figure 15.4 A laser-welded stainless steel T joint (2 mm thick stainless steel) welded using a 400 W fiber laser. (Image Courtesy of SPI Laser UK Ltd.)

By lowering the power density of laser beams, for example, by reducing the output power or using a larger focused spot, the laser beam can create a region of molten material. By firmly holding two metal pieces together and directing the laser to the interface area, the metals can be welded together in an alloy of the two (see Figure 15.4). Very often a shield gas is applied to exclude oxygen from the molten region to increase the weld strength and avoid surface discoloration. Laser welding is essentially similar to laser cutting, but the application of process gas (co-axial assist gas for cutting, shield gas for welding) is very different. Welding techniques are, of course, more complex than simple cutting or engraving, and the engineering details are beyond the scope of this short summary of applications. It should, however, be clear that having the precise heating offered by lasers means the process can be performed in a far more controlled manner.

Laser welding is typically used in applications where a high quality, repeatable weld is required, for instance, where testing (which may involve a destructive test) cannot be performed on every part but instead parts are type-tested and approved. Applications in the nuclear and aerospace industries are common for these reasons. The automotive industry is a significant user of many different types of laser welding in many production environments, including remote laser welding, which involves delivering the beam through a laser scanner to a workpiece, which that can be held around 1 m away, and the creation of tailor-welded blanks, which involves joining different thicknesses or strengths of metal together before stamping.

Other applications in this process group include laser cladding, where the laser beam acts as a heat source to add a surface-layer alloy onto the base part with an applied (usually blown) metal powder and laser hardening—where the component is heated to a suitable temperature (controlled by the laser beam intensity and the speed of motion) and allowed to cool.

One of the most recent uses of lasers is for adding material in successive layers to build up much thicker layers of material. This can again be done by curing polymers or by fusing or sintering high-melting-point powder coatings. This, when combined with computer-controlled mechanical translation stages, allows the gradual buildup of layers to form a 3D structure, a process termed 3D printing. This production of 3D objects can be done without lasers, for example, by simply spraying on molten or curable material from nozzles; however, lasers allow far more choice in the materials that can be added. Highly-refractory materials, such as ceramics, can be added as powder and sintered in place, as can high-temperature metallic alloys. The ability to manufacture complex 3D objects, even ones with closed-off internal cavities, simply from a software "blueprint" is a huge advantage to many manufacturing concerns. Thus, 3D printing has been considered to be one of the most "disruptive" technologies of the 2010–2020 decade and is liable to find ever increasing use in manufacturing throughout the twenty-first century.

This topic will be dealt with in more detail in a following chapter.

REFERENCES

1. G. Overton, D. A. Belforte, A., and Nogee, C. Holton, Laser Marketplace 2015 Report, Laser Focus World, January 2015.
2. A. Mayer, Industrial laser materials processing market report, The Industrial Laser User, issue 32, September 2003.

FURTHER READING

W. M. Steen and J. Mazumder. *Laser Material Processing*, 4th edition, 2010.

16

Laser and LED systems for industrial metrology and spectroscopy for industrial uses

JOHN P. DAKIN
University of Southampton

16.1 INTRODUCTION

The first lasers, initially discovered by Maiman at Hughes Research Laboratories in Malibu, California, were pulsed light sources, using optically pumped doped-crystalline rods. They were first reported in *Nature* in 1960. As is common with promising new technology, the initial publication was followed by ever-increasing "hype" over what they might be capable of. Although it was a dramatic scientific advance, apart from their use in a few very specialized areas, there was at first little justification for any commercial excitement. In fact, they were described by some as "a solution looking for a problem." However, in the present century, their true impact has now started to be fully felt, and they are becoming invaluable in an ever-increasing number of areas of our life.

Because of their initial complexity, rather fragile nature, and need to align mirrors carefully, their first impact was in specialist scientific areas, particularly in spectroscopy and a few areas in metrology [e.g., finding the distance to a retroreflector on the moon with LIDAR (light detection and ranging)!]. Laser machining, albeit using bulky and power-hungry carbon dioxide lasers, was also a fairly early practical use.

Since then, the evolution of relatively cheap and very compact semiconductor lasers has led to their widespread use in industrial equipment; many use them to pump secondary (also compact) solid state and optical fiber lasers.

Laser use now permeates many areas of industrial, office, and home equipment, for example, in metrology units, CD (compact disc) write/read, and laser printers. They have often been so well integrated into everyday equipment that their presence is often no longer obvious to the user, except in instances where the beam extends outside the unit.

The range of laser applications has now become so diverse that the following chapters, and Table VI.1, can only present a few key areas, and the interested reader should consult specialist texts on the subject to learn more. Here, we will try to indicate some important case studies of their applications in industry, starting with the more direct examples, where the presence of the laser beam is more obvious, rather than trying to list them in chronological order.

16.2 OPTICAL METROLOGY IN INDUSTRY

We shall first consider some applications of optoelectronics in metrology, a key area for the construction and manufacturing industries.

16.2.1 Laser metrology

16.2.1.1 THE LASER "STRAIGHT LINE"

One of the first uses of lasers was in the area of surveying and metrology. In air, at constant temperature and humidity, the collimated beam from a laser provides an excellent "straight line." Even if undesirable thermoclines are present, the beam deflection (mirage effect) due to small refractive index gradients is usually not serious for many applications. Fortunately, in one of the main civil engineering areas of use, that of controlling direction of tunneling, mirage deflections are very low, as the composition and temperature of the atmosphere in underground chambers is generally more uniform than outside, in the area close to the earth's surface. Even outside, over short distances, but particularly inside buildings, thermoclines are less of a problem, so the laser straight line finds many applications.

The simplest laser systems merely deliver a line of laser light from the source unit, but many far more complex units are commercially available. Many of these are equipped with a self-leveling capability, using internally dampened optomechanical systems, such that the external laser beam follows a true horizontal path. In addition, many versions have an optical directional-scanning capability (for example, using a scanned mirror) or contain passive beam-splitting optics (for example, diffraction gratings). With such additional optics, the beam can form horizontal or vertical sheets of light, multiple horizontal beams, multiple vertical beams, or any combinations of these. The value of having true horizontal or vertical optical guidelines is clearly a huge bonus, not just for workers in the construction industry, but even for do-it-yourself (DIY) home workers.

16.2.1.2 LASER DISTANCE MEASURING

As mentioned in the introduction, lasers have been used for LIDAR (laser radar) distance measurement by time of flight, even in their earliest years of use. The fundamental principles of LIDAR, and indeed some longer-range applications, are described in far greater detail in the earlier Chapter 12, Volume II on "Remote Optical Sensing by Laser" so only a brief review of the principles is presented here.

The LIDAR technology involves measuring the delay of an amplitude modulation pattern of the laser due to finite transit time of light from the instrument when traveling to and from a reflective surface. This can involve measurement of the delay of a single pulse, an orthogonal pulse train or the phase of a sinusoidal modulation imposed on the laser amplitude.

The first LIDAR instruments were extremely complex and expensive (for example, astronomical telescope–mounted systems to measure lunar orbital distance). However, the ready availability of cheap, "eye-safe" semiconductor lasers (rendered relatively eye safe because of very low power operation and/or operation at 1500 nm, where water absorbs strongly) has now made such systems much simpler and cheaper. This is particularly so when required for short-range applications. Then the return signals are many orders of magnitude higher, due to lack of significant atmospheric absorption and/or lower losses from the inverse square law dependence of return signal with distance.

Cheap LIDAR ranger instruments are now available from most major tool manufacturers at costs that a self-employed builder or even the DIY home "handyman" can afford. A lower-range instrument enables a builder or maintenance man to measure wall-to-wall and floor-to-ceiling distances by simple timing of optical propagation delays. Instruments typically have a range up to 100 m, and distance resolution of a few mm or less. As with longer-range surveying instruments, higher-range instruments can contain a self-leveling capability for the beam and also provide room-volume calculations based on 3D ranging.

As will be described in the following subsections, some simple distance measuring instruments may use LEDs or edge-emitting LEDs (ELEDs) instead of lasers, but lasers are usually the preferred choice whenever high directionality is desired, when there is a need for ultra-high-speed amplitude modulation, or whenever very narrowband optical filtering of returning signals is desired, to reduce the effects of ambient light on detectors.

16.2.1.3 LASER MEASUREMENT OF 3D PROFILES

The types of instruments described earlier have been developed with optical scanning systems, such that they can be projected onto solid objects to measure 3D profiles and even determine finer surface textures. For larger objects, the same technique of time of flight of an amplitude modulated laser beam can be used, and the beam is scanned in azimuth and elevation to cover the surface of the object, enabling distance to each point in the field of scan to be measured by the same LIDAR method.

Provided there is a direct line of path to each point, computer processing can then yield a 3D image of the profile of the object. To form a complete image, in particular, to "see" the back and sides of the object, more than one system is required, unless the object can be rotated such that it can be viewed from several directions.

There is an alternative rather simpler technology, involving cheaper hardware, where a grid pattern of many tiny squares is projected onto the object. This illumination pattern is then monitored by a TV camera at an observation angle away from the light projection path. Computer analysis of the distortion of the grid pattern then yields information on the approximate 3D profile, provided all parts of the grid are visible to the camera. To form a complete image, the object may have to be measured in several orientations, and to cover the full surface, with no missed "gaps," the illuminator may also have to be moved to take more measurements.

The grid method was originally much cheaper to use, as projecting an unmodulated grid image only requires a laser with very simple passive optics and low-cost TV cameras are, of course, readily available. At the present time, however, reasonably low-cost semiconductor lasers are available, with simple modulation capability, so this, with high-speed detector technology has enabled high-accuracy LIDAR-based 3D scanning systems to be produced very economically.

16.2.1.4 PRECISION DISTANCE RESOLUTION AND SURFACE PROFILING USING WHITE-LIGHT INTERFEROMETRY

If it is desired to measure 3D profiles to higher (sub-mm) accuracy, for example, to examine precise distances within optical components such as camera lenses, examine surface texture, quality of machining, or surface flatness of objects, etc., the time-of-flight method using amplitude modulation is less attractive, and other more precise methods are needed. To perform such measurements, the coherent nature of light can be used.

This is now most often performed using the method of white-light interferometry or optical coherence tomography. This concept was already discussed in the optical fiber sensors chapter (and indeed many instruments use optical fibers in their construction) and will be discussed in more detail again in the medical applications chapter, so it will only be briefly reviewed here. The essence of the method is that light from a fairly broadband source, such as an LED or ELED, is split into two beams. One beam is directed to form a tiny (typically ~ 1 μm diameter) focused point on the object and light is then reflected or scattered back to interfere with another reference beam from the same source. This latter beam has been arranged to travel along a linearly variable (time-scanned) reference path within the instrument. Close to the precise point, when the optical paths exactly match, visible inference fringes occur. The path length scan results in a rapid intensity modulation of the light *only* when the path lengths are closely matched, allowing the distance to the surface to be determined precisely from the position of the reference path scanner at that time. With suitable processing of the signals, this can be done to a fraction of the optical wavelength.

16.2.1.5 COHERENT LASER MEASUREMENTS

Many important industrial measurements use the light from a highly coherent (single wavelength) laser source. Such sources include gas lasers, and semiconductor lasers or optical fiber lasers, provided the latter types are equipped with some form of diffraction grating device to produce the desired narrowband optical output. To take advantage of the precision offered by the extremely short wavelength of light, the measurements can again use the technique of interferometry. This again involves splitting the light into two or more optical paths, and then recombining the light to create optical interference fringes. In this case, the detected light intensity varies sinusoidally with optical path difference, changing by a period (360°) for every wavelength (total) path change. If the beam wavefronts and their polarizations are each coplanar, these fringes maintain good visibility (high amplitude modulation index) provided the path length difference remains within the coherence length of the laser.

Interferometry can be used to measure dimensions of objects in many commercial types of optical calipers and micrometers, provided the fringes can be counted without error. This usually means that a zero-position reading has first to be taken, and then, by carefully counting fringe changes, high measurement accuracy is achievable.

With more sophisticated optics, sine and cosine fringe position measurement is possible, enabling the interferometer to unambiguously cross multiple fringes in a homodyne interferometer. These also allow subfringe measurement resolution by processing the two signals, to give precision levels even below the wavelength of the light. With careful design, such homodyne interferometers can be designed for industrial applications (Figure 16.1).

Another frequently used type of interferometric industrial measurement involves setting up a dynamic situation to allow rapidly changing fringe patterns to be measured, meaning that there is then no need to count individual fringes without error. A common means of achieving this is known as heterodyning, which involves mixing beams of different frequency to produce a difference or "heterodyne" signal. Such mixing processes were discussed in more detail in the earlier optical fiber sensors chapter. As mentioned in that chapter, instead of using two frequencies, a single laser can be frequency swept, changing its output frequency with time, such that any optical path difference causes a propagation delay then results

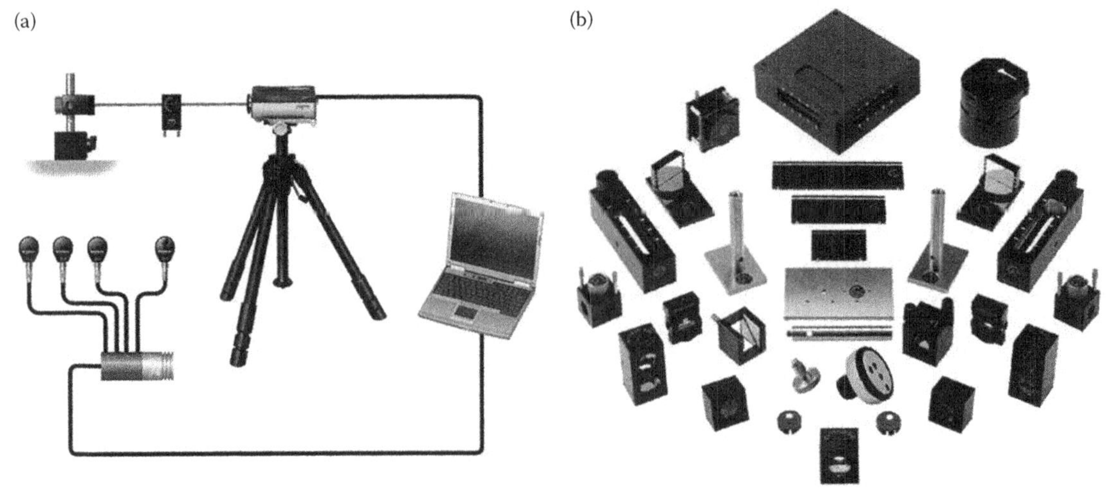

Figure 16.1 (a) An interferometric measurement laser and (b) various add-on optical elements for measuring different mechanical parameters. (Photo Courtesy of Renishaw plc., Wotton-under-Edge, UK.)

in a frequency difference at the detector, which is related to path difference.

Although the simplest measurement possible using interferometry is distance or wavelength, instruments are now commercially available to measure a much wider range of dimensional parameters, such as orientation/alignment, surface flatness, etc. Companies specializing in laser metrology can offer a wide variety of attachments with optical elements to facilitate such measurements (see Figure 16.1). These elements can be attached to industrial machines such as lathes, numerically controlled cutters (NMC machines), and robotic arms.

16.2.1.5.1 Compensation for atmospheric properties

Although the velocity of light is a well-known important physical "constant," its value in air can vary slightly because of the dependence of the refractive index of air. This varies with barometric pressure, temperature, and humidity and even depends on the true gaseous composition of the air. For the highest accuracy measurements, an additional sensor module can be employed to measure these factors and apply a correction either by feeding in data to correct the readout data or by changing the laser wavelength to compensate for the refractive index changes.

Figure 16.2 shows such a device for monitoring the air in the region of the measurement.

16.2.1.5.2 Laser doppler velocimetry

Laser Doppler velocimetry (LDV) involves measuring the velocity of moving objects or particles from the Doppler optical frequency shift that occurs when the target object is moved towards or away from the light source and detector, causing the path in one of the interfering light beams to vary with time. As with LIDAR, the theory of laser Doppler methods was discussed in more detail in the earlier Chapter 12, Volume II on "Remote Optical Sensing by Laser," so, again, we shall only give a short review of the principles here.

As mentioned earlier, varying an optical delay linearly with time, representing constant velocity in a path directly towards (or away from) the light source and detector, causes a phase shift decreasing (or increasing) linearly with time. This results in an increase (decrease) in the optical frequency. If this light returning after reflection from the target object is then mixed (interfered) with a portion of light traveling through a fixed optical reference path from the same source, the detected signal contains a difference frequency (heterodyne signal) component that is directly proportional to the velocity of the object and inversely proportional to the wavelength of the source.

Figure 16.2 Compensation module for an interferometric distance monitor. (Photo Courtesy of Renishaw plc., Wotton-under-Edge, UK.)

LDV measurements have formed the basis of many sensors, ranging from sensors for monitoring moving parts (e.g., pistons, turbine blades) in many types of engines, machines, fluids and even in atmospheric weather systems (where scattering particles are present). There is also, of course, the well-known application of interferometry in laser speed sensing pistols and fixed roadside units for traffic enforcement, although the latter usually tend to use longer wavelength radar signals!

One of the most useful implementations for industrial use is the optical fiber remote LDV system. Such a system uses monomode optical fibers to confer a number of useful advantages:

1. Light guided within fibers can be easily split into two paths with a compact all-fiber component (fused coupler) and then later recombined (mixed or heterodyned) using a similar component before being guided to the detector.
2. Optical fibers can conveniently guide light to (and from) the area where it is most needed. This is highly advantageous in complex areas such as engines and generators, where there is no obvious direct optical path. Without the fiber, countless carefully aligned mirrors would be needed to direct the beam.
3. A monomode optical fiber, particularly a polarization maintaining one, ensures that the modal purity of the laser source is maintained during propagation to the measurement area.
4. If a monomode optical fiber receiving head is used to recollect the reflected or backscattered light, then only a single spatial mode is collected, avoiding the speckle pattern interference, which would lead to fringe ambiguity and signal fading when the beams are combined.
5. A number of compact fiber-compatible optical frequency-shifting (frequency-biasing) components are available. These can add a known fixed frequency shift to the signal in one arm of the interferometer, so that the direction of motion can, without ambiguity, be determined from the heterodyne frequency.

Clearly, when measuring low velocities in severe environments, care must be taken to package the optical fiber carefully to avoid any mechanical or thermal influences on the fiber itself. Otherwise, a linear fiber stretch (or a rapid temperature change of the fiber with time) would also have similar effect to a Doppler shift. Fortunately, there are methods in which the effects of such uplead/downlead sensitivities can be compensated for. For this, please see the earlier discussion of optical fiber hydrophone technology in the earlier optical fiber sensors chapter, as it is also a potential problem in this sensing area.

When LDV is used to monitor movement of gases (e.g., gas flow in turbines) or highly transparent liquids, the only reflected light will be from small suspended particles, if present, or otherwise only from very weak Rayleigh scattering from the fluid itself. There is fortunately a detection method that enables LDV to still be used in low-light circumstances. At very weak light levels, the desired low-noise pure-sinusoidal heterodyne signal is no longer present, as, due to the quantum nature of light, it is reduced to a stream of occasional photons, arriving at seemingly random intervals. Fortunately, however, all is not lost, as the arrival rate is not truly random, each photon having a time-of-arrival probability that is proportional to the expected amplitude of the heterodyne signal that would have otherwise occurred, had the light levels been higher. It is possible to recover a signal that is effectively equivalent to the heterodyne signal by correlating the arrival times of all the detected photons. It is beyond the scope of this application chapter to describe the theory of this in detail. Suffice it to say that suitable correlators are commercially available, and they have lead to the availability of many practical commercial LDV systems, all designed for use where the scattered light is very weak.

16.2.1.5.3 Photon correlation spectroscopy

Photon correlation spectroscopy (PCS) is a method used for the classification of particulate suspensions. It is most valuable for monodisperse (all particles of the same size) or nearly monodisperse particles. The method is very similar to LDV in essence, but the particle suspension is illuminated by a coherent source via a single-spatial-mode filter (a monomode fiber is a useful form of this) and collected by a similar mode filter. Using a single optical fiber for both uplead and downlead, with a fiber directional coupler to isolate the energy in two light directions, we can achieve this very conveniently. The interference then takes place between scattered light components from many particles.

This results in a series of heterodyne signals, one from each of many interfering paths, when the particles move, all of which are seen at the detector. If the fluid is stationary, larger particles tend to move very little, but very small particles are subject to Brownian motion, as they are continually buffeted by the liquid molecules undergoing thermodynamic motion. The smallest particles move the most, as they have much lower inertia, so they have larger heterodyne frequencies. Thus, the Fourier frequency components of the heterodyne signals at the detector depend on the particle size (to be more specific, the hydrodynamic radius) and the temperature and viscosity of the liquid.

As with LDV in gases, the signals can be very weak, usually so weak that the detected signal is just a set of arriving photons, so the processing is again conveniently performed using a photon arrival time correlator.

PCS is established as a valuable tool for examining many industrial suspensions, such as those in the paints, biochemistry, and photographic film industries. As said earlier, it is usually most useful for sizing near-monodisperse suspensions, as otherwise the much larger signals from very large particles can swamp signals from the smaller ones.

16.2.1.5.4 Speckle interferometry

This is a method for determining movement of objects from the changes in the optical "speckle" pattern observed when they are illuminated by a coherent laser. The speckle is due to patterns of constructive (bright regions) or destructive (dark regions) interference, when an object is illuminated by laser light from a diffusely scattered screen. The seemingly random changes in the pattern when an object moves can be detected by cameras and processed to provide information on dimensional changes. The changing pattern gives an indication of stress or strain in mechanical systems. Extremely small dimensional changes, of order less than the wavelength of the light, cause dramatic changes in intensity of the pattern, which can be processed to recover information on the nature of the stresses and strains. Apart from the analytical data, the pictorial information provided can form a very valuable 2D image of static or dynamic distortion of the surface of structures, for example, a strained mechanical part, a medical prosthesis, or the outside surface of a whole automobile or aircraft engine.

For further reading, the review paper from EPFL institute in Switzerland (http://nam.epfl.ch/pdfs/113.pdf) gives an excellent description of this method.

16.2.1.6 OPTICAL SAFETY BEAMS AND CURTAINS

A wide variety of beam-interruption schemes can be constructed using lasers, taking advantage of their narrow focused beam and straight-line propagation. If laser beams are directed across an open path to a detector, the optical link formed can, by sensing beam obstruction, be used to detect objects, vehicles, or personnel entering an area. This can be for purposes of safety (e.g., to warn of, or prevent access to, dangerous areas) or for perimeter security (to detect undesirable intruders). It can then be set to shut down potentially dangerous machines and/or processes or to operate alarm systems. Using various simple optical components, the beams can be reconfigured very easily. For example, a single beam can be redirected, using sets of mirrors, to totally enclose an area. In addition to single beams, "to and fro" zigzag multiple beams can be generated from a single laser using sets of reflecting mirrors to fold back the light. The light can simply be split into multiple beams using diffraction gratings or partial reflectors. Continuous sheets of light with essentially no gaps can be generated from a single laser using cylindrical lenses and pseudo-continuous sheets of light can be formed using beam scanners, although in these latter cases, it will usually be necessary to use several detectors to detect obstruction of any part of the pattern.

16.2.1.7 BAR CODE READING AND PRODUCT SCANNING

Simple bar code reading, which we all know from supermarket checkouts, is of one the most common uses of visible lasers. The light source is a low power laser which is scanned across an area to read a black/white bar code which has been printed on items to be scanned. This can be used to identify products or components from warehouses, on production lines, points of final inspection or at point of sale. More recently, two-dimensional black and white dot patterns are being used to provide more detailed information, for example, on airline boarding cards.

The information content of black and white bar codes, although of great value in many applications, is still very simple compared to the greater information that could be carried by printed labels such as these in future. Clearly, additional degrees of freedom (bits of information) are offered by 3D code patterns, multicolor coding, and even fluorescent labeling. For the latter, not only the fluorescent spectrum but also the fluorescent lifetime of the pigments could be used to carry further information.

16.2.1.8 INDUSTRIAL ROBOTICS

Lasers can assist robotic production in many ways. The first is to allow profiling of objects to be handled or processed by robotic arms, using LIDAR and/or projected grid methods, as described earlier. Optical proximity sensors can be formed simply by launching light into optical fibers (from lasers or from LEDs) and measuring the level of reflected light coupled back into the fiber when an object is close. This is a very simple sensor, invented ~50 years ago, but it is useful as it has a very large change in intensity over very short distances of order 10–100 μm, depending on the fiber core diameter.

Apart from these simpler sensors, a number of tactile pressure-sensitive optical fiber sensors are being developed to attempt to replicate the subcutaneous sensors in the human hand. For example, if optical fibers are bent, light is coupled out, but there are also many other ways of measuring strain in optical fibers that may eventually be used, and there are also many other potential methods using optoelectronic sensor arrays that may not use optical fibers at all.

Of course, for many aspects of robotic guidance and control, the ubiquitous TV camera is already being widely used, due to its well-established nature. Such uses are likely to grow as image processing methods become ever more sophisticated.

Apart from coded labels, CCTV cameras, line-scan cameras, and scanned lasers can all be used to monitor presence, size, profile, color, and many other parameters of products as they travel through the field of view on production lines and conveyor belts. As industrial complexity advances, ever-increasing levels of detection capability and pattern recognition are being used to assist the advanced automation of manufacturing and processing plants.

16.2.2 LED-based sensing systems for industry

Low-cost LEDs and detectors can form the basis of many types of simple industrial sensors with the advantage that no mechanical contact is needed with the target object. Although they are simple, they perform a wide variety of commercially important functions, and so are a valuable aid to modern industry.

For economic reasons, most industrial LED-based sensors rely on detection of light using low-cost silicon detectors, which will also, particularly if not covered by filters, sense ambient light. Fortunately, many important sensors, such as proximity sensors, are very short range in operation, so the LED signal easily dominates the ambient light. If ambient light is significant, optical filters can be used to block much of the ambient light energy.

Longer-range sensors may require narrower band light sources and detector filters and possibly traffic light-type hoods or baffles to reduce ambient light. For the longest paths, the systems may require the light source to be amplitude modulated, so that the intensity modulation pattern can be detected at the optical receiver, clearly distinguishing it from the cross talk signal ambient light. For the highest accuracy or for very longer ranges, only laser sources are really suitable, as they can provide faster modulation rates, better collimation, and they allow narrower band filters to be used to reduce ambient light cross talk on the detector.

A few examples of the types of sensor available are described now.

16.2.2.1 OPTICAL SAFETY INTERLOCK SYSTEMS

Safety interlocks are a vital system component in many industrial areas. For example, it may be that a door, hatch, valve, or even multiple doors/hatches/valves must be closed before a process can be safely started or a system be worked on for maintenance. A wide variety of electromechanical switches exists for such purposes, but electrical switches are notoriously unreliable, failing in both short-circuit and open-circuit mode. This can occur if contaminated with dirt, corrosive chemicals, or even flooded with water. Optical beam-breaking switches, provided the light is modulated in a defined manner, can only fail in "dark" mode, so fail-safe systems can be designed more easily. It can be arranged that a potentially dangerous operation can only be started or continued when light passes through the optically operated switch. This can be, for example, through a physical hole, which is only open when a door or hatch is fully closed, or vice versa, if the alternative safety condition is desired. In many cases, a set of such optical switches can be cascaded, such that it is only in "safe" mode when light is transmitted through the entire system. A further advantage of using light is that it can be arranged for a portion of the light to be available for direct visual inspection by the eyes of an operator, if a "belt and braces" extra level of safety is desired. Using optical fibers to guide light through the arrangement clearly increases the ease of achieving cascaded optical beam-blocking switches and, if desired, guidance of a portion of the light for direct visual verification.

16.2.2.2 SHORT DISTANCE PROXIMITY DETECTORS

The technology used in short distance proximity detectors depends on whether the distance needs to be accurately measured or whether merely a simple indication of close proximity is needed (e.g., to detect the presence of a close object, or for a limit switch to prevent mechanical contact). For the latter, a very simple arrangement is possible, using a single LED chip mounted close to a detector on a common substrate. The light from the LED diverges in a forward direction, such that it illuminates a target object with ever-decreasing irradiance as distance increases. Light scattered back from the object returns back to the detector, and the signal again reduces very rapidly as distance increases. This means that even a very simple amplitude measurement gives a good indication of distance, provided the nature of the surface of the object is known. Clearly, specularly reflective (mirror) surfaces behave very differently from diffusely scattering ones, and blackened objects will, in general, reflect back less light.

If the light source and detector are equipped with collimating lenses, such that the emission is in a narrow beam, and light is only detected from a narrow detection, the response can be tailored to detect light only from a small position in front of the optics, where the two beams (the real beam from the LED and the virtual one, which defines regions from which light can be focused back onto the detector) intersect. This type of "crossed-beam" proximity detector is very useful for detecting the presence of objects in a defined region in front of the detector unit. It is now commonly used to operate water taps in washrooms and automatic flushes in public toilets.

16.2.2.3 MEDIUM-RANGE DISTANCE-MEASURING DETECTORS

Medium-range distance-measuring systems can no longer rely accurately on sensing the amplitude of reflected light, as the distance errors become progressively larger. They therefore need to rely on other methods. Essentially, there are four main methods for this:

1. LIDAR, which was mentioned earlier
2. Stereoscopic camera systems
3. Multiple "crossed-beam" sensors (using arrays of LEDs and arrays of detectors to form multiple "crossed-beams") of the type discussed earlier
4. LED-based time-of-flight sensors.

Stereoscopic camera systems, as the name implies, measure distance much as animals do, by looking at objects from two displayed vantage points, with a computer used as the "brain" to determine distance. This is probably the only satisfactory method, apart from 3D profiling LIDAR, to measure objects having complex shapes.

If multiple "crossed-beam" sensors are used, distance measurement can be achieved with arrays of LEDs and detectors close to the focal plane of lenses. Only one lens is required in front of each array to form the desired set of multiple beams

To obtain quantitative information of objects of known shape and orientation, the time-of-flight LIDAR method is a good engineering solution, and LEDs can be used as cheaper alternatives to lasers for medium distance monitors. The light source is modulated with a time-repetitive signal, which may be a simple pulse train, an orthogonally coded multiple pulse train, or a sinusoidal signal, the latter two providing the best signal-to-noise ratio. All these LIDAR methods rely on measuring the optical delay of light from the light source to the detector (although a second reference detector can be used to detect the true transmitted light pattern and avoid errors due to delays in the light source and also in detection systems, assuming these are matched).

In the case of pulsed systems, a boxcar-type correlator is used to determine the delay, whereas with a sinusoidal system, the phase of the signal is monitored. In the latter, an alternative is to use a phase-locked loop to vary the frequency of the amplitude modulation until the returning signal is in quadrature (90° phase difference) with the transmitted signal and measure the frequency at which this occurs.

16.2.2.4 LIQUID LEVEL SENSORS

The simplest liquid level sensors are hand-operated metal dipsticks and glass-viewing tubes mounted on the side of liquid-containing vessels. These, of course, are merely passive means of viewing the level with the human eye, so no optoelectronics is involved, unless the tubes were to be viewed with a charge-coupled device camera.

There are, however, two commonly used types of optoelectronic liquid level sensors. The simplest uses what would normally be a retroreflective prism, attached to the far end of a simple glass or polymer light guide. If light from an LED is launched into the guide, back-reflected light would normally be reflected, representing a "light-on" state. If, however, the prism becomes immersed in a liquid, internal reflection is prevented by index matching to give a "light-off" state. (Note: A transmissive variant on the retroreflective prism is to have a glass or polymer light-guide rod bent into a "U" shape, where the light is guided round the bend by total internal reflection unless the bottom of the "U" is immersed in the liquid.)

These simple sensors are essentially on/off types only, and so are binary digital in nature. Thus, unless many such sensors are installed at different depths, it can only really detect whether a vessel is correctly filled to the desired level.

A more sophisticated quantitative type of level can be constructed using a medium-range optoelectronic distance sensor, as described earlier. This is mounted above the liquid, pointing downward, and so, provided the system is not violently shaken, the distance to the partially reflective liquid surface can be measured.

16.2.2.5 OPTICAL POSITION ENCODERS AND TACHOMETERS

Optical position encoders are optoelectronic devices that can sense translational or rotational position using light transmission through coded masks. The masks are usually created by evaporating coded metal films on glass (or by etching thin metal sheets or continuous metal films deposited on glass). The idea is to produce a coded binary (on–off) spatial filter (rather like a regular bar code pattern) for transmitted or reflected light. Linear encoders can have parallel sets of coded bands extending along the length of a rectangular mask, with a successively finer pattern for each set of lines, allowing the position to be encoded in a multibit binary form, using light passing to separate detectors for each of the bands. This simple on–off approach can be used for optical micrometers, provided fringes are counted from a known zero position.

For encoding rotational information, a series of concentric annular code bands are produced on a circular disk, and the binary code is read out optically along a radial line.

Simpler forms of the same idea are used for optical tachometers, where the mask on the rotating disk has a simple radial pattern with a number n of lines, such that the transmitted light goes on and off n times for each revolution of the disk. The same concept can be used for linear motion with the light going on and off n times after traversing n lines of the code.

In both cases, there is an alternative using the Moiré fringe concept, where two very fine overlapping grids of slightly different spatial frequency are arranged to move over each other to produce bright–dark patterns, which change from dark to bright even with very small movements.

These are all simple devices, and it should be recognized that much higher precision could always be achieved by counting fringes from such a known zero position in a true optical interferometer, as discussed earlier.

16.3 INDUSTRIAL APPLICATIONS OF OPTICAL SPECTROSCOPY

Spectroscopy is a powerful tool in the industrial environment and an essential one in the chemical, pharmaceutical, and petrochemical industries. The methods of spectroscopy were discussed at length in earlier chapters, so we shall not describe them in detail here, but rather summarize some of the most frequently used methods in the industrial arena.

16.3.1 Absorption spectroscopy

Absorption spectroscopy is an important analytical tool for the chemical and material (metallurgy, paints, varnishes, lubricants, etc.) industries in particular. Analysis of materials is based on their electronic absorption bands (using visible and near-infrared (IR) instruments) or their vibrational ones (using IR spectrophotometers). If many chemical components with different absorption bands are present, a signal recovery process called multispectral analysis can attempt to separate out the detection of each band. This is done by computer analysis of the spectral data and comparison with spectra of expected constituents. Unfortunately, this method rapidly becomes far less effective as more compounds are present and is not really useful at all if the possible compounds have very similar (e.g., same chemical family) or very broad absorption spectra.

There are many manufacturers making instruments, from simple colorimeters to very large laboratory instruments, and many of these offer optical fiber probes to measure samples online. The method cannot easily be used with highly scattering (turbid) samples, unless they can be immersed in refractive index-matching oils.

Most early absorption spectrophotometers were based on broadband incandescent sources (or gas discharge lamps for the ultraviolet region) and used a scanned grating spectrometer to filter the light to a narrow linewidth signal of variable wavelength. These are still the norm for the great majority of applications, but a few advanced instruments improve signal-to-noise ratio by using other more intense tunable-laser sources or new synthetic broadband sources, such as super-continuum, laser-pumped fiber sources.

Almost all current spectrophotometer instruments also improve the signal-to-noise ratio by measuring many wavelengths simultaneously, using one of two main methods:

1. A grating spectrometer with a low-noise detector array in the focal plane, each detector therefore measuring simultaneously at a different wavelength.
2. A scanned interferometer filter, usually of a Fabry–Perot or Michelson type, with the entire throughput light then incident on a single detector. This produces an intensity-modulated signal, which requires a Fourier transform to be performed on the temporal variation of the signal. This is given the name of "fast Fourier transform spectroscopy" or FFT.

Method 1 can only be used to significant advantage in the visible and near-IR region, as it is only there that really sensitive, high-resolution detector arrays are readily available. Thus, FFT tends to be the preferred method for the IR region, which often needs more sophisticated cooled detectors.

The reader is referred back to the chapter on spectroscopic analysis for more details on such instrument aspects.

Absorption spectroscopy is also very important for optical gas detection, a topic dealt with another part of this same applications in Chapter 6.1.

16.3.2 Reflection spectroscopy

Reflection spectroscopy is useful for monitoring the external surface of materials and is particularly useful when examining opaque or translucent items. It can be performed by monitoring specular reflection from flat surfaces, or diffuse reflection from rough or powdery surfaces. Therefore, it has many uses for quality control in the food and pharmaceutical industries. A variation, frustrated total internal reflection spectroscopy is where reflection from an optical interface between a glass surface and the analyte material is measured. Light is usually launched into a prism and the exiting light, which would normally undergo complete internal reflection is analyzed. Measurable spectral variations occur with very highly absorbing compounds. The method is particularly useful for analyzing compliant materials such as polymers or food, where the material can be pressed against a hard glass prism (or vice versa) to provide close optical contact over the measurement area.

Both these types of reflection spectroscopy can be performed with color cameras. Normal reflection spectroscopy is from a distance, for example, to examine items as they pass by on conveyor belts in production lines. This can enable faulty or contaminated products to be identified by their color (including IR or UV spectral variations, if desired), and so is a very useful tool for quality control.

16.3.3 Fluorescence spectroscopy

Fluorescence spectroscopy is a very useful tool for analysis of some types of materials, as it can be used with diffusely scattering samples. Thus, it can be used for surface analysis of materials, regardless of whether they are rough or smooth. It involves exciting the material with short wavelength light (usually at wavelengths well below 1 μm) and observing the reemitted fluorescent light.

One of the big practical attractions of fluorescence is that it is not affected by elastic scattering (light scattered at the same wavelength as the incident laser light) due to the internal optical filtering of the detection system in the instrument that rejects incident light wavelengths. Because of the short wavelengths usually used (visible and near-IR), fluorescence analysis is compatible with optical instruments such as microscopes and telescopes.

A big disadvantage is that only a limited number of materials exhibit strong fluorescence, for example, chlorophyll and several other biological specimens; several natural and mineral oils, particularly aromatic compounds with benzene rings, various organic dyes; some gemstones, such as ruby; glasses containing certain transition and rare earth compounds; and many common semiconductors.

Because of this disadvantage, many fluorescent-monitoring systems involve the use of added tracers, usually highly fluorescent organic dyes such as fluorescein or rhodamine, to materials that would otherwise not fluoresce. These dyes can, for example, be used to help to track flow of water in streams and rivers, detect leaks in plant, mark organic or biological specimens for microscopic analysis, and place identification markers on products on a production line to facilitate easy tracking through the factory.

A useful advantage of fluorescence is that the reemitted light has a significant time delay before it is re-emitted, and this delay can vary from nanoseconds (e.g., for many organic dyes and semiconductors) to several hundreds of microseconds in chromium-doped alexandrite crystals and some Nd-doped laser glasses. It can extend to several milliseconds in ruby crystals (chromium-doped sapphire). Some fluorescent phosphors have yet longer lifetimes, particularly, the phosphorescent materials used in luminous watches, which can store optical energy for hours.

Because of this huge time variation in the fluorescent reemission, individual components in an analyte can often be separated in the time domain, even if they might otherwise have similar excitation and absorption spectra. This separation process is called time-resolved fluorescent spectroscopy. Using the three separate parameters of absorption spectrum, fluorescence spectrum, and fluorescent time delay, it is possible to separate out most fluorescent components likely to be in an analyte in an effective manner.

16.3.4 Raman spectroscopy

Raman spectroscopy observes laser light that is scattered inelastically from compounds. It involves a photon energy change due to photon–phonon interactions, resulting in light being reemitted at shorter (anti-Stokes) and longer (Stokes) wavelengths than the incident light.

Until recently, Raman spectroscopy was a highly expensive and sophisticated analytical procedure, which was only feasible in well-equipped analytical laboratories, as it needed very large and expensive high-power lasers (usually argon ion lasers) and similarly large and expensive grating spectrometers with photomultiplier detectors. However, the rapid evolution of compact high-power semiconductors, solid state and optical fiber-based lasers along with compact spectrometers using sensitive semiconductor detectors has totally changed the practicality of the method.

One of the big practical attractions of Raman scattering is that it is, like fluorescence spectroscopy, not affected by elastic scattering (light scattered at the same wavelength as the incident laser light). It can therefore be used to analyze turbid and translucent materials, medical tablets, many powders, etc., as any elastically scattered (non-wavelength-shifted) incident light is easily removed with internal optical filtering.

It can also be used for surface analysis of materials, again regardless of whether they are rough or smooth.

Unlike fluorescence, almost all compounds exhibit Raman scattering, so its analytical application is far wider in scope. There are other fundamental physical advantages enabling a visible or near-IR laser to probe vibrational bands, which would normally correspond to much longer IR wavelengths. It also has different "selection rules" for the types of absorption that can be monitored, making it complementary to the normal method of absorption spectroscopy.

Raman spectroscopy is now a widely used industrial tool, extensively applied to analyze all types of material from semiconductors, pharmaceutical tablets and powders, polymers, food and dairy produce, and many more. Because it operates at shorter wavelengths than IR absorption methods, it is compatible with optical microscopes, and so can perform microanalysis of samples containing many different types of particles or microstructures.

16.3.5 Photon correlation spectroscopy

This method was discussed in detail earlier because of its similarity (not only in its operating principles but also in its hardware and signal processing equipment) to the technique of LDV. Hence, it will not be discussed further here. (It is mentioned again only because of use of the word "spectroscopy" in the name for this technique.) As stated earlier, it is a valuable tool for examining many industrial suspensions, such as those in the paints, biochemistry, and photographic film industries.

16.4 CONCLUSIONS

We have explained the ways that lasers and LEDs (and for spectroscopy, some other broadband light sources) can be used in industrial applications. As component availability improves and costs reduce, such optoelectronic systems are becoming ever more commonplace. As humans, we use our eyes as our primary sensing method, so it is hardly surprising that optics is so valuable for instrumentation.

ACKNOWLEDGMENTS

The author wishes to thank Renishaw plc. for kindly allowing the use of photographs of some of their industrial instrumentation products.

17

3D Printing applications

CANDICE MAJEWSKI
University of Sheffield

17.1 INTRODUCTION

Three-dimensional (3D) printing (3DP) is an umbrella term for a variety of additive manufacturing processes, where a 3D object is built up one layer at a time (see Figure 17.1). The layers are programmed from a computer-aided design (CAD) model, which consists of many discrete two-dimensional (2D) layers to represent the object. Part manufacture is carried out without

Alternative Terminology

Although this group of technologies is most commonly known as 3DP, a number of other terms are often used interchangeably, for example,

- Additive manufacturing (formal American Society for Testing and Materials [ASTM] definition)
- Additive layer manufacturing
- Direct digital manufacturing
- Freeform fabrication
- Layer manufacturing
- Rapid prototyping/manufacturing
- Tool-less manufacturing

APPLICATIONS

Applications of 3DP can be found throughout almost every industry. Although there are too many of these to include here, examples include the optimization of part geometries within the aerospace and automotive industries (e.g., to reduce weight and/or improve performance) (Figures 17.2 and 17.3), the production of personalized medical aids (Figure 17.4) and implants (prostheses), production of tools and devices to assist manufacturing operations, or production of fashion items such as clothing or jewelry (Figure 17.5). In some cases the CAD models for these can be produced directly from optoelectronic scanning of real objects.

Figure 17.1 Layer-by-layer process of 3DP.

Figure 17.2 Lightweight, optimized, engine block. (Image courtesy of Autodesk, Inc.)

Figure 17.3 3D blade geometry via 3DP. (Image courtesy of EOS GmbH.)

the requirement for any form of additional tooling, opening up possibilities for cost-effective production of personalized parts. The layer-by-layer nature of the process also allows efficient production of substantially more complex geometries (including ones with internal closed cavities) than possible with more traditional techniques such as injection molding.

17.2 3DP PROCESSES USING OPTOELECTRONICS

A variety of methods fall under the umbrella of 3DP, including a variety of different material types, in a range of forms. Traditionally, there has been no unanimous method of classifying systems, with various methods of categorization including feedstock type (powder, resin, laminates, etc.), processing method (e.g., laser-based, jetting, lamination…), speed, cost, and numerous other distinctions. The ASTM recently provided a standardized set of classifications (Table 17.1).

Within these categories, a number of processes involve the use of optoelectronics. The most prominent of these are discussed here.

17.2.1 Material jetting

Material jetting processes involve the selective dispensation of droplets of liquid material, normally through an ink-jet print-head, onto a substrate. These droplets then solidify to form the final part.

Some processes allow the use of materials (e.g., waxes), which are deposited in a semimolten state and solidify as they cool, whereas others involve the use of photopolymer materials, which are optically cured by an ultraviolet lamp, directly following their deposition. This latter method causes a photochemical change within the part to produce the desired properties. Subsequent layers are built up until the final part is produced. Some systems in this category offer the advantage of jetting two separate materials simultaneously, allowing mixing of constituents in varying proportions. This allows the production of support structures for overhanging areas (subsequently removed through a variety of methods) or parts having variable materials properties and/or colors throughout.

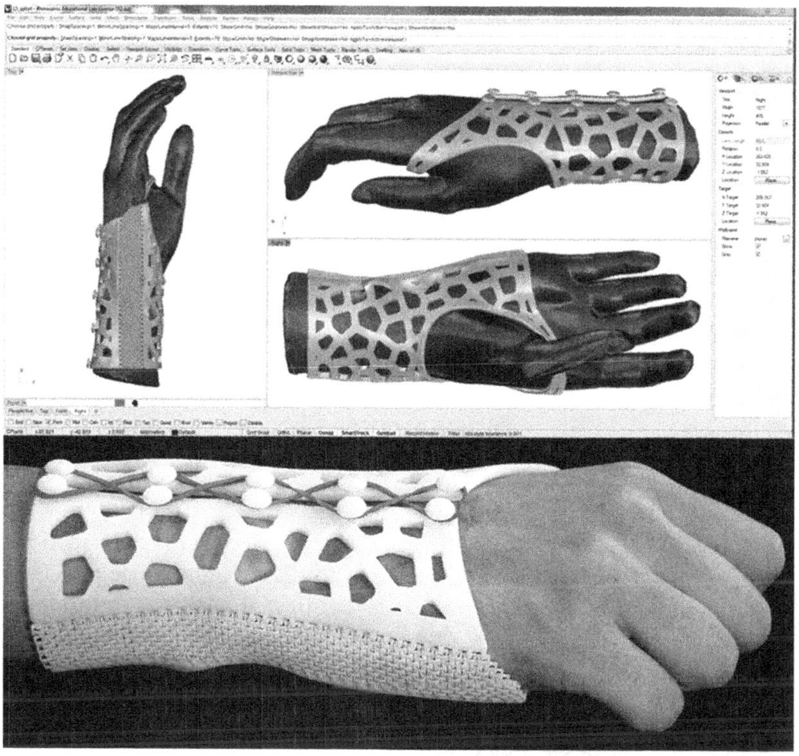

Figure 17.4 Personalized 3DP wrist splint (CAD modeling and final part). (Image courtesy of Dr. Abby Paterson, Loughborough University.)

Figure 17.5 3D printed pendants produced on a Cooksongold Precious M080 3D printer and the Jewellery Industry Innovation Centre, Birmingham City University. (Parts designed by Lionel T Dean and polished by Finishing Techniques, Ltd., and the Jewelle.)

17.2.2 Powder bed fusion

Powder bed fusion relies on inputting thermal energy to melt selected cross-sections of a powdered material (both polymers and metals can be processed within this category). These melted regions solidify upon cooling, whereas the remainder of the build volume remains as a powder surrounding the part(s).

The majority of powder bed fusion processes, including Laser Sintering and Selective Laser Melting, use a directed laser beam in order to provide the required input energy in the desired cross-section (see earlier subchapter by Paul Harrison). Infrared energy can also be used to achieve the desired melting, as in the case of the High Speed Sintering process (Figure 17.6). This process prints a radiation-absorbing "ink" onto the powder bed, immediately followed by heating using an infrared lamp. Printed areas absorb enough energy to melt the powder beneath them, whereas unprinted areas remain as powder.

In addition to these methods, an electron beam can also be used to provide the desired thermal input (Electron Beam Melting).

Table 17.1 Classification of 3DP systems

Category	Description
Binder jetting	Liquid bonding agent jetted onto required cross section of powdered build material
Material jetting	Liquid build material jetted to create cross section directly
Powder bed fusion	Thermal energy (e.g., laser) selectively fuses powdered build material
Directed energy deposition	Thermal energy melts build material directly as it is deposited
Vat photopolymerization	Liquid photopolymer selectively cured by laser or light source
Material extrusion	Build material is extruded through a nozzle or other orifice
Sheet lamination	Sheets of material bonded or fastened to form finished part

Figure 17.6 High Speed Sintering process. (Image courtesy of The University of Sheffield.)

17.2.3 Directed energy deposition

Directed energy deposition also relies on the use of thermal energy; in this case, a high-energy beam is used to melt the build material during its deposition. Processes such as laser-engineered net shaping and direct-metal-deposition produce parts by melting a metal powder as it is deposited to form a melt pool, which then solidifies as the focused beam is moved on. Although there are a number of companies producing equipment in this category, the use of these processes is substantially lower than that of powder bed fusion processes.

17.2.4 Vat photopolymerization

Processes in this category produce parts by inducing a chemical change within a vat of photopolymeric resins. Several of these processes utilize lasers to scan each 2D cross section, whereas others use 2D digital light image projection to cure whole layers in a single pass. Multiphoton polymerization is an emerging method within this category, capable of producing nanosized features (e.g., for use as nerve guides).

17.2.5 Sheet lamination

Sheet lamination involves the stacking and joining of cross-sections of sheet material. Processes within this category range from stacking and bonding layers of paper through to ultrasonically welding metal foils. The majority of these processes use mechanical methods to cut an outline, but in some cases a laser may be used to produce the desired cross-section.

17.3 OTHER APPLICATIONS RELEVANT TO 3DP

Outside of the processing method itself, optoelectronics can be found within a range of other stages of the overall manufacturing process. Although it is not practical to include every situation in which they are used, the following sections provide a range of relevant examples of the most predominant areas.

17.3.1 Preprocess

A number of operations take place prior to the manufacturing stages of any 3DP process, including data preparation and quality checks on the raw materials for the process.

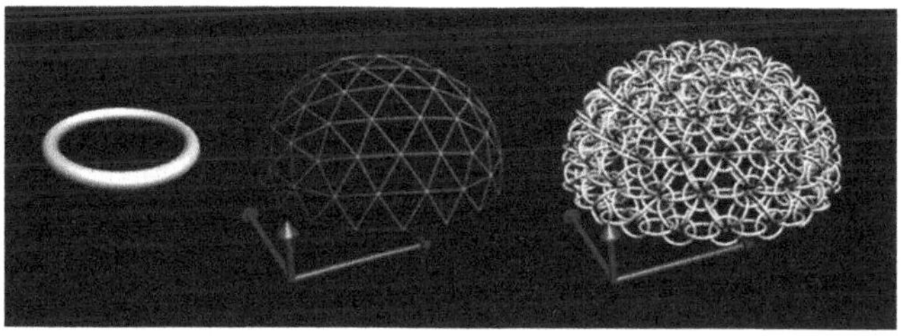

Figure 17.7 Generation of conformal textile geometry-individual "links" is mapped to a uniform, equidistant mesh, draped over a scanned geometry. (Image courtesy of Dr. Guy Bingham, Loughborough University.)

Figure 17.8 CAD model created from full-body 3D scan. (Image courtesy of Ahsan Khan, Code3D.)

17.3.1.1 PRODUCTION OF 3D DATA

The standard method of producing 3D data for these processes is through the use of specific CAD software to design an object from scratch. However, the ability to produce a reliable model from an existing master object has benefits in a wide range of applications; for example, the ability to produce representations of existing components that have no CAD data available or representations of hand-crafted objects. Further benefits arise when considering the personalization of clothing or other objects to be fitted to the human body.

Although it is possible to obtain a fairly accurate representation of an object through physical measurement of all dimensions, this becomes highly inefficient when considering highly complex components or organic geometries such as the human body. In some cases (e.g., rare or culturally significant historical objects), physical handling must remain at a minimum, and noncontact methods must be used. In these situations, the use of noncontact scanning methods to produce a 3D CAD replica of the object is invaluable.

Once the scan data have been captured and processed, it may then be used to create an exact replica of the original item, or to add, subtract, or modify features as required. In the case of objects personalized to the human body, the data would be used as a base for the design of the overall object. Applications of this are varied, ranging from personalization of toys and novelty items through to the production of conformal textiles (Figure 17.7) for both fashion and personal protection scenarios, and the manufacture of personalized prosthetics and medical footwear, etc.

A variety of scanners and methods are available, ranging in terms of price and complexity. An increasing number of hand-held scanners are now available for the general public, including a clip on structured-light scanner for iPhones, which makes use of the phone's in-built camera to capture the 3D data. At the opposite end of the size range, large scanners are available to capture full-body scans of individuals (see Figure 17.8).

17.3.1.2 TESTING OF RAW MATERIAL

The quality of the base material, which feeds into a 3DP process, is critical to the resultant part quality achieved. A variety of different characteristics may be measured, via a number of techniques, in order to ensure the quality and consistency of a particular material batch.

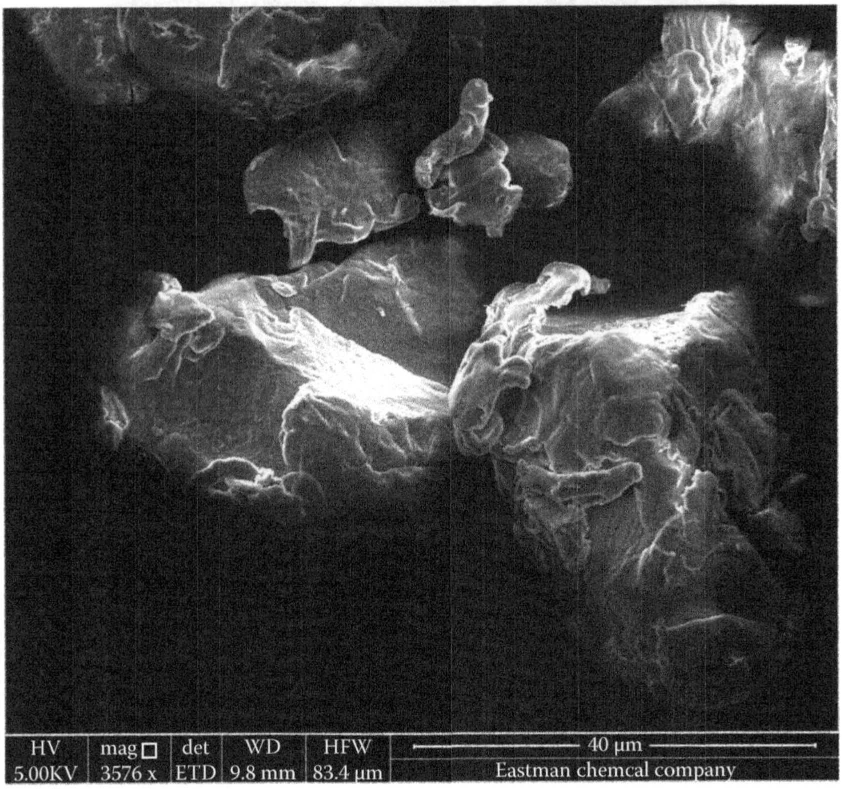

Figure 17.9 Scanning electron microscope image of laser sintering powder. (Image courtesy of Eastman Chemical Company and produced by Dr. Thomas Smart.)

Of particular interest here is the use of lasers in particle size analysis for powder-based systems. An incorrect size distribution can lead to a lack of part strength (e.g., through formation of voids within the final part), or in some cases build failure (e.g., excessively large particles causing disruption during recoating). Laser diffraction analysis involves monitoring of diffraction patterns from a laser beam passed through a suspension of the particles and subsequent calculations of size range and distribution.

Scanning electron microscopy, which produces extremely high-resolution images through the use of a focused beam of electrons, can also provide crucial information, in terms of both examining particle morphology (Figure 17.9) and understanding the internal structure of a finished part.

17.3.2 Process monitoring

A number of current 3DP processes use opto-electronics for process monitoring, with the use of infrared sensing particularly prevalent for processes requiring preheating of raw material prior to, and during, part production. For example, the Laser Sintering process uses a pyrometer to measure powder temperature during its preheat and build stages, whereas in High Speed Sintering, an infrared camera performs the same task. The use of these sensors to control the quantity and uniformity of heating can provide substantial benefits in terms of repeatability and reliability. Future developments are likely to focus on use of sensors and cameras to analyze part quality on a real-time basis throughout a build, with closed-loop control to rectify any errors during the process.

17.3.3 Postprocess

17.3.3.1 PART FINISHING

There are several reasons why some form of post-production part-finishing is required when using 3DP. For many 3DP techniques, the layer-by-layer process, in combination with the build process itself, can lead to a relatively rough surface in one or more orientations. Certain processes may use optical energy to provide some form of surface treatment. For example, it is possible to smooth the surface of a metal part by scanning of the surface with a high-energy laser. This remelts a small layer of material, following which surface tension effects assist in producing a smooth surface.

Other processes, such as stereolithography, require posttreatment in an ultraviolet oven in order to complete the curing process and provide the required properties.

17.3.3.2 QUALITY CONTROL AND TESTING

As with the majority of manufacturing processes, certain checks (geometric accuracy, part strength, etc.) may be required in order to ensure the final part is fit for purpose. Specific examples may include the following:

- *Tensile testing* Tensile testing involves gripping a specimen between two clamps and "stretching" it until it breaks. Various parameters are recorded, including the extension at break. Laser extensometers can be used as an alternative to the standard method of measuring extension through the physical attachment of two devices to the test specimen. This can provide an effective noncontact method of measuring materials, which may damage, or be damaged by, a physical "clip-on" method, as well as preventing inaccuracies, which can occur when a specimen "slips" during the testing process.
- *Surface roughness* Traditional methods of measuring the roughness of a part or component involve the measurement of a single line by lightly dragging a stylus along a section of a surface. Although there are some noncontact methods of achieving this same result, this remains a relatively inefficient and inaccurate method for assessing large areas. Optical techniques such as interferometry or structured light scanning are now increasingly prevalent in 3D surface imaging systems, providing faster and more accurate coverage of much larger surface areas.
- *Accuracy* Depending on the exact 3DP technique used, inaccuracies in the process itself, or other factors such as shrinkage during cooling, can lead to differences between the geometry specified in the original CAD file and the part itself. Scanning techniques such as those described in Section 3.1.1 may also be used to analyze the geometric accuracy of a finished part (Figure 17.10). Scan data are captured from the object, from which the differences between these data and the original CAD data are calculated. This in turn allows the

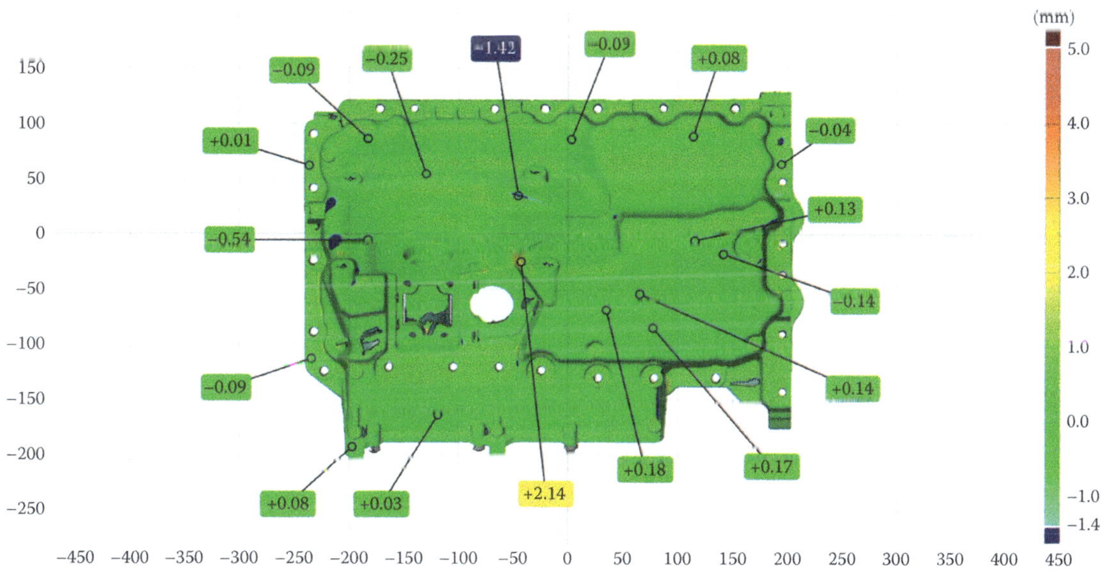

Figure 17.10 Comparison of actual and intended part geometry from scan data. (Image courtesy of Physical Digital Ltd.)

user to identify whether the part falls within acceptable tolerances and any particular areas of concern.

- *Microstructural analysis* In many cases, it is desirable to perform nondestructive testing of parts or components. The internal microstructure of a part can provide a good indication of its mechanical strength, and a number of methods can be used to examine this. A key example of how optoelectronics can assist in this task is through X-ray computed tomography (CT), whereby a series of 2D X-ray images can be combined to give a 3D model of an object. Although this process is predominantly used as a form of medical imaging, the use of micro-CT (higher-resolution imaging than standard CT) can enable a close-up inspection of the internal structure of a component without the need for destructive testing methods.

17.4 CONCLUSION

It is clear that 3DP is a highly disruptive manufacturing technology, the importance of which is growing rapidly. It can be seen that optoelectronics plays an increasingly important role in many 3DP-related areas, within the manufacturing processes themselves, in postproduction surface treatments, and in other applications to assess the overall quality of the parts they produce. It can be expected that this trend will continue as improvements are made to existing methods and new techniques are developed. In particular, the use of optoelectronics in process monitoring and control is likely to be a key area for the integration of these two areas in the future.

18

Fiber optical sensors for monitoring industrial gas turbines

RALF D. PECHSTEDT AND DAVID HEMSLEY
Oxsensis Ltd.

18.1 INTRODUCTION

Gas turbine (GT) engines are widely used for aircraft and marine propulsion, electrical power generation, and fuel gas compression and distribution. There continues to be intensive industrial and academic interest in optimizing GT design and operation to increase fuel efficiency and to reduce pollution together with increased fuel flexibility under greater variable loads and ramp rate conditions [1]. The combustion concepts introduced to respond to these demands are prone to generating instabilities in the GT compressor and combustion systems [2,3], which can lead to potentially catastrophic events such as surge and stall. Fiber optical sensor (FOS) technology is now becoming available that is capable of providing critical real-time information regarding the onset of combustion instabilities in the hitherto inaccessible region of the GT combustor.

The environmental challenges in GT combustors are considerable. Sensors have to withstand temperatures that may exceed 1000°C at pressures up to 60 bar, while having the sensitivity to detect pressure fluctuations of sub-millibar levels at gas pressure oscillation frequencies up to and beyond tens of kilohertz. To date, industry standard sensing solutions for dynamic pressure measurements in GT combustors have mainly relied on piezoelectric transducers, predominantly mounted at ambient temperatures and linked to the combustor via a "semi-infinite" tube. This very practical solution, however, suffers from dynamic pressure frequency response and sensitivity limitations when used to detect the onset of these combustion instabilities and is impracticable in the engine nacelle of an aircraft where electromagnetic immunity (EMI) considerations add to the environmental challenge imposed on these electronic sensors. Maintenance and reliability issues can also arise from condensed water freezing in the semi-infinite tube and potentially creating a leak path.

GT exhaust temperatures are normally measured using thermocouples. However, the need for radiation shielding and mechanical protection reduces their spatial resolution and sensitivity due to the thermal mass of the sleeves to a point where the data are of limited use to compare, for example, the combustion temperature profiles between cans. Indeed, if it were possible, direct combustor gas temperature measurement would be of great value in both the development and commercial use of GTs, but this is often beyond the capabilities of the ubiquitous thermocouple, particularly with respect to its maximum operating temperature. Measuring the exhaust gas temperature instead creates uncertainties regarding which can is responsible for temperature variation due to the angular shift of the temperature profile as the gas travels down the engine. There are further concerns regarding the drift behavior of thermocouples at higher temperatures [4] and the increased number of wires required for high-density profiling.

FOS has an intrinsic advantage in that it spatially separates the optoelectronic readout unit (interrogator) and the passive optical sensor head of the gauging system using fiber optical cables. The nonelectrical sensor head, manufactured from super resistant materials, can be placed in the gas boundary layer for direct measurement of single-point dynamic pressure, static pressure, and temperature using Fabry–Perot (FP) based transducers. For multipoint temperature measurements, silica-based fiber Bragg grating (FBG) sensors systems can be employed, albeit at somewhat lower temperatures [5]. Immunity to EMI is facilitated by the dielectric nature of the fiber optical cable links. These emerging FOS solutions provide the potential for robust multi-measurand measurement capabilities for *in situ* real-time measurement of key combustion parameters in commercial GT applications. This chapter outlines the measurement principles and provides a summary of selected applications.

18.2 THE SENSOR

The operational temperature requirement is very restrictive on the material choice for the sensor head. Indeed, sensor solutions are restricted to passive optical devices that can be fabricated from super resistant materials such as sapphire (aluminum oxide crystal) and inconel (proprietary nickel chromium alloy). These materials will retain adequate mechanic properties at elevated temperatures, as demonstrated recently for a monolithic sapphire-based transducer element [6]. The optical transducer structure based upon an FP etalon provides a practicable scheme for monitoring dynamic pressure, static pressure, and temperature, all from the same sensor head, simultaneously if required. Schematically, the transducer element is depicted in Figure 18.1a.

Typically, pressure sensors consist of an optical structure that has a pressure-sensitive diaphragm with a vacuum cavity formed behind by parallel boundaries. Light incident from the right-hand side (the back) of the sensor reflects from the various optical boundaries and can be made to interfere. The intensity of the interference fringes depends on the optical path difference (OPD) between the reflected beams; the maximum intensity occurs when this OPD is an even number of wavelengths of the incident light.

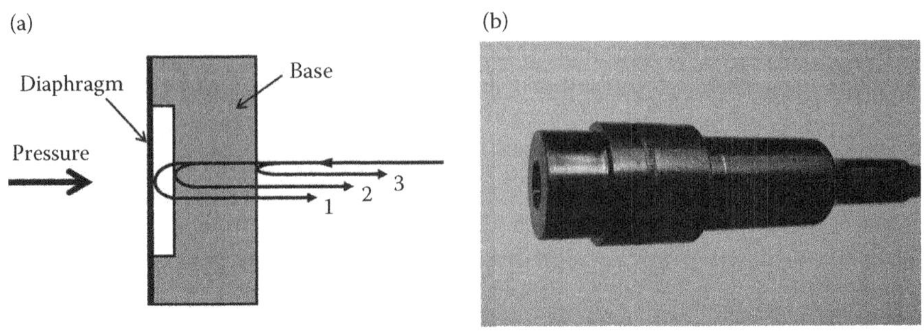

Figure 18.1 (a) Schematic of transducer. (b) Photograph of sensor head.

It can be seen from Figure 18.1a that pressure can be measured from the interference between beams 1 and 2 (because the intensity will depend on the diaphragm deflection in relation to the base) and that temperature can be measured from beams 2 and 3 (because the intensity will now depend on the thermal expansion of the base). A photograph of a packaged sensor structure designed for GT combustors that embodies these principles is shown in Figure 18.1b. The optical transducer element is located in an inconel casing and held in place by a high-temperature seal. The optical fiber is introduced from the back to illuminate the FP cavity and to collect and transmit the reflected light. The entire light path is internal so the sensor is immune to contamination of the pressure-sensitive diaphragm on the front face. There is, however, a temperature limitation of ~750°C when using fiber optical cable for light transmission and data communication. This limitation is overcome by making use of the natural temperature gradient outside the GT combustion chamber provided by the cooling air, enabling hot gas pressure measurements of up to 1000°C at the tip of the sensor.

18.3 THE INTERROGATOR

The interrogator is the optoelectronic readout unit that is connected to the sensor via a fiber optical cable. It illuminates the sensor with infrared light, detects the reflected signal, and transforms this information into a calibrated analogue electrical signal whose voltage is proportional to the parameter being measured.

The precise nature of the returned signal is determined by the OPD variations between the various optical boundaries within the sensor in relation to the incident wavelengths of light. One possible approach is to make use of the intensity of the reflected interference pattern, which changes in proportion to the change in OPD as the diaphragm moves in response to dynamic pressure changes. The corresponding reflected intensity as a function of cavity size is shown in Figure 18.2a. At the steepest part of the curve, the intensity varies approximately linearly with diaphragm deflection, enabling a calibrated intensity to be used as a pressure gauge. However, any intensity changes not due to diaphragm movement introduce measurement errors, especially when used outside a laboratory environment. For example, in GT applications, additional losses due to fiber bending can be caused by the usually severe vibration levels present.

This limitation may be overcome by using the signal processing scheme shown in Figure 18.2b. In this scheme, the sensor is illuminated by two different wavelengths of light simultaneously [7]. The light reflected back from the sensor vacuum cavity is separated into the two wavelength components and the amplitude of each is measured separately. The ratio of the two intensities is then calibrated against pressure. Because many of the factors that will affect the reflected intensity other than diaphragm movement will equally affect both wavelengths, the intensity ratio remains largely unchanged. In addition, due to its simple setup and minimal signal processing, a large signal bandwidth can be achieved. A simultaneous

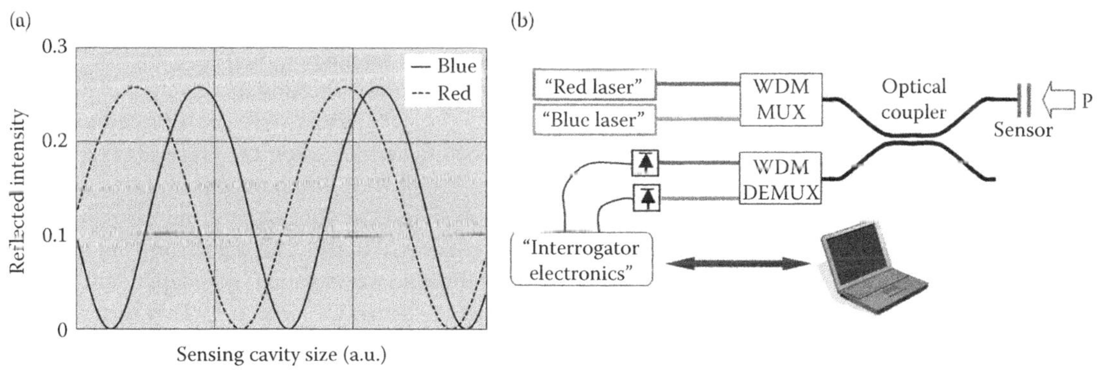

Figure 18.2 (a) Return intensity from sensor. (b) Schematic of interrogator.

temperature measurement function can be added, providing additional static pressure information and a more accurate dynamic pressure calibration factor [8]. Other interrogation schemes are also viable such as those reviewed [9].

18.4 APPLICATIONS

Four GT applications of FOS are presented. They are all aimed at either demonstrating the viability of the fiber optical sensing approach or gathering data to demonstrate the long-term reliability of the sensor solutions. The three pressure measurement applications include a high-pressure combustor test facility and 7–300 MW GT electrical power plants. A summary of the diverse applications of FBG-based temperature sensors for multipoint measurements throughout the engine is also given.

18.5 FIBER OPTICAL DYNAMIC PRESSURE SENSOR ENDURANCE TESTS IN A SIEMENS 300 MW SGT5–4000F COMMERCIAL GT COMBUSTOR

The FP optical structure as the sensitive element for dynamic pressure measurement is presented to the GT using the configuration shown in Figure 18.3.

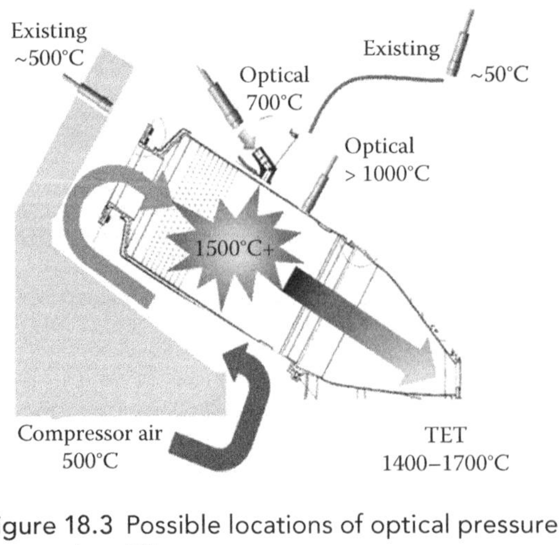

Figure 18.3 Possible locations of optical pressure sensors in a GT.

Optical dynamic pressure sensors are linked to a remote optoelectronic interrogator by fiber optic cables, and the system confers the following advantages:

1. Combustion instability measurements are made closer to the event. Indeed, the front face of the sensor can be mounted flush to the inside of the combustor for maximum sensitivity.
2. Higher frequency content can be measured because the acoustic signal is not attenuated by a connecting tube.
3. Immunity to electromagnetic noise.

An example of such an installation is on a Siemens SGT5–4000F (V94.3) at RWE's 1.3 GW Combined Cycle Gas Turbine CCGT Station at Didcot, Oxfordshire, United Kingdom (Figure 18.4). Optical pressure sensors were fitted to two locations: in the exhaust thermocouple lance and on the burner flanges. The sensor location temperatures are between 400°C and 550°C.

The availability of dynamic pressure data obtained directly from the GT compressor, combustors, and exhaust stages provides the potential to develop instrumentation to mitigate combustion instabilities caused by the interaction of fluctuating heat release of the combustion process with naturally occurring acoustic resonances. These interactions can produce high-frequency pressure oscillations within the combustor that can lead to very costly mechanical failure.

The 3-year sensor system deployment at Didcot Power Station in their Siemens SGT4000 engines has proved that the sensors can survive engine conditions for considerable periods of time and that recorded dynamic pressure data agree well with legacy sensors and expected results. An example of the dynamic frequency map as a fuction of time during the engine shutdown is given in Figure 18.5.

The optical sensors were recalibrated and then returned to service when the GT was taken offline for overhaul, with different sensors fitted at different times and the longest continuous run lasting in excess of 10,000 h. The data from the pre- and postcalibration performed by independent organizations are given in Table 18.1. This series of FOS trials forms part of the process of building confidence in the durability and fidelity of the optical technology.

Figure 18.4 FOS installation at Didcot power station.

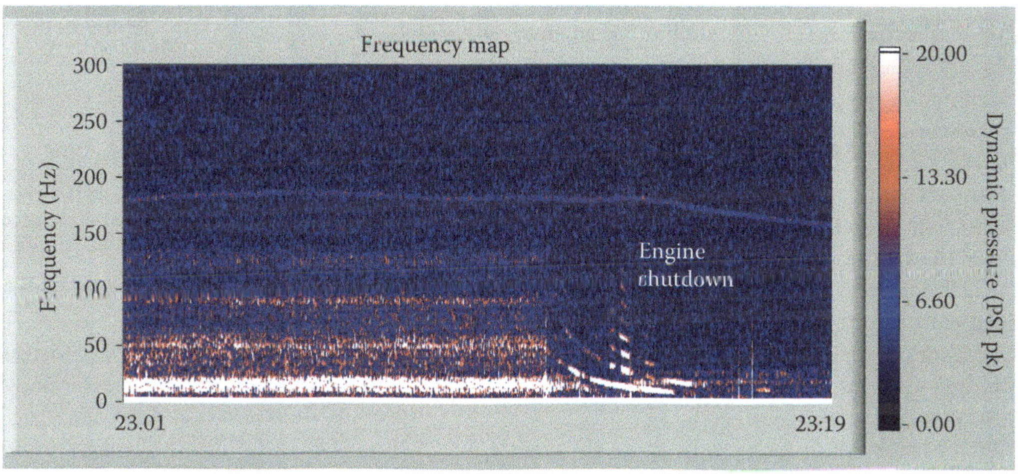

Figure 18.5 Recorded dynamic frequency map during engine shutdown.

Table 18.1 Long-term variation of calibration factor

Sensor serial number	Fired hours between calibrations	Pre- and post-didcot calibration variation (oxsensis) (%)	Pre- and post-didcot calibration variation (independent) (%)
EG192	10,580	4.8	4.2
EG360	7,100	2.5	4.6
EG363	7,100	2.4	1.7

18.6 FOS FOR CARBON EMISSION REDUCTION ON 3.9–5.3 MW COMMERCIAL GT GENERATORS

GT combustion instrumentation necessarily varies with plant size and complexity. Large engines, with high capital values, can support more expensive solutions than small engines, of which the Centrax KB5/7 class is a competitive example. The sub-6 MW class of industrial GTs is frequently not fitted with production combustion instrumentation. A conservative approach is taken to control settings on the production test bed, and then the "safe-set" engine is dispatched. The consequence of this is that performance and emissions are not optimized, and if combustion instability does occur in service, it may not be noticed until engine damage occurs. Figure 18.6 shows a comparison between an intact burner assembly and a damaged burner assembly on another, larger, industrial GT type due to combustion instabilities [10].

The availability of dynamic pressure sensor instrumentation is therefore essential to ensure and maintain the reliability and safety, as well as the efficiency of new and existing engine components. However, piezoelectric transducer systems cannot, as yet, be relied upon to provide reliable long-term dynamic pressure data at high temperatures. Effort is therefore being aimed at FOS to bridge the temperature capability gap in pressure measurement above 700°C, and robust optical pressure sensor systems operating at 1000°C with multi-measurand capability (ΔP, P, and T) are beginning to emerge.

These FP-based optical pressure transducers have been deployed on medium-sized industrial GTs with the aim at demonstrating longer life and higher fidelity measurements than are possible with equivalent piezoelectric transducers. Long-term deployment of the new sensor on commercial Lean Premix Combustion systems will deliver better control of the combustion systems, cleaner combustion with consequent reduction in NO_X emissions. The FOS technology addresses sensor durability by forming the sensor element from high melting point (2053°C) Sapphire material possessing excellent mechanical and anticorrosive properties.

An example of dynamic pressure data that were extracted from a GT engine test-bed rig is presented in Figure 18.7.

Close agreement is observed between the low-frequency dynamic pressure measurements using both fiber optical and piezoelectric transducers, noting that the optical system is not picking up the 50 Hz electromagnetic noise signal from the mains power supply. Long-term testing will provide the data required to support the commercial and safety case for commercial deployment of this new sensor technology.

18.7 DYNAMIC PRESSURE FREQUENCY RESPONSE IMPROVEMENT IN A HIGH-PRESSURE COMBUSTOR TEST RIG

The experiments were carried out in the high-pressure combustor test rig at the DLR Institute of Combustion Technology in Stuttgart, Germany [11].

(a) (b)

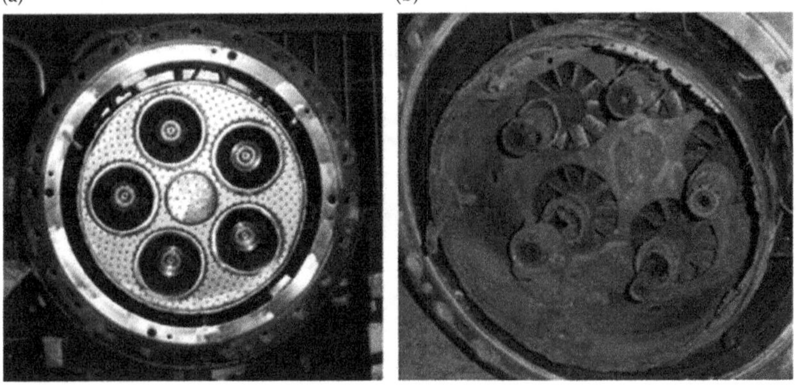

Figure 18.6 Burner assembly: (a) intact and (b) damaged.

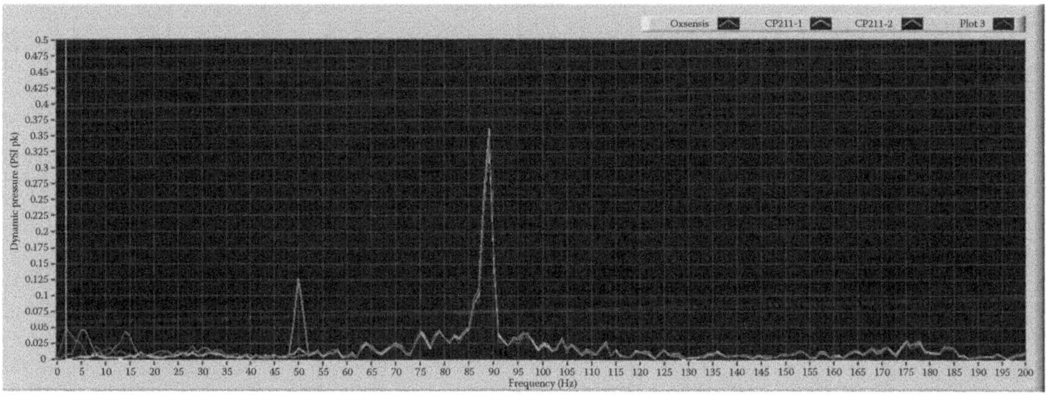

Figure 18.7 Frequency analysis of dynamic pressure data captured.

The combustors can be operated with different types of fuel such as natural gas (NG), "off-spec" NG, syngases, and hydrogen, as well as blends of these. The main air can be electrically preheated up to 725°C. Several pressure transducers are normally installed to record the static pressure in the combustion chamber and pressure drops across the intake and the burner. The data acquisition system is employed to monitor the conditions of the test rig as well as emissions.

This rig was used to undertake back-to-back testing of piezoelectric and optoelectronic dynamic pressure sensors with measurement points at two liner locations, indicated as sites 1 and 2 in Figure 18.8a. Figure 18.8b shows the location of the optical dynamic pressure sensor with a measuring range of 0–50 bar that is mounted at an angle of 50° to the main cavity. The piezoelectric sensor is connected via a long metal tube to reduce the temperature at the sensor and to improve access for replacement. Both sensor outputs were recorded at a sampling rate of 10 kHz for periods of 1 s intervals.

Enriching hydrogen with NG decreases the reactivity of the fuel/air mixture, and therefore changes the flame shape and position because of different chemical kinetics. This transition regime is very often accompanied by excited acoustic instability modes.

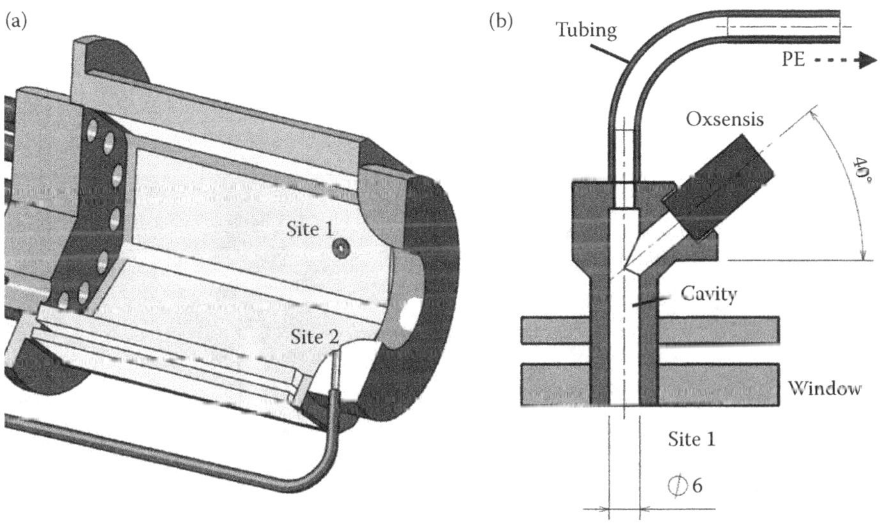

Figure 18.8 (a) Schematic of FLOX® combustor and locations of measurement points and (b) cross section of instrumented side wall showing cavity and sensor locations at site 1.

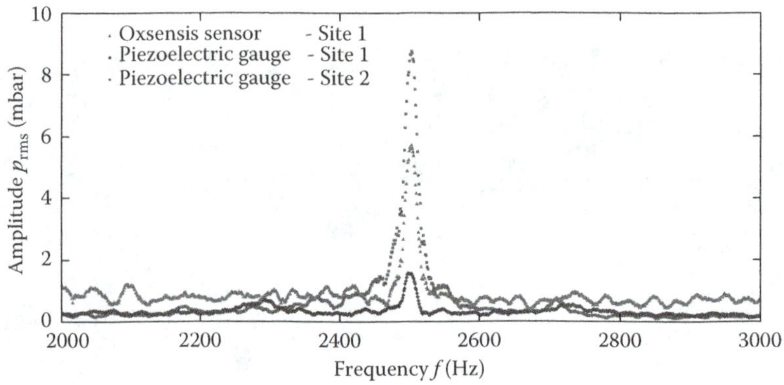

Figure 18.9 Close-up view of the first pressure pulsation peak around 2.5 kHz.

Several distinct peaks are discernible in the frequency spectrum recorded by the optical and piezoelectric sensors at sites 1 and 2. Figure 18.9 is a close-up view of the first pulsation peak around 2.5 kHz, thought to correspond to a longitudinal mode of the combustion chamber. Thus, sites 1 and 2 are located at different longitudinal node and antinode locations of this mode and therefore register different pulsation amplitudes. Further, it can be seen that the peak amplitudes measured by the piezoelectric and the optical sensors at site 1 are 1.7 and 5.7 mbar, respectively. This difference may be explained by the frequency dependent attenuation of the tubing system. The measured amplitudes differ by a factor of 0.3, indicating typical attenuation values for tubes in agreement with atmospheric calibration tests at ambient temperature using calibrated microphones [12].

18.8 HIGH-DENSITY FOS AND INSTRUMENTATION FOR GT OPERATION CONDITION MONITORING

Variations in GT exhaust temperature are used to provide timely information on combustor fault conditions, but the method is prone to false alarms. Such control systems are normally based on thermocouple arrays that provide a discrete number of measurement points with limited spatial resolution due to radiation shielding and bulky packaging, making it difficult to diagnose annular can-to-can temperature variations. Improvements in reliability of this approach have been demonstrated using an annular array of FBG sensors to accurately measure the annular static and dynamic exhaust temperatures at both startup and steady operation conditions [13]. Figure 18.10 shows the installed

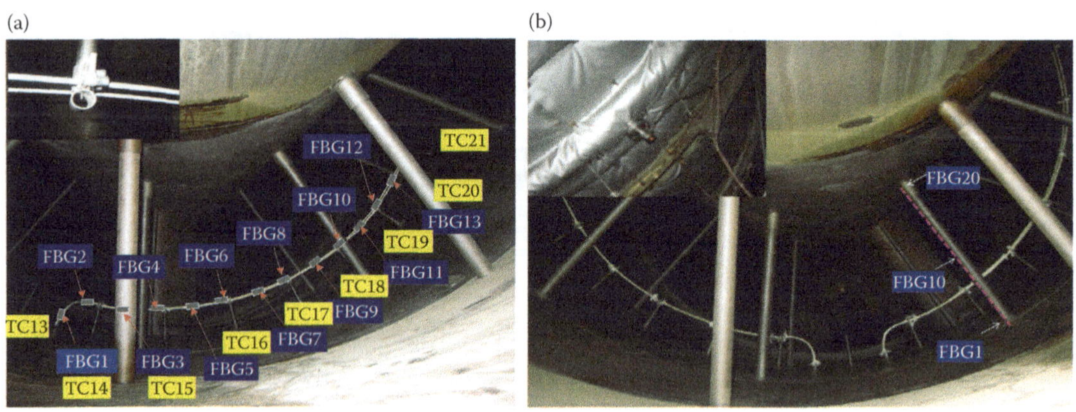

Figure 18.10 Installed FBG array for (a) circumferential and (b) radial temperature measurements.

FBG array in the exhaust gas duct for both circumferential and radial sensor distribution.

A 1–5 Hz interrogator with four-channel capability was used to detect the static signal, and the dynamic temperature signals were recorded at a 1 kHz update rate. The FBG-based sensor arrays have demonstrated their unique advantages in providing high-density, multipoint, and multifunction capability in measuring static and dynamic responses from a GT.

The ultimate goal would be to provide sensor instrumentation for direct combustor temperature measurement but normally this requirement is beyond the capabilities of most thermocouple sensors. However, the recent developments in fiber optical temperature sensors have the potential to alleviate many of the limitations of thermocouples, especially as new sapphire fiber-based sensors increase the upper temperature limit of the sensors [14].

FBG sensors have great potential for multipoint temperature measurement along a fiber optical cable and FP sensors for single-point measurements. Applications of these sensors within GT engines are diverse. Optical temperature sensors can be installed at precombustor flow path, turbine inlet, downstream of the combustor or exhaust, and interstage locations for measuring dynamic temperature anomalies.

REFERENCES

1. Docquier, N., Candel, S., Combustion Control and Sensors: A Review, *Progress Energy Combustion Science*, Vol. 28, 107 (2002).
2. Lieuwen, T. C., Yang, V., eds., *Combustion Instabilities in Gas Turbine Engines: Operational Experience, Fundamental Mechanisms, and Modeling*. Progress in Astronautics and Aeronautics, Vol. 210. Reston, VA, AIAA, 2005.
3. DeLaat, J. C. et al., Active Combustion Control for Aircraft Gas-Turbine Engines-Experimental Results for an Advanced, Low-Emissions Combustor Prototype, NASA/TM-2012-217617, *50th Aerospace Sciences Meeting*, Nashville, TN, January 9–12, 2012.
4. Anderson, R. L. et al., Decalibration of sheathed thermocouples. In *Temperature, It's Measurement and Control in Science and Industry*, J. F. Schooley, ed., Vol. 5, American Institute of Physics, New York, pp. 977–1007, 1982.
5. Mihailov, S. J., Fiber Bragg Grating Sensors for Harsh Environments, *Sensors*, Vol. 12, 1898 (2012).
6. Pechstedt, R. D., Fibre Optic Pressure and Temperature Sensor for Applications in Harsh Environments, *Proc. SPIE*, Vol. 8794, 879405 (2013).
7. Dakin, J. P., Withers, P. B., Optical fibre pressure or displacement sensor, UK Patent Application GB2202936A.
8. Maillaud, F. F. M., Pechstedt, R. D., High-Accuracy Optical Pressure Sensor for Gas Turbine Monitoring, *Proc SPIE*, Vol. 8421, 8421AF (2012).
9. Fiber Optic Sensors for Aerospace Applications, Aerospace Information Report AIR6258, SAE Aerospace (2014), http://standards.sae.org/wip/air6258/.
10. Goy, C. J. et al., *Monitoring Combustion Instabilities: E.ON UK's Experience*. Progress in Astronautics and Aeronautics, Vol. 210. pp. 163–178, Reston, VA, AIAA, 2005.
11. Winterburn, A. et al., Extension of an optical dynamic pressure sensor to measure temperature and absolute pressure in combustion applications, *Proceedings of the 6th International Gas Turbine Conference*. Brussels, October 2012.
12. Lourier, J.-M. et al., Numerical Analysis of the Acoustic Transfer Behavior of Pressure Ducts Utilised for Microphone Measurements in Combustion Chambers. ASME Paper No. GT2010-22805, 2010.
13. Xia, H. et al., High-Density Fiber Optical Sensor and Instrumentation for Gas Turbine Operation Condition Monitoring, *Journal of Sensors*, Vol. 2013, 206738 (2013).
14. Elsmann, T. et al. Advanced Fabrication and Calibration of High-Temperature Sensor Elements Based on Sapphire Fiber Bragg Gratings, *Proc. SPIE*, Vol. 9157, 91572N-1 (2014).

19

Raman gas spectroscopy

ANDREAS KNEBL, JÜRGEN POPP, AND TORSTEN FROSCH
Leibniz Institute of Photonic Technology

19.1 INTRODUCTION

In recent years, Raman spectroscopy has emerged as a powerful analytical tool to identify and quantify the components of gas mixtures in continuous and nonconsumptive fashion. Although the Raman effect has been known since 1928,[1] its use has mainly been constrained to fundamental research regarding the structure and characteristics of molecules for a long time. This had been due to the inherently weak signal and technical limitations leading to high instrumentation costs and very long measurement times. The advent of the laser in the 1960s was a turning point, but only recent advances in laser as well as detector technology and new enhancement techniques enabled researchers to develop Raman spectroscopy into a comprehensive technique for multigas analysis.

Common methods to analyze gas mixtures in an industrial, medical, or research setting include electrochemical sensing, gas chromatography–mass spectrometry (GC–MS), and infrared (IR) absorption spectroscopy. Electrochemical gas sensors are only sensitive to a single or few gas species. Also, cross sensitivities can cause serious problems. Although GC–MS is highly sensitive and accurate, expensive and bulky equipment and the time necessary for measurements are the major drawbacks. For separation, the gas has to travel through and is retained in a capillary column for up to 20 min. This does not allow rapid and continuous monitoring of processes without sampling and gas consumption. IR absorption spectroscopy offers high-resolution, high-sensitivity gas measurements but is not sensitive to diatomic homonuclear molecules such as oxygen (O_2), hydrogen (H_2), or nitrogen (N_2). With these techniques, a comprehensive multigas analysis would require a combination of multiple methods. This also means that different gas volumes are probed at different times leading to inaccuracies and errors. Raman gas analysis provides the simultaneous identification and quantification of all but noble gases and does not require sample preparation, making it the ideal tool for time-resolved, nonconsumptive, and noninvasive analysis of complex gas mixtures. The only limiting factor is the weak signal intensity making it necessary to take advantage of sophisticated enhancement mechanisms to measure gaseous components down to low concentrations of single-digit ppm.

In the following, the theoretical background of Raman gas spectroscopy and the measurement principle will be introduced. Additionally, exemplary applications of Raman gas spectroscopy in medicine, environmental research, and industry will be given.

19.2 THEORY OF RAMAN GAS SPECTROSCOPY

When monochromatic laser light is guided into a gas sample, most photons pass the volume without any interaction and remain unaltered. A small fraction of the photons is scattered by the gas molecules. The scattered radiation mainly consists of two parts: (1) The major part results from elastic scattering and has the same frequency as the excitation light. The responsible process is called Rayleigh scattering. (2) With approximately three orders of magnitude smaller intensity, inelastically scattered and thus frequency-shifted light occurs as the minor part. The energy transfer is due to changes in the rotational, vibrational, or rotational–vibrational state of the molecule. This inelastic process is called Raman scattering or the Raman effect.

IR absorption and Raman scattering are complementary effects: While the IR absorption depends on a change in dipole moment due to the altered rotation/vibration, Raman scattering relies on a change in the polarizability of the molecule. For molecules with a center of inversion, the two effects are mutually exclusive for the eigen modes of the molecule, such that a rotational/vibrational mode is either IR active or Raman active. As the

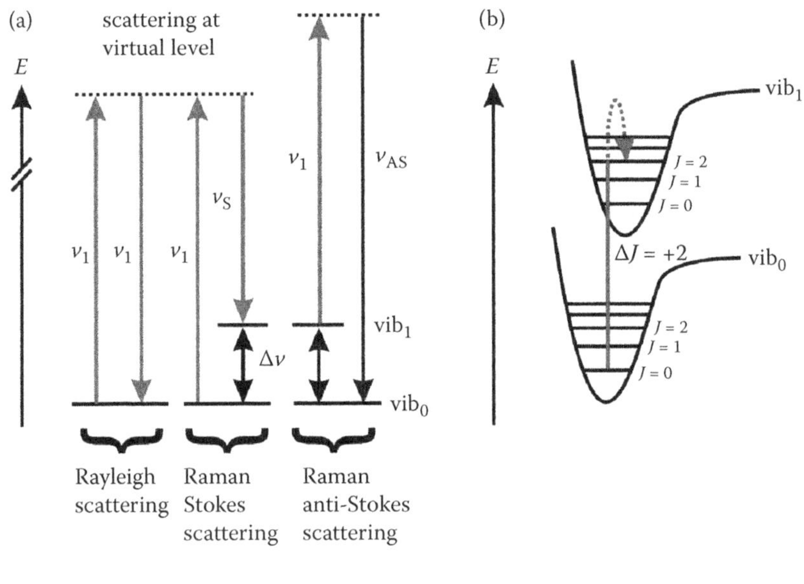

Figure 19.1 Schematic diagram of the Raman scattering process. (a) Energy-level diagram of different scattering processes. Light with a frequency ν_1 interacts with a molecule. For the case of Rayleigh scattering, there is no energy transfer and the frequency remains ν_1. If part of the energy of the incoming light is transferred and the molecule is excited into a higher vibrational state (vib$_1$), we speak of Raman Stokes scattering. The frequency of the scattered light is then reduced to $\nu_s = \nu_1 - \Delta\nu$. If energy is transferred from the molecule and the molecule drops into a lower vibrational state (vib$_0$), the frequency of the scattered light grows to $\nu_{AS} = \nu_1 + \Delta\nu$. This process is called Raman anti-Stokes scattering. (b) Energy diagram of a rotational–vibrational transition. The transition takes place from vibrational level vib$_0$ to level vib$_1$ and the respective rotational levels J=0 to J=2 (ΔJ = 2). This transition can thus be found in the S-branch on the Stokes side of the spectrum.

Raman effect does not require a permanent dipole moment, it occurs for all but noble gases. In particular, the IR-inactive diatomic, homonuclear molecules such as nitrogen, hydrogen, and oxygen yield a Raman signal.

Spectral position and intensity of the Raman signal depend on the structure of the gas molecule. The Raman peaks show up symmetrically on both sides of the laser line. This shift results from the difference in energy of the two involved energy levels. The blue shift (toward higher frequencies) is called anti-Stokes shift denoting a transition from a higher energy level to a lower energy level, and the red shift (toward lower frequencies) is called Stokes shift denoting a transition from a lower energy level to a higher energy level. Usually, the Stokes signal is more intense because for gas temperatures around 300 K, the lower, nonexcited energy levels are more populated than the upper levels. The ratio of the Stokes to the anti-Stokes peak may be used to measure the temperature of the sample. Figure 19.1a shows the Stokes as well as the Rayleigh process in an energy-level diagram.

Quantum mechanical selection rules govern the transitions. The occurring bands and branches and their spectral positions are the characteristics for the scattering molecule. Thus, the molecule can be unambiguously identified taking advantage of this spectral "fingerprint." The scattering is basically independent of the initial laser wavelength (disregarding resonance effects). Hence, all gas species can be identified using one light source. The determination of allowed and active transitions, energy levels, and resulting Raman frequencies quickly becomes complicated, according to the complexity of the molecule. That is why the following theoretical consideration only covers linear molecules to point out the important principles and essential features.

Linear molecules composed of N atoms have $3N-5$ fundamental normal modes of vibrations. These modes are grouped into parallel (along the internuclear axis, vibrational angular momentum $l=0$) and perpendicular vibrations (perpendicular to the internuclear axis, $l=1$). With J, the total angular momentum (exclusively the nuclear spin), the general selection rules for Raman scattering at linear molecules are[2]

$$\Delta J = 0, \pm 2 \qquad \text{if } l = 0 \tag{19.1}$$

$$\Delta J = 0, \pm 1, \pm 2 \qquad \text{if } l \neq 0 \tag{19.2}$$

$$+ \not\Leftrightarrow - \qquad s \not\Leftrightarrow a \tag{19.3}$$

meaning that positive rotational levels (+) do not combine with negative ones (−), and symmetric rotational levels (s) do not combine with antisymmetric ones (a). $\Delta J = J' - J''$ denotes the change in rotational level with the single prime standing for the upper and the double prime for the lower level. According to these selection rules, several branches can occur in a Raman spectrum: (1) the central Q-branch denotes no change in rotation ($\Delta J = 0$), consisting of several closely spaced lines with smaller wavenumber shifts than the pure vibrational line. (2) The R- and S-branches denote a transition into a higher rotational level ($\Delta J = +1$ and $\Delta J = +2$, respectively), consisting of a series of lines with larger wavenumber shifts than the pure vibrational line. (3) The O- and P-branches denote a transition into a lower rotational level ($\Delta J = -2$ and $\Delta J = -1$, respectively), consisting of a series of lines with smaller wavenumber shifts than the pure vibrational line. In many cases, the lines of the Q-branch overlap or are superimposed leading to a single strong line. In Figure 19.1b, a Stokes process with $\Delta J = +2$ is displayed in an energy-level diagram.

For a parallel vibration, the Raman shifts of the fundamental rotation–vibrations are given by[2]

Q-branch: $\Delta J = 0$

$$\tilde{\nu}_{Q(J)} = \tilde{\nu}_{vib} + (B' - B'')J(J+1) - (D'_J - D''_J)[J(J+1)]^2 \tag{19.4}$$

O-and S-branches: $\Delta J = -2$ and $\Delta J = +2$, respectively

$$\begin{aligned}\tilde{\nu}_{O(J),S(J)} = {} & \tilde{\nu}_{vib} + \frac{3}{4}(B' - B'') - \frac{9}{16}(D'_J - D''_J) \\ & + \left[(B' + B'') - \frac{3}{2}(D'_J - D''_J)\right]m \\ & + \frac{1}{4}\left[(B' - B'') - \frac{11}{2}(D'_J - D''_J)\right]m^2 \\ & + \frac{1}{2}(D'_J - D''_J)m^3 - \frac{1}{16}(D'_J - D''_J)m^4\end{aligned} \tag{19.5}$$

where $m = -2J + 1$ for the O-branch and $m = -2J + 3$ for the S-branch, $\tilde{\nu}_0$ is the wavenumber of the pure vibrational transition. The molecular constants B (rotation constant) and D_J (centrifugal distortion constant) are specific functions of the vibrational state, often additionally labeled with the vibrational quantum number as a subscript. The primes again denote the upper and lower level.[2,3]

Neglecting the centrifugal distortion ($D' = D'' = 0$) and considering the rotational constants B' and B'' to be equal, we get

$$\begin{aligned} &S\text{-branch}, \tilde{\nu}_{S(J)} = \tilde{\nu}_{vib} + 4BJ + 6B, && \Delta J = +2 && J = 0,1,2,\cdots \\ &Q\text{-branch}, \tilde{\nu}_Q = \tilde{\nu}_{vib}, && \Delta J = 0 && J = 0,1,2,\cdots \\ &O\text{-branch}, \tilde{\nu}_{O(J)} = \tilde{\nu}_{vib} - 4BJ + 2B, && \Delta J = -2 && J = 2,3,4,\cdots \end{aligned} \tag{19.6}$$

Therefore, the first lines of the S-and O-branches ($J = 0$ and $J = 2$, respectively) are $6B$ away from the primary line; all subsequent lines have a spacing of $4B$ to the previous line. Here, B has the unit cm^{-1}.

The uncertainty of the energy levels leads to natural line broadening. Together with other homogeneous broadening effects such as the pressure broadening, this would lead to Lorentzian-shaped spectral lines. However, inhomogeneous broadening such as Doppler broadening contributes a Gaussian line shape. In total, homogeneous and inhomogeneous broadening add up to build a Voigt profile (convolution of Lorentzian and Gaussian).[4]

The intensity of the lines is governed by the population of the respective energy levels. Thus, the lines envelope can be described by a Boltzmann distribution. Additionally, each rotational state consists of $2J+1$ coinciding levels. This degeneracy leads to a weighting factor for each line of $2J+1$.[5] Also, the nuclear spin statistical weight can introduce intensity alterations of the peaks[6]:

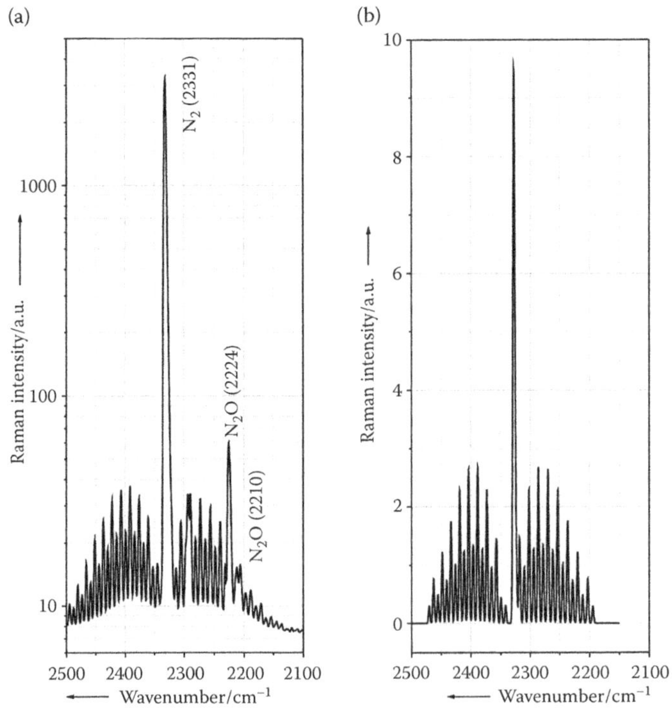

Figure 19.2 Comparison of measured and calculated nitrogen spectra. (a) Spectrum of nitrogen (N_2) and nitrous oxide (N_2O) measured with fiber-enhanced Raman spectroscopy. Zoomed in section of a broader spectrum is shown. (Modified from Hanf, S. et al., *Anal. Chem.* *87*(2), 982–988, 2015.) (b) Calculated Raman spectrum of N_2 (Data derived from Bendtsen, J., *J. Raman Spectrosc.* *2*(2), 133–145, 1974.), considering Gaussian line broadening ($\sigma = 1.3$ cm^{-1}), level occupation according to Boltzmann distributions, rotational level degeneracy, and nuclear spin statistical weight.

$$\frac{g_a}{g_s} = \frac{I_q}{I_q + 1} \tag{19.7}$$

where I_q is the nuclear spin quantum number, and g_a and g_s are the statistical weights for antisymmetric and symmetric quantum states, respectively.

Nitrogen (N_2) shows a typical spectrum with the described features. Figure 19.2b shows a calculated Raman spectrum of N_2. Gaussian line broadening, level occupation according to Boltzmann distributions, and nuclear spin statistical weight have been considered in the calculation; see Equation 19.8. For N_2, $I_q = 1$, and therefore, the symmetric states (for N_2 the even J values) are twice as likely and thus intense as antisymmetric states (for N_2 the odd J values). The relative intensity of a nitrogen Raman peak I_P can be calculated as

$$I_P(\nu) = g \cdot (2J+1)\exp\left(-\frac{hcBJ(J+1)}{kT}\right) \frac{1}{\sigma\sqrt{2\pi}}\exp\left(-\frac{1}{2}\left[\frac{\nu - \nu_P}{\sigma}\right]^2\right) \tag{19.8}$$

with $g = 1$ for even J values and $g = 1/2$ for odd J values (according to Equation 19.7). The last term denotes the Gaussian line broadening: σ stands for the standard deviation or better speaking the line width, and ν_P is the spectral peak position calculated according to Equations 19.4 and 19.5.

The measured Raman spectrum of N_2 in Figure 19.2a is very similar and shows the same characteristics: the unresolved Q-branch is centered at 2331 cm^{-1}, and the O- and S-branches are clearly visible and display the Boltzmann envelope as well as the intensity alternation due to the spin statistical weight. There is a slight shift in relative wavenumbers of the peak positions between data from the works of Bendtsen et al.[7] and from Hanf et al.[8]

In a highly resolved Raman spectrum of nitrogen, the frequency shifts of the rotational lines can be seen as described in Equation 19.6. As displayed in Figure 19.3, the measured spectrum again compares well with a calculated spectrum with a small line width, simulating a high-resolution device. The lines in the O- and S-branches can easily be ascribed to the respective rotational transitions (Figure 19.3a).

For multigas mixtures, the resulting spectrum is the superposition of the spectra of the components. Depending on the components and the spectral resolution of the setup, this can lead to overlaps and ambiguities. Multigas spectra can be analyzed using multicomponent analysis tools, fitting the experimental data with weighted reference spectra of several gases (see also Section 19.3).

The overall intensity I of the Raman signal depends on several factors:

$$I \propto NI_0\sigma_R \propto NI_0(\nu_0 \pm \Delta\nu)^4 |\alpha|^2 \tag{19.9}$$

where N is the number of gas molecules in the base level, I_0 is the intensity of the laser light, σ_R is the Raman scattering cross section of that specific transition, ν_0 is the frequency of the laser light, $\Delta\nu$ is the Raman frequency shift, and $|\alpha|$ is the magnitude of the polarizability tensor. Thus, the Raman intensity is linearly proportional to the concentration of the probed gases. To actually be able to determine the concentration of a component of the sample gas, a calibration measurement with a reference gas of known concentration is necessary. For the evaluation, temperature and pressure have to be considered too.

This chapter introduced details specific for Raman gas spectroscopy.

19.3 MEASUREMENT PRINCIPLE: EXPERIMENTAL SETUP AND DATA ANALYSIS

The setup that is necessary for Raman spectroscopic measurements of gas samples basically consists of four parts: (1) the laser, (2) the sample container, (3) the spectrometer with a dispersive element, and (4) a highly sensitive detector. The inherently weak Raman signal of gas samples represents a major challenge if short measurement times, high sensitivity, and a fairly small setup are required. The setup or device for continuous, noninvasive Raman gas analysis is preferably portable. Thus, high-power lasers or ultrashort pulse lasers are impractical, and small, energy-efficient diode or diode pumped lasers are the way to go.

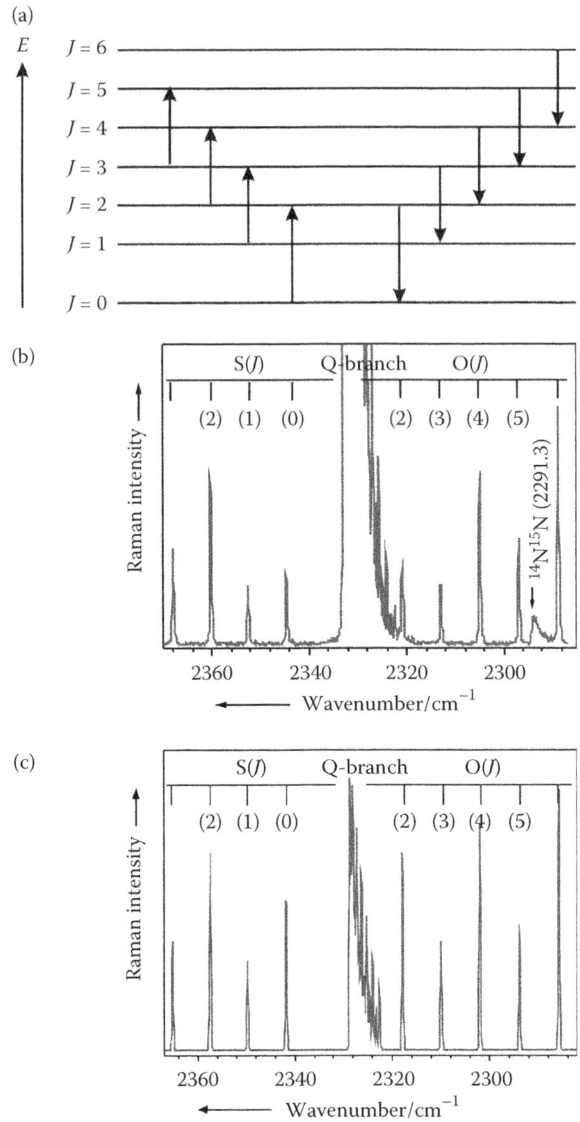

Figure 19.3 (a) Energy diagram of rotational transitions. (b) Highly resolved Raman Stokes spectrum of nitrogen. (Modified from Hanf, S. et al., *Anal. Chem.* *86*(11), 5278–5285, 2014.). S-and O-branches are clearly split up in single lines. For each line, the respective transitions are displayed in the energy diagram. (c) Calculated Raman spectrum of N_2 (Data derived from Bendtsen, J., *J. Raman Spectrosc.* *2*(2), 133–145, 1974.), considering Gaussian line broadening ($\sigma = 0.15$ cm^{-1}), level occupation according to Boltzmann distributions, rotational level degeneracy, and nuclear spin statistical weight.

Consequently, it is necessary to increase the interaction between laser light and sample gas. To solve that problem, several sophisticated enhancement strategies have been invented, developed, and employed.

One strategy is based on a multipass cavity used as sample container. The laser light is forced to pass the cavity and thus the sample gas multiple times. Consequently, the interaction is increased. A range of different multipass cavity architectures has been developed and employed to Raman gas analysis. These include retro-reflective cavities,[9] near confocal cavities,[10] and resonant cavities.[11] A good overview of cavity-enhanced techniques is given by Gagliardi and Loock.[12] A resonant cavity is a high-finesse cavity built up by at least two

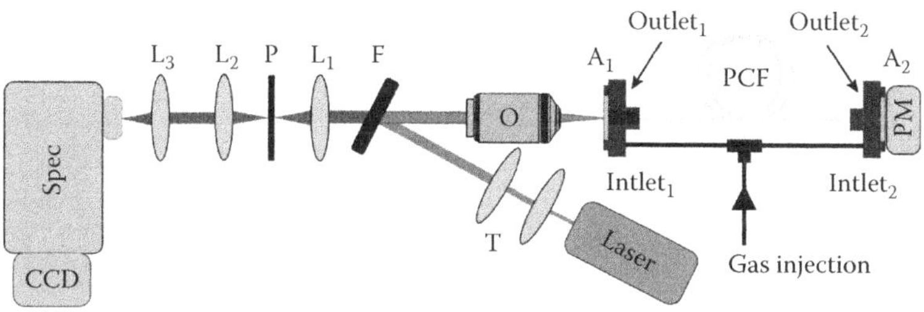

Figure 19.4 Schematic sketch of the setup for fiber-enhanced Raman spectroscopy. PCF, hollow-core photonic crystal fiber, used as a sample container; T, telescope; F, edge filter; OL, objective lens; A_1, A_2, fiber adapter assemblies; PM, power meter; L_1, L_2, L_3, aspheric lenses; P, pinhole; Spec, spectrometer. (Modified from Hanf, S. et al., *Anal. Chem.* *87*(2), 982–988, 2015.)

mirrors. When laser light is coupled into the cavity, the narrow resonance of the cavity leads to mode selection and frequency locking of the laser frequency. Moreover, the laser light is reflected a great number of times in the cavity, effectively storing it and making the cavity an "optical capacitor." This leads to a huge power buildup. The enhancement can be expressed in terms of an effective resonator length, which is the resonator length times the enhancement factor. For high-finesse cavities, the enhancement can be in the order of 10^4. The technique of cavity-enhanced Raman spectroscopy is often abbreviated as CERS.

Another enhancement technique is based on optical waveguides, namely, metal-coated capillaries[13] or hollow-core photonic crystal fibers.[14,15] As the hollow core serves as a sample container, only miniscule sample volumes are necessary. The laser light is coupled into the light guide. Due to the guiding mechanism (total reflection; photonic bandgap or antireflection), the intensity is confined to the core mode(s) over the whole length of the waveguide. This has the effect of an extended focus, increasing the length of light–gas interaction with high intensity. The waveguide has to be chosen carefully because low attenuation is needed for the excitation as well as the Stokes-shifted Raman signal light. The signal can be gathered at either the front (same side as laser light coupling) or the end of the fiber. Enhancement factors up to the order of 10^3 have been reported.[8] An exemplary setup for fiber-enhanced Raman spectroscopy (FERS) is displayed in Figure 19.4.

For multigas mixtures, a multicomponent analysis is necessary for identification and quantification. For that, the measured spectrum is fitted with calibrated reference spectra of gases that are expected to constitute the mixture. The weight of each single reference spectrum tells the concentration of that gas in the mixture. The calculations are performed with the help of an overdetermined linear equation system with the calibration gas, g; measured gas, a (mixture of gases); intensity, $I(\tilde{\nu})$; concentration, c; detector pixel, n; and number, m, of extracted gases[16]:

$$\begin{bmatrix} I(\tilde{\nu})_{g_1 1} & \cdots & I(\tilde{\nu})_{g_m 1} \\ \vdots & \ddots & \vdots \\ I(\tilde{\nu})_{g_1 n} & \cdots & I(\tilde{\nu})_{g_m n} \end{bmatrix} \begin{bmatrix} c_1 \\ \vdots \\ c_m \end{bmatrix} = \begin{bmatrix} I(\tilde{\nu})_{a1} \\ \vdots \\ I(\tilde{\nu})_{an} \end{bmatrix} \tag{19.10}$$

Also, peaks that are not accounted for by the used reference spectra are highlighted using this procedure: if the difference between the measured spectrum and the deconvoluted individual spectra differs from a zero baseline, information about additional gases can be gained.[16]

19.4 APPLICATIONS: RAMAN GAS SPECTROSCOPY IN MEDICINE, ENVIRONMENTAL RESEARCH, AND INDUSTRY

Raman spectroscopic measurements allow the continuous, nonconsumptive, and highly sensitive identification and quantification of gases. The power of Raman gas spectroscopy can be seen in Figure 19.5: due to their spectral shift, a multitude of gases including the IR-inactive homonuclear species can be detected and quantified simultaneously. This capability is required in a broad range of applications, from medicine to environmental

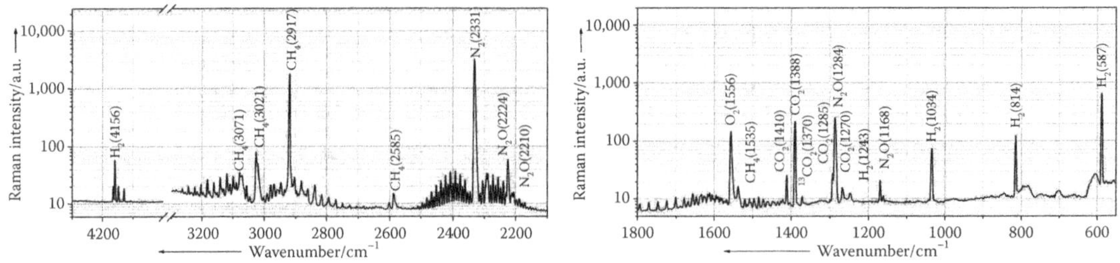

Figure 19.5 Enhanced Raman spectrum of the rotational and rotational–vibrational bands of a complex gas mixture: rotational Raman bands of H_2 and rotational–vibrational Raman bands of CO_2, N_2O, O_2, CH_4, N_2, and H_2. (Modified from Hanf, S. et al., *Anal. Chem.* *87*(2), 982–988, 2015.)

sciences and industry. Although there are already commercial products for Raman gas spectroscopy, a lot of work is still ongoing in research institutions and many setups and devices are unique or in the prototype stage.

19.4.1 Medical applications

Breath analysis has garnered lots of interest and attention in medicine and analytical sciences.[17] The human breath carries a multitude of biomarkers that could be used as early stage indicators for metabolic, lung, or airways diseases. Several analytical techniques are investigated for clinical use for breath analysis, in particular GC–MS.[17a,18] However, most of these techniques—such as GC–MS—do not allow a direct sampling of the exhaled air. Raman spectroscopy offers direct sampling with fairly short measurement times. The limits of detection and quantification of enhanced Raman spectroscopic techniques are coming towards the medically relevant range but must be further improved. Recent publications about fiber-enhanced Raman gas spectroscopy of human breath show great potential.[8,14,17b]

Molecular hydrogen and methane are biomarkers in the human breath for malabsorption disorders. Hanf et al.[8] showed that FERS is suited well as highly sensitive and selective point-of-care examination method. They demonstrated the ability to monitor the levels of H_2 and CH_4 indicative for oligosaccharide intolerances or small intestinal bacterial overgrowth syndrome (SIBO); see Figure 19.6. In a hydrogen breath test—a diagnostic tool for different malabsorption disorders—an increase of 20 ppm above basal is considered positive. An

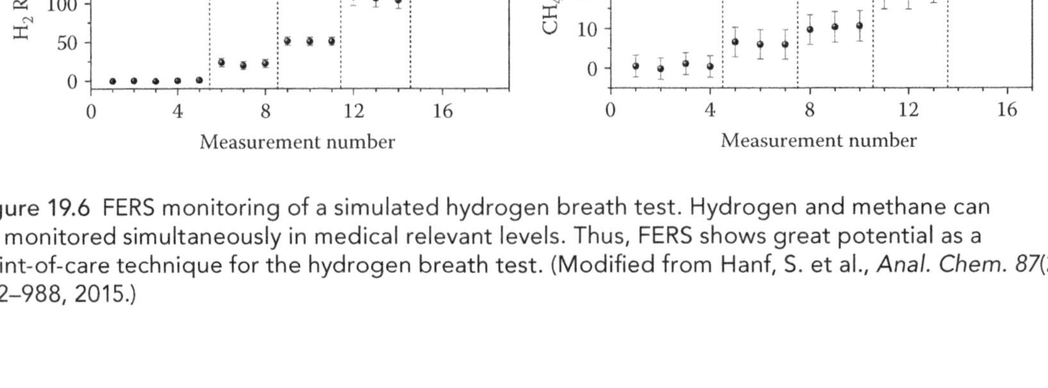

Figure 19.6 FERS monitoring of a simulated hydrogen breath test. Hydrogen and methane can be monitored simultaneously in medical relevant levels. Thus, FERS shows great potential as a point-of-care technique for the hydrogen breath test. (Modified from Hanf, S. et al., *Anal. Chem.* *87*(2), 982–988, 2015.)

elevation above 12 ppm is already an indicator for SIBO. As H_2 can be converted to CH_4 by bowel bacteria, the latter also is an important parameter that should be checked in addition[8] to increase the diagnostic sensitivity.

Another application in the field of medicine is the monitoring of anesthetic gases. With Raman gas spectroscopy, the simultaneous monitoring of nitrogen (N_2), oxygen (O_2), and carbon dioxide (CO_2) as well as the anesthetic gases nitrous oxide (N_2O), sevoflurane, isoflurane, and desflurane is possible with a time resolution comparable to the breath cycle.[9,10,19] This enables a continuous monitoring, giving the anesthetist the chance to adjust the artificial respiration as well as the dosage of the anesthetic gases. Raman spectroscopy could thus help to increase patient safety during surgeries requiring anesthesia.

19.4.2 Environmental applications

Another field of application is the environmental sciences. The understanding of gas flows, sinks, and wells, and the formation, spread, and exchange as well as storage of gases is crucial for climatology and biogeochemistry.[20] Raman spectroscopic methods can be employed in mesocosm experiments and in clima stations to monitor changes in gas concentrations. Raman spectroscopy is particularly suited to monitor the gas exchange of atmosphere, soils, and plants as N_2, O_2, CO_2, and methane (CH_4) can all be monitored at the same time and isotopes of those gases can be discriminated. Consequently, Raman spectroscopy could be applied in multiscale approaches to contribute to the understanding of biogeochemical processes and climate change.

Enhanced Raman spectroscopy made it possible to investigate the gas exchange processes in peat bog ecosystems.[11] O_2, CO_2, CH_4, and N_2 were simultaneously analyzed in real time. This allowed the continuous observation of the dynamics of greenhouse gases evolving from the climate-sensitive peat bog ecosystem. The Raman gas measurements were performed in the head space of a water-saturated, raised peat bog ecotron. Various important ecosystem parameters were determined for different light regimes, characterizing plant and soil effects. Raman gas spectroscopy thus has proved to be an extremely versatile analytical technique for the monitoring of climate-sensitive ecosystems and the quantification of greenhouse gases.[11]

The discriminatory power of Raman spectroscopy facilitates stable isotropic tracer experiments. The discrimination between $^{12}CO_2$ and $^{13}CO_2$ allows experiments examining plant respiration, especially leaf dark respiration, and to investigate the effects of drought and pest infestation.[16,21] The simultaneous and continuous monitoring of $^{12}CO_2$, $^{13}CO_2$, and O_2 enables researchers to study the photosynthetic gas kinetics, leaf dark respiration, and the respiratory quotient. Due to the ^{13}C-labeling, the resource flow can be analyzed additionally.[16] An exemplary Raman spectrum of the atmosphere during a leaf dark respiration experiment and the Raman gas spectra of $^{12}CO_2$ and $^{13}CO_2$ are displayed in Figure 19.7. Using the same technique, the effects of drought and pest infestation on European beech seedlings were examined. The respirational gas exchange was monitored and analyzed for periods of drought as well as after the plants were exposed to aphids. Thus, Raman spectroscopy enables the investigation of stress effects on trees.[21,22]

Also, Raman spectroscopy can be used to monitor the respirational activity of cave bacteria[23] and soil microbes.[24] Microbes are the driving forces in biodegradation and biomineralization processes. Jochum et al.[24] used CERS to investigate the effects of soil contamination with benzene. For that, ^{13}C-labeled benzene was spiked on soil, and the heterotrophic (microbial) soil respiration was monitored in the headspace above the soil sample. The respiratory quotient was analyzed over several days with high time resolution. Together with the evolution of the $^{13}CO_2$ concentration, the data clearly showed the microbial benzene degradation.[24] Besides degradation, bacteria are also responsible for biomineralization. The respiratory activity associated with the growth of bacteria such as *Arthrobacter sulfonivorans* cannot be observed using conventional turbidity-based optical density measurements due to concomitant mineral formation in the medium.[23] The Raman-based gas analysis, however, allowed studying the life cycle and respiratory activity of important cave bacteria.[23]

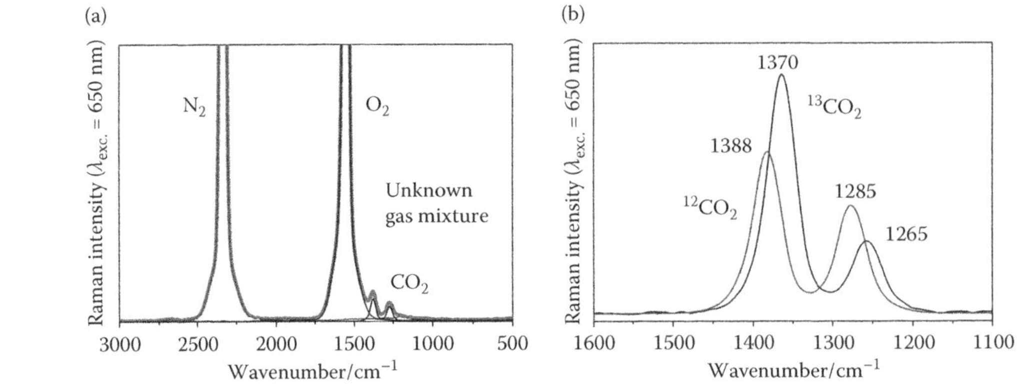

Figure 19.7 (a) Example of a Raman gas spectrum (λ_{exc} = 650 nm) during a typical leaf dark respiration measurement. Raman spectrum of the unknown gas mixture and the Raman spectra of the individual gaseous components are shown (N_2 and O_2 are rotational–vibrational spectra, with unresolved O- and S-branches). The concentrations of the individual gases can be deconvoluted from the experimentally acquired envelope. (b) The Raman gas spectra of $^{12}CO_2$ and $^{13}CO_2$ can be distinguished and simultaneously quantified due to their spectral shift and differences in the intensity distribution of the Fermi diad. (Modified from Keiner, R. et al., *Analyst*, *139*(16), 3879–3884, 2014.)

The understanding of the nitrogen cycle is extremely important in environmental research; however, the comprehensive investigation of N gases, such as N_2 and N_2O, is very challenging so far. Keiner et al.[25] applied CERS and traced the stepwise reduction of ^{15}N-labeled nitrate by the denitrifying bacteria *Pseudomonas stutzeri*. CERS allowed the simultaneous measurement of all relevant gases to trace the fate of the ^{15}N-labeled substrate and to understand the formation and

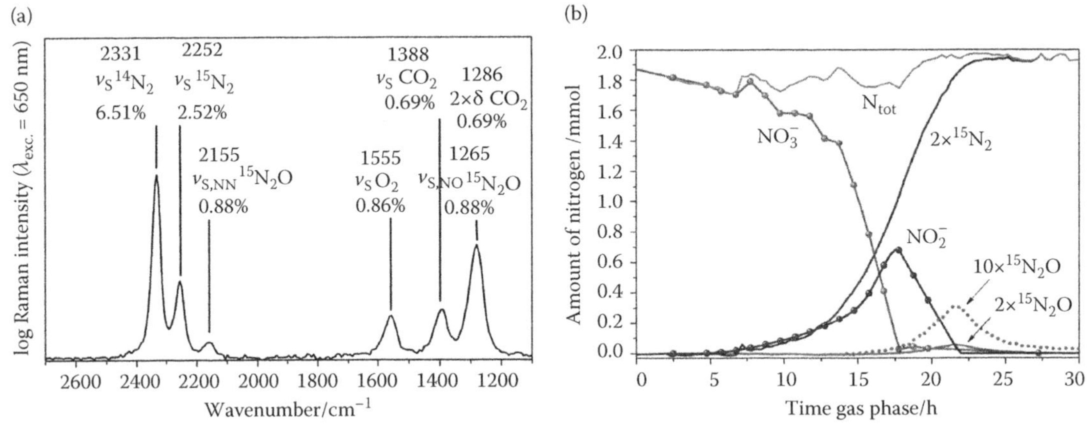

Figure 19.8 (a) Closed cycle Raman multigas spectrometry of the head space gas volume of a batch denitrification experiment with ^{15}N-labeled substrate. The five gases $^{14}N_2$, $^{15}N_2$, $^{15}N_2O$, O_2, and CO_2 are identified; their modes are assigned; and the gas abundance is quantified in percent volume. (b) Raman spectroscopic multigas monitoring of a denitrification experiment with ^{15}N-labeled substrate. The concentration courses of nitrate (NO_3^-), nitrite (NO_2^-), nitrous oxide (N_2O), and dinitrogen (N_2) during the successive reduction $NO_3^- \rightarrow NO_2^- \rightarrow NO \rightarrow N_2O \rightarrow N_2$. The continuously measured gas concentrations of $^{15}N_2$, $^{15}N_2O$, and CO_2 are displayed in solid lines. The amounts were corrected for the number of *N* atoms present in the respective gas molecules. To account for the low concentration of N_2O, the amount is scaled up by a factor of 10. The concentrations of nitrate and nitrite are depicted as dots, each representing a taken sample (acquired from a parallel culture to rule out any disturbance to the closed cycle experiment). N_{tot} represents the calculated total nitrogen balance from all nitrogen components. (Modified from Keiner, R. et al., *Anal. Chim. Acta*, *864*, 39–47, 2015.)

degradation processes of the gaseous compounds; see Figure 19.8. Recently, Jochum et al. showed direct Raman spectroscopic measurements of biological nitrogen fixation under natural conditions.[26] They measured the nitrogen fixation rates of *Rhizobium* bacteria living symbiotically on the roots of legumes with the help of CERS. Raman gas spectroscopy could thus contribute substantially to the understanding of nitrogen cycling in both natural and agricultural systems.[25]

19.4.3 Industrial Applications

Raman spectroscopy also has a high potential for industrial applications. With a Raman spectroscopic device, the composition of natural gas[27] or biogas[27b,28] can be determined on-site and continuously. This is of great interest at the site of extraction or production as well as in the plant to better control the combustion process.

Hippler[27d] shows that CERS with low-power diode lasers is suitable for online monitoring of natural gas mixtures, including H_2, H_2S, N_2, CO_2, and alkanes. Kiefer et al.[27b] demonstrate the advantages of Raman spectroscopy over gas chromatography: a Raman spectroscopic device is integrated in a gas turbine power plant, and the data are compared to the power plant's integral GC device. Although the accuracy is comparable, the significantly higher time resolution of the Raman spectroscopic device allows seeing short-time fluctuations that are not resolved in the GC measurements. The derived information is valuable for purposes of controlling the gas turbine operation.[27b] Even without the use of an enhancement technique, the content of all molecular natural gas components whose content exceeds 0.005% could be reliably determined with a measurement time of 100 s by Petrov and Matrosov.[27c]

Besides natural gas, Raman spectroscopic devices are also used in applications related to biogas. Numata et al.[28a] demonstrated that Raman spectroscopy is capable of identifying fermentation gases and determining their partial pressures in the gas mixture. Taking it one step further, Eichmann et al.[28b] impressively show that a Raman gas analyzer based on a retro-reflective gas cell successfully detects and quantifies all relevant gas components in a biogas plant, that is, CH_4, CO_2, N_2, and H_2O, and reports their individual concentrations over time.[28b]

Francisco and Rich[29] show that CERS could also be used to monitor industrial emissions. They present a multigas analyzer based on intracavity Raman scattering able to monitor pollutants, such as SO_2, as well as CO, CO_2, N_2, and O_2.[29]

One major advantage of Raman spectroscopy for industrial applications is pointed out by Eichmann et al.[28b] and Francisco and Rich[29]: IR absorption based gas analyzers cannot detect homonuclear diatomic gas species, making it necessary to use multiple devices, whereas with Raman spectroscopy, one device is sufficient to detect all relevant components of a multigas.

Another possible usage is the monitoring and control of the food ripening process during storage and transport.[30] With the help of a FERS gas sensor, a gas mixture typical for fruit chambers was monitored: O_2, CO_2, and ethylene (C_2H_4), which is used to trigger fruit ripening, could be detected at the same time. Also, the commonly used cooling agent ammonia (NH_3) can be detected by FERS, which potentially could be used to trigger an alarm for too high or low concentrations. Thus, FERS has the potential to be used as a versatile gas sensor throughout the complete postharvest production chain, including storage, transport, and industrial fruit ripening.[30]

19.5 CONCLUSION AND OUTLOOK

Raman spectroscopic methods allow the simultaneous identification and quantification of various components of a sample gas with one measurement. The gas mixture can be analyzed continuously in a nonconsumptive fashion and with high sensitivity and specificity. Thus, Raman gas spectroscopy has great potential in research as well as industrial applications. However, several steps have to be taken before Raman gas spectroscopy will become a broadly used, standard technique. On the one hand, the devices have to be miniaturized and produced at comparable cost to competing instruments. On the other hand, the device should be easy to use for a non-expert may it be a biogeochemist or a clinician. This requires the device to be robust to withstand an employment outdoors, in the clinic, or a power plant. Additionally, calibration and alignment as well as peak identification should be automated. Cavity as well as fiber-enhanced techniques are promising, and further developments will lead towards more practicality.

REFERENCES

1. (a) Smekal, A., Zur quantentheorie der dispersion. *Naturwissenschaften* 1923, *11* (43), 873–875; (b) Raman, C. V.; Krishnan, K. S., A new type of secondary radiation. *Nature* 1928, *121*, 501–502; (c) Landsberg, G.; Mandelstam, L., Eine neue Erscheinung bei der Lichtzerstreuung in Krystallen. *Die Naturwissenschaften* 1928, *16* (28), 557–558; (d) Rocard, Y., Les nouvelles radiations diffusées. *CR Academy of Sciences* 1928, *190*, 1107–1109; (e) Cabannes, J., Un nouveau phénomène d'optique: Les battements qui se produisent lorsque des molécules anisotropes en rotation et vibration diffusent de la lumière visible ou ultraviolette. *CR Academy of Sciences* 1928, *186*, 1201–1202; (f) Kramers, H. A.; Heisenberg, W., Über die streuung von strahlung durch atome. *Zeitschrift für Physik A Hadrons and Nuclei* 1925, *31* (1), 681–708.
2. Weber, A., Raman spectroscopy of gases. In *Handbook of Vibrational Spectroscopy*. Chalmers, J. M.; Griffiths, P. R., eds. John Wiley & Sons: Chichester, 2002, 176–195.
3. Long, D. A., *The Raman Effect: A Unified Treatment of the Theory of Raman Scattering by Molecules*. John Wiley & Sons: Chichester, 2002.
4. Steinfeld, J. I., *Molecules and Radiation an Introduction to Modern Molecular Spectroscopy*. Dover Publications Inc.: Mineola, 2005.
5. Herzberg, G., *Molecular Spectra and Molecular Structure I. Spectra of Diatomic Molecules*, 2nd ed., D. Van Nostrand Company, Inc.: Princeton, NJ, 1963.
6. Haken, H.; Wolf, H. C., *Molekülphysik und Quantenchemie-Einführung in die experimentellen und theoretischen Grundlagen*. Springer-Verlag: Berlin and Heidelberg, 2006.
7. Bendtsen, J., The rotational and rotation-vibrational Raman spectra of 14N2, 14N15N and 15N2. *Journal of Raman Spectroscopy* 1974, *2* (2), 133–145.
8. Hanf, S.; Bögözi, T.; Keiner, R.; Frosch, T.; Popp, J., Fast and highly sensitive fiber-enhanced Raman spectroscopic monitoring of molecular H_2 and CH_4 for point-of-care diagnosis of malabsorption disorders in exhaled human breath. *Analytical Chemistry* 2015, *87* (2), 982–988.
9. Schlüter, S.; Popovska-Leipertz, N.; Seeger, T.; Leipertz, A., Gas sensor for volatile anesthetic agents based on Raman scattering. *Physics Procedia* 2012, *39*, 835–842.
10. Schlüter, S.; Krischke, F.; Popovska-Leipertz, N.; Seeger, T.; Breuer, G.; Jeleazcov, C.; Schuttler, J.; Leipertz, A., Demonstration of a signal enhanced fast Raman sensor for multi-species gas analyses at a low pressure range for anesthesia monitoring. *Journal of Raman Spectroscopy* 2015, *46* (8), 708–715.
11. Frosch, T.; Keiner, R.; Michalzik, B.; Fischer, B.; Popp, J., Investigation of gas exchange processes in peat bog ecosystems by means of innovative Raman gas spectroscopy. *Analytical Chemistry* 2013, *85* (3), 1295–1299.
12. Gagliardi, G.; Loock, H.-P., *Cavity-Enhanced Spectroscopy and Sensing*. Springer: Berlin and Heidelberg, 2014.
13. (a) Rupp, S.; Off, A.; Seitz-Moskaliuk, H.; James, T. M.; Telle, H. H., Improving the detection limit in a capillary Raman system for in situ gas analysis by means of fluorescence reduction. *Sensors (Basel)* 2015, *15* (9), 23110–23125; (b) Buric, M. P., Gas phase Raman spectroscopy using hollow waveguides. Dissertation, University of Pittsburgh, 2010; (c) James, T. M.; Rupp, S.; Telle, H. H., Trace gas and dynamic process monitoring by Raman spectroscopy in metal-coated hollow glass fibres. *Analytical Methods* 2015, *7* (6), 2568–2576.
14. Hanf, S.; Keiner, R.; Yan, D.; Popp, J.; Frosch, T., Fiber-enhanced Raman multigas spectroscopy: A versatile tool for environmental gas sensing and breath analysis. *Analytical Chemistry* 2014, *86* (11), 5278–5285.
15. (a) Benabid, F., Hollow-core photonic bandgap fibre: New light guidance for new science and technology. *Philosophical Transactions a Mathematical Physical Engineering Sciences* 2006, *364* (1849), 3439–3462; (b) Benabid, F.; Roberts, P. J.; Couny, F.; Light, P. S., Light and gas confinement in hollow-core photonic crystal

fibre based photonic microcells. *Journal of the European Optical Society: Rapid Publications* 2009, *4, 09004-1–09004-9.*
16. Keiner, R.; Frosch, T.; Massad, T.; Trumbore, S.; Popp, J., Enhanced Raman multigas sensing—A novel tool for control and analysis of (13)CO(2) labeling experiments in environmental research. *Analyst* 2014, *139* (16), 3879–3884.
17. (a) Buszewski, B.; Grzywinski, D.; Ligor, T.; Stacewicz, T.; Bielecki, Z.; Wojtas, J., Detection of volatile organic compounds as biomarkers in breath analysis by different analytical techniques. *Bioanalysis* 2013, *5* (18), 2287–2306; (b) Boegoezi, T.; Popp, J.; Frosch, T., Fiber-enhanced Raman multi-gas spectroscopy-what is the potential of its application to breath analysis? *Future Science Bioanalysis* 2015, *7* (3), 281–284.
18. Buszewski, B.; Kęsy, M.; Ligor, T.; Amann, A., Human exhaled air analytics: Biomarkers of diseases. *Biomedical Chromatography* 2007, *21* (6), 553–566.
19. Schlüter, S.; Seeger, T.; Popovska-Leipertz, N.; Leipertz, A., Atemzyklusgenaues Anästhesiegas-Monitoring mit einer laserbasierten Raman-Sonde unter klinischen Bedingungen. *tm-Technisches Messen* 2016, *83* (5), 289–299.
20. (a) Ciais, P.; Sabine, C.; Bala, G.; Bopp, L.; Brovkin, V.; Canadell, J.; Chhabra, A.; DeFries, R.; Galloway, J.; Heimann, M.; Jones, C.; Le Quéré, C.; Myneni, R. B.; Piao, S.; Thornton, P., Carbon and other biogeochemical cycles. In *Climate Change 2013: The Physical Science Basis. Contribution of Working Group I to the Fifth Assessment Report of the Intergovernmental Panel on Climate Change.* Stocker, T. F.; Qin, D.; Plattner, G.-K.; Tignor, M.; Allen, S. K.; Boschung, J.; Nauels, A.; Xia, Y.; Bex, V.; Midgley, P. M., eds. Cambridge University Press: Cambridge and New York, 2013; (b) Schulze, E. D.; Luyssaert, S.; Ciais, P.; Freibauer, A.; Janssens, I. A.; et al., Importance of methane and nitrous oxide for Europe's terrestrial greenhouse-gas balance. *Nature Geosci* 2009, *2* (12), 842–850; (c) Reichstein, M.; Bahn, M.; Ciais, P.; Frank, D.; Mahecha, M. D.; et al., Climate extremes and the carbon cycle. *Nature* 2013, *500* (7462), 287–295; (d) Brevik, E. C., Soils and climate change: Gas fluxes and soil processes. *Soil Horizons* 2012, *53* (4),doi:10.2136/sh12-04-0012.
21. Keiner, R.; Gruselle, M. C.; Michalzik, B.; Popp, J.; Frosch, T., Raman spectroscopic investigation of 13CO2 labeling and leaf dark respiration of *Fagus sylvatica* L. (European beech). *Analytical Bioanalytical Chemistry* 2015, *407* (7), 1813–1817.
22. Hanf, S.; Fischer, S.; Hartmann, H.; Keiner, R.; Trumbore, S.; Popp, J.; Frosch, T., Online investigation of respiratory quotients in *Pinus sylvestris* and *Picea abies* during drought and shading by means of cavity-enhanced Raman multi-gas spectrometry. *Analyst* 2015, *140* (13), 4473–4481.
23. Keiner, R.; Frosch, T.; Hanf, S.; Rusznyak, A.; Akob, D. M.; Kusel, K.; Popp, J., Raman spectroscopy—An innovative and versatile tool to follow the respirational activity and carbonate biomineralization of important cave bacteria. *Analytical Chemistry* 2013, *85* (18), 8708–8714.
24. Jochum, T.; Michalzik, B.; Bachmann, A.; Popp, J.; Frosch, T., Microbial respiration and natural attenuation of benzene contaminated soils investigated by cavity enhanced Raman multi-gas spectroscopy. *Analyst* 2015, *140* (9), 3143–3149.
25. Keiner, R.; Herrmann, M.; Küsel, K.; Popp, J.; Frosch, T., Rapid monitoring of intermediate states and mass balance of nitrogen during denitrification by means of cavity enhanced Raman multi-gas sensing. *Analytica Chimica Acta* 2015, *864*, 39–47.
26. Jochum, T.; Fastnacht, A.; Trumbore, S.E.; Popp, J.; Frosch, T., Direct Raman Spectroscopic Measurements of Biological Nitrogen Fixation under Natural Conditions: An Analytical Approach for Studying Nitrogenase Activity. *Anal Chem* 2017, *89* (2), 1117–1122.
27. (a) Buldakov, M. A.; Korolkov, V. A., Matrosov, II; Petrov, D. V.; Tikhomirov, A. A.; Korolev, B. V., Analyzing natural gas by spontaneous Raman scattering spectroscopy. *Journal Optical Technology* 2013, *80* (7), 426–430; (b) Kiefer, J.; Seeger, T.; Steuer, S.; Schorsch, S.; Weikl, M. C.; Leipertz, A., Design and characterization of a Raman-scattering-based sensor system

for temporally resolved gas analysis and its application in a gas turbine power plant. *Measurement Science and Technology* 2008, *19* (8), 085408; (c) Petrov, D. V., Matrosov, I. I., Raman gas analyzer (RGA): Natural gas measurements. *Applied Spectroscopy* 2016, *70* (10), 1770–1776; (d) Hippler, M., Cavity-enhanced Raman spectroscopy of natural gas with optical feedback cw-diode lasers. *Analytical Chemistry* 2015, *87* (15), 7803–7809.

28. (a) Numata, Y.; Shinohara, Y.; Kitayama, T.; Tanaka, H., Rapid and accurate quantitative analysis of fermentation gases by Raman spectroscopy. *Process Biochemistry* 2013, *48* (4), 569–574; (b) Eichmann, S. C.; Kiefer, J.; Benz, J.; Kempf, T.; Leipertz, A.; Seeger, T., Determination of gas composition in a biogas plant using a Raman-based sensor system. *Measurement Science and Technology* 2014, *25* (7), 075503.
29. Francisco, T. W.; Rich, R. R.; Society of Photo-optical Instrumentation Engineers, Intracavity Raman spectroscopy for industrial stack gas analysis, in Proceedings-SPIE the International Society for Optical Engineering, Environmentally conscious manufacturing IV, Philadelphia, PA, 2004, Vol. 5583, pp. 68-75.
30. Jochum, T.; Rahal, L.; Suckert, R. J.; Popp, J.; Frosch, T., All-in-one: A versatile gas sensor based on fiber enhanced Raman spectroscopy for monitoring postharvest fruit conservation and ripening. *Analyst* 2016, *141* (6), 2023–2029.

PART VII

Oil, gas, and mineral exploration and refining

Table VII.1 Optoelectronic applications in the oil and gas industry

Application	Technology	Advantages	Disadvantages	Current situation (at time of writing)	More reading
Permanent reservoir monitoring (seabed seismic) using active sonar. Air-gun sound source, with large fiber hydrophone sensor arrays.	Multiplexed interferometric optical fiber sensors. Usually fixed seabed arrays or vessel-towed arrays of sensors. Spin-off from naval research.	No subsea electronics, improved reliability. Very compact cable structure for large arrays.	Needs care to handle very high amplitude broadband signals, which can easily overload sensor interrogation systems. Still relatively new technology.	Many examples of towed arrays. Has been used in two fixed installations (240 km installed cable). Still competing with electrical systems.	See this section.
Downhole seismic point sensors.	Multiplexed interferometric optical fiber sensors.	No downwell electronics, high temperature and pressure ability, good seismic performance.	Very restrictive space for sensor packages. Still expensive compared to electrical systems. Still relatively poor spatial resolution.	Has been used in a few niche applications, >90% of market still electrical.	See this section.
Downhole distributed seismic sensors.	Distributed acoustic optical fiber sensors on single optical cable.	High spatial resolution and over 1000 sensing points. Easy to install and can even use existing downhole fibers. High total cost, but low cost per sensor.	Relatively poor seismic performance compared to point sensors. Nature of acoustic interaction with a linear cable is more complex.	Still a new technology, but rapidly growing, use in a number of experiments.	See this section.
Downhole flow-noise sensors.	Distributed acoustic optical fiber sensors on single optical cable	High spatial resolution, simple to install and can use existing fibers, to give low cost per sensor	Nature of acoustic interaction is more complex and noise floor can be high	Becoming quite widely adopted	See this section

(*Continued*)

Table VII.1 (*Continued*) Optoelectronic applications in the oil and gas industry

Application	Technology	Advantages	Disadvantages	Current situation (at time of writing)	More reading
Downhole temperature sensors	Distributed optical fiber temperature sensor, most using Raman scattering	High spatial resolution, simple to install and can also use existing fibers, low cost	Lower accuracy/ resolution than point-measurement sensors	Has been adopted worldwide, so probably the most mature optical technology in oil wells.	See also Volume II, Chapter 11 (Optical Fiber Sensors).
Downhole pressure	Bragg grating optical fiber sensors	No downhole electronics. High temperature capability	Difficult to deploy and install	Quite widely adopted	See also Volume II, Chapter 11 (Optical Fiber Sensors)
Flammable gas sensing around petrochemical plants and oil rigs	Spectrographic sensors, mostly in infrared region. Can use continuous wave (CW) or pulsed lasers for measurement over extended airborne paths (absorption, LIDAR or DIAL).	Electrically safe in explosive environments. Can measure over long distances with line-of-sight. Can be networked using optical fibers.	Many use relatively new LED/laser technology, so more expensive than simple electrical (electrochemical and pallister) types.	Some infrared sensors are becoming well established, and ever more common as component developments continue	See also Volume II, Chapter 12 (Remote Optical Sensing by Laser) and Chapter 15 (Spectrographic Analysis)
Other gas sensing (toxic gases and CO_2 monitoring).	As above	Can measure over long distances with line-of-sight. Can be networked using optical fibers.	Many use relatively new LED/laser technology, so more expensive than simple electrochemical types.	Some infrared sensors are becoming more common as component developments continue.	See also Volume II, Chapter 12 (Remote Optical Sensing by Laser) and Chapter 15 (Spectrographic Analysis).

20

Fiber optics in the oil and gas industry

ANDRE FRANZEN
Shell International Exploration and Production B.V.

20.1 OPENING REMARKS

As the handbook covers the broader subject of optoelectronics, we feel there is a need to emphasize our great bias in this chapter towards optical fiber systems. This choice has been made because of the nature of the petrochemical and mining industries, which often requires equipment to operate in remote and particularly severe environments, namely, deep-sea, downwell, and in mine, usually very far from traditional terrestrial infrastructure and hence only accessible via very long (multi-kilometer) cables. Optical fiber cables, being thin, extremely robust when packaged, low loss, and, if needed, having a very high bandwidth are the preferred options for these areas.

For completeness, it is appropriate, before moving on, to briefly list some of the nonfiber uses of optoelectronics:

1. Solar panels to power remote equipment [1]
2. Optoelectronic displays such as LCD displays [2]
3. Optical gas sensors [3]
4. Optical spectroscopy of samples taken to the laboratory [4]
5. Line-of-site optical communication between offshore oil installations such as drilling rigs and remote satellite platforms as well as subsea communication [5,6]

All of these other fiber optic technologies are presented in Table VII.1. Many of these are fairly generic technology, used widely across many applications, and are discussed in more detail in other chapters of the book, so we will not discuss them further here.

20.2 INTRODUCTION TO THE OIL INDUSTRY

Before delving into fiber optic applications in the oil industry, the driver of these developments shall be explained here. The rapid population growth in combination with increasing prosperity in developing countries is expected to lead to a surge in energy consumption over the coming decades [7]. Despite implementation of energy-saving measures, forecasts indicate an increase in energy demand in 2035 ranging from 55% up to 95% compared to 2000 [8,9]. Renewable energy sources should provide a substantial part of the required energy. However, it is likely that a range of sources will be needed to supply this vital energy over the coming decades. To do this in a sustainable way, a part of the world's *energy mix* could come from *renewables*, with *fossil fuels* and *nuclear power* providing the rest. Moreover, oil is the basis for many products in today's industrialized world, and it is unlikely that sufficient alternatives will become available over the coming decades [10].

As a result, the challenge for oil and gas companies is to supply sufficient amounts of oil and in a sustainable way. This needs continuous improvements throughout the production cycle. The integrated supply chain that ranges from upstream to downstream is displayed in Figure 20.1. Activities to explore, that is, find, and produce hydrocarbons are collectively referred to as upstream activities. Production locations include onshore oil and gas wells, offshore platform-based or subsea production systems, as well as production of, for example, sugarcane as an ingredient for biofuels. Downstream activities include all efforts to process and refine the hydrocarbons, eventually resulting in products that can be sold. Products include fuels, and also chemicals, for example, plastics, as a feedstock for other industries. The fiber optic sensing technologies discussed in this chapter are tailored to upstream applications, although they also offer significant potential for application for downstream as well as in other industries. In upstream applications, the main challenge is to increase hydrocarbon production; although the era of easy oil is over, that is, easy drillable sources of light oil are becoming increasingly scarce. Thus, the driving factor in the oil industry to use fiber optics is the enablement of improved economical exploitation

Figure 20.1 Modern hydrocarbon recovery and processing. The onshore field (1) uses fracturing techniques in long horizontal wells to extract gas or oil from shales. Steam is injected in an onshore field (2) to lower the oil viscosity and consequently increase the recovery factor. New resources are increasingly found in challenging areas, such as deep sea environments (3) or in arctic conditions (4). All these complex production conditions require increasing level of monitoring to ensure efficient operation. (Courtesy of Shell.)

Table 20.1 Typical examples for various sensor types

Point sensors	Quasi-distributed sensor	Distributed sensor
Pressure gauges	Multipoint pressure gauges	Temperature sensing
Temperature gauges	Array temperature sensing	Acoustic sensing
Chemical gauges	Seismic sensors	
Strain gauges		

of natural resources as well as improved health, safety, and environment (HSE) [11]. Fiber optic technologies provide certain properties that cannot be achieved by conventional technologies, such as resistance to high temperatures, electromagnetic interference (EMI), and distributed measurements, just to name a few. Various environments, as presented in Figure 20.1, require different solutions and thus pose a very specific challenge to the required solution. There are in principle three categories of sensors that are widely deployed in the oil industry: point sensors, quasi-distributed sensors, and truly distributed sensors (Table 20.1).

Besides the sensor developments, the interrogator development has a special place in the industry as often the off-the-shelf solutions do not give the required solution, for example, Fiber Bragg grating (FBG; explained more in Section 20.4) interrogator development to achieve higher strain resolution or truly distributed systems to achieve unique measurements along the whole length of the fiber. In order to comprehend the environmental challenge that the oil industry faces, a few but major aspects will be presented here and explained to such an extent that the reader should be able to grasp the overall challenge and understand the implemented solution.

20.3 TYPICAL WELL CONSTRUCTIONS

Most permanent sensors in the oil industry are deployed inside boreholes. These boreholes can be used for either the production of hydrocarbons (i.e., oil, gas or gas condensate) or the injection of gas, water, or chemicals to drive hydrocarbons in certain directions from where those hydrocarbons can be produced, or there are boreholes just for monitoring purposes. These are the basic three types of wells that therefore determine the potential applications, namely, production monitoring, injection monitoring, and field monitoring.

When looking at the generic well structure all types of wells have a common setup: they all have a *casing* (see Figure 20.2), which is required to ensure that the borehole does not collapse, they all have a *well head*, which confines the access from the reservoir to the surface, and they also have a form of *tubing*, which reaches usually the reservoir section. For *production wells*, a *packer* is required to guarantee the isolation from the reservoir to the upper section of the well. The well head is seen as a second barrier, and it is standard to have two barriers in most counties to avoid integrity issues. The (production) tubing is a tube in which the hydrocarbon is produced up to the surface. All wells need a connection to the reservoir, that is, the hydrocarbon bearing layer; this is usually achieved by perforating, that is, shooting through the casing at the correct depth. This setup limits the placement of any sensor downhole; ultimately only two locations are available for sensing: along the tubing and along the casing (see Figure 20.2; blue and green indicate the fiber cable positions). Although one is closer to the environment one wants to monitor, the other allows much easier deployment of sensors. Also, at times there are legislative constraints on where to place the sensor. Without delving into the various production methods such as artificial lift [beam pumps, electrical submersible pump (ESP)], steam production (cyclic steam stimulation), or injection techniques (water, steam, and chemicals), the focus can now be shifted to the applications and their deployments.

20.3.1 Field architecture structure basics

To complete the overview, it is important to understand that the measurements taken in the field are

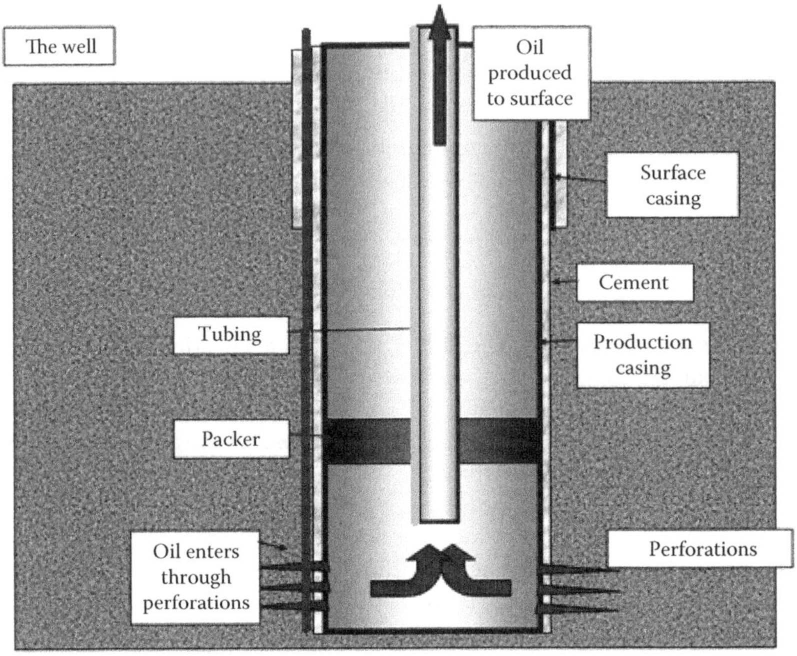

Figure 20.2 Typical well schematic. (From Wikipedia-Well, http://upload.wikimedia.org/wikipedia/en/7/7f/Oil_Well.png (Accessed January 18, 2015). With permission.)

potentially hundreds of kilometers away from the person analyzing them.

Therefore, the data transport is a vital stage in this development and so is the storage of information.

Usually, a deployment can be separated in four building blocks:

downhole,
deployment,
surface deployment,
transmission, and
storage.

Once the data are available in the database, various people, such as production technologists,

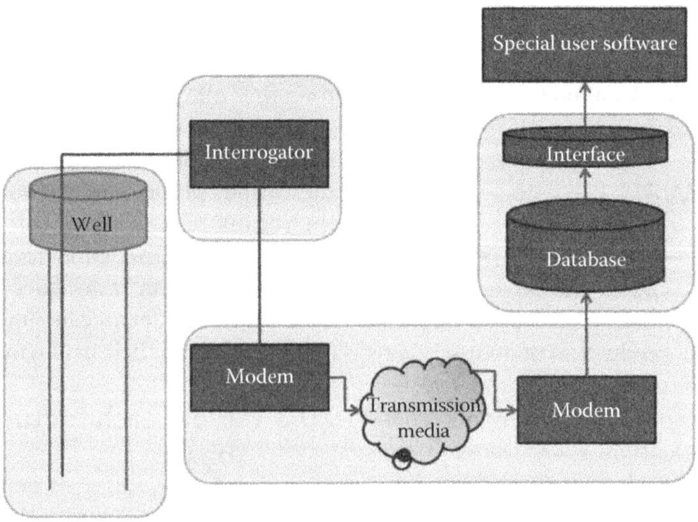

Figure 20.3 Data architecture.

petrophysicists, reservoir engineers, geologists, and seismologists, can access the data and use them for further processing using specialized software. The schema in Figure 20.3 only shows the main elements; in reality, there are elements such as firewalls and proxy servers in use, but this generic architecture does not change.

20.4 FBG INTERROGATOR

Usually, a sensor is connected to a fiber that in turn is connected to an appropriate interrogator. If a commercial solution is available, it is always the preferred path due to the fact it is proven and commercial aftercare can be relied on. At times, it is still required to develop specific interrogators to achieve best-in-class performance as the requirements change compared to standard interrogators (Table 20.2).

20.4.1 Requirements

Unlike other commercial environment, for example, rotating equipment or airplane wings, where very fast changes require to be measured, downhole conditions usually change very slowly; therefore, the requirement of fast interrogation is usually not given, but instead a higher wavelength resolution is preferred. Another aspect of field instrumentation is that the interrogator needs to be able to survive ambient temperatures of up to 60°C without active cooling, a condition that is seen in desert scenarios. As power consumption is another challenge in desert environments, often the only supply of electricity is by the use of solar panels; due to the vast distances, low-power consumption stipulates another stringent requirement.

20.4.2 Developments

An analysis of various interrogation schemes leads to tunable filter-based interrogator, and a prototype was realized as depicted in Figure 20.4. A hydrogen cyanide ($H^{13}CN$) gas cell was used to achieve temperature correction especially for the temperature dependency of the Fabry Perot tunable filter and other components [13]. Additionally, it allowed high resolution to be obtained as the molecular absorption lines are well defined within an inaccuracy of 0.5 pm; see Figure 20.5 for the molecular spectrum at different temperatures. Combining a high-temperature processing unit and a self-developed robust algorithm resulted in an interrogator that could face the set requirements.

20.4.3 Results

The designed interrogator was tested in an oven where temperature swings could easily be emulated, and the FBGs to be observed were placed outside in a thermally stabilized chamber as presented

Table 20.2 Building blocks of the field data architecture

1. Downhole installation	The part where fiber optic equipment is installed downhole, either behind casing if it is a newly drilled well or on tubing possible for both newly drilled well and worked over wells
2. Surface installation	Is the placement of the appropriate interrogator and all connection to the well site, conversion of measurement data to a digital format
3. Data transport	Is concerned about the transfer of the data via a suitable medium, for example, TCP/IP, ModBus over Ethernet, WiFi, WiMax, microwave link, and fiber optic backbone from the well site to the office
4. Data storage	Focuses on the collection and storage of field data; depending on the data type that usually happens in the process control domain with the exceptions of specific array data where the data end up in the office domain; in either case, the data will be stored in a database

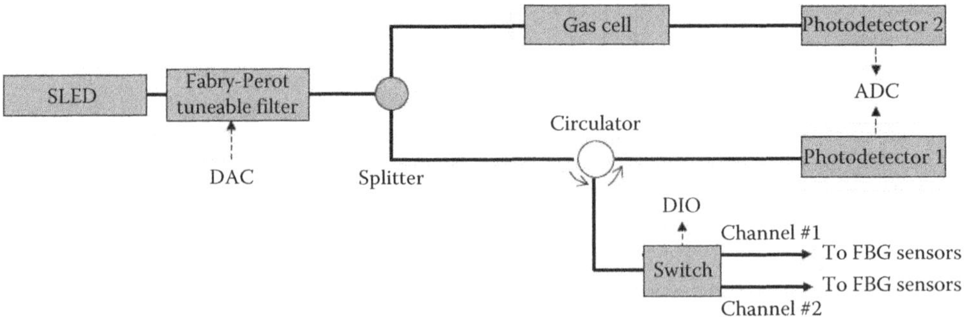

Figure 20.4 Tunable filter-based interrogator with two channels. (From Lumens, P.G.E., *Fibre-Optic Sensing for Application in Oil and Gas Wells*, Eindhoven University of Technology, Eindhoven, 2014.)

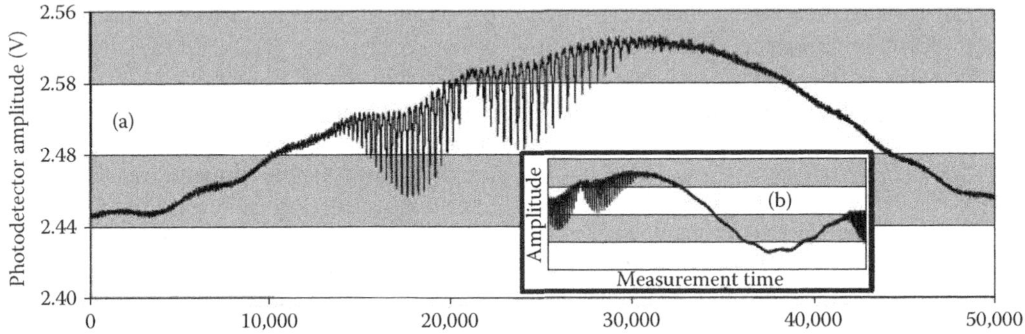

Figure 20.5 Spectral measurement of molecular absorption lines of hydrogen cyanide (H13CN) in a gas cell at a filling pressure of 100 torr, as measured by recording the transmission through the cell while sweeping the tunable filter at (a) room temperature and (b) at a temperature of 70°C. (From Lumens, P.G.E., *Fibre-Optic Sensing for Application in Oil and Gas Wells*, Eindhoven University of Technology, Eindhoven, 2014.)

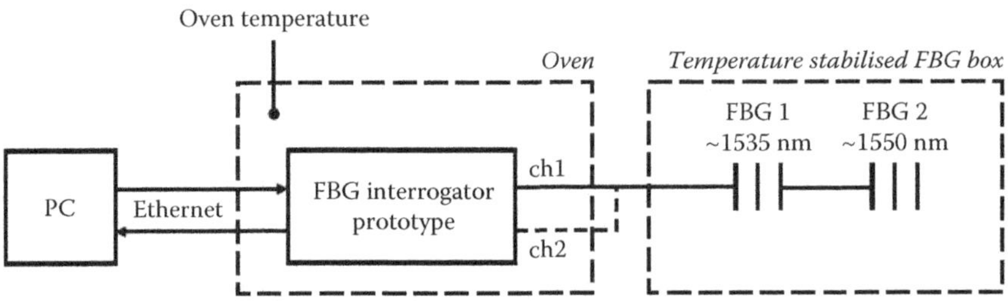

Figure 20.6 Test setup to simulate desert condition.

in Figure 20.6. The acquisition time is 8 s, which is rather slow for an FBG interrogator; in turn, the unit could operate up to 63°C and showed a repeatability of 0.6 pm, which is approximately a factor of 2 better than commercial interrogators at that time. Also, the temperature stability is better than 2.5 pm over a range from 20°C to 63°C, thus ensuring good performance in the field environment. An additional factor for success was that this development delivered a low-cost solution, which was proven as well.

20.5 FIBER OPTIC PRESSURE GAUGES

One of the longest standing measurements in the oil industry is the pressure measurement. For instance, pressure buildup and drop-down tests were developed to estimate the reserves and continuing pressure measurements to measure natural flow to surface, just to name a few applications. Most traditional pressure transducers use a membrane that

Figure 20.7 Downhole pressure gauge. (From Smartfibres, *Smartfibres—Oil and Gas*. http://smartfibres.com/oil-and-gas (Accessed November 12, 2014); emphBen Hunter, Technology overview of permanent downhole distributed pressure sensing with fiber bragg grating sensing technology. http://smartfibres.com/docs/DPS_Overview.pdf (Accessed November 12, 2014). With permission.)

deflects due to the pressure and that deflection can be measured in turn with a strain gauge. In fiber optics, an equivalent element to the strain gauge is known as the FBG; see also the section 1.1.2 on FBG.

Besides strain, the FBG is sensitive to temperature. Once the FBG substitutes the standard strain element and is combined with a second FBG for an additional temperature measurement, a perfect pressure sensor is designed. The temperature impact of the strain sensing FBG can be compensated by the second one, leaving the true strain value. This sensor type has the advantage that it does not require power and hence is EMI resistant. These are two rather important aspects in downhole deployments as other components may cause issues along those lines, for example, submersible pumps. Another advantage is that these gauges can be cascaded and the limit is determined by the interrogator used. Also, it is fully operational at 150°C with the prospect to achieve even higher temperatures and is rated for 350 bars. In terms of pressure accuracy, the sensor performs at 0.1% FS and has a resolution of 0.05% FS [15]. Looking at the temperature performance, an accuracy of 0.1°C with a resolution of 0.001°C was achieved. The long-term stability was determined to be 0.05% FS/year. These are typical values for qualified gauges in the industry (Figure 20.7).

20.5.1 In-well level monitoring

Having the access to sufficiently accurate pressure gauges as presented previously, a new application opened up in the oil market: in-well level monitoring.

The fluid level in observation wells is related to the oil rim in reservoir. Oil rims can be controlled by pushing the oil up or down by injecting or pumping off water below the oil and additional injection of gas above the oil or by decreasing/increasing the production. Thus, the rim can be confined in a certain depth, which is where the actual producing wells are drilled to. The rim can always be shifted up or down to remain at this depth with the help of the water-gas injection, but it is vital to know where this depth is. To make matters worse, the height of the rim can change as well; therefore, it is of paramount importance to monitor the oil rim behavior. With the help of the pressures measured in each of the fluid gas, oil, and water, we can identify the contacts between gas–oil [free gas–oil contact (FGOC)] and oil–water [free oil–water contact (FOWC)] in the well due to simple gradient calculation. When there are at least two gauges in the same medium, it is a straightforward calculation to determine the gradient in that medium. Then we only have to identify the location where the gradients intersect, and the FGOC and FOWC contacts are established. This information in turn can be used to control the injection of the gas and water. To be certain that the gradients are calculated with enough precision, there is a trade-off due to the uncertainty of the gauges, and the amount of gauges in one fluid a cut-level algorithm was developed to obtain more precise contacts.

20.5.2 Pressure gauge deployment

In 2008, 2012, and 2014, the in-well level monitoring system as described earlier was deployed in the Middle East in all cases in an identical way. Up to 15 pressure gauges were deployed downhole in a redundant way as depicted in Figure 20.8.

Several gauges are placed in nylon inserts and connected to each other and then placed in a specially designed gauge carrier; see Figure 20.8. This

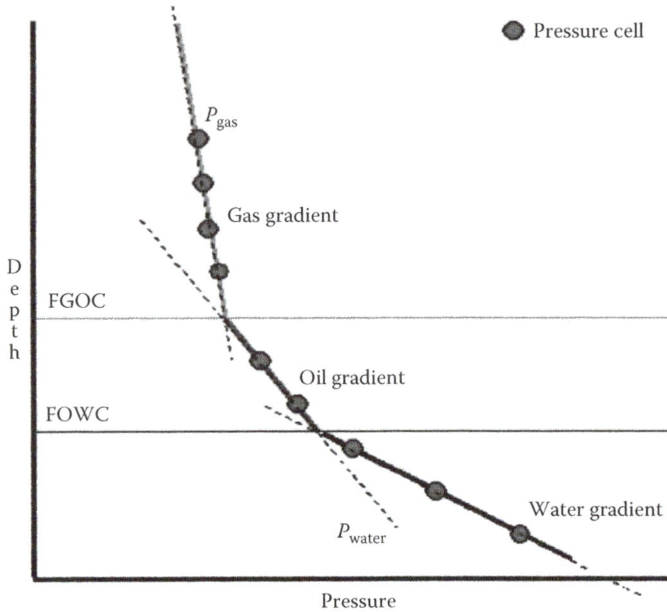

Figure 20.8 Level gradient diagram.

way clamps can easily be used to fix the gauges against the tubing, avoiding expensive mandrels, see Figure 20.9. In order to have a timely deployment, the gauges are placed in a special carrier system that allows easy attaching of the gauges with the clamps onto the tubing (Figure 20.10).

One can see that several sections need to be connected on the rig floor but that it is a relatively minor activity. Once the gauges are deployed at the correct location, only downhole cable needs to be installed, which is clamped onto the tubing until one reaches the well head. Here the cable is brought

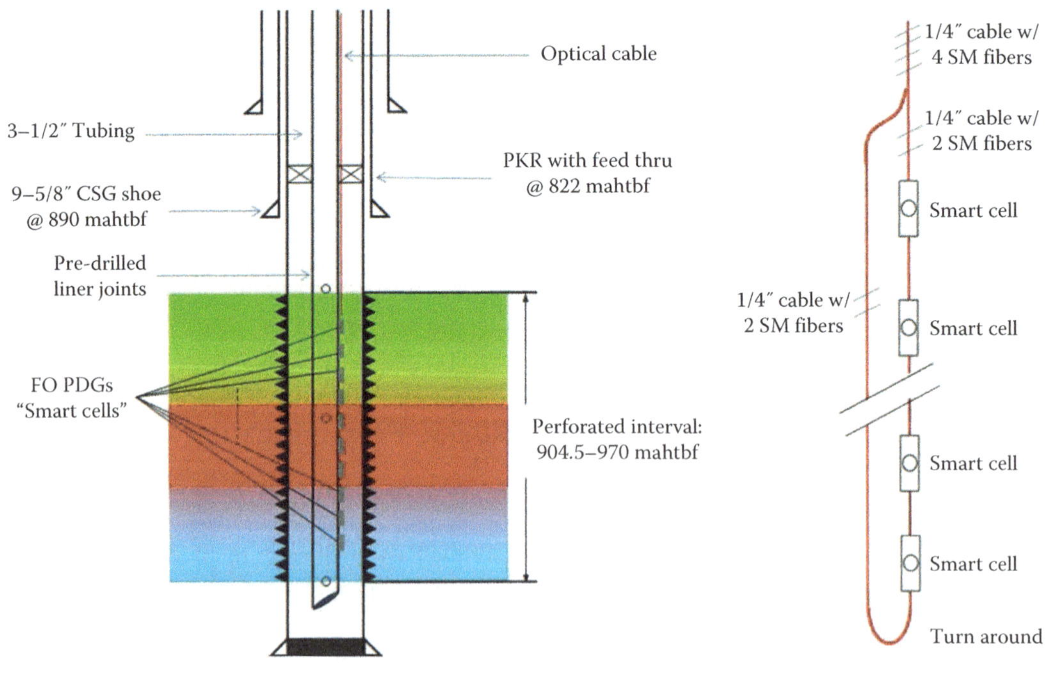

Figure 20.9 Generic schematic and optical equivalent.

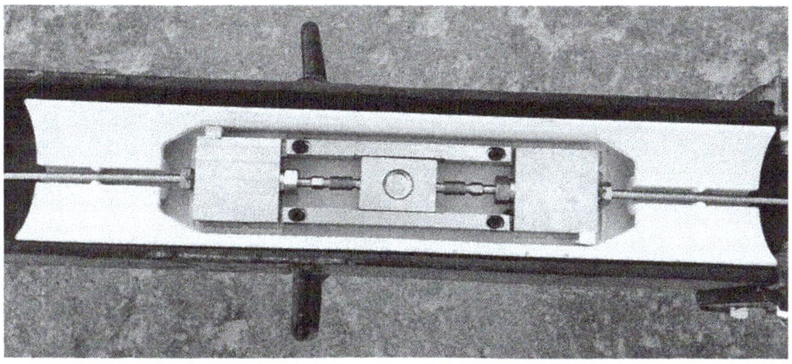

Figure 20.10 Fiber optic downhole pressure gauge in insert and clamp.

Figure 20.11 Gauges and gauge carrier assembly at workshop (left) and gauges on catwalk (right).

out through the wellhead exit and is terminated in a junction box via a pressure barrier for which standard oil field components are used. The cable is then connected to the interrogator system comprising an FBG interrogator with an optical scrambler, and a small form-factor industrial PC with an appropriate modem, which is then connected to a WiMax system where Modbus/Transmission control protocol over Ethernet is facilitated for the actual transmission (Figure 20.11).

The WiMax system allows the transmission from the well site via several relays to the headquarters (HQs), some 260 km away. The field cabinet is rather simple but very effectively built up, as one objective here is to obtain as much shadow as possible to provide cooling to the surface equipment. It should additionally be power efficient (as there was no power supply) and then also be desert sand proof (IP66), the full cabinet setup can be viewed in Figure 20.12.

The wavelength data are received in HQ and stored in an appropriate database. Once the data arrive, a first algorithm converts these wavelengths into pressure and temperature information, which are then stored separately. As the first algorithm finishes, it automatically triggers a second algorithm, the aforementioned cut-level algorithm (CLA). This software takes all pressure information and evaluates it; as a result, the gradients in the three media, that is, gas, oil, and water, are determined, and from there the gas–oil contact and the oil–water contact. This information is made visible via a specially developed user interface for ease of use to the end user (Figure 20.13).

The system runs about six times in 24 h and generates a wealth of information in comparison with the traditional Gradio Logging Tool, which were only run three or four times a year. This now allows the end user to adjust and therefore optimize water and gas injection leading to a better oil recovery rate and additionally avoids any well intervention, thus minimizing HSE exposure.

20.6 DISTRIBUTED SENSING

A technique that can be called the first truly distributed measurement besides optical time domain reflectometrys (OTDRs) is distributed temperature sensing (DTS), developed in the 1980s at Southampton University in the United Kingdom;

Figure 20.12 Field cabinet with solar panel.

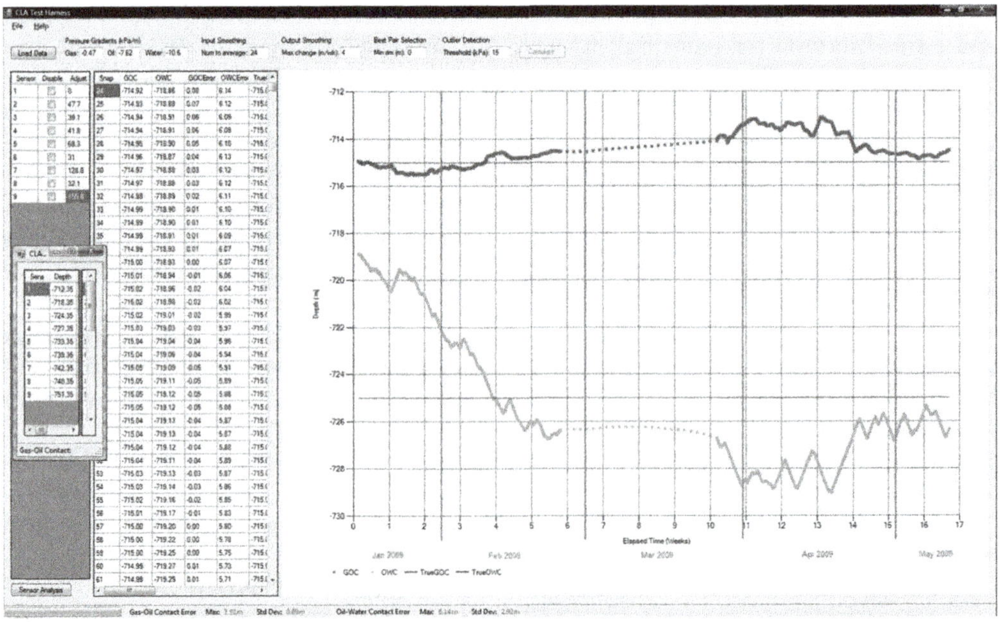

Figure 20.13 Output of the CLA running in HQ.

it measures the temperature usually every meter along the fiber. The details are covered in the Section 1.1.4 on DTS [16] and therefore will not be discussed here further. The original applications were on the surface, for example, fire protection in tunnels, power cable monitoring, transformer monitoring, monitoring of hot pipes, and/or vessels and leakages in pipes. However, soon applications were found in the downhole environment as well, such as leak detection (well integrity) or warm-back measurements for the identification of hydrocarbon bearing layers. It is now one of the widest deployed fiber optic technologies in exploration and production (E&P) of hydrocarbons [17,18]. The usual way of executing measurements in the oil field is by deploying the fiber along the well, preferably from the toe the deepest part of a well to the surface. This way the reservoir section can be monitored as well as artificial lift elements such as ESPs or gas lift valves and other potential heat or cooling (gas leaks, etc.) sources.

20.6.1 Types of installation

There are a few ways of deploying fibers in a well. First of all, for the readers understanding, the fibers are usually packaged in control lines; the most common size is 1/4 in. but also 1/8 in., and occasionally, smaller diameters can be used in order to withstand the challenging operating conditions (Figures 20.14 and 20.15).

Another way to achieve ruggedization of the fiber is by jacketing various polymers around the fiber. Standard polymers for this procedure are PVDF, hytrel, polyethylene (PE), nylon, PVC, fire-retardant PE, and polyurethane.

The two mechanical structures lead to different deployments: control lines can be deployed in a permanent manner by clamping the control line to the tubing or casing during installation.

The rod solution is very suitable for a deployment inside the tubing after the well was completed. There is also a third way, here control line is

• Warm-back analysis	To identify injection zones
• In/outflow analysis	To identify coarsely the inflow of hydrocarbons or outflow of injection water/polymers
• Steam breakthrough	To ensure optimal production in steam-assisted environments
• Out-of-zone-injection	To identify zones that are injected into wrongly, that is, outside the intended injection zones
• Gas lift monitoring	To optimize the production
• Well integrity	To identify leaks in the production tubing

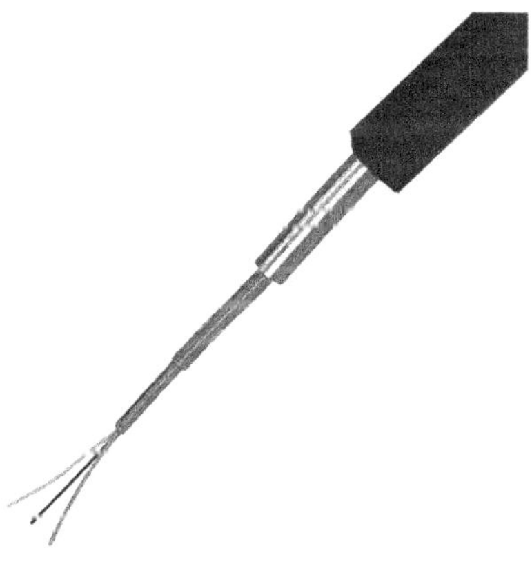

Figure 20.14 Traditional downhole cable from AFL. (From AFL. https://www.aflglobal.com/Products/ Fiber-Optic-Cable/Downhole/Low-Profile/Low_Profile_Downhole_Cable.aspx (Accessed January 18, 2015). With permission.)

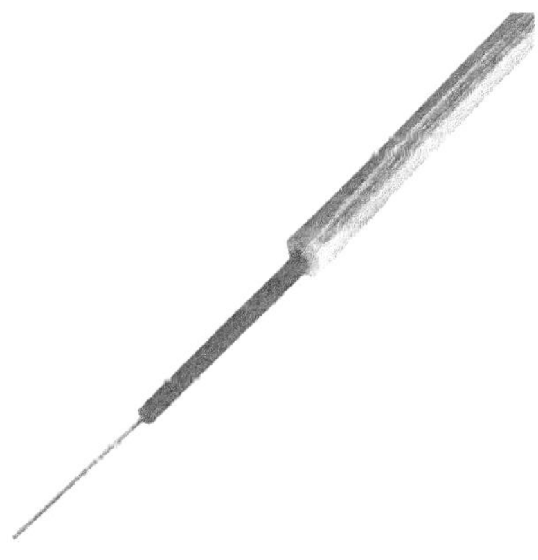

Figure 20.15 Fiber rod from AFL. (From AFL. https://www.aflglobal.com/Products/ Fiber-Optic-Cable/Downhole/FiberRod/Fiber-Rod.aspx (Accessed January 18, 2015). With permission.)

Table 20.3 Installation types

Type	Installation	Measurement time	Benefits	Application area
Control line	CL with fiber must be installed during well deployment	Continuous	Low-cost, permanent monitoring	Any type of well, can be deployed on tubing and on casing. Usually not inside tubing as diameter reduces incurring deferment
Fiber rod		During installation	Limited time duration monitoring	Any type of well, only inside tubing diameter reduction incurring deferment
Pumped fiber	CL must be installed during well deployment	Continuous	Replacement of fiber possible, permanent monitoring	Any but usually high-temperature wells, see control line

installed during deployment, and fiber is pumped in at a later stage, but this is only used for rare cases, for example, steam-assisted gravity drainage (SAGD) wells. All three deployment strategies have their own advantages, as there is always a trade-off among the deployment strategy, cost, and measurement required (see Table 20.3).

20.6.2 Applications

There are quite a few applications, but it would go beyond the purpose of this chapter to discuss them all. Therefore, we do not delve into this topic too deep. Still, an idea should be given what DTS can be used for; in general, any technology needs to be proven by an application, and the real test and success is the commercial exploitation of it. Hence, it should be mentioned that typical thermal monitoring applications in the E&P world are interity monitoring, gas-lift monitoing, and inflow monitoring; just to name a few.

The list does not mean to be complete but just to give a few examples of DTS applications. Further reading can be found in Refs. [16,17].

20.7 HIGH-TEMPERATURE FIBERS

As explained in Section 20.6, there are different types of deployment of fibers in wells, and they all depend on the type of application.

An even more challenging downhole environment is found when hydrocarbons are extremely viscous and require additional help to become more movable. One way to achieve that is using steam; here a larger downhole area will be heated up for a prolonged time, usually between 3 and ≥12 months, and then produce a process called SAGD [22].

This is carried out either by using the same well for both steam injection and back-production, so-called huff-and-puff, or by using two wells that are drilled in parallel, one above the other (Figure 20.16). Although the upper one is used to inject steam, the lower one is used to produce the hydrocarbons. *Huff-and-puff* is a method in which only one well is used for both steaming and production [23,24]. There are also other environments the industry operates, in which are at elevated temperatures [25]. As can be seen from these descriptions, the environment becomes elevated in temperature, some ≥280°C, and thus, the fibers need to be able to withstand the steam injection temperatures.

The main challenge that is found is not the temperature but the hydrogen that is found in abundance in such an environment. It is the lightest and most common element found on the earth, and it exists in atomic and molecular forms. Although molecular H2 can stay in a glass matrix (i.e., does not recombine with any other element in this form), atomic H combines with, for example, germanium, phosphor, and other core materials, resulting in the permanent loss of light.

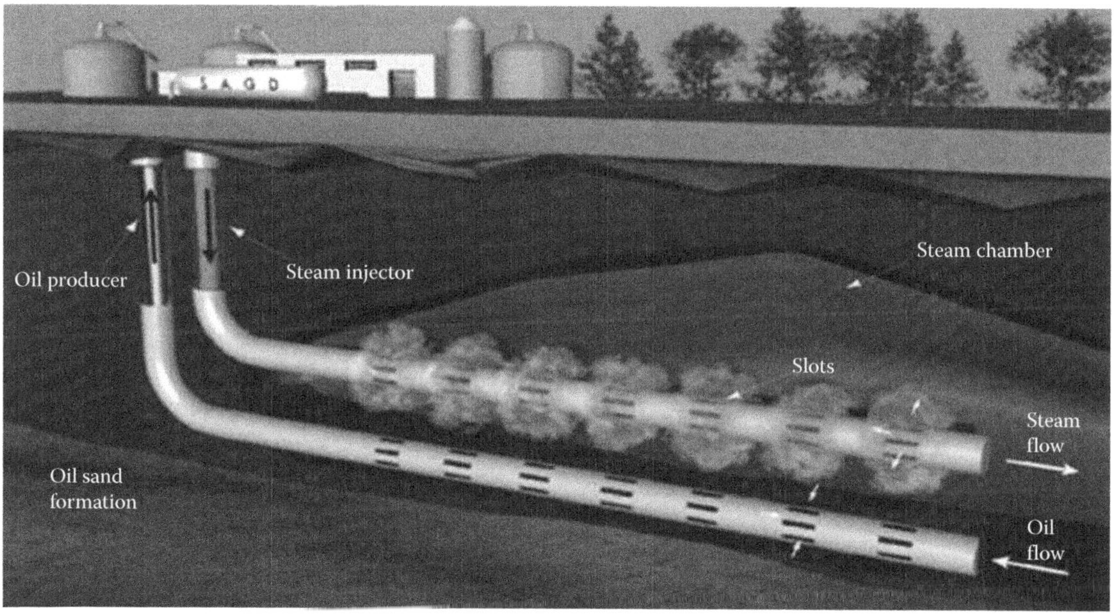

Figure 20.16 SAGD principle. (From Harvest Operations Corp, Blackgold Oil-Sands Project. http://www.harvestenergy.ca/operating-activities/oil-sands/bg-enviro-assessment.html# SAGD-101 (Accessed August 8, 2014). With permission.)

Therefore, a means must be found to inhibit the atomic reaction within the fiber. One way to do that is to use specialized coatings; metal coatings have shown promising results but are very costly, so other coatings and fiber structures are under investigation.

The most promising one is based on polyimide–carbon coating that allows building a hermetical wall around the fiber, which is even for hydrogen difficult to penetrate. In order to evaluate what is commercially in the market, a test setup was designed and a test matrix established. The fibers under test were placed in a specially designed and outgassed cylindrical container as depicted in Figure 20.17.

In a second step, the best performing fibers were packaged in a fiber in metal tube (FIMT) that then was placed in a larger cylinder and exposed to the same procedure, temperature, and combined temperature with hydrogen. The test was designed

Figure 20.17 Container holding the fiber under test during set up. (Courtesy of Shell.)

Figure 20.18 Oven with container holding fibers under test. (Courtesy of Shell.)

such that first the temperature impact up to 300°C was observed, and then in a second run, the impact of hydrogen together with the temperature was observed; the test setup is shown in Figure 20.18. In the figure, one can see the impact of temperature and hydrogen on the loss of high-temperature fiber, an arbitrary example but reflecting most fibers. One can see that after 620 h, approximately 26 days, the loss has come to equilibrium from which it does not deviate in the usual area of interest, that is, 1550 nm window. The traditional water peaks can still be seen, but their widths have been narrowed allowing the region of interest to be used (Figure 20.19).

This obviously needed to be carried out carefully in small steps as internal stresses in the fiber needed to be minimized.

When investigating the wavelength regime at 1550 nm, various fibers show different performances: some almost immediately show huge losses rendering that particular fiber useless, whereas others show an increased loss which almost levels out resulting in an unsuitable fiber for high-temperature downhole conditions due to attenuation loss (Figure 20.20).

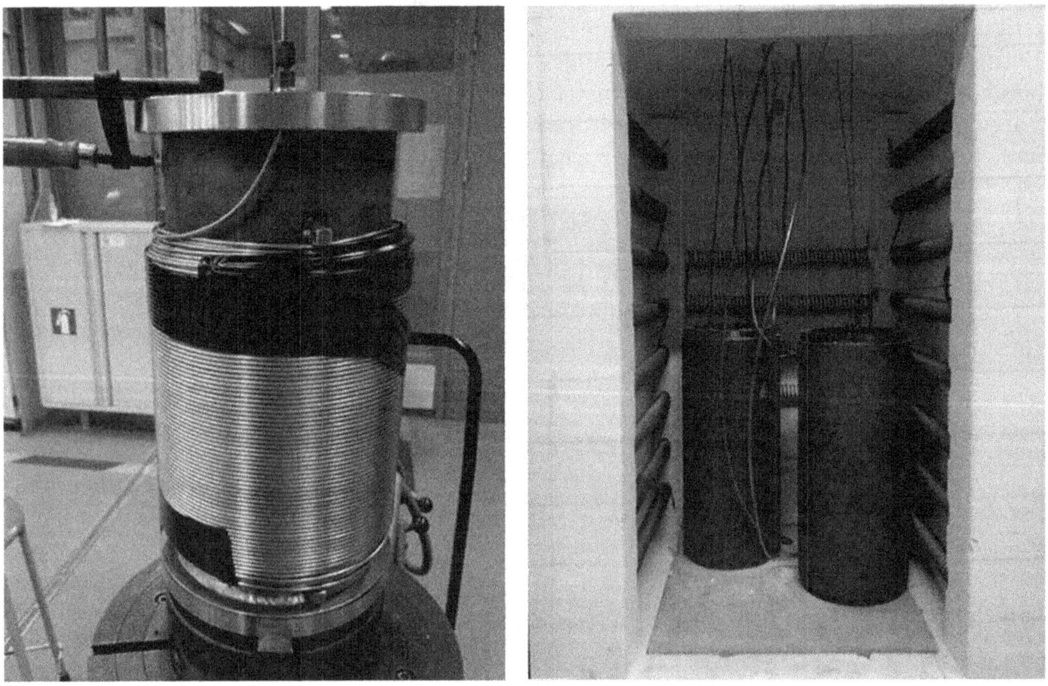

Figure 20.19 FIMT test. (Courtesy of Shell.)

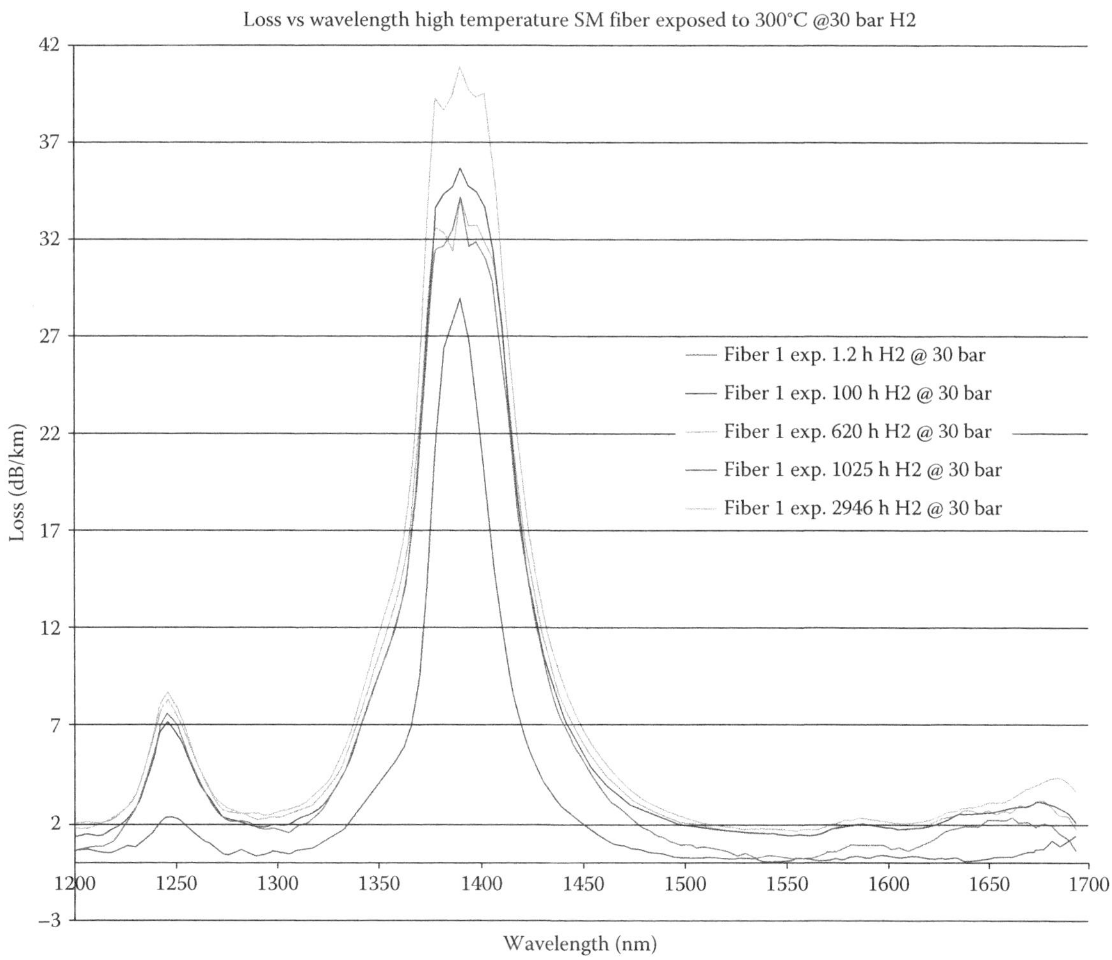

Figure 20.20 Impact of hydrogen at 300°C over time.

However, there are a few fibers that show losses that are relatively small and are also constant over time; these fibers are the ones required for reliable downhole measurements when high temperatures are seen (Figure 20.21).

20.8 DISTRIBUTED ACOUSTIC SENSING

With the arrival of distributed acoustic sensing (DAS) [26], a new chapter in fiber optics research was opened and the resulting applications are numerous nowadays such as perimeter protection, geohazard monitoring, leak detection, or the downhole environment fracture monitoring [27], flow measurements [28], and vertical seismic profiling (VSP) [29]. The detailed working principle and additional details are described in Chapter 2 and will not be further detailed here. However, ultimately, it is based on coherent OTDR [30], so optical pulses are sent down the fibers, which interact with the impurities in the fiber. The change of the locations of impurities due to acoustical impact between two consecutive pulses is proportional to the change of the strain seen in that fiber section; this can be observed along the whole length of the fiber, hence distributed. There are other parameters to observe such as channel length, gauge length, and sampling rate, and their details and use are given in Chapter 2.

20.8.1 Cable types for DAS

In an effort to determine whether commercially available downhole cables can be deployed for DAS measurements, several cable designs were tested in

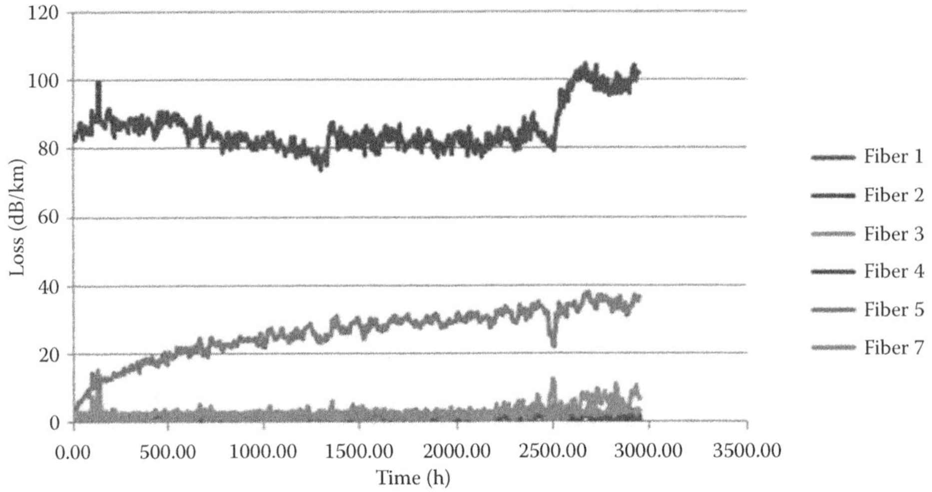

Figure 20.21 Loss versus time for various fibers at 300°C and at 1550 nm.

a test well in Houston, Texas [31]. The cable designs varied in wall thickness (0.028″, 0.049″) and FIMT design (1/8″ metal cable), encapsulated and bare, gel-filled and nonfilled plus a fiber rod as well as single-mode (standard and bend-insensitive) and multi-mode fibers (MMFs).

The test conditions of the well were dry and water filled, which pose the extreme conditions in a well; two vibroseises were used as source generators. The cables were all bundled together, lowered into the well, and held in place by specially designed springs.

When looking into the recording of a DAS channel under wet condition (see Figure 20.22),

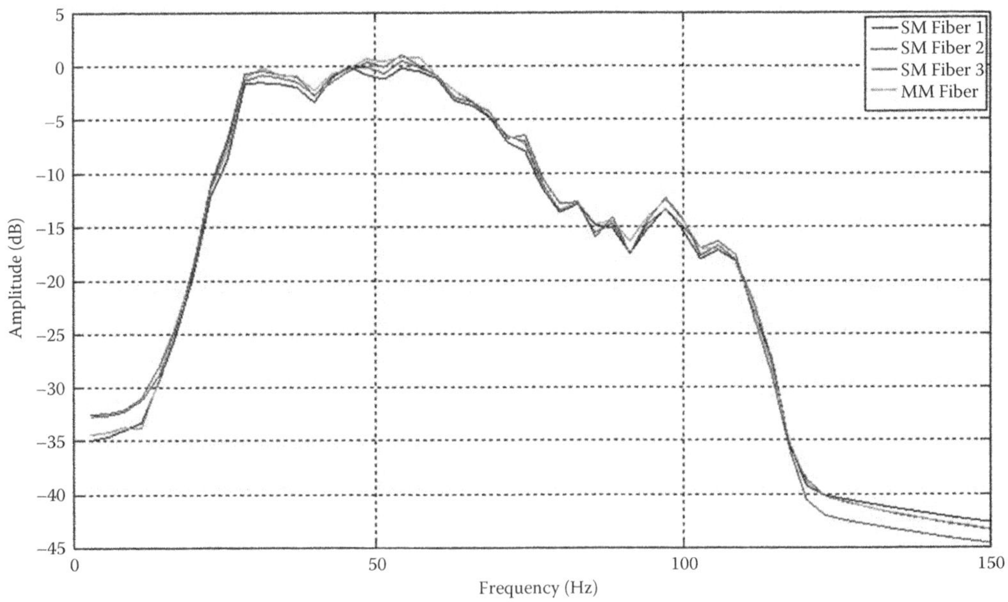

Figure 20.22 Channel response for three different SMFs and an MMF. (From La Follett, J. R. et al., Evaluation of fibre-optic cables for use in distributed acoustic sensing: Commercially available cables and novel designs. *SEG Conference Paper*, October 26–31, Denver, CO, SEG Paper No. 2014-0297, 2014. With permission.)

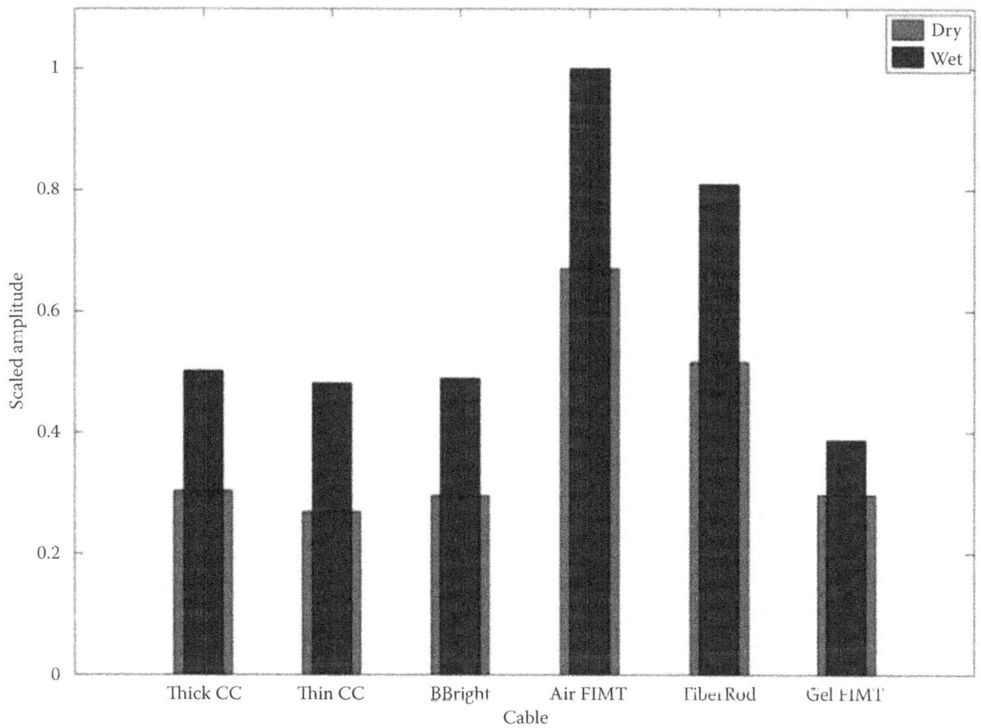

Figure 20.23 Amplitude response of various cable constructions. (From La Follett, J. R. et al., Evaluation of fibre-optic cables for use in distributed acoustic sensing: Commercially available cables and novel designs. *SEG Conference Paper*, October 26–31, Denver, CO, SEG Paper No. 2014-0297, 2014. With permission.)

the channel responses are very similar and hardly any difference can be observed. Of course, this is also due to the limited length of the fibers, so the impact of the attenuation is minimal. However, it shows that for certain fiber lengths, the type of fiber has a limited impact. When looking at the different cable structures, the Air FIMT (Figure 20.23) outperforms all other designs under both dry and wet conditions. When evaluating the typical commercial 1/4″ control line design (different wall thickness), the difference among them is rather negligible, indicating that the ultimate performance is independent from the structure. The fiber rod shows an interesting performance as it is between the FIMT and 1/4″ control line design; unfortunately, this design is not suited for the real downhole deployment.

20.8.2 Fracturing

One of the first applications in which DAS was used, in the oil industry, was the monitoring of fracturing operations, a technique used to break open rock downhole to allow hydrocarbons to flow to the well and be produced from there. This was successfully monitored in Groundbirch, British Columbia, Canada, in 2011 [32], so that additional work was undertaken to further this application and ensure its commercialization. For this kind of deployment, a control line is used containing a suitable fiber. The control line needs to be deployed over the reservoir section (area of interest). This means that the control line needs to be deployed all along the well: from the toe (furthest away point) through a packer and the well head outlet to the interrogator box. The fracturing procedure is fairly simple but effective for the hydrocarbon production [33]; fluid, usually water with additives, is pumped into the target zone, exits the well through the local perforations (holes shot to allow fluid penetration), breaks open the rock, and then creates small fractures through which the water transports the sand or additives, which help to keep these fractures open once the injection pressure and fluid are gone. DAS data are acquired during the time of fracturing and often afterward to monitor where these fractures are being created. The generated noise gives information, which is an indication how successful the fracturing job was and whether any additional fracturing is required.

20.8.3 Vertical seismic profiling

A powerful seismic tool is VSP. As explained in Chapter 2, DAS allows detecting minute acoustical signatures; in the E&P world, these signals are referred to as seismic waves or acoustic waves, which are of low frequency and usually low energy [34]. The picked up acoustic signal contains only very limited information, but if additional information is added, it becomes a rather powerful tool.

If a source, for example, a vibroseis or an air gun or similar, is used as a seismic source and then the data such as GPS coordinates are correlated, a lot of information can be deducted from the measurements such as speed of the traveling wave, and whether and where it reflected at an interface layer. Traditionally, this is carried out by using geophones as they are extremely sensitive, but DAS is nowadays approaching this sensitivity as well. However, the biggest advantage comes from the fact that the measurement is distributed, and hence the aperture becomes much larger compared to the usually 8–12 geophones that are deployed in a string (Figure 20.24).

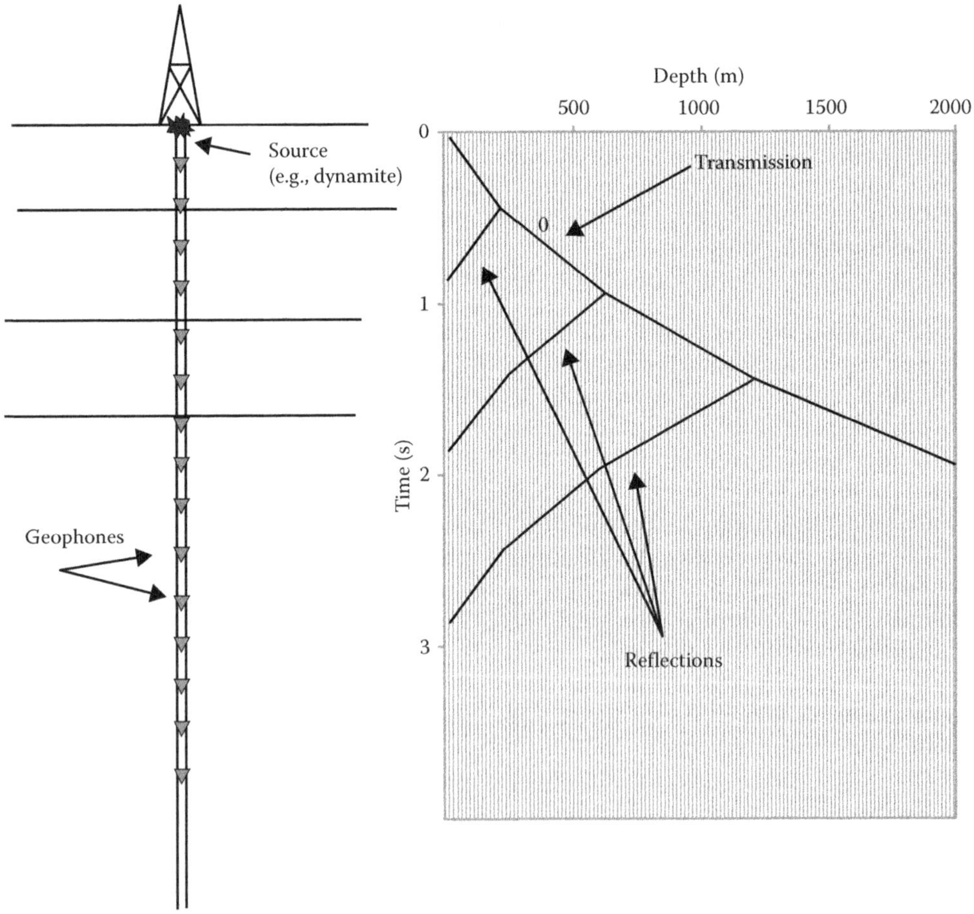

Figure 20.24 VSP principle. (From Dr. Ir. G. Drijkoningen, Introduction to reflection seismics. http://www.slideshare.net/DelftOpenEr/chapter-6-vertical-seismic-profiles (Accessed August 14, 2014). With permission.)

20.8.4 Surface seismic

To determine the subsurface structure, acoustical signals are launched in the surface and the reflections are detected via fiber optical cable. The usual sources in this scenario are vibroseises; the pickup can be via patch cord, control line, or surface cable. There are other systems such as the permanent reservoir monitoring systems that usually are dedicated subsea installations and not considered here because they do not use a distributed sensor but point sensors. The way most distributed acoustic interrogators operate is to determine the length change at a given location, and this local imprint together with GPS coordinates and some additional processing can be used to get good quality seismic data.

20.8.5 Helically wrapped cable

As the DAS interrogators measure the minute differential dynamic strain along the fiber, two questions came up. One question was on sensitivity: How sensitive are we and where is it in comparison with, for example, geophones? The other question was on the broadside detection: Can we measure another direction other than the inline direction of the fiber effectively? In order to address these questions, various solutions were considered; the one here presented is the *helically wrapped cable* nicknamed *HWC*. It allows waves impinging perpendicular or any other angle to contribute to the measured signal in the DAS interrogator.

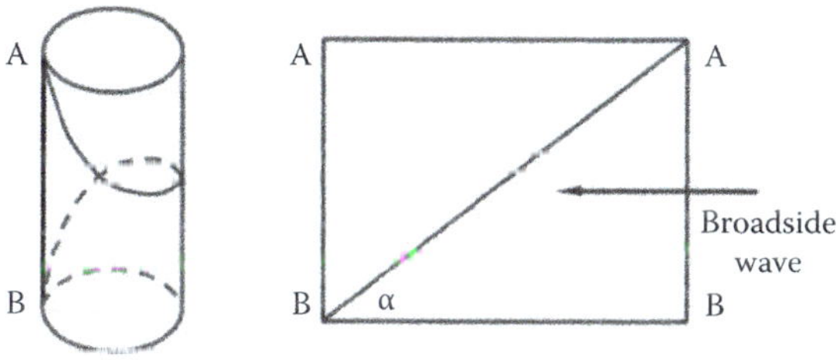

Figure 20.25 Left: A helically wound fiber (blue) in the cable (red). Right: when cut along line AB and flattened, a surface is obtained, where the wrapping angle α is defined. (From Hornman, K. et al., Field trial with of a broadside sensitive distributed acoustic sensing cable for surface seismic, SPE. London: SPE, June 10–13, 2013. 75th EAGE. Vol. SPE EUROPEC 2013. With permission.)

A wrapping angle α of 30° has nearly isotropic sensitivity (Figure 20.25), and with a linear combination of a straight fiber, it is theoretically possible to generate a response without in-line sensitivity.

At this wrapping angle of 30°, the cable contains ~2 m of fiber per 1 m of cable. This means an increase in the overall sensitivity with respect to a straight fiber, which was the other challenge. The response depends now only on the soil properties. The solid curves in Figure 20.26 relate to a soft ground, and the dotted curves to a harder ground. Spatial or seasonal variations in ground properties will lead to a changing response, which needs to be taken into account in the analysis of time lapse surveys.

20.8.5.1 FIELD TRIAL HWC

An 800 m HWC, as described previously, was manufactured and deployed in a field trial in Schoonebeek, the Netherlands. The cable was buried in 0.8 m depth and the soil was relatively soft. Three-dimensional accelerometer was placed in alignment with the chosen DAS channel spacing to measure the seismic signal.

A vibroseis source was used on several neighboring roads, generating a distribution of offsets and azimuths per receiver station. A shot record with a cross-line offset, recorded with vertical accelerometers; a straight fiber; and the HWC is shown in Figure 20.27. It can be seen that the HWC is sensitive to reflection energy, in comparison with the straight fiber. The reflection data are not visible on the horizontal accelerometers (not shown), proving that the HWC is indeed broadside sensitive. More details are found in Ref. [36].

20.9 STATUS OF FIBER OPTICS IN THE OIL AND GAS INDUSTRY

The world of E&P was introduced to the reader and the associated field architectures and infrastructure to get measurements to people in the office that can use those to make informed decisions. Also, well-chosen fiber optic applications were presented that went through iterative development stages from fundamental ideas over breadboard designs to real-world E&P applications such as oil rim monitoring, VSPs, and surface seismic. My aim is to show the reader that developing fiber optic solutions is not only an academic exercise but can lead to real-world applications and

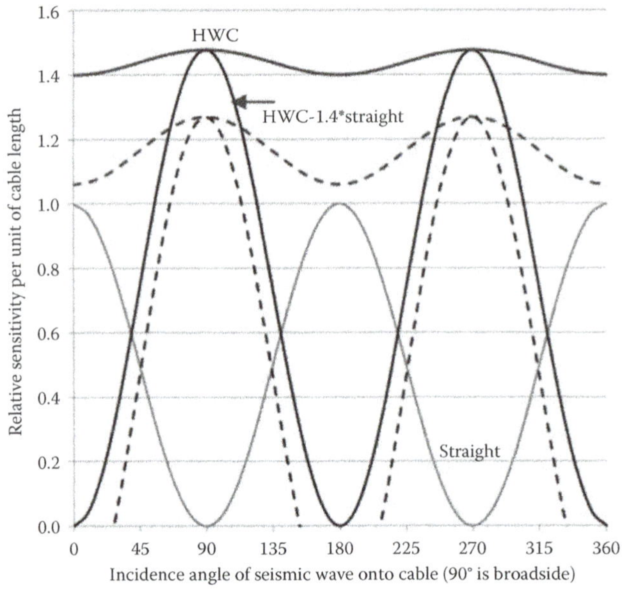

Figure 20.26 Response as a function of incidence angle φ for a cable with a straight fiber (green), HWC with wrapping angle α of 30° (red), and a linear combination of the two (purple) for ground with $V_P = 1500\ \mathrm{m\,s^{-1}} = V_S = 250\ \mathrm{m\,s^{-1}} = \rho = 1600\ \mathrm{kg m^{-3}}$. For a soil with $V_s = 500\ \mathrm{m\,s^{-1}}$. (From Hornman, K. et al., Field trial with of a broadside sensitive distributed acoustic sensing cable for surface seismic, SPE. London: SPE, June 10–13, 2013. 75th EAGE. Vol. SPE EUROPEC 2013. With permission.)

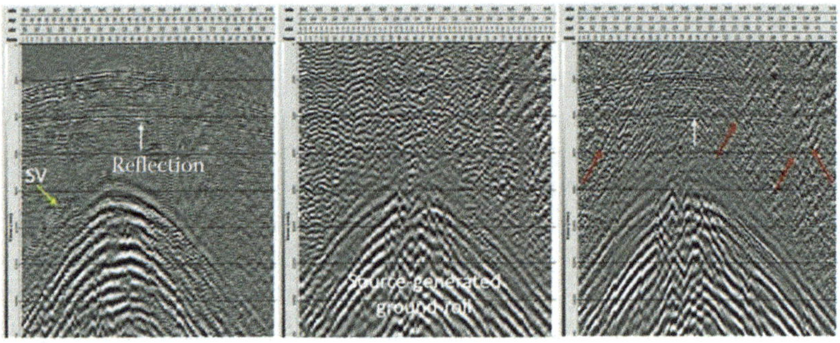

Figure 20.27 Shot records recorded in three collocated detector systems: (1) vertical accelerometers (left), (2) a straight fiber (center), and (3) the HWC (right). The white arrows point at reflection energy, absent on the straight fiber; the red arrows on the HWC record (right) point at ground-roll from ambient noise sources. The yellow arrow points at a direct SV wave, used for the analysis of azimuthal sensitivity. An AGC was applied to the data. (From Hornman, K. et al., Field trial with of a broadside sensitive distributed acoustic sensing cable for surface seismic, SPE. London: SPE, June 10–13, 2013. 75th EAGE. Vol. SPE EUROPEC 2013. With permission.)

deployments, even at a commercial level nowadays. One item that needs to be in any system designers mind is the fact that information needs to be presented to the end user; this requires to have accurate measurements in the field, a reliable transport of those measurements into a central data storage and appropriate processing. Only then these data can be visualized by the end user, providing information, not data that can be used in the optimization process. Applications that are currently on the edge of research are chemical sensing and low-cost point acoustic sensing, whereas commercialization for DAS flow sensing seems to be well on its way. The advantages, besides the technical advantages of fiber optics mentioned in the introduction, are as follows:

- Reduced HSE risk as after the permanent installation of the equipment during run-in it can be reused at any moment in time without well intervention
- Reduced deferment as there is no time required to deploy any of the sensors for any measurement
- One cable for multiple measurements; usually several fibers are deployed in a cable, so a combination of measurements can be carried out, for example, DTS and DAS, pressure and DTS or DAS, etc.
- No EMI issues

Therefore, fiber optics brings a wealth of new opportunities to the oil world and will continue to do so.

This will give E&P the opportunity to optimize on production and produce safer and better than ever before.

ACKNOWLEDGMENT

The author expresses his gratitude and thanks to Lex Groen, Arthur van Rooyen, Roel Kusters, Dan Joinson, Gijs Hemink, Bill Birch, Hans den Boer, Peter in 't panhuis, Paul Lumens, Brandan Wyker, Jon La Follett, Juun van der Horst, Kees Hornman, Albena Mateeva, Samantha Grandi, Boris Kushinov, Paul Zwartjes, Maartje Koning, Keith Hunt, Philip Holweg, Sheikhan Al-Kadhuri, Evert Moes, and many other people in various Shell Operating Units in Oman, Brunei, Canada, and other places around the world who all contributed to work that led to this publication.

REFERENCES

1. Solar Power, Wikipedia. http://en.wikipedia.org/wiki/Solar_power (Accessed January 24, 2015).
2. https://en.wikipedia.org/wiki/Liquid-crystal_display.
3. https://en.wikipedia.org/wiki/Gas_detector.
4. https://en.wikipedia.org/wiki/Spectroscopy.
5. Subsea Wireless Group (SWiG), OTM Consulting Ltd. http://www.subseawirelessgroup.com/ (Accessed January 24, 2015).
6. J. Mulholland and D. McStay, FMC Technologies Ltd. http://www.offshoremag.com/articles/print/volume-71/issue-1/subsea/wireless-communication-enhancessubsea-production-monitoring.html (Accessed January 24, 2015).
7. P. G. E. Lumens, *Fibre-Optic Sensing for Application in Oil and Gas Wells*, Eindhoven University of Technology, Eindhoven. ISBN 978-90-386-3567-5, 2014.
8. International Energy Agency (IEA), 2012 Key World Energy Statistics, Paris, 2012. http://alofatuvalu.tv/FR/12_liens/12_articles_rapports/IEA_rpt_2012_us.pdf.
9. C. Rühl, P. Appleby, J. Fennema, A. Naumov, and M. Schaffer, Economic development and the demand for energy: A historical perspective on the next 20 years, *Energy Policy*, Vol. 50, No. 3, 109–116, 2012. http://www.foresightfordevelopment.org/sobipro/55/870-economic-development-and-the-demand-for-energy-a-historicalperspectiveon-the-next-20-years.
10. Shell, New lens scenarios. http://www.shell.com/global/future-energy/scenarios/newlens-scenarios.html (Accessed November 12, 2014).
11. Occupational Safety and Health. Wikipedia. http://en.wikipedia.org/wiki/Occupational_safety_and_health (Accessed January 24, 2015).
12. Wikipedia-Well. http://upload.wikimedia.org/wikipedia/en/7/7f/Oil_Well.png (Accessed January 18, 2015).
13. P. G. E. Lumens, Low-Cost High-*Temperature FBG Interrogator*, Eindhoven, ISBN 978-90-444-0920-8, 2010.
14. Smartfibres, *Smartfibres—Oil and Gas*. http://smartfibres.com/oil-and-gas (Accessed November 12, 2014).
15. emphBen Hunter, Technology overview of permanent downhole distributed pressure sensing with fiber bragg grating sensing technology. http://smartfibres.com/docs/DPS_Overview.pdf (Accessed November 12, 2014).

16. A. van der Spek and J. J. Smolen, Distributed temperature sensing—A DTS primer for oil and gas production, shell. http://w3.energistics.org/schema/witsml_v1.3.1_data/doc/Shell_DTS_Primer.pdf (Accessed June 20, 2017).
17. A. Ukil, H. Braendle, and P. Krippner, Distributed temperature sensing: Review of technology and applications, *IEEE Sensors Journal*, Vol. 12, No. 5, 885–892, 2012.
18. K. Johannessen, B. Drakeley, and M. Farhadiroushan, Distributed acoustic sensing—A new way of listening to your well/reservoir. *Intelligent Energy International Conference*, and Exhibition, March 27–29, Utrecht, The Netherlands, Paper SPE 149602, 2012.
19. AFL. http://www.aflglobal.com/Products/Fiber-Optic-Cable/Downhole/Low-Profile/Low_Profile_Downhole_Cable.aspx (Accessed January 18, 2015).
20. AFL. http://www.aflglobal.com/Products/Fiber-Optic-Cable/Downhole/FiberRod/Fiber-Rod.aspx (Accessed January 18, 2015).
21. Harvest Operations Corp, Blackgold Oil-Sands Project. http://www.harvestenergy.ca/operating-activities/oil-sands/bg-enviroassessment.html# SAGD-101 (Accessed August 8, 2014).
22. Wikipedia-SAGD, Steam-assisted gravity drainage. http://en.wikipedia.org/wiki/SAGD (Accessed August 8, 2014).
23. M. Cuiyu, L. Yuetian, L. Peiqing, W. Chunhong, and L. Jingli, Study on steam huff and puff injection parameters of herringbone well, *The Open Petroleum Engineering Journal*, Vol. 6, No. 1, 69–75, 2013. https://benthamopen.com/contents/pdf/TOPEJ/TOPEJ-6-69.pdf.
24. J. Alvarez and S. Han, Current overview of cyclic steam injection process, *Journal of Petroleum Science Research*, Vol. 2, No.3, 2013. http://www.jpsr.org/Download.aspx?ID=5755.
25. Offshore Operations Subgroup of the Operations and Environment Task Group, *Subsea Drilling, Well Operations and Completions*, NPC North American Resource Development Study. September 15, 2011.
26. Wikipedia-DAS, Distributed acoustic sensing. http://en.wikipedia.org/wiki/Distributed_acoustic_sensing (Accessed August 8, 2014).
27. Marathon Oil Corp, Animation of hydraulic fracturing (fracking). https://www.youtube.com/watch?v=VY34PQUiwOQ (Accessed August 16, 2014).
28. P. In't panhuis, H. den Boer, J. van der Horst, R. Paleja, D. Randell, D. Joinson, P. B. McIvor, K. Green, and R. Bartlett, Flow monitoring and production profiling using DAS. *Annual Technical Conference and Exhibition (ATCE)*. SPE, October 27–29, 2014. SPE 170917.
29. A. Mateeva, J. Mestayer, B. Cox, D. Kiyashchenko, P. Wills, J. Lopez, S. Grandi, K. Hornman, P. Lumens, A. Franzen, D. Hill, and J. Roy, Advances in distributed acoustic sensing (DAS) for VSP, *SEG Technical Program Expanded Abstracts 2012*, Vol. 30, 1–5, 2012.
30. Wikipedia-OTDR, Optical time-domain reflectometer. http://en.wikipedia.org/wiki/Optical_time-domain_reflectometer (Accessed August 16, 2014).
31. J. R. La Follett, B. Wyker, G. Hemink, K. Hornman, P. Lumens, and A. Franzen, Evaluation of fibre-optic cables for use in distributed acoustic sensing: Commercially available cables and novel designs. *SEG Conference Paper*, October 26–31, Denver, CO, SEG Paper No. 2014-0297, 2014.
32. M. M. Molenaar, D. Hill, P. Webster, E. Fidan, and B. Birch, First downhole application of distributed acoustic sensing (DAS) for hydraulic fracturing monitoring and diagnostics. *Society of Petroleum Engineers, SPE Hydraulic Fracturing Technology Conference*, 24–26 January, The Woodlands, TX, Paper No. SPE-140561-MS.
33. E. Rowley, Shale oil "could add 50bn to UK economy." February 14, 2013, http://www.telegraph.co.uk/finance/newsbysector/energy/9868680/Shale-oil-could-add-50bn-to-UK-economy.html (Accessed September 07, 2014).
34. Wikipedia-Seismic wave. http://en.wikipedia.org/wiki/Seismic_wave (Accessed September 6, 2014).

35. Dr. Ir. G. Drijkoningen, Introduction to reflection seismics. http://www.slideshare.net/DelftOpenEr/chapter-6-vertical-seismicprofiles (Accessed August 14, 2014).
36. K. Hornman, B. Kuvshinov, P. Zwartjes, and A. Franzen, Field trial with of a broadside sensitive distributed acoustic sensing cable for surface seismic, SPE. London: SPE, June 10–13, 2013. 75th EAGE. Vol. SPE EUROPEC 2013.

21

Oilfield production monitoring with fiber-optic sensors

PHILIP NASH
Stingray Geophysical

21.1 INTRODUCTION

In this chapter we describe the use of fiber-optic sensors for permanent reservoir monitoring (PRM) of oilfields.

One of the most important uses of optoelectronics in the oil and gas industry is for oilfield production monitoring, as part of a process to enhance production and yields. Much of this monitoring involves insertion of various sensors into oil wells, as described in Section 20. Such sensors measure a range of parameters including temperature, pressure, flow, vibration, and chemical composition. These sensors give very valuable information about the local conditions, but they are essentially a one-dimensional measurement, in that they give information about a particular depth in the well, which itself represents a single point in the field.

An additional and complementary approach is to image the entire oilfield in three dimensions. This is most commonly achieved using active seismic imaging, a technique by which an acoustic source is used to produce downward propagating seismic energy within the field, and the reflections of this from the various geological features of the field are detected by a network of seismic sensors. This allows a three-dimensional image of the field to be built up, in much the same way as a magnetic resonance imaging scanner produces an image of the human body (an alternative approach uses EM (Electromagnetic) fields to achieve a somewhat different image based on rock conductivity). A schematic of a PRM system is shown in Figure 21.1.

The image so formed is an image of the "hard" structures within the field (that is, the rock strata) and does not give a direct representation of the presence of oil or other fluids within the rock. Rather, this can be inferred from changes in the details of the image as fluids move around the field. These changes are caused by changes in the impedance of the rock due to the presence of fluids (which cause amplitude and velocity changes) or by dimensional changes as the rock layers settle.

If a series of time lapse images are taken of a producing field (say at 6 monthly intervals), then changes between the images can be used to infer information about the movement of oil (and other fluids such as gas and water) within the field, and

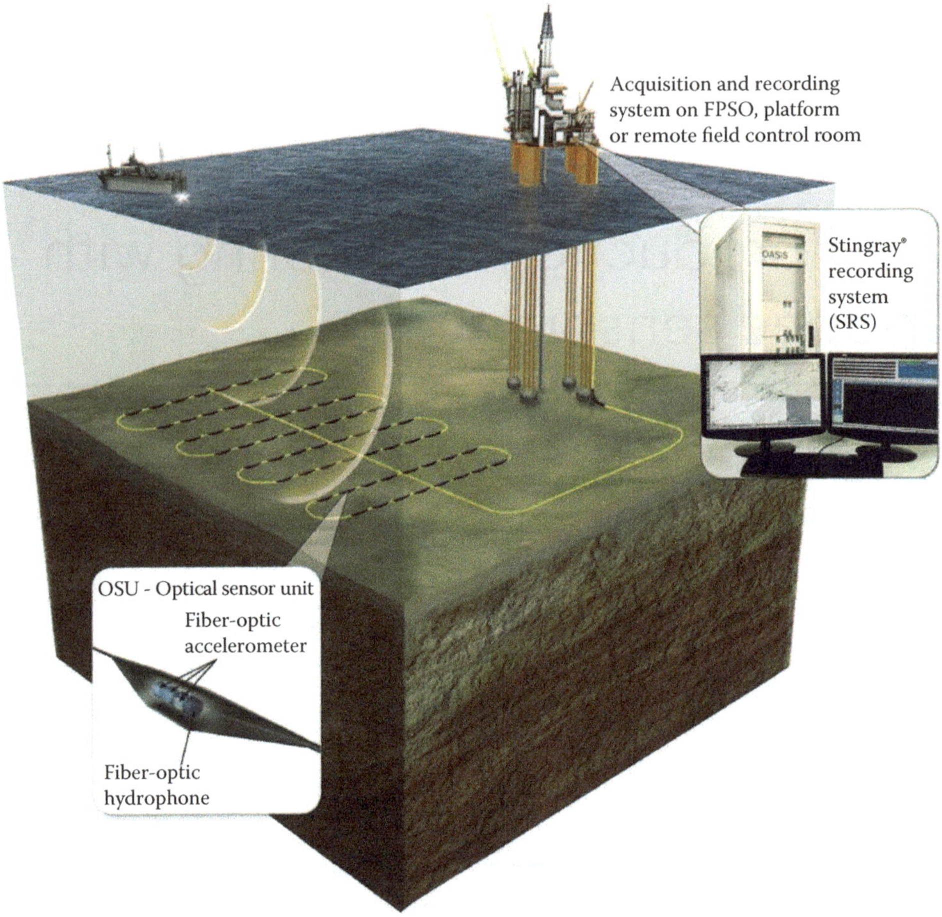

Figure 21.1 Schematic of PRM system.

this information, either by itself or in conjunction with well information, can be used to optimize production from the well.

The active seismic source for such surveys is normally a towed air gun array. The receiver network can be realized in a number of ways, but the highest performance and most repeatable approach is to use a network of seismic sensors, permanently installed on the seabed above the field. Such a system is known as a PRM system or sometimes a life of field seismic system. Use of such a system, in conjunction with enhanced recovery techniques, can increase total recovery from an oilfield from 40% to in excess of 50%, which represents a huge amount of additional hydrocarbon.

The seabed sensor networks which can act as the acoustic receivers are amongst the largest and most complex sensor systems ever constructed. They can cover around 100 km^2 of seabed with over 400 km of cable, along which are spaced in excess of 40,000 individual seismic sensors. The sensors are typically a combination of accelerometers and hydrophones, grouped as three orthogonally mounted accelerometers and one hydrophone into what is known as a 4-c (four-component) seismic package, which allows full reconstruction of the seismic wave field.

The performance, environmental and reliability requirements for such sensor networks are very demanding. They may be required to operate in water depths of up to 3000 m and to give a high level of seismic performance without failure for in excess of 25 years. Furthermore, as the sensors are being used to measure very small changes in the field, their performance must be extremely stable throughout the field lifetime.

Early PRM systems were typically based on electrical sensors, usually piezoelectric hydrophones and accelerometers or geophones. The first such system was installed by BP on the Valhall field in the North Sea in 2003, and is still operational today (Kammedal et al. 2004). These systems give good performance, but require a large amount of subsea electronics, and have been known to suffer from issues of reliability and subsea power distribution.

More recently (starting in 2008), PRM systems based on fiber-optic sensors have started to be introduced (Nash et al. 2009). These sensor networks are electrically completely passive, relying on simple robust fiber-optic sensors which require no underwater electronics and which are designed to be highly stable for periods in excess of 25 years.

This is a classic application for fiber-optic sensors, and an interesting case study because of the demanding requirements, the large scale and the very powerful economic benefit of such systems. Here, we describe three approaches to a fiber optic solution, which have been implemented in recent years. The three approaches are a coupler-based reflectometric interferometric system (Nash et al. 2000), a Fibre Bragg Grating (FBG)-based reflectometric interferometric system (Nakstad 2011), both using a combination of time and wavelength division multiplexing, and a coupler-based interferometric system using a combination of frequency and wavelength division multiplexing (Bunn 2008).

21.2 SENSING PRINCIPLES

All three of the systems discussed here are based on interferometric principles; that is, they measure the optical phase change included in a coil of optical fiber by the seismic signal acting on the coil. The mechanical packaging of the coil determines whether the phase change is induced by acceleration (accelerometer) or pressure (hydrophone). It should be noted that, other than this packaging, the different sensor types are optically identical. The principles of interferometric fiber optic sensing are discussed in Chapter 11, Volume II.

Figure 21.2 shows a schematic of a hydrophone. The hydrophone comprises a compliant mandrel supported on a metallic structure over an air gap. The fiber coil (typically in the range 10–150 m length) is wound onto the mandrel and encapsulated in a material such as epoxy resin. An acoustic signal acting on the coil causes a localized pressure change which compresses the coil, and causes a phase change in light travelling through the coil (which is a combination of a physical strain effect and a refractive index change caused by the elasto-optic effect). A typical hydrophone is shown in Figure 21.3.

Figure 21.4 shows a schematic of an accelerometer based on a typical sensing principle. The accelerometer is a mass-spring device, where the mass is a steel cylinder and the two springs above and below the mass may be mandrels made of a compliant material, whereas the can is a lighter hollow metallic cylinder. The fiber coil is wound around one of the springs. If the accelerometer is accelerated along its axis the mass moves relative to the can and this movement compresses the spring. This is converted into an outward expansion of the compliant cylinder, which in turn produces a change in length of the fiber coil, and therefore a phase change in light passing through the coil. By careful design the optical phase change can be

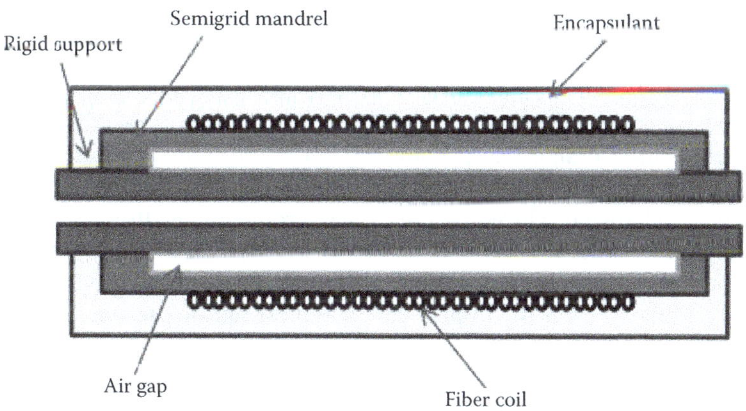

Figure 21.2 Schematic of hydrophone.

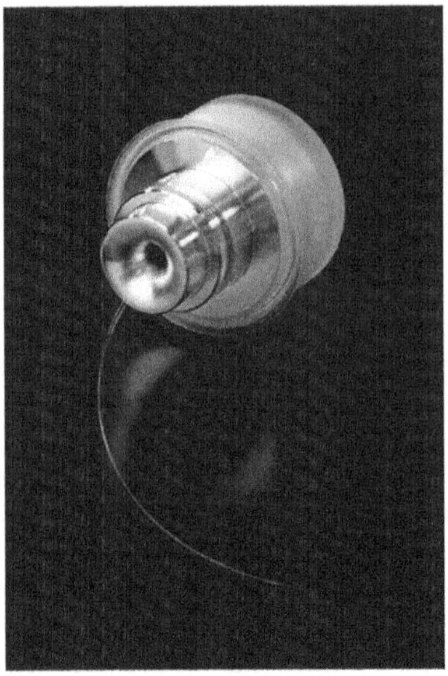

Figure 21.3 Typical optical hydrophone.

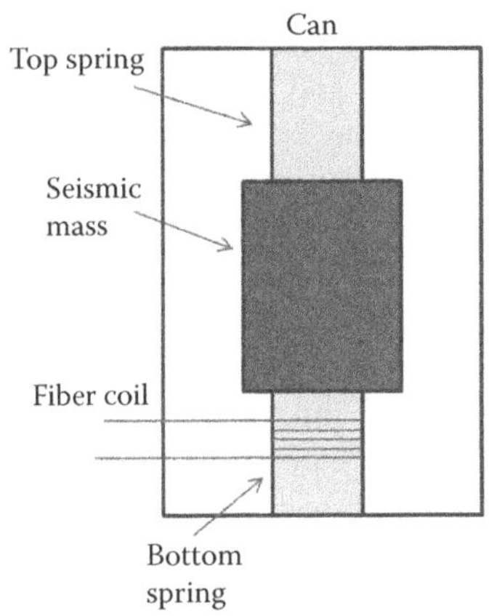

Figure 21.4 Schematic of an accelerometer.

made linear with respect to the axial acceleration. A typical accelerometer is shown in Figure 21.5.

All three commercial systems described here use variants of these sensor designs. Important design considerations include the sensor responsivity (the optical phase change produced by unit acceleration/pressure), the senor linearity (and/or harmonic distortion), and directivity (which should be omnidirectional for the hydrophone, but with a high on/cross-axis response ratio for the accelerometer).

Various techniques have been used for extraction of the optical phase change from these sensors, as

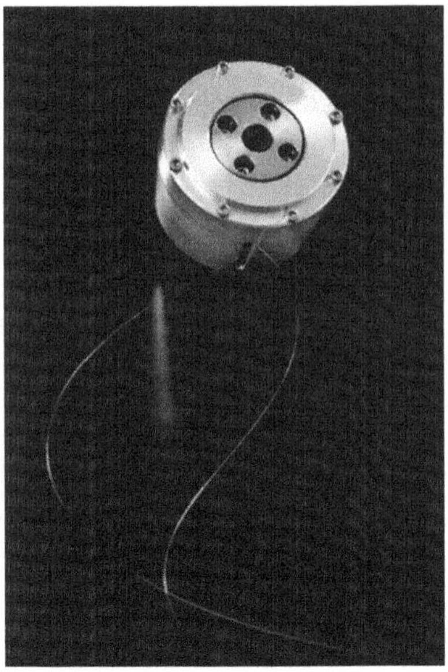

Figure 21.5 Typical optical accelerometer.

shown in Figure 21.6. Approaches 1 and 2 both use a version of the reflectometric approach described in Chapter 11, Volume II. In this approach, partially reflecting mirrors are placed on either side of the coil, and the coil is interrogated with an optical pulse pair. The reflection of pulses from mirrors in either side of the coil is compared, and the phase change between these mirrors is a measure of the length of the optical coil. In Approach 3, a Michelson interferometer is used. The sensor coil forms one arm of the interferometer and an insensitive reference coil, typically located inside the sensor coil, forms the other arm (or alternatively a push-pull arrangement with positive and negative sensing coils forming the two arms can be used).

In the four-component seismic package required for PRM, three accelerometers are located in an orthogonal arrangement to measure the components of the seismic field. The accelerometers are typically located in a pressure housing to shield them from ambient pressure, although in some designs the housing is pressure balanced and the accelerometers are directly exposed to the pressure. The hydrophone sits outside the pressure housing and is always directly exposed to ambient pressure. A typical sensor package is shown in Figure 21.7. Within the sensing package the sensors are typically multiplexed together onto a single fiber. In Approaches 1 and 2, the sensors are time multiplexed, so that time of flight of optical pulses to the

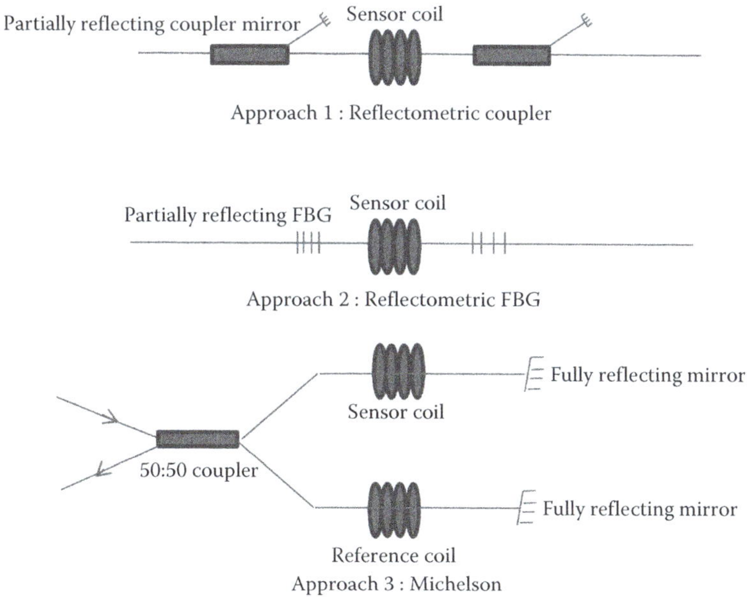

Figure 21.6 Alternative interferometric sensor approaches.

Figure 21.7 Optical sensor unit containing three accelerometers and one hydrophone.

sensors is used to distinguish them. In Approach 1 the mirrors are implemented using fiber splitters with a mirror on one arm of the splitter (the mirror may be a simple silvered fiber end, or it may be a Faraday rotator mirror (FRM) to reduce polarization effects). In Approach 2, the mirrors are implemented using low reflectivity FBGs. These have the advantage of being intrinsically imprinted into the fiber, but their reciprocal reflective nature means that such systems are susceptible to multi-path crosstalk.

In Approach 3, the sensors are multiplexed using frequency division multiplexing (FDM), as described in Chapter 11, Volume II. In this case the interrogating light is continuous, not pulsed. The mirrors at the end of the interferometer arms can again be slivered mirrors, or FRMs (this type of approach can also be used in a pulsed TDM multiplexing architecture).

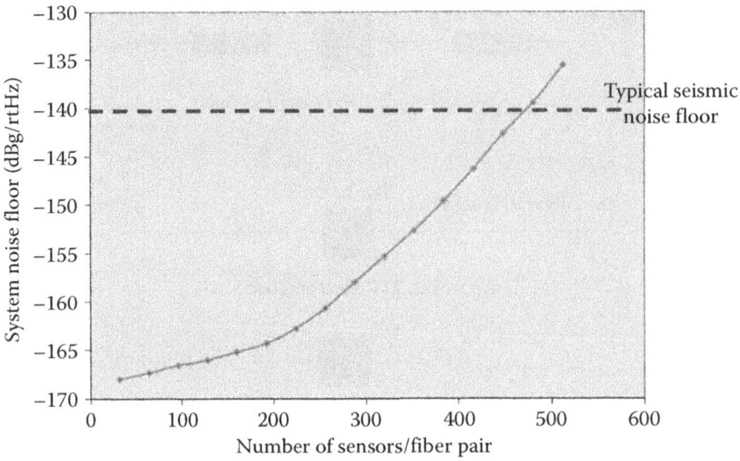

Figure 21.8 Achievable system noise floor vs. number of sensors.

Multiple sensor packages can be multiplexed onto a smaller number of fibers by extension of these multiplexing techniques. In Approach 1, four sensor units are time interleaved using Time Division Multiplexing (TDM). Wavelength division multiplexing (WDM) is then used to combine larger numbers of sensors (Nash et al. 2009). This is typically achieved by using a number of lasers, spaced on the International Telecommunications Union (ITU) grid, and combining their outputs using standard wavelength multiplexing components. As an example, with 16 sensors per wavelengths, and 16 optical wavelength spaced on the ITU grid, 256 sensors in 64 four-component sensor packages can be multiplexed onto one optical fiber (or more correctly, one optical fiber pair, as usually separate input and return fibers are used). Approach 2 also uses a TDM/WDM combination, although in this case TDM is only used within the 4-C package, so for 16 wavelengths the multiplexing ratio is 64, or 16 sensor packages.

Approach 3 uses a combination FDM/WDM approach, in which the number of FDM sensors is determined by a matrix arrangement (Kersey 1990, see Figure 1b). Typically in this arrangement for N frequencies there are 2N fibers and N^2 sensors, so for example for 8 frequencies 16 fibers are required and there will be 64 sensors. If WDM is added then the multiplexing ratio is increased by the number of wavelengths used (there may be some confusion here between frequency and wavelength multiplexing. In this context, FDM means that the optical signals are closely spaced in frequency (typically separated by several GHz) and are distinguished by filtering in the electrical domain. WDM means that the optical signals are widely spaced in the frequency domain (typically by THz) and are filtered in the optical domain).

There is a trade-off between number of multiplexed sensors (i.e., multiplexing ratio) and system performance, in particular system noise floor and dynamic range. Figure 21.8 shows a plot of allowed system noise floor (expressed in units of acceleration) against number of multiplexed sensors. The figure shows that for typical seismic noise levels of around-140 dB re 1 g/rtHz, over 400 sensors can be multiplexed together, and larger numbers if a higher noise floor is acceptable.

21.3 MECHANICAL DESIGN ISSUES

Irrespective of the sensor operating principles and optical multiplexing architecture, mechanical design is a key factor in ensuring a stable and reliable system. The major environmental issues that a PRM system are likely to encounter are as follows:

- Transport by road, air, and/or ship over long distances
- Storage for periods of possibly several months on docksides where temperatures may be as high as +70°C (in direct sunlight) or as low as −40°C
- Deployment from surface vessel onto the seabed in water depths of up to 3000 m
- Survival up to 25 years at 3000 m water depth

Of these items the last two are particularly challenging. One objection often raised to the use of fiber in these applications is the supposed fragility of fiber in the tough environment of the oil field. The reality is that, provided it is properly packaged, fiber sensor approaches can be at least as tough, and probably more durable, than other approaches. Some of the key factors to allow for in the mechanical design of PRM systems are

Mechanical simplicity: Minimize number of mechanical components, especially those which are put under strain or which are expected to exhibit some form of mechanical motion

Material and component selection: Good material selection is a key factor in determining long-term reliability and stability. The key materials (for instance, mandrel materials for the active parts of the sensor) should be selected to have properties which are highly stable with time. Likewise, components should be used which have long track records and significant evidence of good field reliability. This is not always compatible with using state-of-the-art components, so when new components are introduced they need to be thoroughly qualified, ideally using recognized industry standards such as Telcordia.

Strain minimization: Minimize strain applied to fiber in sensor coils by shielding sensors from pressure wherever possible e.g., accelerometers should be housed in a pressure housing to protect from pressure-induced strain.

Corrosion resistance: Consideration should always be given to the short and long term impact of corrosion. In particular it is good practice to minimize exposure of metals directly to seawater, through the use, for example, of polyurethane overmolds.

Fiber protection in cable: Fiber in cables is vulnerable to damage due to strain and also to hydrogen darkening, especially in deep water. Use of welded steel fiber tubes containing hydrogen scavenging gel has been proven to significantly reduce vulnerability in these areas.

An Optical Sensor Unit (OSU) designed to the above criteria is shown in Figure 21.7.

21.4 SYSTEM QUALIFICATION AND RELIABILITY ASSESSMENT

Rigorous qualification of a technology is required to prove that it will be fit for purpose in the operating environment—this is especially true in an environment such as the oil industry. Qualification of a PRM system is likely to include tests such as

- Subjecting a section of sensor cable to its full expected life cycle, including transportation, storage, and simulated deployment.
- Extended depth testing to at least the intended operating depth (plus a significant safety margin).
- Accelerated temperature testing of sensors to demonstrate long-term stability. As a PRM system is expected to be highly stable for 25 years, it is an interesting question how that can be proven during product qualification. Fortunately, accelerated temperature testing, although not definitive, does give some indication of long-term stability over a shorter duration test. The concept is based on accelerating the ageing properties of materials at low temperatures by exposing them to higher temperatures. For instance, exposure of a rubber to 50°C for 1 month, approximately simulates the ageing of the same rubber at 5°C for 2 years.

Figure 21.9 shows a hydrophone that has been exposed to 50°C temperature for a period of 1 year, simulating 30 years of operation at 5°C (a standard seabed temperature in deep water). It can be seen that over the simulated 30-year period there is no measurable shift in the sensor sensitivity or frequency response. This is crucial in a system that is being used to measure small changes in the field over an extended time period.

Information about system reliability can be obtained through the qualification process, but the scale of the qualification testing, both in terms of numbers and sensors and duration, gives only limited information about the long-term reliability of a full scale system. This is normally assessed through a quantitative reliability analysis, which takes into account the system architecture, together with reliability information on individual components, ideally obtained from real field data (there is much field data, for instance, on optical components obtained from the telecoms industry). The analysis typically uses a Monte Carlo simulation to predict how many sensor failures will occur over a specified lifetime. Analyses of these types show that for a typical optical-based PRM system, >98% of sensors can be expected to be operational after 25 years. Failures will likely

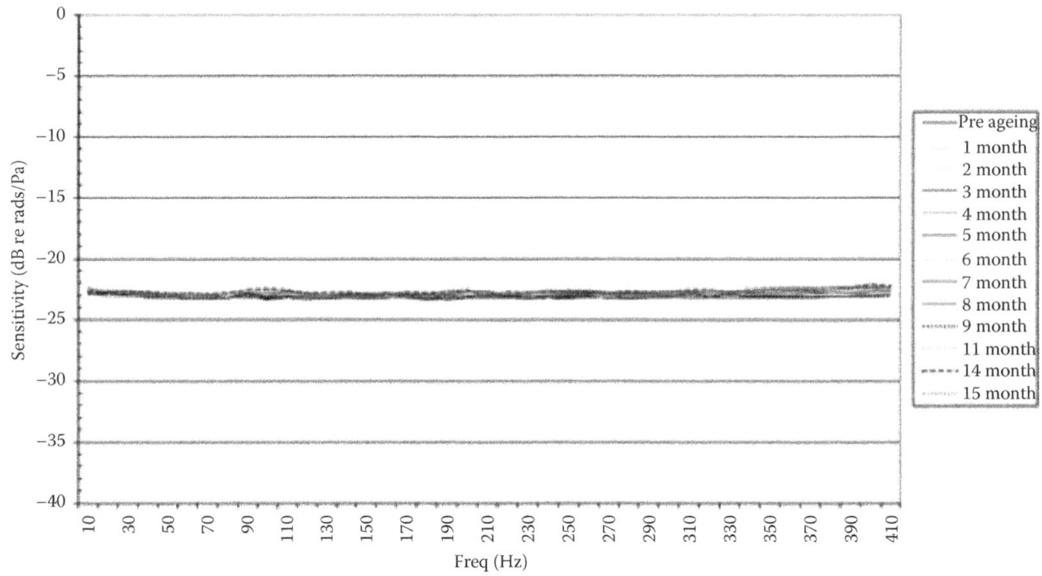

Figure 21.9 Hydrophone response over equivalent 30 years period.

be due to random failures of optical components, connectors, or pressure seals.

The main evidence on the reliability for RM systems comes from the relatively small number of systems so far deployed. For instance, the system deployed at Ekofisk in 2008 (Nakstad et al. 2011) had 16,000 sensor channels, and based on published figures all are still working in 2014, 6 years later.

21.4.1 Performance issues

In addition to the demanding environmental requirements, PRM systems are required to achieve a very high level of performance, often significantly more demanding than other sensor applications. A few of the key performance requirements are discussed below.

Noise performance. As PRM systems are detecting very small seismic reflections against background noise, it is important that the system noise floor is as low as possible, and ideally below the environmental noise level. As background seismic noise at deep water locations can be very low, this puts a demanding requirement on system noise, and this requirement extends to ever lower frequencies, e.g., performance down to 0.5 Hz or even 0.1 Hz is often required. A typical system noise specification would be <100 ng/rtHz for an accelerometer, which for a typical sensor sensitivity equates to a system noise floor of tens of microradians, even down to <0.5 Hz. As system noise is often dominated by laser phase noise, careful design and selection of lasers is needed, in conjunction with self-noise referencing techniques.

Dynamic range. In addition to exhibiting very low system self-noise, PRM systems are also required to provide an accurate reconstruction of the very high amplitude wideband direct arrival from the seismic air gun. The nature of this signal is such that the optical sensors are subject to an extremely rapid phase change, which can be of the order of 1000 radians in less than 10 ms. Many optical interrogation schemes struggle to cope with this, either due to undersampling of the signal (in the time domain) or due to multiple sidebands spreading into adjacent frequency bins (in the frequency domain). Various proprietary techniques have been developed to deal with these effects, and these have resulted in some systems being able to achieve a dynamic range in excess of 180 dB.

Operational issues. The PRM application is, as discussed previously, interesting because of the high demands made in terms of scale,

Figure 21.10 Prototype system being deployed during field test.

performance, and reliability. However, these systems also need to operate reliably and with minimum disruption to their surrounding environment (typically a platform on an operating oilfield). This leads to a number of considerations.

The system must be capable of deployment from a standard installation vessel and integration with the often complex infrastructure of an existing oilfield.

Size, weight, and power of the interrogation system on the platform should be minimized. Consideration also needs to be given to minimizing acoustic and Electromagnetic emissions from the system, and various safety and emission standards will need to be met.

There will need to be maintenance and repair plans for the system. Repair of the interrogation unit is likely to be required at times and will typically be by module replacement on the platform. Repair of the subsea system is much more onerous, and typically systems will be designed such that no planned subsea maintenance/repair will be required during a system lifetime.

Field testing. The overall performance of a PRM technology is best characterized through field testing, ideally in a representative underwater environment using a seismic source to excite the sensors. Figure 21.10 shows a prototype system being deployed in 140 m water depth during a field test.

21.5 THE FUTURE

We have seen that PRM is a demanding application for optoelectronic systems, in terms of scale, performance, and reliability. Another issue we have not discussed is that of cost. The cost of building, installing, and operating a PRM system can be in the many tens of millions of dollars.

Despite this, the potential benefit of a PRM system is very large. This can be in reduced operational cost of fields (e.g., smaller number of dry wells) and/or enhanced production rates, or enhanced final production volumes. The benefits of these may vary significantly outweigh the costs.

Despite all these factors the take-up of PRM into the oil/gas market has been slow. The first installation took place in 2003 and in the succeeding 11 years there have only been eight installations. The scale of these installations has been very large, totaling over 1000 km of sensor cable and over 100,000 sensors in total. However, considering the number of fields under development or production over that period, and the number of those that have the potential to significantly benefit from use of PRM, the degree of take-up has been surprisingly small.

There have been a number of reasons for this. The high up-front build/installation cost is a barrier, because although the eventual payback may be many times the initial cost, the cost is a significant extra factor in field development at a time when capital expenditure is under tight control, and the benefit will only be fully realized over a number of years. This issue is made worse by short-term fluctuations in the oil price such as the dramatic price drop seen in late 2014. The operational and organizational difficulties of design, build, and installation are also a significant factor. Finally, proving the long-term reliability of these expensive systems, which only deliver benefits if they operate reliably for many years, has also been a concern.

Although this may sound a little negative, experience of the introduction of optical sensor technology into other areas of the oil industry suggests a more positive outcome. Two good examples are distributed temperature sensing (DTS) and distributed acoustic sensing (DAS). Both DTS and DAS were research technologies, which showed considerable promise but slow adoption for a number of years. However, both technologies eventually reached the point where they overcame the barriers to adoption, and are now widely used, and with very beneficial results to the industry. There is no reason why PRM should not follow the same path.

Perhaps the key question is whether PRM will always be a niche application used in certain very specific instances, or whether it will experience widespread adoption, at least in a significant minority of field cases. The answer to this will lie partly in technical developments, partly in cost, and partly in establishing in quantitative terms, the clear benefit of PRM to the industry.

In terms of technical developments, we can expect to see further increases in multiplexing efficiencies. Current multiplexing schemes allow in excess of 250 sensors per fiber, but studies and laboratory work have shown that well in excess of 1000 sensors per fiber is theoretically possible (Liao et al. 2012). This requires the use of techniques such as distributed optical amplification to balance the losses associated with multiplexing. Other improvements may include lower noise lasers with an emphasis on very low frequency (sub 1 Hz) performance and wider bandwidth sensors, especially for passive applications. Finally, improvements in processing power and in miniaturization of integrated optics components will help to reduce the size of the platform-based interrogation system, which can be another bar to PRM installation.

On the cost side, the biggest issue is to break the vicious circle, which means that optical system costs were initially higher than more conventional electrical system because so far they have been built in relatively small quantities. However, the use of components and production facilities developed for telecoms applications will help here, and significant cost reductions have already been seen, such that optical approaches are now comparable or cheaper than electrical approaches.

All these factors will help. However, the biggest difference will be made if the promised PRM benefits in terms of enhanced resort performance can be conclusively demonstrated. Significant progress has been made in this area also, and, based on the experience in the introduction of other optical sensor technologies to the oil industry, we can expect to see much more widespread adoption, so that PRM will become both technically and commercially one of the success stories of the optical fiber sensing industry.

REFERENCES

Bunn, B., Fiber-optic permanent installed sensor cable for seismic reservoir monitoring. *Offshore Technology Conference*, Houston, TX, OTC-19445-MS, 2008.

Kammedal, J. H. et al., Initial experience operating a permanent 4C seabed array for reservoir monitoring at Valhall. *SEG2004 Extended Abstracts*, Society of Exploration Geophysicists, pp. 2239–2242, 2004.

Kersey, A., Multiplexed interferometric fiber sensors. *Proceedings of the 7th Optical Fibre Sensors Conference*, Sydney, 1990.

Liao, Y. et al., Highly scalable amplified hybrid TDM/WDM array architecture for interferometric fiber-optic sensor systems, *Journal of Lightwave Technology*. 31(6), pp. 882–888, 2012.

Nakstad, H., Langhammer, J. and Eriksrud, M., Fibre optic permanent reservoir monitoring breakthrough, *12th International Congress of the Brazilian Geophysical Society*, Society of Exploration Geophysicists, pp. 1440–1443, 2011.

Nash, P., Cranch, G. and Hill D., Large scale multiplexed fibre-optic arrays for geophysical applications, *SPIE Industrial Sensing Systems*, Vol. 4202, pp. 55–65, 2000.

Nash, P. et al., High efficiency TDM/WDM architectures for seismic reservoir monitoring, *Proc SPIE 7503*, International Society for Optical Engineering, 2009.

22

Applications of visible to near-infrared spectroscopy for downhole fluid analysis inside oil and gas wellbores

GO FUJISAWA, OLIVER C. MULLINS, AND TSUTOMU YAMATE

Schlumberger

22.1 INTRODUCTION

Timely assessment of hydrocarbon reservoir fluid is very important for upstream oil/gas companies to assess their reservoirs and make optimum plans for development and production of their underground resources. Visible to near-infrared (VIS/NIR) spectroscopy is a powerful analysis tool and utilized in many industrial settings, and it has been used to analyze fluid inside oil/gas wellbores for over 20 years. Today, downhole fluid analysis (DFA) by VIS/NIR spectroscopy is widely accepted by upstream oil/gas companies as a valuable and indispensable method to understand their reservoir immediately after a well is drilled.

Implementation of DFA by VIS/NIR spectroscopy has always been a very challenging scientific and engineering task. The nature of this challenge becomes obvious when contrasting the environmental requirements between oil/gas wellbores to normal laboratories. Maximum wellbore temperatures can reach 175°C or higher for a deep wells; moreover, the temperature is by no means stable during measurements. Wellbore pressure can reach 100 MPa or much higher for deep-water wells. For example, DFA spectrometers may experience ~200 MPa pressure in very deep wells in the Gulf of Mexico. DFA spectrometers may experience severe mechanical shocks and vibration during transportation to a well site, often over unpaved road for several hours, and during conveyance into a bottom of wellbore, especially if DFA spectrometers are integral as a part of a drill string. Space is a premium for DFA instruments and hardware needs to be designed so that it can fit into a cylindrical tool string of approximately 10 cm or less in diameter. Once DFA spectrometers are dispatched from the base to wells for measurement, there is no chance of any easy calibrations. Thus, it is required that the calibration of the spectrometer will last a minimum a few months to years. Instrument reliability is critically important. Lost time due to instrument failure is prohibitively expensive in some operating environment. For example, operations of a deep water offshore rig can cost as high as 1 million US dollars per day. It is no exaggeration that oil/gas exploration is a most challenging environment for technology and is comparable to demands of space exploration and military requirements.

In this review, we first explain the nature of fluid spectra in oil/gas wells, and the design considerations for DFA spectrometers, and we provide an example of spectrometer hardware and measurement performance. Then, we discuss actual applications including the setting of measurement, fluid sampling optimization, fluid sampling quality assurance, *in situ* fluid property measurements, and briefly advanced applications.

22.2 VIS/NIR SPECTROSCOPY FOR DFA

This section will present the nature of fluid spectra in oil/gas wells, design considerations for DFA spectrometers, and principles of measurements and interpretations.

22.2.1 Fluid spectra in oil/gas wells

The majority of molecules constituting fluid in oil/gas reservoir are various types of hydrocarbons and water. Generally, small amounts of carbon dioxide, hydrogen sulfide, and nitrogen are also present in some formation fluid, although in some cases, these nonhydrocarbon components can dominate. Figure 22.1 shows several spectra of crude oil and water.

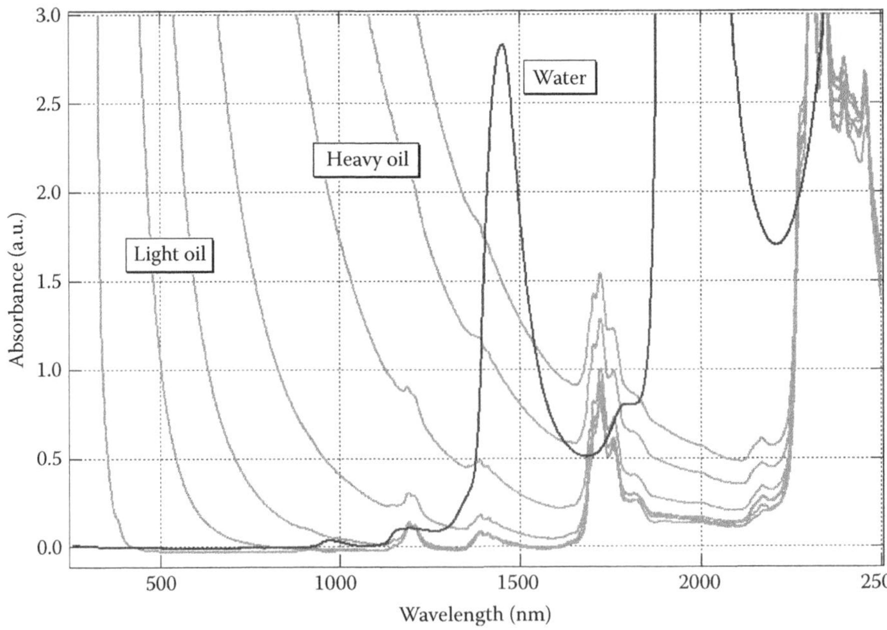

Figure 22.1 VIS/NIR spectra of crude oil and water.

Broadly speaking, three types of information can be extracted from VIS/NIR spectrum.

First, water has very distinctive absorption peaks at about 1440 and 1940 nm. The peak at 1440 nm corresponds to the two-stretch vibrational overtone (symmetrical or asymmetrical) of the oxygen–hydrogen (OH) bond. The peak at 1940 nm corresponds to vibrational modes of combination of bend and stretch modes of the OH bond. Oil has a group of absorption peaks at around 1725 nm that corresponds to vibrational overtone of carbon–hydrogen stretch mode. The absorption bands at 2350 nm correspond to the CH stretch plus bend combination bands. These water and oil peaks are easily detectable by a basic spectrometer, and these measurements were realized in the first generation of DFA instrument about 20 years ago [1]. The magnitude of the overtone and combination bands reduces at higher energies. These bands are all formally "forbidden" yielding rather small absorption coefficients in comparison to the fundamental vibrational bands in the mid-infrared, which are roughly at least 1000 times bigger. This fact is quite useful; relatively long path lengths can be used enabling bulk sample analysis with the NIR.

Second, the oil spectrum shows increasing absorbance toward shorter wavelength. Generally, heavier crude oil tends to absorb more at longer wavelength than lighter crude oil. This mode of absorption is attributed to excitation of delocalized π-electrons found in aromatic molecules. Larger fused aromatic ring systems have greater delocalization of the π-electrons, thereby requiring less energy for excitation, thus absorbing longer wavelength light. This absorption corresponds to electronic excitation, which is quite distinct from the vibrational excitation discussed earlier. Two extreme examples are benzene with its lowest electronic transition at ~250 nm and graphite, which is a zero bandgap semiconductor. The π-electrons in benzene molecules are delocalized only within a confinement of one aromatic ring, thus it still has a relatively large energy gap and excited only in ultraviolet light, resulting in transparent color to human eyes. On the other hand, the π-electrons in graphite are completely delocalized and readily excited at both visible and NIR light, resulting in black color. In crude oil, there is a group of molecules known as asphaltenes that have an aromatic core consisting of about seven fused rings, with several peripheral aliphatic hydrocarbon chains. It is the delocalized π-electrons in these asphaltenes molecules that absorb in VIS/NIR light and give the brown to black color to crude oil. Asphaltenes impart high viscosity to the crude oil (and asphalt!), as well as decrease the yield of liquid fuels in refining. Thus, crude oil color is an important indicator of oil type in reservoir.

Third, a closer look at the details of the oil spectrum around 1725 nm shows that it depends on the composition of hydrocarbon fluid. The resonance frequency of carbon–hydrogen vibration mode is slightly different when it is in methane (CH4) or in ethane (–CH3) or in larger hydrocarbon molecules (dominated by –CH2–). Essentially, the different masses of these mechanical oscillators cause a shift in the resonance frequencies. As a result, methane has a different absorption spectrum from ethane, and ethane has a different absorption spectrum from propane, and so forth. However, this difference diminishes soon as a size of hydrocarbon molecules increases. This similarity among hydrocarbon molecules sets a practical limit to a number of independently extractable composition groups. Measuring these subtle differences among hydrocarbon peaks requires a better quality spectrometer compared to simple oil–water detection or oil color measurements. Several studies have demonstrated that it is still possible to determine grouped composition in hydrocarbon fluid by spectroscopic measurement of relatively limited number of wavelength and resolution [2–5].

22.2.2 Hardware design considerations

Several environmental factors demand special considerations for designing DFA spectrometer hardware.

22.2.2.1 TEMPERATURE

High-temperature impacts spectrometer hardware designs in three ways. First, it may simply destroy or significantly reduce the lifetime of optoelectronics and electronics components by excessive heat. Failure rates of typical plastic packaged components increase dramatically above 150°C. At 175°C or above, packaging of optoelectronics and electronics components need to be carefully selected in high temperature plastic, metal, or ceramic package or risk instrument failures [6]. Second, elevated temperature may significantly alter the characteristics and performance of optoelectronics components. Light-emitting diode (LED) and laser emission typically decrease in intensity and shift in wavelength at high temperature, even if they survive and remain operational. Photodiodes sensitivity decrease and their cutoff shifts toward longer wavelength. Third, elevated temperature may slightly change optomechanical properties of spectrometer hardware. Fabry–Perot interferometers and band-pass filters shift in wavelength due to thermal expansion, thermal stress, and refractive index change. Thermal expansion and stress may distort the spectrometer unit and its optical alignment. While these effects may not be as dramatic as component failures, they are, nevertheless, sources of considerable measurement errors.

22.2.2.2 PRESSURE

The only way to handle extreme high pressure is to seal the sensitive spectrometer from high pressure. Only optical window(s) should be facing high pressure fluid extracted from the formation. Sapphire is a perfect material for DFA spectrometer optical windows as it is hard, tough, chemically stable, transparent in the entire VIS/NIR wavelength range, and affordable. Spectroscopic measurements can be made with one window for reflection measurements or with two windows for more conventional transmission geometry.

22.2.2.3 MECHANICAL SHOCK AND VIBRATION

Any mechanically moving or movable components are most likely to fail or misalign when subject to severe shock and vibration. Chopper wheels, rotating diffraction grating shafts, and interferometers are key components of conventional and Fourier transform infrared spectroscopy (FTIR) spectrometers, but they can be a significant source of instrument failure or measurement error in the DFA spectrometer. If these moving components are necessary, utmost considerations should be taken for their mechanical designs. Repetitive mechanical shocks and vibrations also cause electronics component failures, particularly when the temperature is high.

22.2.2.4 SPATIAL LIMITATION

A spectrometer using a dispersing element, such as a grating or prism requires a longer optical path or smaller element detector array to achieve greater wavelength resolution. This may be difficult in a DFA spectrometer as it has to have a small foot print to fit into the high-pressure tool housing.

22.2.2.5 CALIBRATIONS AND MAINTAINABILITY

We measure background with an empty cell or reference cell before the sample measurement to ensure data quality in laboratory. We take it for granted that this condition remains essentially unchanged between background measurement and sample measurement. This is not the case for realistic DFA spectrometer measurements. Once a spectrometer unit is dispatched from a base, there is little opportunity to calibrate it. It is transported to a well site by land or air, and then it is lowered into a wellbore as a part of a suite of measurement instruments suspended by an electrical cable, called Wireline in oilfield terminology, or as a part of a drill string, called logging while drilling (LWD). Either way, a spectrometer is conveyed to a designated depth at various temperatures and pressures. In between, it may experience severe shock and vibration. Embedded in this hostile environment, the DFA spectrometer hardware needs to be stable during all these processes to make reliable measurements.

22.2.3 DFA spectrometer example

An example of DFA spectrometer is given below with hardware architecture, interpretation software, and typical accuracy.

22.2.3.1 HARDWARE

Designs of DFA spectrometers are restricted by several environment factors described in the previous section. Nevertheless, a reasonably accurate VIS/NIR spectrometer can still be built by carefully selecting a specific hardware architecture and design details. Figure 22.2 shows a schematic of a DFA spectrometer hardware configuration. A tungsten halogen incandescent lamp operating around 3000 K gives a blackbody radiation with a peak around 1 μm and covers the entire VIS/NIR wavelength range. Unlike lasers or LEDs, the temperature change from room temperature to over 175°C has negligible impact to emission of a 3000 K radiation source. A rotating chopper wheel enables reduction of noise. The chopped source light is selectively guided into a measurement cell and reference path via fiber optics. A

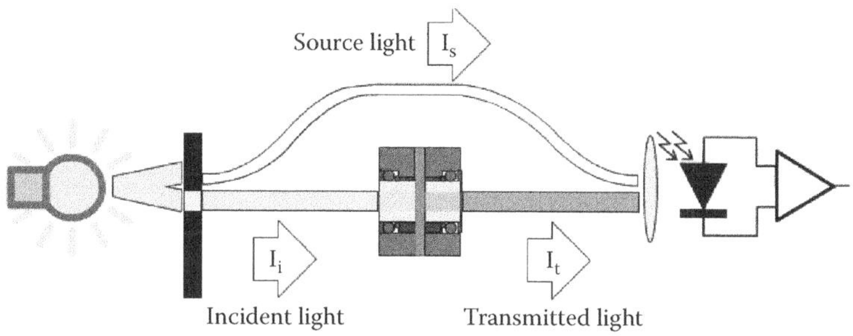

Figure 22.2 A schematic of DFA spectrometer hardware. This figure is drawn for only one wavelength channel for simplicity. An actual spectrometer has multiple individual photodetectors coupled with narrow band-pass filters, each pair for one wavelength channel.

measurement cell consists of two thick sapphire crystal optical windows separated by a fluid flow channel where actual formation fluids flow. The transmitted light will be distributed to individual photodetectors each coupled with a designated band-pass filter (only one channel is drawn in Figure 22.2). The number of wavelength channels can be 10, 20, or any numbers considered necessary for intended applications. Fiber bundles consisting of several thousands of optical fibers connect each optical component and coupling. The bundles are branched and randomized as needed. All components are designed and tested to withstand high temperature and mechanical shock and vibration. More recent DFA spectrometers have a grating spectrometer to cover information-rich 1600–1800 nm wavelength along with a filter-array spectrometer described here [7].

22.2.3.2 INTERPRETATION

As the measured fluid is at such high pressure and temperature, its spectrum is substantially different from those at a standard laboratory condition often found in spectrum database library. Thus, specialized databases for hydrocarbon spectra measured at high pressure and high temperature are needed for accurate composition analysis on the acquired sample spectrum. Once the database is built, interpretation schemes can be developed by standard chemometrics approaches such as principal component regression or partial least squares regression method. The latest DFA spectrometer can analyze hydrocarbon fluid and interpret its composition into five groups; methane, ethane, propane-to-butane, hexane+, and carbon dioxide as long as no appreciable water is present in the sample. For the upstream oil and gas industry, there is a very important parameter named gas-to-oil ratio (GOR) to characterize hydrocarbon fluid. Inside high-pressure hydrocarbon reservoir, crude oil may contain a large amount of dissolved methane, ethane, and other smaller molecules that will be released as gas once the fluid is produced to environmental conditions of one atmosphere and surface temperature. The GOR gives the ratio of gas volume over liquid volume once the single-phase fluid is brought to surface. For example, a crude oil of GOR 100 m^3/m^3 means 100 m^3 of gas is produced along with 1 m^3 of crude oil. Thus, the surface production facility requirements depend dramatically on the GOR. DFA spectrometers give GOR from derived compositions by either solving equations of state or neural network approaches.

22.2.3.3 ACCURACY

The accuracy of DFA spectroscopy measurements and its interpretation depends on multiple factors including hardware, interpretation methods, sample types, and measurement conditions. Figure 22.3 is an example of comparisons of several DFA measurement and interpretation results against laboratory composition analysis results by gas chromatography (GC). Agreements are surprisingly good considering all environmental difficulties described earlier.

22.3 APPLICATIONS INSIDE OIL AND GAS WELLBORES

This section will present typical applications of DFA spectrometer instruments inside oil/gas wellbores.

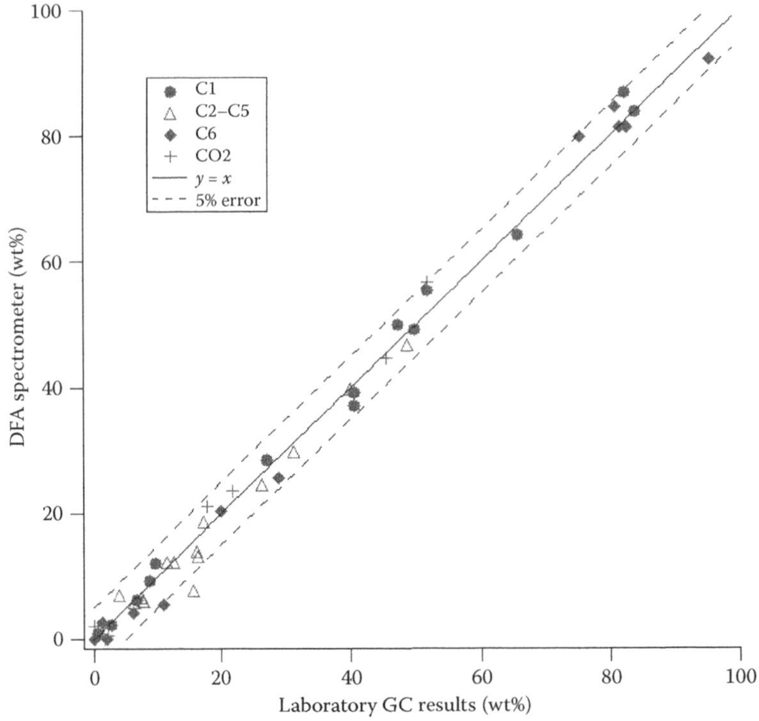

Figure 22.3 Comparison of measured fluid compositions of DFA spectrometer analysis and laboratory GC analysis.

22.3.1 Formation tester

Properties of the rock formation penetrated by a wellbore are first measured by several instrumented tools relying on nuclear, electrical, and sonic measurements. Properties typically measured include formation density, porosity, basic lithology, resistivity, and stress. Then, a fluid sampling program by a formation sampling tool or "formation tester" is planned based on the formation evaluation. The DFA spectrometer instrument is conveyed to a wellbore as a part of formation tester tool by Wireline tool string or LWD tool string.

Formation tester tool strings typically includes one or more probe or inlet modules, pump modules, DFA modules, and sample chamber modules [8]. A probe or inlet module has a tube or opening that forms a seal with the formation wall allowing extraction of formation fluids by use of a pump. It is essential to avoid contamination of the formation sample with mud in the borehole or with mud filtrate that leaks into the formation. The inlet module can be a simple probe or guarded probes that can sample uncontaminated formation fluid more efficiently, or a pair of packers that seals a section of a wellbore vertically. A pump module creates a necessary differential pressure to move fluid from the formation through the flowline of the formation tester tool. Fluid can be expelled from the formation tester into the wellbore until the desired fluid is obtained. The DFA modules analyze the flowing fluid in real time to evaluate fluid properties and to enable the operators to make necessary decisions such as whether and when to acquire a sample. If the fluid flowing inside the formation tester is judged as worthy of further analysis in laboratory, the fluid is captured in a sample bottle in a sample chamber module.

22.3.2 Sampling program optimization

The first type of DFA spectrometer application is sampling program optimization. Although formation properties are analyzed and understood to some extent prior to formation testing, real-time DFA spectrometer analysis on the flowing fluid often reveals reservoir information that is not available from other measurements. As a result, the fluid sampling program can be altered from the original plan and optimized. An illustrative example may be an initial assumption that a reservoir contains homogeneous hydrocarbon fluid. Once the DFA spectrometer starts to provide evidence that the fluid is not homogeneous, the operator will naturally change a sampling program to capture more fluid samples from different parts (depths) of the wellbore than originally planned so that complexities of reservoir fluid along the wellbore are fully understood.

22.3.3 Sampling quality assurance

At each wellbore point (depth) that an operator decides to take a formation fluid sample, it is critically important that the fluid flowing into the formation tester is truly representative of the fluid inside the formation. This problem may not seem very obvious but nonrepresentative sampling can happen for two different reasons. First, formation oil almost always contains some dissolved gas. Similarly, formation gas also may contain light-end liquids. As an example, a bottle of coca-cola contains dissolved gas, which is released with pressure decrease. If the pressure is reduced to mobilize formation fluid and the pressure drops below saturation pressure (i.e., bubble point or dew point) of formation fluid, phase segregation will take place inside the formation and resulting in multiphase fluid flowing into formation tester, which is no longer representative due to different phase mobility and volumetrics. Second, oil wells are drilled with drilling mud, which is pumped down in the center of the drill pipe and travels up the annulus between the pipe and the borehole wall. The mud maintains wellbore pressure to avoid blowouts and avoid wellbore collapse, carries the rock cuttings to surface, lubricates the drill string, and forms a mud cake at permeable zones avoiding excessive fluid loss. When drilling mud contacts with permeable formation, the liquid part of drilling mud (called mud filtrate) will invade into formation while the solid part of drilling mud will stay and plug pore space of the borehole wall. When fluid flows from formation into formation tester, the first arriving fluid is heavily contaminated by invaded mud filtrate. Typically, fluid flow for hours as required before it becomes sufficiently free from contamination, given limitations of pump rate and pressure drawdown. The DFA spectrometer instrument is used to monitor the spectrum of the flowing fluid continuously and analyze if it is in single phase and sufficiently free from contamination. The DFA spectrometer literally provides eyes to fluid sampling process [1,9,10].

22.3.4 *In situ* fluid property measurement

Fluid worthy of detailed analysis is captured in a sample bottle and analyzed in laboratory later. Pressure–volume–temperature analysis measures fluid GOR and other fluid properties. GC measures gas and liquid compositions in detail. Some fluid properties are, however, not measured reliably in laboratory because they can change irreversibly from reservoir to laboratory. For example, water pH is reliably measured at *in situ* condition by DFA spectrometer [11]. A small amount of pH-sensitive color dye is mixed *in situ* in the formation tester with formation water and its color change is measured by spectrometer. In contrast, a typical laboratory pH measurement is done at room temperature and atmospheric pressure conditions after dissolved gas is liberated and minerals precipitated. Naturally, its pH value is different from the original pH value in reservoir. Another example is detection of low concentration of hydrogen sulfide. Hydrogen sulfide in producing fluid has significant impact on operator because it is hazardous and corrosive. However, it is well known that laboratory measurement of low-concentration hydrogen sulfide (H_2S) is not reliable because H_2S is consumed by metal and mud filtrate before the sample is analyzed in the laboratory. *In situ* measurement inside borehole is the only way to detect low concentration of hydrogen sulfide accurately.

22.3.5 Advanced applications

DFA at each measurement point (depth) along the wellbore gives the property of fluid at that particular point of wellbore in real time. These measurements are very valuable themselves as already explained. Recently, there have been several studies that relate property of formation fluid at one point to others by using DFA spectroscopy and equation of state (EoS). For example, some reservoirs are known to have compositional gradient that can be measured and interpreted by DFA spectrometer and described by an EoS. Gradients in GOR can be evaluated with the cubic EoS. Some reservoirs are known to have asphaltene concentration gradient that can be measured as fluid color by DFA spectrometer and successfully characterized by a newly developed EoS, the Flory–Huggins–Zuo EoS, for dissolved (or colloidally suspended) asphaltenes. This EoS is based on the resolution of the molecular and colloidal species of asphaltenes codified in the Yen–Mullins model. These advanced applications give spatial distribution of fluid properties across the reservoir compared to basic DFA that gives fluid properties at a measurement point. The details of these advanced applications are beyond the scope of this introductory review. Interested readers are encouraged to further readings [12,13].

22.4 CONCLUSIONS

This review discusses the application of VIS/NIR spectrometers to operable in hostile environment inside oil and gas wells. DFA by VIS/NIR spectroscopy provides vital information for operators to understand their hydrocarbon reservoirs and optimize their development plans. Indeed, many hundreds of these DFA spectrometers have been commercially produced and annually run in thousands of wells to perform *in situ* spectroscopic measurements of reservoir fluids in oil and gas wellbores worldwide.

REFERENCES

1. A. R. Smits, D. V. Fincher, K. Nishida, O. C. Mullins, R. J. Schroeder, and T. Yamate, In-situ optical fluid analysis as an aid to wireline formation sampling, *SPE Formation Evaluation*, vol. 10, no. 2, pp. 91–98, 1995.
2. C. W. Brown, Trading wavelength for absorbance resolution: predicting the performance of array detection systems, *Applied Spectroscopy*, vol. 47, no. 5, pp. 619–624, 1993.
3. O. C. Mullins, T. Daigle, C. Crowell, H. Groenzin, and N. B. Joshi, Gas-oil ratio of live crude oils determined by near-infrared spectroscopy, *Applied Spectroscopy*, vol. 55, no. 2, pp. 197–201, 2001.
4. M. A. van Agthoven, G. Fujisawa, P. Rabbito, and O. C. Mullins, Near-infrared spectral analysis of gas mixtures, *Applied Spectroscopy*, vol. 56, no. 5, pp. 593–598, 2002.
5. G. Fujisawa, M. A. van Agthoven, F. Jenet, P. A. Rabbito, and O. C. Mullins, Near-infrared compositional analysis of gas and condensate reservoir fluids at elevated pressures and temperatures, *Applied Spectroscopy*, vol. 56, no. 12, pp. 1615–1620, 2002.
6. C. Avant, S. Daungkaew, B. K. Behera, S. Danpanich, W. Laprabang, I. De Santo et al., Testing the limits in extreme well conditions, *Oilfield Review*, vol. 24, no. 3, pp. 4–19, 2012.
7. G. Fujisawa and T. Yamate, Development and applications of ruggedized VIS/NIR spectrometer system for oilfield wellbore, *Photonic Sensors*, vol. 3, no. 4, pp. 289–294, 2013.
8. J. Creek, M. Cribbs, C. Dong, O. C. Mullins, H. Elshahawi, P. Hegeman et al., Downhole fluids laboratory, *Oilfield Review*, vol. 21, no. 4, pp. 38–54, 2009/2010.
9. O. C. Mullins, J. Schroer, and G. F. Beck, Real-time quantification of OBM filtrate contamination during openhole wireline sampling by optical spectroscopy, *Presented at SPWLA 41st Annual Logging Symposium, SS*, Houston, TX, June 4–7, 2000.
10. S. Betancourt, G. Fujisawa, O. C. Mullins, A. Carnegie, C. Dong, A. Kurkjian et al., Analyzing hydrocarbons in the borehole, *Oilfield Review*, vol. 15, no. 3, pp. 54–61, 2003.
11. B. Raghuraman, M. O'Keefe, K. O. Eriksen, L. A. Tau, O. Vikane, G. Gustavson et al., Real-time downhole pH measurement using optical spectroscopy, *SPE Reservoir Evaluation and Engineering*, vol. 10, no. 7, pp. 302–311, 2007.
12. O. C. Mullins, *The Physics of reservoir fluids: discovery through downhole fluid analysis*. Sugar Land, TX: Schlumberger, 2008.
13. O. C. Mullins, A. E. Pomerantz, J. Y. Zuo, and C. Dong, Downhole fluid analysis and asphaltene science for petroleum reservoir evaluation, *Annual Review Chemical Biomolecular Engineering*, vol. 5, pp. 325–345, 2014.

23

Mid-infrared spectroscopy for future oil, gas, and mineral exploration

CHRISTIAN M. MÜLLER, FLORIAN RAUH, THOMAS SCHÄDLE, MATTHIAS SCHWENK, ROBERT STACH, AND BORIS MIZAIKOFF
University of Ulm

BOBBY PEJCIC
CSIRO, Energy Flagship

23.1 VOLATILE ORGANIC COMPOUNDS AND OIL MONITORING

Given that volatile organic compounds (VOCs) are among the most commonly detected organic contaminants in water, the general public becomes increasingly aware of the problems associated with the contamination of ground and surface water resources. During the past few decades, the worldwide production and consumption of oil and petroleum products have increased dramatically. Concurrently, this leads to a higher risk of environmental pollution. Aromatic hydrocarbons from petroleum-based products may enter the marine environment from a number of different sources, e.g., oil spills due to leaking tankers, accidents at vessel and drilling operations, or industrial discharge. Industrial oil production and offshore delivery globally accompany these oil spill events. From such oil spills, VOCs, especially benzene, toluene, ethyl benzene, naphthalene, and xylenes (or "BTEX+N"), may dissolve to a small yet relevant extent within the water phase. Therefore, they cause contaminations at concentrations ranging from ppb (μg/L) to ppm (mg/L) levels. For detection of contaminants emanating from accidents during industrial oil spill into the marine environment, various hydrocarbon sensors have been developed, which are suitable for marine monitoring and surveys [1].

Besides the established state-of-the-art analytical systems, infrared (IR) spectroscopy is an emerging technology in terms of rapid and miniaturizable sensing techniques currently available in this domain. Methods such as gas chromatography coupled with mass spectrometry (GC-MS) usually require extensive laboratory equipment and time-consuming sample preparation steps. With few exceptions, such methods are expensive and difficult to operate during in-field deployment; thus, efficient environmental monitoring and online water quality control require more compact sensing technologies enabling ubiquitous application. Regarding water quality control, Fourier-transform infrared (FTIR) spectroscopy is capable of directly analyzing liquid-phase samples in a continuous or quasicontinuous fashion. However, conventional FTIR techniques are limited by strong background absorption interferences particularly from water and usually require sizeable spectrometers derived from laboratory equipment.

Given recent innovations in IR sensor technologies, i.e., specifically mid-IR (MIR) waveguides and light source technologies such as quantum cascade lasers, which are capable of emitting high-intensity radiation throughout the entire MIR regime, measurement strategies based on the fundamental principles of internal total reflection are of increasing importance and offer a useful alternative to conventional analysis methods. Based on these technologies, analytical systems providing robust and reliable in-field sensing capabilities may be scaled to lower dimensions, and therefore offer the potential for in-field applications even at harsh conditions taking advantage of microfabricated devices and components.

23.1.1 VOC "fingerprinting" in aqueous environments

Today, it has readily been demonstrated that advanced attenuated total reflection (ATR) spectroscopy techniques are capable of reliable hydrocarbon identification in geochemically relevant fluids [2]. They have successfully been used for the detection of a wide variety of organic constituents, as well as for the analysis of oil in water and water in oil.

This section reviews current IR-ATR-based methods for the detection and quantification of hydrocarbons partitioning into the water phase from oil matrices by evaluating characteristic spectral features within the so-called fingerprint region (i.e., 600–800 cm^{-1}). In this regime, each aromatic hydrocarbon constituent is characterized by a distinctive pattern of absorption features, which may be used to identify specific compounds even in complex mixtures comprising a variety of organic species [3].

To analyze VOC "fingerprints" resulting from and attributed to distinctive oil types, recent studies [3] have demonstrated methods where aqueous solutions containing dissolved hydrocarbons are pumped across a polymer-coated zinc selenide (ZnSe) waveguide (i.e., the ATR crystal

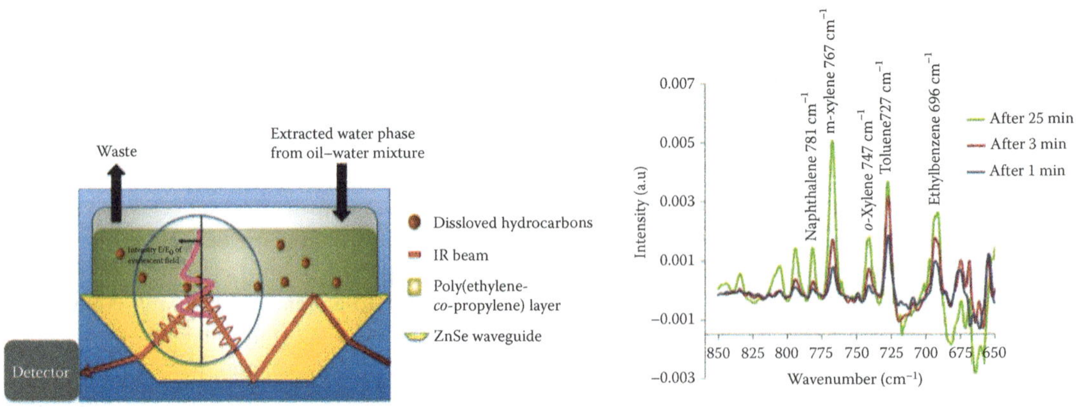

Figure 23.1 Left: Schematic measurement concept for MIR VOC monitoring using ATR techniques. Right: IR spectra of the aqueous phase from a crude oil–water mixture using a polymer-coated ZnSe waveguide after enrichment. (Adapted from Schaedle, T. et al., *Anal. Chem.*, *86*(19), 9512–9517, 2014. With permission.)

serving as the active sensing element based on several internal reflections producing an according evanescent field emanating from the waveguide surface; Figure 23.1), while continuously recording IR-ATR spectra of the VOCs partitioning into the polymer membrane. Next to the quantitative enrichment of VOCs amplifying the analytical signal, using a hydrophobic polymer membrane has the additional effect that water is effectively excluded from the analytically probed volume defined by the penetration depth of the exponentially decaying evanescent field (i.e., a few micrometers at MIR wavelengths), which is generated at the waveguide/polymer interface.

Collecting IR spectra during this enrichment process reveals clearly separated absorption bands of relevant VOCs (naphthalene, xylenes, toluene, and ethyl benzene) in the spectral region of 650–800 cm^{-1} (Figure 23.1). These absorption features may readily be identified and quantified from crude oil-in-water samples at ppb levels of concentration, thereby leading to a promising strategy for online monitoring of hydrocarbon pollutants in marine environments.

Figure 23.1 demonstrates that characteristic concentration patterns of VOCs may be used for in situ quantification via polymer-coated IR-ATR waveguides, and that using waveguide-based and miniaturized IR sensing systems for in situ oil fingerprinting in marine environments is already within reach, given the current state of IR technology development.

23.1.2 Quantification of VOCs in marine environments

Besides in situ fingerprinting, in situ quantification of petroleum-based VOCs plays an important role in marine monitoring. A major factor impeding the progress in understanding and controlling petroleum-based contaminants in the coastal environment is the difficulty associated with sampling aqueous systems containing VOCs. Current analytical methods (e.g., GC-MS) readily quantify petroleum-based aromatic hydrocarbons in marine environments; however, they are expensive, time consuming, and error prone, given the potentially large sampling errors resulting from the volatility of the sampled constituents.

In contrast, advanced IR sensor technologies enable VOC fingerprinting and quantification within a single measurement after appropriate calibration routines have been established. Additionally, sampling errors are avoided due to the possibility of direct in situ measurements without further sample preparation steps [4].

Recent studies have impressively demonstrated the suitability of IR-ATR sensor systems utilizing polymer-coated ZnSe ATR crystals for the direct quantification of BTEX+N constituents in water matrices, and that continuously operating quantitative measurement routines are indeed feasible [4].

In Figure 23.2, IR spectra of a polyisobutylene-coated ZnSe ATR crystal exposed to aqueous solutions of naphthalene are shown at various

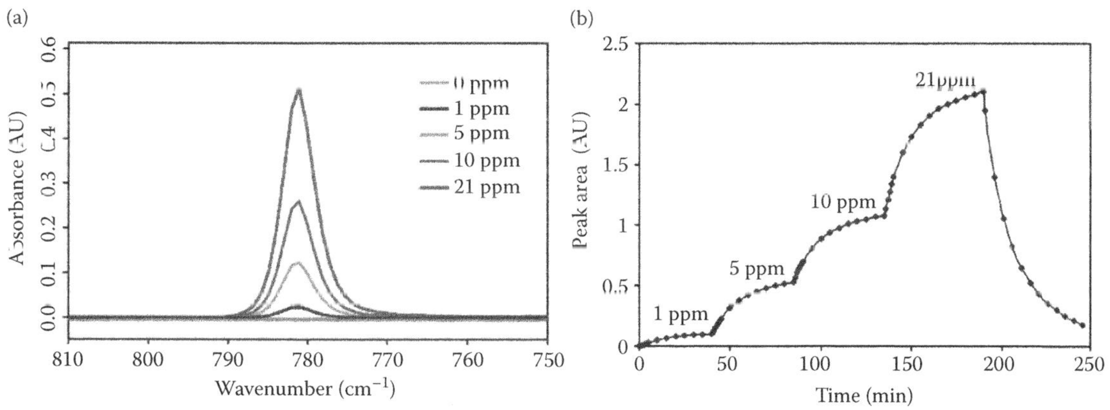

Figure 23.2 (a) IR spectra of naphthalene at different concentrations dissolved in deionized water. (b) Response time of the ZnSe waveguide when exposed to different concentrations of naphthalene dissolved in deionized water. (Adapted from Pejcic, B. et al., *Org. Geochem.*, 55, 63–71, 2013.)

concentrations (1–21 ppm) along with the respective diffusion curves.

In general, such polymer membranes require ~40 min to equilibrate with the surrounding sample and ~60 min to completely regenerate, which compares favorably to discontinuous GC-MS measurements. It should also be noted that the response of a continuously operating sensor system to concentration changes is significantly faster (see also Figure 23.2b), i.e., within several seconds due to shifts of the partition equilibrium. Hence, such sensors are ideally suited for monitoring threshold levels. The response time may be further decreased by using membrane-coated optical fibers in lieu of the rather large conventional ATR crystals. Last but not least, quantitative signals may also be derived from the slope of the diffusion curve, which also encodes the concentration information rather than using equilibrium evaluation routines. Thereby, quantitative readings may be established—albeit with larger standard deviations—after few minutes of exposure.

Considering the spectra shown in Figure 23.2a, it is evident that the intensity of the naphthalene vibration band increases with increasing concentrations. Further experiments revealed that the intensity of absorption features of all investigated BTEX+N compounds increased linearly with increased concentrations in solution (i.e., from the lower ppm region to ~80 ppm). Hence, calibration functions may be established enabling the quantification of minute amounts of petroleum-based aromatic hydrocarbons using simple least squares regression techniques. If more complex mixtures with strongly overlapping absorption features or unknown constituents have to be analyzed, multivariate calibration and data evaluation strategies have to be augmented.

For sensor testing, solutions of crude oil in deionized water were used as sample, which simulates the case of an oil spill.

Results obtained by Pejcic et al. [4] demonstrate the capabilities of IR-ATR techniques operating as quantitative sensors for rapidly obtaining data on dissolved hydrocarbon contaminants without the need of sample preparation, and that the obtained results compared well to established yet discontinuously operating methods such as GC-MS. Given the current trends in miniaturizing FTIR- and laser-based sensor technologies, it is evident that in-field deployment appears suitable for these methods.

Next to liquid-phase analysis, there is a high demand for trace gas sensors capable of rapid, sensitive, and selective detection of vapor-phase contaminants. Similar to liquid-phase VOC sampling, state-of-the-art gas detection at ppb levels is based on gas chromatographic techniques. Yet, recent progress in IR sensor technology, specifically the fabrication of practically applicable nanostructures, provides an emerging analyzer platform proven competitive for in situ and online gas sensing.

23.2 MIR SPECTROSCOPY OF GAS HYDRATES

Gas hydrates are clathrate compounds consisting of a rigid framework of water molecules bonded by two hydrogen bonds each and encasing usually small guest molecule inside the established cavities. In natural gas hydrates, most commonly methane (CH_4) is the guest molecule, frequently accompanied by other short-chained aliphatic molecules along with gases such as hydrogen sulfide (H_2S) or carbon dioxide (CO_2).

To date, three main structures of gas hydrates have been identified depending on the guest molecule: structure I (sI), structure II (sII), and structure H (sH) (see Figure 23.3). The radius of the cavities ranges from 3.91 to 5.79 Å with internal volumes of 0.25–0.81 nm^2 (i.e., if the cavities are considered to be spherical).

The majority of gas hydrates is found at continental margins (i.e., in water depths ranging between 300 and 800 m), and within the permafrost areas, which renders them a potential future source as fossil energy carrier. Even the most conservative estimations assume that there is almost seven times more CH_4 immobilized within gas hydrates than in all conventional gas reserves at a global scale (i.e., $1 \times 10^{15} m^3$ compared to $0.15 \times 10^{15} m^3$, respectively).

Besides its potential as fossil energy reserve, the impact of gas hydrates on the climate is a matter of ongoing discussion because methane has a global warming potential that is 21 times higher than that of CO_2 and could therefore potentially contribute to an unknown extent to the climate change.

Because CO_2 may also be encapsulated within clathrate hydrates, current efforts are on the way investigating the substitution of CH_4 within natural gas hydrates by carbon dioxide, thus (1) gaining a source of fossil fuel and (2) potentially sequestering and immobilizing CO_2 from the atmosphere, which would ideally render methane retrieved from gas hydrates a zero-emission energy source.

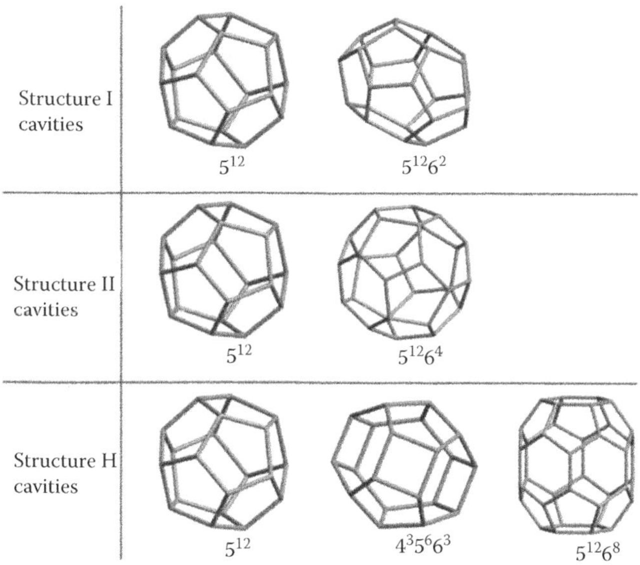

Figure 23.3 Cavities provided by different gas hydrate structures.

Also, the capability of clathrate hydrates to encage different gases makes them an interesting strategy for storing gaseous media (e.g., fuels) at low temperatures and moderate pressures, which is safer than the storage of liquefied gases at high pressures in heavy-duty steel bottles, as practiced nowadays.

Most commonly today, gas hydrates are an issue during gas/oil recovery because they may cause blockage of pipelines, which in turn leads to a reduced flow and therefore a reduced gas/oil production. In the worst case, even destruction of the pipeline may be induced. Low dosage kinetic inhibitors can be used to avoid plugging [5]. In order to optimize their function knowledge about gas hydrate formation and dissociation kinetics are crucial.

MIR spectroscopy offers unique insight into the mechanisms of gas hydrate formation and dissociation at a molecular level, as each involved molecule shows a distinct IR absorption pattern such that gases may be analyzed as well as various additives including, e.g., tetrahydrofuran (THF) or sodium dodecyl sulfate (SDS). Furthermore, IR spectroscopy may be used to study molecular changes during phase transitions, i.e., here, from liquid water into solid gas hydrate. Last but not least, IR sensors are nowadays readily miniaturized and may thus provide robust in-field sensing technologies for real-world applications in pipeline monitoring, deep sea studies, or analyses in permafrost environments.

23.2.1 MIR spectroscopy of gas hydrates

This section will present selected examples for the application of MIR spectroscopy for the advanced characterization of gas hydrates.

23.2.1.1 INFLUENCE OF DETERGENTS ON GAS HYDRATE FORMATION

Luzinova et al. [6] investigated the effects of a surfactant, namely SDS, on the formation and growth of gas hydrates via fiber optic evanescent field spectroscopy (FEFS).

It is a known fact that surface-active molecules accelerate the formation of gas hydrates. Therefore, propane hydrates were grown in the presence and absence of SDS in a stainless steel pressure vessel that was equipped with a polycrystalline IR-transmitting optical fiber (i.e., silver halide ATR element) in order to collect spectroscopic information on the growth of the hydrate. The phase change was monitored via spectral changes, as well as changes in temperature and pressure.

It is supposed that the nonpolar SDS molecules act as a carrier for the likewise nonpolar propane molecules, thereby actively transporting them to the surfaces inside the steel vessel (i.e., the steel walls and the optical fiber surface), where the nucleation starts first, given the presence of few

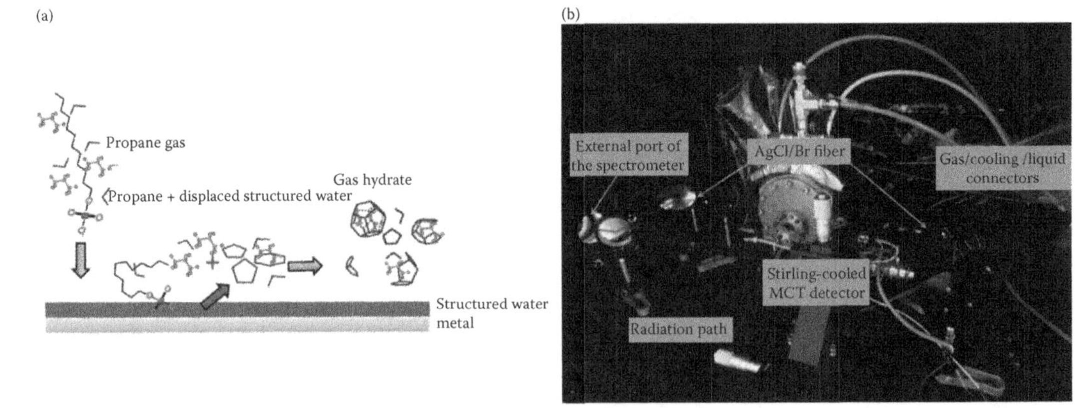

Figure 23.4 Left: Hypothesized mechanism of gas hydrate growth at surfaces promoted by SDS (schematic). (Reproduced from Luzinova, Y. et al., *Chem. Eng. Sci.*, *66*(22), 5497–5503, 2011. With permission.) Right: MIR-FEFS setup for the in situ analysis of gas hydrates.

structured monolayers of water. Thus, structured water fragments (i.e., potentially preconfigured components of larger water cages forming clathrates) that are displaced from the surface by the detergent molecules may facilitate the nucleation of gas hydrates at or near such surfaces. The hypothesized mechanism is shown in Figure 23.4.

23.2.2 Carbon dioxide hydrates

Accumulation of CO_2 gas in the atmosphere is one of the key problems that need to be solved in order to counteract the so-called greenhouse effect and associated climatic changes. Among the currently discussed and evaluated solutions is also the storage of CO_2 as gas hydrates in either the deep sea sediments or the injection into methane hydrate deposits provided an accompanying exchange of the gases. Similar to CH_4, CO_2 initiates the formation of gas hydrate structure I [5]. In order to gain a more detailed insight into the relevant mechanisms, suitable monitoring techniques and analysis methods need to be applied. Here, we summarize recently emerging MIR-based measurement techniques for the evaluation and monitoring of CO_2 gas hydrates.

Kumar et al. [7] could successfully apply ATR IR (IR-ATR) spectroscopy to evaluate thin films of CO_2 hydrate. The deconvolved spectrum clearly showed two separate peaks for CO_2 molecules in the small (2347 cm^{-1}) and large (2336 cm^{-1}) cages of the hydrate structure, respectively.

The advancement of an IR-ATR-based technique to analyze hydrates in the bulk is clearly based on the progress in FEFS. This method uses a MIR transparent optical fiber in lieu of a conventional ATR crystal, here, consisting of silver halides (i.e., polycrystalline AgCl/AgBr mixtures) serving as the waveguide and transducer. IR radiation is guided within such fibers via total internal reflection, thereby giving rise to an evanescent field emanating at the interface of the core-only waveguide with the surrounding medium. Using this technique, it is possible to analyze bulk gas hydrates in situ within the penetration depth of the evanescent field. This measurement principle has recently been adopted for studying CO_2 hydrates [8].

Figure 23.4 depicts a CO_2 hydrate measurement setup, which consists of a high-pressure autoclave surrounded by the optical components necessary to guide the IR radiation from the light source (i.e., FTIR spectrometer), to the detector. In order to guide the fiber through the pressure cell, feed-through ports equipped with custom-made polytetrafluorethylene chucks were implemented. CO_2 gas hydrate was formed inside the cell using appropriate additives (here, THF and SDS), and the obtained spectra were correlated with the recorded pressure and temperature traces. The absorption features have been evaluated via integration of the relevant peak areas or through application of appropriate peak fitting routines. Dissociation of the hydrate was induced via slowly reducing the pressure within the cell [8].

The evaluation of the absorption bands via IR-FEFS, however, adds the additional perspective of also sensing additives and their unique absorption signals for providing further insight into the molecular processes involved during gas hydrate formation and dissociation due to the increased sensitivity compared to conventional ATR methods. Furthermore, this monitoring technique can be of assistance in regard to the exchange reaction between CH_4 and CO_2 hydrates.

23.3 IR SENSORS FOR MINERAL EXPLOITATION

Ever since the first studies using IR spectroscopy on minerals in the middle of the 20th century, IR spectroscopy has been established as a routine method in the field of mineral research. The utility of IR spectroscopy in mineral research ranges from (1) identifying minerals and deriving information regarding composition, bonding, and structure, irrespective of whether the material is crystalline or amorphous, to (2) obtaining qualitative and quantitative information on complex mineral mixtures and rocks, and (3) remote sensing studies for mineral exploration [9]. This section will provide a brief summary on the potential of IR techniques offered to geosciences in the field of mineral exploitation.

23.3.1 Oil shale characterization via IR-based methods

The utility of IR spectroscopy in mineral research becomes apparent when studying complex mineral matrices such as shales, and in particular oil shales. Oil shales are a type of sedimentary rock rich in kerogen. Kerogen, frequently also referred to as insoluble organic matter (OM), is derived from decomposed plants or animals and is finely dispersed within the shale rock. It is considered the most abundant form of OM on earth, and therefore plays an increasingly important role as unconventional fossil energy source. Crude oil may be exploited from oil shales, as hydrocarbons are produced upon heating of the kerogen either within the earth crust or via pyrolysis in a facility. Natural gas may also occur as shale gas, which is trapped within the shale rock or likewise may be produced upon heating.

Apart from the organic content, oil shales are mainly comprised of clay minerals (i.e., illite, smectite, kaolinite, and chlorite), quartz, feldspars (i.e., albite and orthoclase), and carbonates (i.e., calcite, dolomite, and siderite) in varying particle sizes, thus rendering oil shales a demanding sample for characterization. Both the mineral and organic components show distinct vibrational features in the MIR spectral range, e.g., the hydroxyl vibrations of the clay minerals at ~3600 cm^{-1}, the aliphatic vibrations at ~2900 cm^{-1}, or the carbonate feature at 1400 cm^{-1}. However, oil shales are frequently complex mixtures, and several major absorption features overlap in the region at 1200–400 cm^{-1}. Hence, an unambiguous assignment of the silicate minerals often proves difficult.

Due to the cumbersome sample preparation and issues arising from overlapping IR signatures, geoscientists usually resort to more established techniques such as X-ray diffraction for mineralogy and Rock-Eval analysis for kerogen characterization, respectively.

However, with the increased adoption of multivariate data analysis techniques, including principal components analysis (PCA) or partial least squares regression in combination with advanced IR techniques that require less sample preparation (e.g., ATR spectroscopy or photoacoustic IR spectroscopy), both the experimental effort and the data analysis could be simplified and improved. This applies to both the mineral composition of the shale, as presented by Müller et al. [10] among others, and kerogen characterization. These contributions have paved the way for advancing the usage of IR spectroscopy in oil shale research, especially if one considers that IR spectrometers may readily deploy in the field, thus offering the opportunity to characterize samples on location.

23.3.2 IR sensors in the field: From laboratory-based systems to space-borne hyperspectral imaging for mineral exploration

Remote sensing offers the opportunity of surveying large areas of the earth surface from aircrafts or satellites. Geoscientists have been interested in remote sensing ever since these techniques have emerged using the acquired images and associated data for mapping, studying geological settings, or prospecting sites of interest. As most mineral deposits are not close to the surface and the penetration depth of most sensor systems is only a few centimeters, an immediate and direct detection of exploitable deposits is not possible. However, some minerals are associated only with a certain rock type or occur only along certain geological formations. Hence, remote sensors may offer that type surface information, and with the correct interpretation of the geological background, relevant mineral deposits may be located.

Most remote sensors rely on the detection of radiation that is reflected by the earth surface and are therefore spectrally limited by the transmissivity of the atmosphere. Water, ozone, and carbon dioxide as the main constituents absorb radiation in the visible-near infrared (VNIR) and MIR regions and thus strongly reduce the wavelength bands available for remote sensing to the so-called atmospheric window. Early satellite-based sensors such as the Multispectral Scanner on the first Landsat satellites were limited to four channels, i.e., small spectral bands within the atmospheric window in the visual and NIR regions (0.5–1.1 μm), whereas newer satellites, e.g., Terra or Landsat 8, are equipped with sensors such as the advanced spaceborne thermal emission and reflection radiometer (ASTER) with up to 15 or even 185 (Hyperspectral Imager Suite) channels, including the MIR.

Sensor systems with higher number of bands are called hyperspectral instead of multispectral sensors, such as the airborne visible IR imaging spectrometer or the spatially enhanced broadband array spectrograph system, which are more commonly used on board of airborne instruments rather than satellites, however, not exclusively. There are several methods to evaluate the gathered information, and the selected approach is largely dependent on the nature of the respective problem of interest. In the context of mineral exploration, the data obtained from the ASTER sensor can be analyzed by different algorithms, i.e., by the calculation of indices or channel ratios, multivariate analysis such as PCA, and powerful shape-fitting algorithms such as mixture-tuned matched filtering (MTMF). An example of an MTMF analysis from an ASTER image is shown in Figure 23.5, where mineral components

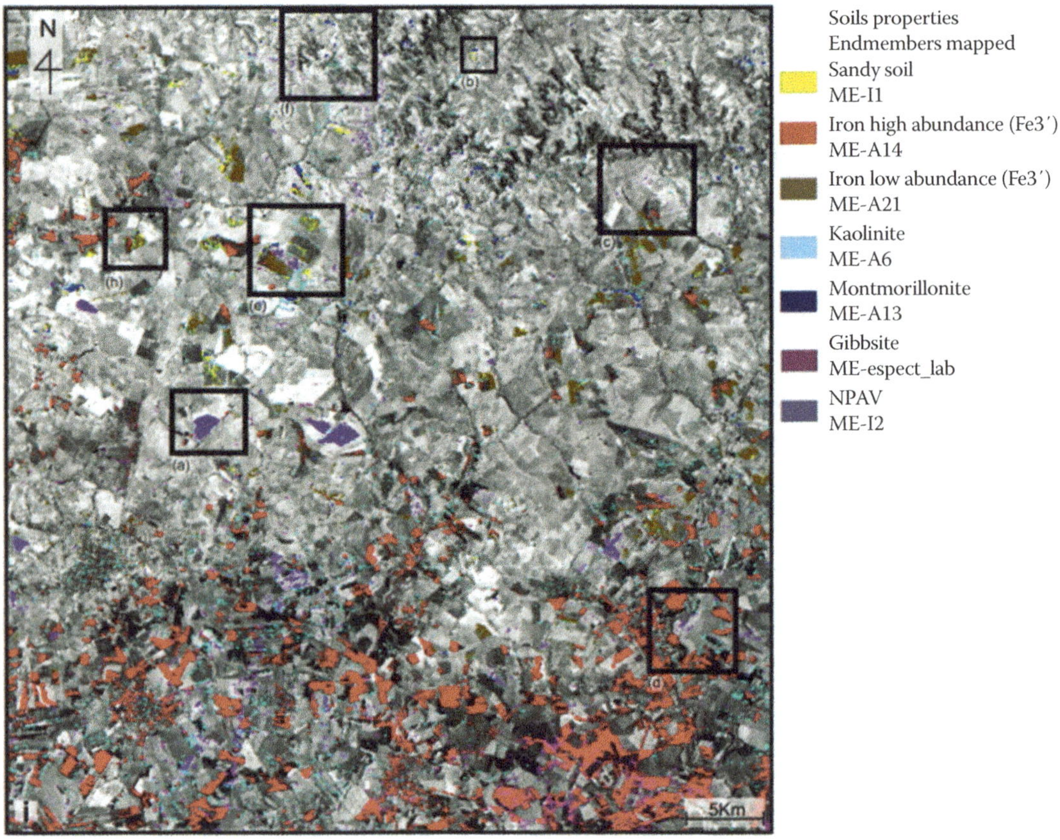

Figure 23.5 Mineral classifications in a vegetated area in Brazil. (Partly adapted from Vicente, E. and de Souza Filho, C. R. *Remote Sens. Environ.*, *115*(8), 1824–1836, 2011.)

of soil were characterized despite dense vegetation covering the surface [11].

23.4 CONCLUSIONS

The presented case studies provide a brief overview on the effectiveness of IR spectroscopy and sensor technologies for mineral, oil, and gas exploitation operating in the MIR (3–15 μm) spectral regime.

Devices introduced herein for monitoring contaminants such as VOCs using MIR spectroscopy provide promising platforms towards robust analytical data recording. Hence, MIR sensors for water and air quality monitoring/control are a rapidly growing field of application that offers a variety of future opportunities to close the gap between laboratory analysis, in-field studies, fingerprinting and quantification.

The described examples of IR spectroscopic studies on different clathrate hydrates reveal the potential of these optical technologies in gas hydrate research, which may play a pivotal role within future energy supply strategies. MIR sensors offer a detailed insight into the kinetics and dynamics of gas hydrate formation and dissociation. The influence of additives such as detergents or additional gases may be studied in molecular detail and will most certainly be expanded into studies at real-world or in situ conditions (e.g., influence of salt ions, bacteria, etc.) during future applications. Because IR spectroscopic sensors can be readily miniaturized, they may be applicable as in situ deep sea analyzers or for spacecraft or remote lander applications for analyzing gas hydrates at the surface of comets or planets.

Last but not least, MIR spectroscopic sensors are also being applied for the identification, quantification, and exploitation of economically valuable rock types such as oil shales and provide important information on the mineralogy of the rock and the included OM. This information along with the capabilities of state-of-the-art remote sensor systems is a powerful tool for the exploration of mineral deposits on a large scale.

In summary, IR spectroscopic sensors have the potential to not only play an important role in future energy and resources production but also monitor and prevent the impacts of mining and other retrieval activities on the aqueous and atmospheric environments.

REFERENCES

1. P. Lambert, A literature review of portable fluorescence-based oil-in-water monitors, *J. Hazard. Mater.*, vol. 102, no. 1, pp. 39–55, 2003.
2. A. Gonzalvez, S. Garrigues, M. de la Guardia, and S. Armenta, The ways to the trace level analysis in infrared spectroscopy, *Anal. Methods*, vol. 3, no. 1, p. 43, 2011.
3. T. Schädle, B. Pejcic, M. Myers, and B. Mizaikoff, Fingerprinting oils in water via their dissolved VOC pattern using mid-infrared sensors, *Anal. Chem.*, vol. 86, no. 19, pp. 9512–9517, 2014.
4. B. Pejcic, L. Boyd, M. Myers, A. Ross, Y. Raichlin, A. Katzir, R. Lu, and B. Mizaikoff, Direct quantification of aromatic hydrocarbons in geochemical fluids with a mid-infrared attenuated total reflection sensor, *Org. Geochem.*, vol. 55, pp. 63–71, 2013.
5. E. D. Sloan and C. A. Koh, *Clathrate Hydrates of Natural Gases*, Third Ed., CRC Press, Boca Raton, FL, 2007.
6. Y. Luzinova, G. T. Dobbs, Y. Raichlin, A. Katzir, and B. Mizaikoff, Infrared spectroscopic monitoring of surface effects during gas hydrate formation in the presence of detergents, *Chem. Eng. Sci.*, vol. 66, no. 22, pp. 5497–5503, 2011.
7. R. Kumar, S. Lang, P. Englezos, and J. Ripmeester, Application of the ATR-IR spectroscopic technique to the characterization of hydrates formed by CO_2, CO_2/H_2 and $CO_2/H_2/C_3H_8$, *J. Phys. Chem. A*, vol. 113, no. 22, pp. 6308–6313, 2009.
8. M. Schwenk, Y. Raichlin, A. Katzir, and B. Mizaikoff, In-situ monitoring of CO_2 gas hydrate formation via mid-infrared fiber optic evanescent filed absorption spectroscopy, The 8th International Conference on Gas Hydrates (ICGH8-2014) Beijing, China, 28 July–1 August, 2014.
9. F. D. van der Meer, H. M. A. van der Werff, F. J. A. van Ruitenbeek, C. A. Hecker, W. H. Bakker, M. F. Noomen, M. van der Meijde, E. J. M. Carranza, J. B. de Smeth, and T. Woldai, Multi-and hyperspectral geologic remote sensing: A review, *Int. J. Appl. Earth Obs. Geoinf.*, vol. 14, no. 1, pp. 112–128, 2012.

10. C. M. Müller, B. Pejcic, L. Esteban, C. D. Piane, M. Raven, and B. Mizaikoff, Infrared attenuated total reflectance spectroscopy: An innovative strategy for analyzing mineral components in energy relevant systems, *Sci. Rep.*, vol. 4, p. 6764, 2014.

11. L. E. Vicente and C. R. de Souza Filho, Identification of mineral components in tropical soils using reflectance spectroscopy and advanced spaceborne thermal emission and reflection radiometer (ASTER) data, *Remote Sens. Environ.*, vol. 115, no. 8, pp. 1824–1836, 2011.

PART VIII

Applications in energy generation and distribution

This section presents applications in the field of energy generation and distribution.

It specifically excludes instrumentation for the extraction of oil and gas, which was dealt with in the previous Part 7. As with many of the other chapters, solar energy generation plays a big and rapidly increasing part in energy generation. However, in this area, it is far more important and even significant enough to qualify as a disruptive technology. Although having been known for many years, it is currently crossing a major threshold. The scenario is changing from being an expensive technology, only being used because of subsidies, to becoming a fully economic solution to many of our future energy needs. It is said that sufficient solar energy strikes the Earth in a few hours to be able to satisfy all the world's annual energy requirements.

It has also been proposed to store electrical energy by using it to pressurize (or even liquefy) air and then use the high-pressure air to regenerate the electrical energy in a gas turbine.

In addition, optical sensors are involved in many more traditional means of energy generation, with the particular advantage of having nonelectrically conducting optical fiber leads. As before, we will now tabulate the main application areas.

Before going on to case studies, it is necessary to give a cross-reference back to the chapter on sensors for gas turbine engines from the industrial section. Similar gas turbines to these are also used extensively in the generation of electrical power from pressurised steam or burning gas or oil.

Table VIII.1 Summary of applications of optoelectronics in energy generation

Application	Technology	Advantages	Disadvantages	Current situation (at time of writing)	More reading
Solar energy generation.	Conversion of solar energy (mainly visible and NIR light components) to electricity, using photovoltaic cells. Technologies include the more common poly-and monocrystalline silicon, plus many advanced thin-film technologies, which potentially save cost by using less active material. Other new materials are being extensively researched, including new high-efficiency inorganic materials and low-cost polymer ones. Electrical inverters are used to convert DC to mains-compatible AC power, and also feed excess power into the national grids.	No moving parts Long life and low maintenance Can be used in remote locations with no infrastructure Well-suited to microgeneration, e.g., on roofs of private houses of small business premises. Can also be used in large field-mounted arrays ("solar farms") for high-power utility generation.	Conversion efficiency of commercial panels is still low (circa 20% max, at time of writing) Still higher cost than conventional methods. Generation is only possible in daylight hours, and power is also very low if cloudy.	Currently requires subsidies to be viable, except in hot countries or in remote areas, where no existing power grid. However, solar is rapidly becoming more and more viable. Costs are reducing rapidly with scale, and more systems are becoming economical without subsidy (vis. they are reaching "grid parity"). The factors which will make it the most attractive future energy source are panel cost reductions and the availability of viable technologies for large scale energy storage.	See this section, and also Volume II, Chapter 16 (Optical to Electrical Energy Conversion).

(*Continued*)

Table VIII.1 (*Continued*) Summary of applications of optoelectronics in energy generation

Application	Technology	Advantages	Disadvantages	Current situation (at time of writing)	More reading
Optoelectronics in wind energy systems. (In sensors for structural integrity and also for flashing safety lights on wind-turbine towers).	Fiber-grating strain sensors in tower structure and in the wind turbine blades Optical tachometers to measure rotation rate. High-intensity flashing LED strobe lights to warn aircraft of presence of tower.	In-fiber gratings form excellent miniature sensors for embedding in glass or carbon fibre composites. They are relatively immune to lightning strike and create less stress concentration points than embedded electrical gauges. LEDs are more efficient and reliable for warning strobe lamps	More expensive than traditional electrical strain sensors.	Grating strain sensors are used in many large wind turbines. They are also potentially very suitable for new-generation submarine turbines and wave-energy generators, as they are nonelectrical and immune to salt-water corrosion. LED strobe lamps for towers are now the industry standard, replacing gas-filled flash lamps.	See Volume II, Chapter 11 (Optical Fiber Sensors) and Volume 1, Chapter 10 (LEDs).
Optoelectronic monitoring of oil, gas, and steam powered turbine generators.	Optical sensors for gas turbine engines and associated electrical generators. These include simple cameras and a diverse range of optical fiber sensors for strain, temperature, and pressure.	Fibre sensors can be made to withstand the very high temperatures in gas turbine engines. In-fiber gratings form excellent miniature sensors for embedding in high-voltage electrical machines.	More expensive than traditional electrical strain sensors, but in many areas the latter cannot be used.	Becoming more accepted in these difficult environments.	See Volume II, Chapter 11 (Optical Fiber Sensors). More discussion on turbine sensors in the industrial section.

(*Continued*)

Table VIII.1 (*Continued*) Summary of applications of optoelectronics in energy generation

Application	Technology	Advantages	Disadvantages	Current situation (at time of writing)	More reading
Optoelectronic monitors for electrical transmission lines.	Simple cameras, to monitor line integrity. Fiber optic current sensors in electrical distribution components. Distributed optical fibre temperature sensors, to check for hot-spots in high power cables.	Cameras are a convenient way to monitor line integrity. Fiber optic sensors can measure current in high voltage lines, because of nonelectrical nature of the fibre cable. Optical fibre current sensors are potentially a lot cheaper than high voltage current transformers. Distributed sensors are well suited to detecting hot-spots in a long power cable, without risk of missing such a fault.	Optical fiber current sensors have required a very high capital investment to develop. The main problems were in developing fiber cables with sufficient insensitivity to vibration and temperature.	Cameras are widely used in many areas. Optical fiber current sensors are finally becoming an economic solution to the problem of measuring current in high voltage lines.	See Volume II, Chapter 11 (Optical Fiber Sensors).
Infrared cameras and optical pyrometers for monitoring surface temperatures.	Infrared cameras can measure temperature of plant, to get an instant view of overheating. Emission of infrared light is used in industrial pyrometers for surface temperature measurement.	Pyrometers are valuable noncontact temperature measuring devices, which can operate from many meters away Very useful for measuring temperatures in electrical machines and gas turbine rotating parts.	The signal, varies with the emissivity of surfaces, although this error can be reduced if two-wavelength readings are taken.	Pyrometers can measure overheating parts in machines, overhead lines, etc.	See Volume I, Chapter 4 (Detection of Optical Radiation).

(*Continued*)

Table VIII.1 (*Continued*) Summary of applications of optoelectronics in energy generation

Application	Technology	Advantages	Disadvantages	Current situation (at time of writing)	More reading
Monitoring of combustion processes in boilers for traditional coal, oil, and gas generation plant.	Optical combustion monitors, measuring the emitted spectrum of flames Optical gas sensors Optical pollution sensors, measuring undesirable gas and smoke emissions with LIDAR and DIAL laser systems	Spectrometers can monitor flames to set optimum combustion conditions. Gas sensors can detect dangerous leaks of CH_4 fuel gas and toxic gases such as CO. LIDAR sensors can detect air quality near the plant	Discrete optical gas sensors are still not as cheap as simple electrical types (e.g., pellistor, chemfet),	Many of these optical sensors (except for LIDAR and DIAL) are in direct competition with electrical sensors, but are becoming more common as component availability and performance improves and cost reduces.	See Volume II, Chapter 11 (Optical Fiber Sensors), Volume II, Chapter 12 (Remote Optical Sensing by Laser), and industrial section.

24

Applications of electricity generation by solar panels

FERNANDO ARAÚJO DE CASTRO
National Physical Laboratory

24.1 INTRODUCTION

The sun is the most abundant energy resource that we have available and delivers about 885 million terawatt hours (TWh) per year to the surface of the Earth. This is more than 5600 times the world's total energy consumption in 2012, according to the International Energy Agency. One very convenient way of harnessing this energy, with low maintenance and no moving parts, is via the photovoltaic (PV) process, where a semiconductor material is used to directly convert light into electricity.

PV solar energy has been used since the 1970s for terrestrial applications but, until recently, it has represented a very small part of the energy mix. The introduction of generous government "green" incentives in various countries from the mid-1990s has led to an impressive market growth, with an average growth rate of cumulative-installed capacity of 49% per annum, from 2003 (2.6 GW) to 2013 (139 GW), to reach an amount capable of generating at least 160 TWh of electricity per year in 2013.

PV schemes now generate 3% of the energy demand in Europe, and even very conservative growth scenarios from the European Photovoltaic Industry Association indicate that this could reach 10% by 2030. The cost of PV electricity (levelized cost of electricity) has reached grid parity in many countries, and at the time of writing, PV development is changing from a policy-driven scenario (dependent on government subsidies) to a market-driven economy. By the time this chapter is read, this will probably already be the case in many countries of the world.

Recent years also have seen a broadening of the PV market growth, which was previously concentrated in Europe, despite having not the best insolation conditions, except in southern European lands. Asia and North America are quickly increasing the installed capacity, and it is expected that the Sunbelt countries will soon follow as PV technology prices decrease. Indeed countries such as Brazil and India have announced large investments and changes in energy market regulations to boost their internal PV markets (Figure 24.1).

Chapter 16 "Optical to Electrical Energy Conversion: Solar" in Volume II describes many aspects of the enabling technology for PV panels. This chapter will describe the main application areas of the technology. The use of PV energy can be separated into two main areas: stand-alone and grid

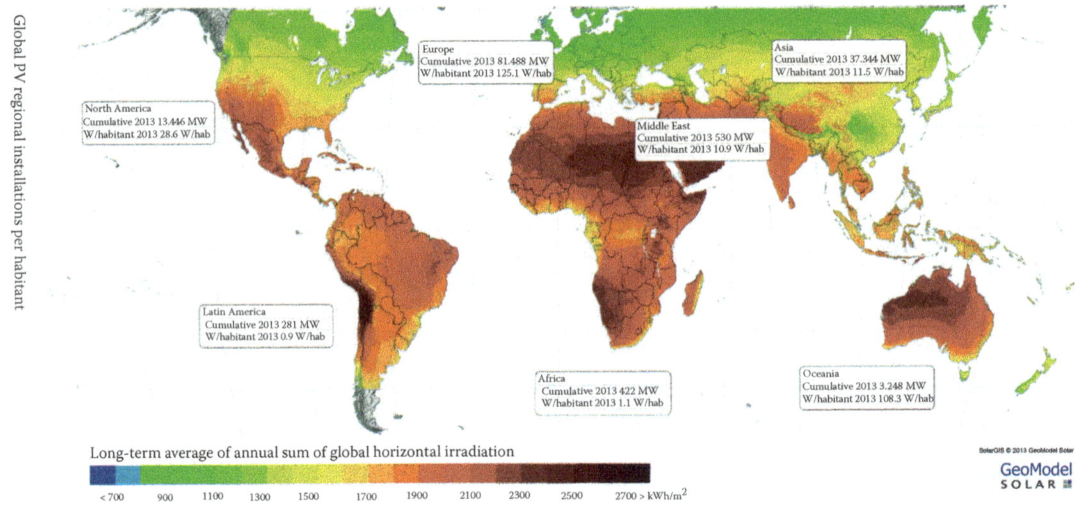

Figure 24.1 World map indicating the annual sum of global irradiation and accumulated installed PV capacity in 2013 in different regions. (Reproduced with permission from "Global Market Outlook for Photovoltaics 2014–2018", EPIA, June 2014.)

connected. We will show examples of the two cases separately and discuss general requirements. Some account of the residential use of PV will, for completeness, be repeated in Chapter 30 "Applications for home and mobile portable equipment."

24.2 STAND-ALONE SYSTEMS (OFF-GRID)

One of the beauties of solar power is the ability to be decentralized, i.e., installed in locations where it is economically and socially desired. This is particularly important in large countries with sparse population and in isolated areas where the cost of building and maintaining a large energy distribution network is too high. For instance, in many developing countries, poor distribution infrastructure means that people from small villages need to travel to larger towns or cities, even to charge up their mobile phones. In East Africa, solar pods are being used to provide cost-competitive energy for household use and to recharge mobile phone batteries. This has a direct impact on the local economy and at the same time reduces environmental pollution by offsetting the use of diesel generators. The solar pods use wireless control, and in-built performance monitoring allows companies to remotely check and assist customers.

Decentralized facilities are also becoming widely used in the industrialized world to provide power to modern urban devices. Figure 24.2 shows a rubbish bin that uses PV cells to provide energy, which is stored to then compact the rubbish at specific intervals of time. The energy can also be used to power sensors and wireless communication systems to inform waste collection managers when the bin is full. This allows one so-called Bigbelly bin to collect five times more rubbish than a conventional one of the same size, drastically reducing the need for collection trucks, which has a direct positive impact on carbon emissions, fuel use, and traffic congestion. It is only recently that these indirect impacts of renewable decentralized energy are starting to be taken into account when calculating the levelized cost of electricity. Taking full life cycle costs and factors such as these into consideration will certainly increase the attractiveness of investments in PV technology.

Other examples of stand-alone applications include powering of LED street lighting, traffic systems (cameras, signal, and information displays), security surveillance cameras, remote weather monitoring stations, seismic monitoring stations, offshore oil platforms, pipeline flow monitoring points, powering of remote wireless sensors and communication points, and container management.

In some cases, these stand-alone applications also need to be portable, for example, for camping and caravanning, for traffic lights at road work sites and for disaster recovery situations. One of the earliest reports of portable PV systems used

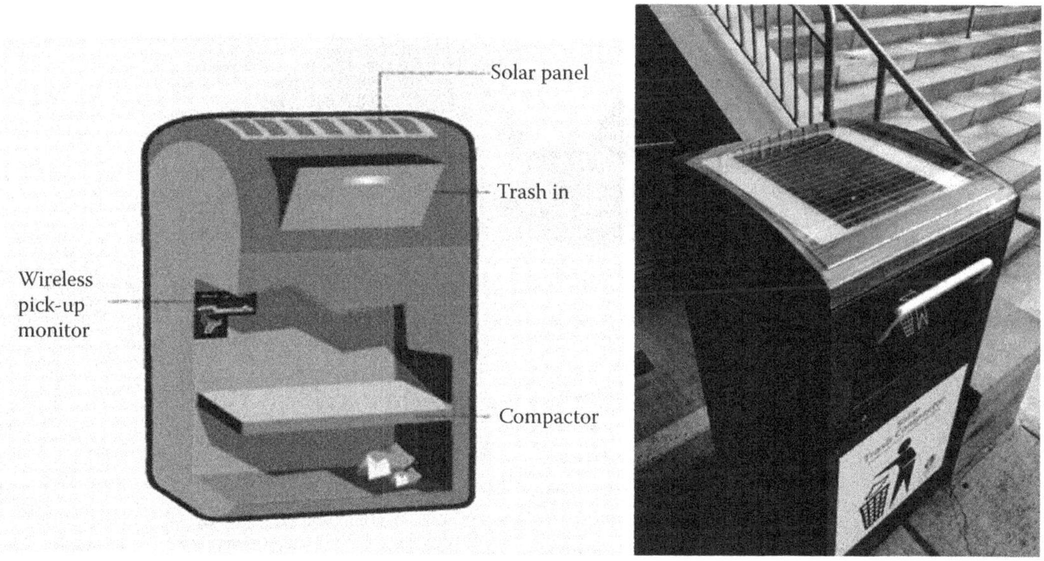

Figure 24.2 Solar-powered bin can compact rubbish and power wireless communication. (Image Courtesy of Bigbelly, Needham, MA.)

during a disaster was in 1998 when Hurricane Hugo struck first the Virgin Islands and then the coast of South Carolina. A trailer-mounted PV system was deployed to, with the aid of storage batteries, power a law enforcement traffic facility and an orphanage 24 h a day for several weeks until utility power was restored. The system supplied 12 V (direct current) and 115/220 V (alternating current) of electrical power, via inverters, from a 2.64 kW peak PV array. Since then, many local governments have invested in mobile PV systems to substitute for, or complement, the use of conventional gasoline or diesel generators for disaster relief. The problems with the latter include high fuel costs, safety, and availability of fuel, making PV generators a better alternative (Figure 24.3).

In full sunlight, these stand-alone PV systems provide typically 100 W to 4 kW of power (but this can reach 15 kW or more in specific cases) and are connected to an electrical power panel or to a device through extension cords. When the

Figure 24.3 Example of deployment of stand-alone PV system for disaster relief situation. (CSM Courtesy of Ecosphere Technologies, Inc., Stuart, FL.)

Figure 24.4 Examples of flexible and lightweight PV systems. (PowerShade military tent image courtesy of PowerFilm Inc. ITO-free organic solar cell module on flexible substrate. ©Fraunhofer ISE, Freiburg, Germany.)

emergency is over, the system is disconnected and redeployed. They can be airlifted, but the most common trend is to mount the solar systems on trailers for road transport. The energy can be used directly or stored in batteries for later use in places such as camp sites, social events, construction sites, clinics, shelters, gasoline stations, and businesses. The maximum power that these PV systems can deliver is limited due to the restrictions of weight and physical size of the PV system (including PV panels, batteries, and electronics) associated with the transportation units. More lightweight products would reduce transportation costs and simplify deployment. Numerous thin film and novel PV technologies promise flexible and lightweight PV panels (see Figure 24.4). They could play an important role in this market if they can demonstrate the same level of performance and reliability of conventional rigid crystalline silicon PV modules.

24.3 GRID-CONNECTED PV SYSTEMS

As the name suggests, grid-connected PV systems are those when the PV generation acts as an energy source that the grid can then distribute to customers. These systems can be large and centralized, where a large PV power station feeds electricity to high-voltage transmission lines. Depending on the size, these then bring energy to power stations or substations and then finally to consumers. In line with the decrease in PV prices, the size of PV power plants has been increasing, and this volume of course leads to yet lower prices.

At the time of writing, the world's largest PV power station is Topaz Solar Farm in the San Luiz Obispo County, California (see Figure 24.5). In contrast to most PV farms that use Si PV, Topaz uses thin-film CdTe technology and with nine million modules, has 550 MW capability, which is enough to power 160,000 average homes in the United States in 2015. It is expected that these will soon be dwarfed by many much larger sites, although the ease of location for PV, and the high cost of distribution, leads to a compromise, which might lead to an optimum size for such sites.

India is currently building the largest solar park complex in the world, Gujarat Solar Park. In 2015, the Gujarat complex has a capacity to generate 500 MW of power and is estimated to save eight million tons of CO_2 emissions per year. At the time of writing, additional power stations are being installed, and, when complete, the complex will have a capacity of more than 850 MW.

In between large-scale power plants and localized, small-sized PV applications, there has been an increase in new concepts of medium-sized solar installations. One such concept is that of solar islands, which, despite the name, can also be built on land. The concept is of building multiple solar panels in a large conglomerate that can be rotated to track the sun using a single tracking system (see Figure 24.6). This conglomerate is built over some water, which naturally provides cooling underneath the panels to keep them operating at optimum conditions. At the time of writing, different versions of floating PV plants (with or without a tracking system) had been announced in many countries, such, as the United Kingdom, Italy, France, Australia, India, and Japan. In hot countries, the solar panels not only produce electricity but can also help to reduce water evaporation by 90% in the covered area. In South Australia,

Figure 24.5 Aerial photo of Topaz Solar Farm, one of the largest PV solar plants in operation in 2015. (Photo Courtesy of BHE Renewables, Des Moines, IA.)

evaporation of uncovered water reservoirs can cause an annual loss corresponding to a reduction of about 2.5 m of water depth!

Another concept that regained substantial interest is that of building-integrated PVs. The novelty has been the change in focus from discrete roof-mounted PV systems fixed to preexisting buildings to new fully integrated systems where the PV device becomes part of the building structure. Making the PV module part of the building envelope (e.g., tiles, windows, walls) contributes to the energy efficiency of the buildings but can also become more aesthetically appealing to the general public than bolted-on panels. The design of tall buildings with integrated PV systems is a challenge for engineers. Shadows from other buildings on building facades, chimneys, etc., depending on the direction of the sunlight, can lead to strong variations in generated voltage and power during the day and from panel to panel. This requires shaded panels to be bypassed with diodes, as they have high impedance in reverse bias, and also needs care in the design of the following electronic inverter. The solution relies on smart electronic systems, combined with active monitoring.

At the other end of the scale, there has been a very large increase in small add-on rooftop installations, such as on individual houses and commercial buildings. Offices and other places of work, particularly factories, use electricity

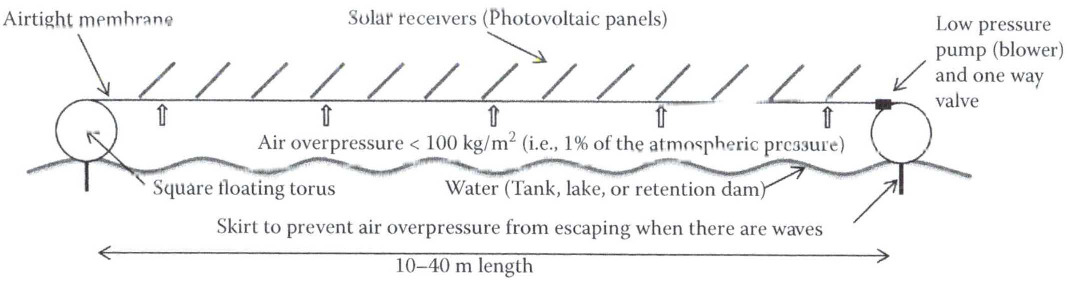

Figure 24.6 Sketch of a floating solar island concept. (Image Courtesy of Novaton, Zürich, Switzerland.)

mostly during the day, thus having a high degree of self-consumption. However, in most residential buildings, there is a mismatch between the period of the day that electricity is produced (e.g., during the day) and the time when it is mostly consumed (e.g., during the evening). Because most households are connected, additional energy that is not used can, with intelligent electronics, be introduced into the grid for others to use. This is advantageous, but it means that, as more and more homes are using PV panels, energy companies and regulators may, at times, have to be able to cope with that increase in supply. Therefore, much effort has been put into researching ways of increasing self-consumption of PV electricity. This problem, which also applies to larger solar farms and wind energy systems, can potentially be resolved by storing energy locally. In the home, only smaller scale energy storage is possible, for example, by heat pumps to store it as heat or to use the electricity to charge batteries. As these additional storage units can be expensive, it is important to assess each project individually to find the more cost-effective solution, which can include the combination of heat pumps and batteries.

24.4 IMPACT ON THE ELECTRIC GRID

The increase of variable PV electricity generation introduces ever-increasing challenges to the management of the electric grid. Knowledge of the available sun energy resource at the location of interest requires long-term measurement of daily temperature, direct beam irradiance, diffuse irradiance from the sky, and irradiance from the ground surface. Irradiance measurements are normally measured using a pyranometer, an instrument that is capable of measuring global solar irradiance. It measures all incoming energy in the hemisphere above the plane of the instrument. It is also possible to measure only the diffuse irradiance by projecting a shadow that blocks the sun's disk only, while the desired diffuse irradiation component is measured. For measurement of direct irradiance, a pyrheliometer can be used. In this case, a thermopile placed at the end of a long tube aimed at the sun allows selective measurement of sunlight radiation with an acceptance angle of about 5°. Measurement of direct irradiance is important for specific applications, such as concentrator PVs, where diffuse irradiance does not generate much energy. For regions with no solar irradiance measurement capabilities, cloud coverage data obtained from local weather stations (and sometimes from satellite data) have also been used to predict solar irradiance levels (Figure 24.7).

With sufficient statistical meteorological data, seasonal and daily variations can be modeled, and the PV plant capacity can be designed accordingly. However, unfortunately, actual sunlight intensity does not follow statistical probability curves, and there are very significant variations in the instantaneous power incident on real panels. Coping with these very large variations in generation (e.g., due to shadows by clouds) is much more challenging. One particular difficulty is that it can be localized to only part of a module (or even of a whole solar farm!), leading to sudden drops in voltage or reversed current polarity through the diodes on individual "dark" cells or groups of cells.

As discussed in Volume II, Chapter 16, a PV module is composed of multiple solar cells that can be connected in series or in parallel to achieve the desired voltage and current output characteristics, providing power in a range from a few watts to hundreds of megawatts. These modules are often subdivided into strings of connected cells, and also use bypass diodes to minimize the effect of localized loss in performance on the overall module output. Even with this precaution, a single underperforming cell will negatively affect the output voltage of the string. More advanced module technology uses power optimizer chips (DC–DC converters) connected directly to smaller elements of the string to allow an increased range of maximum power point and facilitate connection of modules that are exposed to different external conditions. These chips can be combined with intelligent power inverter electronics in the grid to maximize module and array performance. All these features are needed to minimize energy loss and potential damage to the module and to the grid.

There are different possible solutions to tackle the problem of variability in power output, a feature common to other energy generation technologies that rely on intermittent sources (such as wind). For small installations, energy storage (e.g., batteries) during the peak production hours can be used to offset short periods of decreased production. For that to work, fast and intelligent electronics needs

Figure 24.7 Instruments for solar irradiance measurements. Pyranometer with thermal sensor for global irradiance measurements (left top). Two-axis tracked pyrheliometer for direct normal irradiance measurements (left bottom). Pyranometer with shading ball for diffuse irradiance measurements (right). (Reproduced from http://www.volker-quaschning.de; for further information, see Quaschning, V., The sun as an energy source, *Renewable Energy World* 05/2003, pp. 90–93. With permission.)

to be in place to monitor and quickly change the power output of the production plant.

The most obvious storage means for electricity is to charge secondary cells (batteries) and discharge them when needed. Unfortunately, this is still a very expensive option, as lithium batteries are very expensive and the cheaper lead acid batteries have shorter lifetimes, particularly when deeply discharged. Without great care in design of the charging circuitry, they can also easily become overcharged. The voltage can get too high if energy continues to be supplied after battery reaches full charge. This will convert water in the electrolyte to gaseous hydrogen and oxygen, which can only be partially replaced by catalytic conversion, leading, as with discharging too deeply, to faster degradation. Voltage regulators are used to prevent overheating, and control set points can be defined to vary with battery temperature. However, good temperature control of the charger is challenging and relies on temperature probes being positioned in a location where the temperature is as close as possible to that of the battery (ideally directly attached).

In future, the prices of lithium batteries are likely to reduce dramatically due to the expected greatly increased use for electric vehicles, and this also conveniently forms an outlet for excess electricity in the home. It may happen that the family car helps to provide an electricity storage system for the home, if not needed for long journeys the next day! At least one electric vehicle manufacturer has started to market their lithium storage units for home use.

The other form of energy storage in the home is in the use of excess electricity to produce heat, either rather inefficiently to heat the hot water tank or more efficiently with a heat pump to heat the floor of the house, which can form a large storage heater. The latter is, of course, only really useful in winter months.

For energy storage on a larger scale, many more options are available. In hilly areas, the power can be used to pump water to a lake on a hilltop and recover energy later using a hydroelectric plant. The energy can also be used to pump gas into a large high-pressure vessel or underground cavern and recover it with a gas turbine generator, but very large vessels or existing caverns with no gas leaks to the surface are required. Other possibility is to use the power to drive high-efficiency, ultra-low-energy-loss flywheels, storing the energy mechanically, such that the flywheel can later drive an alternator to recover the energy. This is, of course, only a short-term storage method. One of the most recent energy storage ideas appears at first rather bizarre but when examined in detail

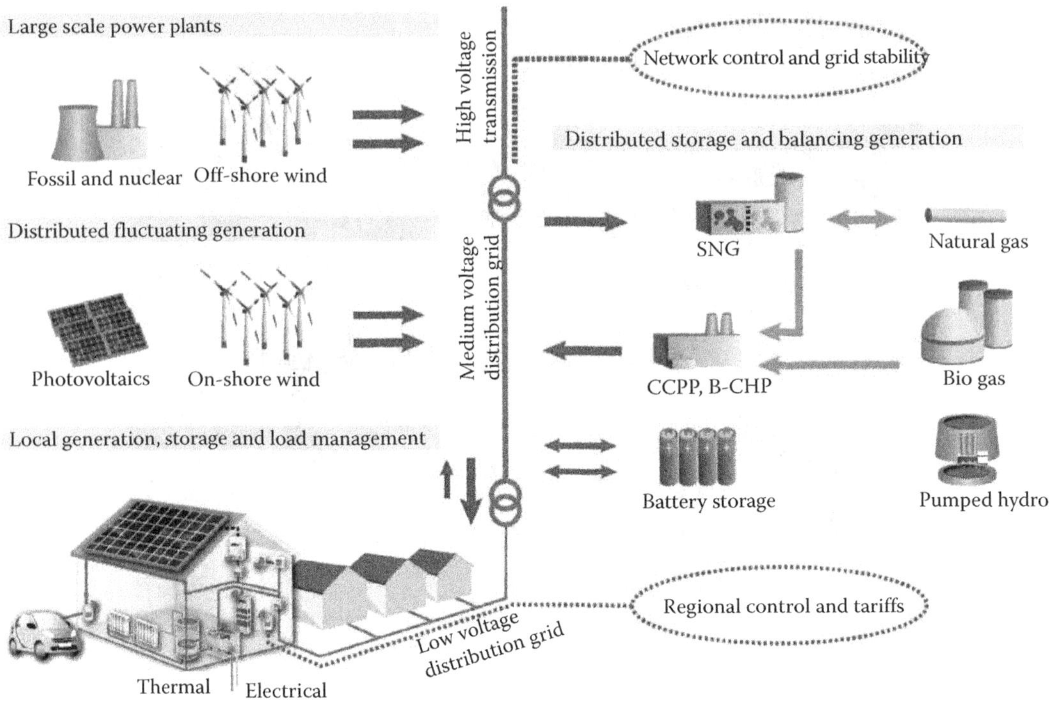

Figure 24.8 Examples of how storage solutions can be used as buffer to balance energy generation and consumption. (Image Copyright ©ZSW, Stuttgart Germany.)

has great engineering attractions. This is to pump air through repetitive nozzles to liquefy it (as is already done commercially to produce liquid air, oxygen and nitrogen), then to store the air as a liquid. Unlike storage of energy as compressed air, this does not need a high-pressure vessel, and the thermal insulation requirements are simpler when scaled up, so liquified air can be stored for very long periods, if desired. Energy is recovered simply by allowing it to boil (it will do this anyway at ambient temperatures of course) and use it to generate electricity with a conventional gas turbine. No novel technology is required, and it appears to become a more attractive engineering concept as it is scaled up in size.

Figure 24.8 shows how the use of storage devices as a buffer can increase self-consumption of generated PV electricity and reduce the required electricity from the grid, throughout the year*.

* For more information see J. Binder, C.O. Williams, T. Kelm, Increasing PV self-consumption, domestic energy autonomy and grid compatibility of PV systems using heat-pumps, thermal storage and battery storage, *EU PVSEC Proceedings*, pp. 4030–4034 (2012).

Another possibility is to use a decentralized power plant system, where power plants are built at different sites but act as a single power plant source. Studies have shown that such system would drastically reduce the effect of unexpected supply conditions as clouds in one location will often not affect the others. It is clearly more attractive when large-scale energy storage systems are available. Despite the requirement for transmission lines between sites, studies indicate that estimated additional costs beyond those of conventional systems would range from \$0.003 to \$0.03/kWh. The more cost-effective option will certainly be region dependent and will need to take into account the impact on the grid structure. For instance, when more energy is fed into the power grid than is removed from it, the grid frequency can increase and render it unstable. To ensure network security, the German's System Stability Act of May 2012 scheduled retrofit of PV systems to include power inverters that are able to reduce output when frequency rises too high or to turn themselves off smoothly. Similar measures were taken in other countries. It is, therefore, clear that high levels of electronic control will be required in future PV systems and that the accuracy of power measurement and timing will be crucial for consumer and investor trust.

Currently, PV electricity is sold in units of actual generated energy (in kWh), but the PV system price is based on *nominal* kWp, which is the peak power generated by the solar module under standard test conditions (as defined in IEC 60904), adjusted by the "performance ratio" of the system. The adjustment is based on models that estimate efficiency losses using available data to predict the generated system output in a specific location. Standard test conditions of 25°C, 1000 W/m2 AM1.5G direct irradiance at normal angle of incidence are rarely encountered in the real world and do not take into account a number of environmental parameters that affect the performance of a PV system in real operation, such as spectral changes, the dependence of solar cell efficiency on incidence angle, low light behavior of the solar device, and the behavior at realistic temperatures. This, combined with the fact that different PV technologies may present different dependencies on the parameters listed previously, indicates that peak efficiency is not a very meaningful parameter when consumers and investors (including governments) need to make informed decisions about where PV power plants should be built and which type of PV technology should be used. The Technical Committee 82 of the International Electrotechnical Commission is in the process of developing an energy-rating standard (IEC 61853), which would be an energy-based metric for PV.

24.5 CONCLUSIONS

Electricity generation by solar energy has become a reality both for stand-alone applications and for grid-connected energy supply, but, at the time of writing, is currently being held aloft by generous subsidies. The future world energy mix will be diverse. Solar, because of its low maintenance requirement, unobtrusive nature, and ever-decreasing cost, is particularly well positioned to play an increasing role. As with many alternative energy schemes, it poses challenges for engineers to integrate intermittent energy sources with the grid, which will require intelligent power electronics, intelligent system design, and, as energy production by this means increases, a viable large-scale method of storing the energy. Perhaps, more than all other technologies, it also brings the real possibility of a future of energy security (abundant energy supply from the sun) with less human impact on our environment.

25

Advantages of fiber optical sensors for power generation

RALF D. PECHSTEDT AND DAVID HEMSLEY
Oxsensis Ltd.

25.1 ADVANTAGES OF FOS FOR POWER GENERATION

Gas turbine engines have been widely used in electrical power generation for decades. More recently, there has been an increasing interest in optimizing gas turbine design and operation to improve fuel efficiency and to reduce pollution together with enhanced fuel flexibility under greater variable loads and ramp rate conditions. The combustion concepts addressing these demands are generally prone to generating instabilities in the gas turbine compressor and combustion systems [1]. As a result, events such as surge and stall could occur with potentially catastrophic damage to the gas turbine plant, risk to lives, and severe disruption of electricity delivery capacity.

Fiber optical sensor (FOS) technology is now becoming available that is capable of providing critical real-time information regarding the onset of combustion instabilities in the hitherto inaccessible region of the gas turbine combustor. A key advantage of FOS for gas turbine applications is therefore their ability to operate at elevated temperatures [2,3]. This allows, for example, a more direct dynamic pressure measurement in close vicinity to the combustion process. Figure 25.1a is a schematic of a gas turbine, indicating possible locations of optical pressure sensors. A picture of a gas turbine driving an electrical generator is shown in Figure 25.1b.

FOS could replace piezoelectric transducers that are commonly mounted via a "semi-infinite" tube that attenuates higher frequencies and thus limits its potential to detect the onset of combustion instabilities.

Figure 25.2a shows a ruggedized fiber optical pressure sensor together with a fiber optical connector box loaded onto a transportation trolley. The sensor head was designed to fit a monitoring port in the combustor of a gas turbine of an industrial power generating plant. A picture of the installed sensor is shown in Figure 25.2b. Also visible in the foreground is the semi-infinite tube arrangement used for mounting conventional pressure sensors.

At ultra-high temperatures, thermocouples suffer from increased drift and reliability problems. Sapphire is being investigated as a base material for optical sensors capable of measuring temperatures up to ~1500°C and beyond [4], and the first successful prototypes are emerging from laboratories. These sensors have the potential for direct combustor temperature measurements, the knowledge of which is critical for design and operation of more efficient gas turbines.

A unique advantage of fiber Bragg grating-based FOS systems is their ability to measure at multiple

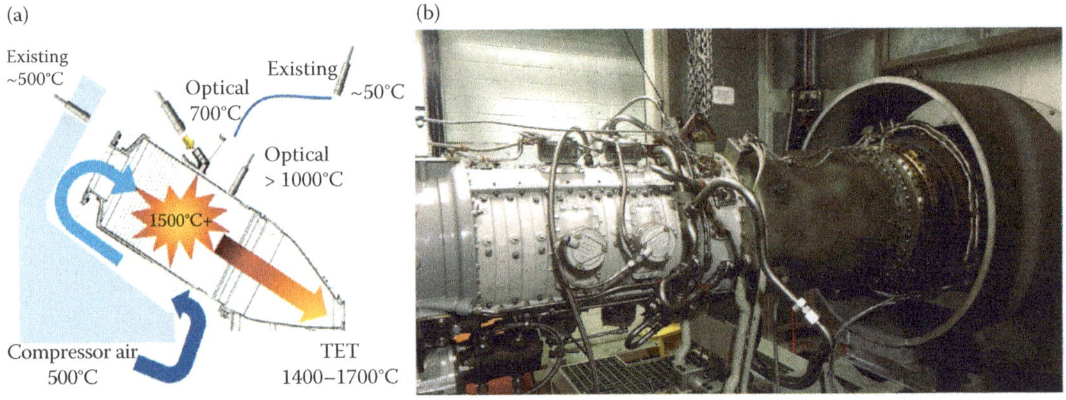

Figure 25.1 (a) Possible locations of pressure sensors. (b) Siemens 501-KB7S 5 MW aero-derivative gas turbine packaged by Centrax Ltd., Newton Abbot, UK.

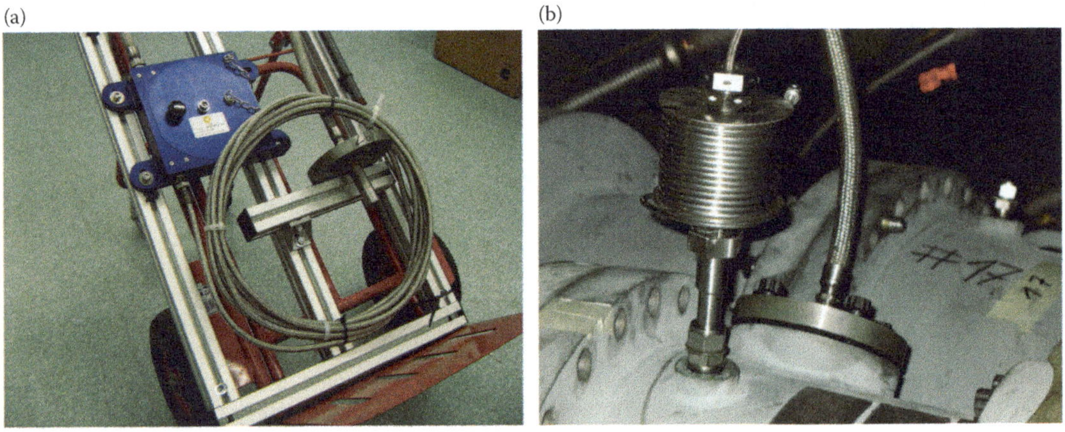

Figure 25.2 (a) Ruggedized fiber optical pressure sensor. (b) Installed optical sensor and semi-infinite tube.

discrete positions along a single fiber strand. This enables, for instance, densely spaced gas turbine exhaust gas temperature measurements [5] that would be difficult to implement using conventional thermocouples due to the significant amount of electrical wiring required. In addition, FOS avoid problems related to electromagnetic pickup and temperature gradients along the electrical wires carrying the thermocouple signals, thanks to their immunity to electromagnetic interference and the dielectric nature of the fiber optical cable.

For a more detailed description, we refer to Chapter 18.

REFERENCES

1. Lieuwen, T., McManus, K., That elusive hum, *Mech. Eng.*, Vol. 124(6), 53–55 (2002).
2. Mihailov, S.J., Fiber Bragg grating sensors for harsh environments, *Sensors*, Vol. 12, 1898 (2012).
3. Maillaud, F.F.M., Pechstedt, R.D., High-accuracy optical pressure sensor for gas turbine monitoring, *Proc SPIE*, Vol. 8421, 8421AF (2012).
4. Busch, M. et al. Inscription and characterization of Bragg gratings in single-crystal sapphire optical fibres for high-temperature sensor applications, *Meas. Sci. Technol.*, Vol. 20, 115301 (2009).
5. Xia, H. et al., High-density fiber optical sensor and instrumentation for gas turbine operation condition monitoring, *J. Sens.*, Vol. 2013, Article ID 206738.

PART IX

Applications for medicine, health monitoring, and biotechnology

Table IX.1 Summary of applications of optoelectronics in medicine, healthmonitoring, and biotechnology

Application	Technology	Advantages	Disadvantages	Current situation (at time of writing)	More reading
External examination of individual body regions, or of the whole body, using cameras.	Visible or infrared cameras to detect signs of illness from optical images. (hot-spot detection and skin coloration)	Simple, large-area, noninvasive technology. Potentially should be more objective than a human observer. Infrared sensors are very good at observing unusual temperatures arising from localized illness or inflammation.	In many cases, humans are still better at visual examination and performing a diagnosis from what they see. Machine vision is still poor at detecting major illnesses such as tumors with low error rate.	Has great potential, as multispectral imaging is possible, and as the associated signal processing algorithms are improved. There should be no reason why the diagnostic capability should not exceed human capability, particularly as a wider spectral range is available.	See Volume II, Chapter 4 (Camera Technology)
External spectroscopic examination of smaller regions through the skin (subcutaneous).	Many types of spectroscopic/colorimetric sensors for oximetry, glucose levels, etc.	Noninvasive.	Usually only measures over small areas.	Sensors for blood oximetry are most developed.	
External laser examination of smaller regions through the skin, or via organs, such as the eye or ear.	Laser Doppler velocimetry to measure blood flow. Optical coherence methods, such as OCT, to resolve in-skin and subcutaneous features.	Noninvasive methods giving very useful information which is difficult to achieve by other means.	Usually only measures over small areas.	Optical coherence methods are becoming a highly valuable tool in medical diagnostics. It can be used to see in-skin and subcutaneous features. In the eye, it can measure and detect many abnormalities.	See this section for OCT

(*Continued*)

Table IX.1 (*Continued*) Summary of applications of optoelectronics in medicine, healthmonitoring, and biotechnology

Application	Technology	Advantages	Disadvantages	Current situation (at time of writing)	More reading
Use of cameras for guidance of robotic surgery tools.	Short-focus camera used with robotic surgery arm, the latter mechanically geared to give higher precision cuts (with laser or knife) than an unaided surgeon could make.	A major development for precision surgery.	None	Very sophisticated (and expensive) commercial devices exist, and are being used in many well-equipped hospitals.	
Fiber optical sensors for temperature, pressure, and strain.	Bragg grating sensors, Fabry Perot sensors and various distributed sensors.	No electrical connections, so much safer using invasive (in-body) sensors. Really tiny sensors (small is "beautiful" in the body!). Bragg grating sensors can be formed into a multisensor array in a single length of fiber. Distributed fiber sensors can measure over extended lengths and distributed Bragg grating sensors could potentially measure at many closely spaced points.	More expensive than traditional electrical sensors. Most distributed sensors require multiple delay loops of fiber to effectively enhance distance resolution.	Still relatively little use, considering the advantages. Bragg grating sensors and Fabry Perot types are probably used the most. The use of multiplexed pressure sensors by Flanders University (this section) may lead to more widespread use of these.	See Volume II, Chapter 11 (Optical Fiber Sensors) and this section.

(*Continued*)

Table IX.1 (*Continued*) Summary of applications of optoelectronics in medicine, healthmonitoring, and biotechnology

Application	Technology	Advantages	Disadvantages	Current situation (at time of writing)	More reading
Optical sensing of oxygen and CO_2	Optical monitoring of sensing layers with active indicator ingredients embedded in gas-permeable polymers. Dissolved oxygen sensors rely on quenching of fluorescent lifetime of reagents such as ruthenium complexes.	Again, no electrical connections when using invasive (in-body) sensors. Really tiny sensors—small is "beautiful" in the body! Dissolved oxygenation sensors using fluorescent dyes do not consume oxygen, so the sensors do not change the environment, and so are less affected by surface-contamination films.	Reaction time relies on diffusion of gas through the polymer. CO_2 sensors are essentially pH sensors, so can be affected by other compounds. Sensors require temperature compensation in some applications.	Commercial dissolved oxygen sensors are manufactured by several companies. They have a big market for bioreactors, again due to not consuming oxygen.	See also: Wolfbeis, OS, *Fiber Optical Chemical Sensors and Biosensors*, CRC Press (1991).
Selective optical detection of metals, ions, and biological molecules (including DNA)	Chemical absorption indicators for ions or molecules. Selective fluorescence, either intrinsic to the biomolecule or via a binding indicator. Selective binding sites to give ultraselective detection	Very many possible methods. Selective binding gives the best fingerprinting. The methods are particularly well-suited for producing very large, close-packed, sensor arrays in small volumes.	Many of the methods need care to prevent contamination of samples. Occupation of binding sites by similar contaminant molecules could be a problem.	Becoming a very important technique for large-detector cell-count sensor arrays.	See also: Wolfbeis, OS, *Fiber Optical Chemical Sensors and Biosensors*, CRC Press (1991).

(*Continued*)

Table IX.1 (*Continued*) Summary of applications of optoelectronics in medicine, healthmonitoring, and biotechnology

Application	Technology	Advantages	Disadvantages	Current situation (at time of writing)	More reading
Photodynamic therapy	Selective bonding of chemical agents to tissue (e.g., staining), such that it is sensitized to high-energy light. This can be used to destroy the undesirable tissue, such as malignant regions of the body, when illuminated with high energy light.	Prospect of treatment of some forms of cancer, etc., without chemotherapy or ionising-radiation treatment.	Need to find suitable selective binding agents, to ensure only unhealthy tissue is affected.	Is finding uses, but, as with many cancer treatments, has yet to advance to be a real "game changer"	
Breath analysis using spectroscopy.	Many diseases or medical conditions give rise to charges in the organic compound composition of human breath. This can most easily be analyzed by absorption spectroscopy, with strongest lines in the Infrared region.	Rapid, noninvasive analysis of breath using spectrometers provides a useful diagnostic tool. Fourier transform IR spectroscopy is the most versatile measurement method.	In some cases, diet may affect the compounds in the breath, but this is not a major problem.	A recent development, having great promise due to the rapid diagnostic capability.	

(*Continued*)

Table IX.1 (*Continued*) Summary of applications of optoelectronics in medicine, healthmonitoring, and biotechnology

Application	Technology	Advantages	Disadvantages	Current situation (at time of writing)	More reading
Spectroscopic sensors for testing of drugs.	Many solid drugs are produced in the form of highly scattering powders or pills, so Raman spectroscopy or fluorescence spectroscopy are very useful tools. For soluble drugs, many other spectroscopic methods are applicable.	Raman spectroscopy is a highly versatile tool for analysis of scattering powders. It can probe for characteristic spectral lines in regions inaccessible to absorption measurements Fluorescence is a far less versatile analytical tool, and also many products do not fluoresce anyway. It can, however, be enhanced using staining of molecules using fluorescent indicators.	Weak signals from Raman spectroscopy. Fluorescence cannot provide such good analytical "fingerprinting".	The viability of Raman spectroscopy has been transformed by the availability of compact high-power semiconductor and solid-state lasers. It is now a far more valuable analytical tool than before.	See Volume II, Chapter 15
Optical coherence tomography.	Location of light reflection points from within tissues based on interfererometric ranging methods. White light interferometry can determine the reflection profile as a function of distance into the tissue.	Particularly useful for ophthalmology, as the lens of the eye is clear and transparent. Very attractive method for analysis of skin and any subcutaneous tissue just below the skin.	Fairly expensive systems.	This is a rapidly developing field of medical diagnostics. Applications are mainly in the ophthalmology and dermatology fields so far.	See Volume II, Chapter 11 (Optical Fiber Sensors) and this section.

(*Continued*)

Table IX.1 (*Continued*) Summary of applications of optoelectronics in medicine, healthmonitoring, and biotechnology

Application	Technology	Advantages	Disadvantages	Current situation (at time of writing)	More reading
Laser surgery	Use of high-power lasers to cut, ablate, or cauterize tissues, and/or to kill undesirable cells, such as malignant ones.	Very precise cutting, and compatible with robotics. Noncontact, so no need to sterilize tools. The heating from the laser can cauterize wounds, so reduces bleeding. Particular organs or parts of organs can be dyed to give selective absorption. The skin is most easily treated, but most parts of the body can be operated on. Possible to correct focal length of cornea lens with lasers, for eyesight correction.	Expensive equipment. Need to control skin penetration at wavelengths where tissue has a lower absorption.	Cost of compact high-power semiconductor lasers (or lasers to pump other solid-state lasers) has fallen dramatically. Laser eyesight correction was initially viewed with skepticism, but is now an established technique.	
3D printing of medical and dental prosthetics.	Use of light projectors and scanned lasers to cure resins to produce 3D objects. In many cases, the basic digital data used to produce the image can also be obtained using optical 3D scanners.	A huge variety of objects of any desired shape (including ones with cavities) can be produced or copied in this manner. Even very strong, hard, and rigid objects can be produced using fillers in polymer or by using lasers to sinter even high-temperature materials. The data necessary for production of parts can be transmitted to anywhere in the world.	The highest precision professional equipment is still expensive, but getting less as the industry expands.	Costs are falling rapidly and the ability of these processes to produce the objects will lead to ever greater use. Equipment to produce simple polymer parts is becoming available for home use.	See https://en.wikipedia.org/wiki/EnvisionTEC as an example of use of the technology.

(*Continued*)

Table IX.1 (*Continued*) Summary of applications of optoelectronics in medicine, healthmonitoring, and biotechnology

Application	Technology	Advantages	Disadvantages	Current situation (at time of writing)	More reading
Human eye-implanted semiconductor image cameras to provide some degree of vision to the blind.	A semiconductor image camera is implanted in the eye and the electrical output is connected to the optic nerve, such that the brain can eventually learn to decipher images from the electrical signals.	Some degree of sight is restored to otherwise totally blind people.	At the time of writing, the quality of sight achieved is still poor. There is a desire to be able to power the sensor for longer periods than currently possible.	Progress in this field is considerable, as both the miniaturization of sensors and their power consumption improves. Research is also aimed at improving the signals from detector to brain, to best allow the brain to learn how to decipher the image from electrical signals.	

26

Medical applications of photonics*

CONSTANTINOS PITRIS
University of Cyprus

TUAN VO-DINH, R. EUGENE GOODSON, AND SUSIE E. GOODSON
Duke University

26.1 INTRODUCTION

Over the past two decades, optic and photonic technologies have become exceedingly ubiquitous in the key enabling devices that are central to many aspects of everyday life. The importance of these paradigm-shifting technologies has so been recognized and appreciated that the United Nations Educational, Scientific and Cultural Organization adopted a resolution declaring 2015 as the International Year of Light. The impact of photonic

* This chapter has been updated by Constantinos Pitris for this Second Edition, using portions of the First Edition version, contributed by Tuan Vo-Dinh.

technologies on medicine and life sciences has also been significant. Light-based innovations have the potential to improve the quality of life and prognosis of patients but also contribute to the decline of the steeply rising cost of health care. Technological advances in the areas of lasers, detectors, and sensors have enabled the introduction and, relatively quick, acceptance of a number of optical technologies into clinical practice. At the same time, the room for further penetration of photonic technologies in medicine, rather than shrinking, seems to be expanding into both established and new areas.

The purpose of this chapter is to introduce scientists and engineers of various disciplines related to photonics to the nature of the interaction of light with biological material and the established and emerging applications of light-based technologies in medicine. Photonics technologies are being used for both diagnostic and therapeutic applications. In the case of diagnosis, this can be performed *in vitro* (i.e., with tissue cultures or samples "in glass"), *ex vitro* (i.e., using samples excised from the human body), or *in vivo* (i.e., in the living human body). *In vitro* techniques have mostly been developed for mass and inexpensive screening of disease. However, *in vivo* photonic-based diagnostics allow a complete investigation of the tissue of interest without the need for biopsies (sample excision and processing) and can be performed in real time. This chapter provides an overview of photonic technologies for *in vitro, ex vivo,* and *in vivo* diagnostic and therapeutic applications. Future prospects are also briefly discussed.

26.2 BASICS OF TISSUE AND ITS INTERACTION WITH LIGHT

In multicellular organisms, the basic building blocks of life are the cells. These membrane-encapsulated units are composed of an abundant number of molecules and molecular complexes, which provide functionality to the cell. The major components of a cell are its membrane, nucleus, cytoplasm, and various types of organelles, which, in a coordinated fashion, orchestrate the processes that maintain life. Cells of similar structure and function are organized into tissues, which serve a specific common goal. Several groups of tissue are organized into organs, and organs are combined into systems, each serving a particular function. Just like any other collection of organic materials, organized in a complex and intricate structure, tissues exhibit common radiation–material interactions. Reflection and refraction occur at the interfaces of different materials. At the level of the molecules and small subcellular structures, incident light can be either scattered (elastically or inelastically) or absorbed, with subsequent reemission in the form of fluorescence [1]. These interactions can be exploited to both provide an understanding of the underlying structure and physiology, leading to the diagnosis of disease, and enable light-based interventions, leading to therapeutic benefits.

26.2.1 Reflection and refraction

The index of refraction of biological structures differs depending on their constituent molecular concentrations. Usually, the index of refraction of tissues varies from close to 1.33 (the index of refraction of water) to 1.70 (the index of refraction of melanin) in the range of visible to near-infrared (NIR) light. As a result, reflection and refraction occur at the surface and the various interfaces of the cells and tissues of the human body.

26.2.2 Scattering

Scattering describes the change in the direction of propagation and, sometimes, energy of light when incident on heterogeneities within a bulk medium. The heterogeneities are usually nonuniform spatial and/or temporal distributions of the refractive index in the medium due to physical inclusions or random thermal motion. Scattering depends on the size, morphology, and structure of the constituents of tissue (e.g., lipid membrane, collagen fibers, nuclei). Changes in the characteristics of these elements, due to disease, affect their scattering properties, thus providing a means for detection and diagnosis of pathophysiology.

Scattering can be elastic or inelastic. In the case of elastic scattering, there is a change in the direction of propagation but no energy shift. Elastic scattering by particles smaller than one-tenth of the wavelength of the incident light is called Rayleigh scattering and is isotropic and proportional to the inverse of the fourth power of the wavelength. Elastic scattering by larger particles is called Mie scattering and is anisotropic (mostly forward). Mie scattering exhibits a complex oscillatory wavelength dependence that is characteristic of the particle size relative to the incident light wavelength. During inelastic scattering, the energy of the incident photons, as well as the direction, changes. The energy is transferred to or from the vibrational energy levels of the molecules, a phenomenon called Raman scattering. Because the vibrational energy levels are usually characteristic of the chemical bonds of the molecules, the energy shifts due to Raman scattering are representative of the biochemical structure of the material irradiated.

26.2.3 Absorption

Absorption is the transfer of energy from light to a molecular species. It occurs when the photon frequency matches the "frequency" associated with the molecule's energy transitions. Electrons absorb the energy of the light and transform it to vibrational motion. The energy is subsequently dissipated as thermal energy via the interaction with neighboring atoms and the molecule returns to its ground state. Unlike the ultraviolet and visible regions of the spectrum, infrared (IR) light is absorbed due to the presence of vibrational energy levels. Transitions between energy levels are well defined for different molecules and can serve as a spectral fingerprint to be exploited for understanding the variations in the chemical composition of cells due to the presence of disease.

26.2.4 Fluorescence

In addition to nonradiative relaxation pathways, energy absorbed by some molecules can be dissipated by emission of radiation. In biological samples, this can occur by fluorescence emission. During this transition, some of the absorbed energy is converted to heat by nonradiative processes. Subsequently, the molecules relax from the lowest vibrational energy level of the excited state to a vibrational energy level of the ground state, simultaneously emitting the energy difference in the form of fluoresce. The resulting photons have lower energy than the incident photons (red shifted). Because for a given excitation wavelength, the emission transitions are distributed over different vibrational energy levels, a broad fluorescence spectrum is generated and can be measured. There are many naturally occurring molecules that can fluoresce, termed endogenous fluorophores. They include amino acids, structural proteins, enzymes and coenzymes, vitamins, lipids, porphyrins, etc. Exogenous fluorescent molecules can also be administered as contrast agents (e.g., cyanine dyes), molecular markers (e.g., green fluorescent protein), etc.

26.2.5 Light transport and photon migration

Light transport in tissues is governed by the relative magnitudes of absorption and scattering. Describing the light propagation through the intricate structures of most tissues is rather complex and usually relies on diffusion theory (applicable when scattering dominates in the region of the visible and NIR) and transport theory or Monte Carlo simulations (when absorption and scattering are of similar magnitudes). These approaches result in complex analytical models or numerical solutions, which can be used to predict the behavior of emitted light, given an incident wave on an object, or, inversely, given that the characteristics of the emitted light predict the structure and optical characteristics of an unknown sample.

26.3 *IN VITRO* AND *EX VIVO* APPLICATIONS

When solid or liquid samples are available, these can be processed away from the patient, *in vitro* or *ex vivo*. Optical and photonic tools can be used to significantly speed the processing time and increase the processing volume while, at the same time, reducing the cost. Such technologies, including DNA sequencers and biochips, are revolutionizing our understanding of biology, health, and disease.

26.3.1 DNA sequencing

DNA sequencing is the delineation of the precise series of nucleotides in a DNA molecule or even an entire genome. The methodology that dominated the DNA sequencing efforts for many years was the detection of radioactive nucleotides inserted in the DNA during the replication process [2]. Electrophoresis of these samples revealed the distribution of radioactive bands of varying lengths from which sequence could be read. A major improvement to this process was the result of labeling of each type of nucleotide with a different fluorescent dye, thus enabling much faster sequencing without radiography. The fluorescent DNA passed through an electrophoresis tube where a detector distinguished the nucleotides by their color [3]. The resulting automation greatly increased throughput, reduced cost, and enabled deciphering large sequences contributing to the success of the human genome project [4,5]. In order to fully expand our understanding of the genomic code and the variations characteristic of each individual, there is now a push for so-called next-generation sequencing. New methods are being developed to allow complete, fast, and inexpensive sequencing of entire genomes. Most of these methods are based on highly parallel processes with the use of either luminescent moieties or fluorescence molecules as the enabling sensors [6].

26.3.2 Biochip and microarrays

The terms "biochip" and "microarray" are sometimes used interchangeably. However, there are important differences in both the design and the concept behind each device. Biochip refers to a material or a substrate that has a two-dimensional array of probes for biochemical assays as well as appropriate measurement and/or recording circuitry [7]. Biochips imply both miniaturization, usually in microarray formats, and the possibility of low-cost mass production. They can be classified either by the nature of their probes or by the type of transducer used. If the probes are nucleic acids, the devices are called DNA biochips, DNA chips, genome chips, DNA microarrays, gene arrays, or genosensor arrays. If the probes consist of antibodies or proteins, the devices are referred to as protein chips or protein biochips. A recently developed system with both DNA and antibody probes on the same platform is referred to as a multifunctional biochip [8] (Figure 26.1).

Biochips can also be classified by the type of transducers used. Typical transducer techniques involve (1) optical measurements (fluorescence, luminescence, absorption, Raman, surface plasmon resonance, etc.), (2) electrochemical measurements, and (3) mass-sensitive measurements (surface acoustic wave, microbalance, etc.). Because of its inherently sensitive detection capability, fluorescence is the most commonly used technique in DNA hybridization assays. Other approaches include a type of spectral label for DNA probes based on surface-enhanced Raman scattering (SERS) for use in cancer diagnostics [9]. The development of a biosensor, also for DNA probes, using visible and NIR dyes has also been reported [10]. Microarrays, however, consist only of arrays of probes but do not include sensor microchips integrated into the system [11]. Microarrays usually have separate, relatively large, detection systems that are more suitable for laboratory-based applications. They can incorporate large numbers of probes (tens of thousands), which could be used to identify multiple biotargets with very high speed and high throughput. These devices are very useful for mass screening for gene and drug discovery applications.

Both microarrays and biochips use probes operating as biological recognition systems, also called bioreceptors. A bioreceptor is a biological molecular species (e.g., an antibody, an enzyme, a protein, or a nucleic acid) or a living biological system (e.g., cells, tissue, or whole organisms) that uses a biochemical mechanism for recognition. Bioreceptors are the key to specificity for biochip technologies, responsible for binding the analyte of interest to the sensor for the measurement. Bioreceptors can take many forms and are as numerous as the different analytes that can be monitored using biosensors. They can generally be classified into five major categories: (1) antibody/antigen, (2) enzymes, (3) nucleic acids/DNA, (4) cellular structures/cells, and (5) biomimetic probes (synthetic probes that mimic receptors of living systems). DNA microarrays, including those in biochips, can be fabricated using high-speed robotics on a variety of substrates. The substrates can be thin plates made of silicon, glass, gel, gold, or a polymeric material such as plastic or nylon or may even be composed of beads at the ends of fiber-optic bundles [12]. Oligonucleotide microarrays are fabricated either by *in situ* light-directed combinatorial synthesis that uses photographic masks for each chip [13] or by conventional synthesis followed by immobilization on glass substrates [14]. Arrays with more than 250,000 different oligonucleotide probes or 10,000 different complementary DNAs (cDNAs) per cm^2 have been produced [15]. These arrays have been designed and used for quantitative and highly parallel measurements of gene expression, to discover polymorphic loci, and to detect the presence of thousands of alternative alleles.

The most noteworthy impact of microarray and biochip technologies, in conjunction with bioinformatics, is the facilitation of an entirely new

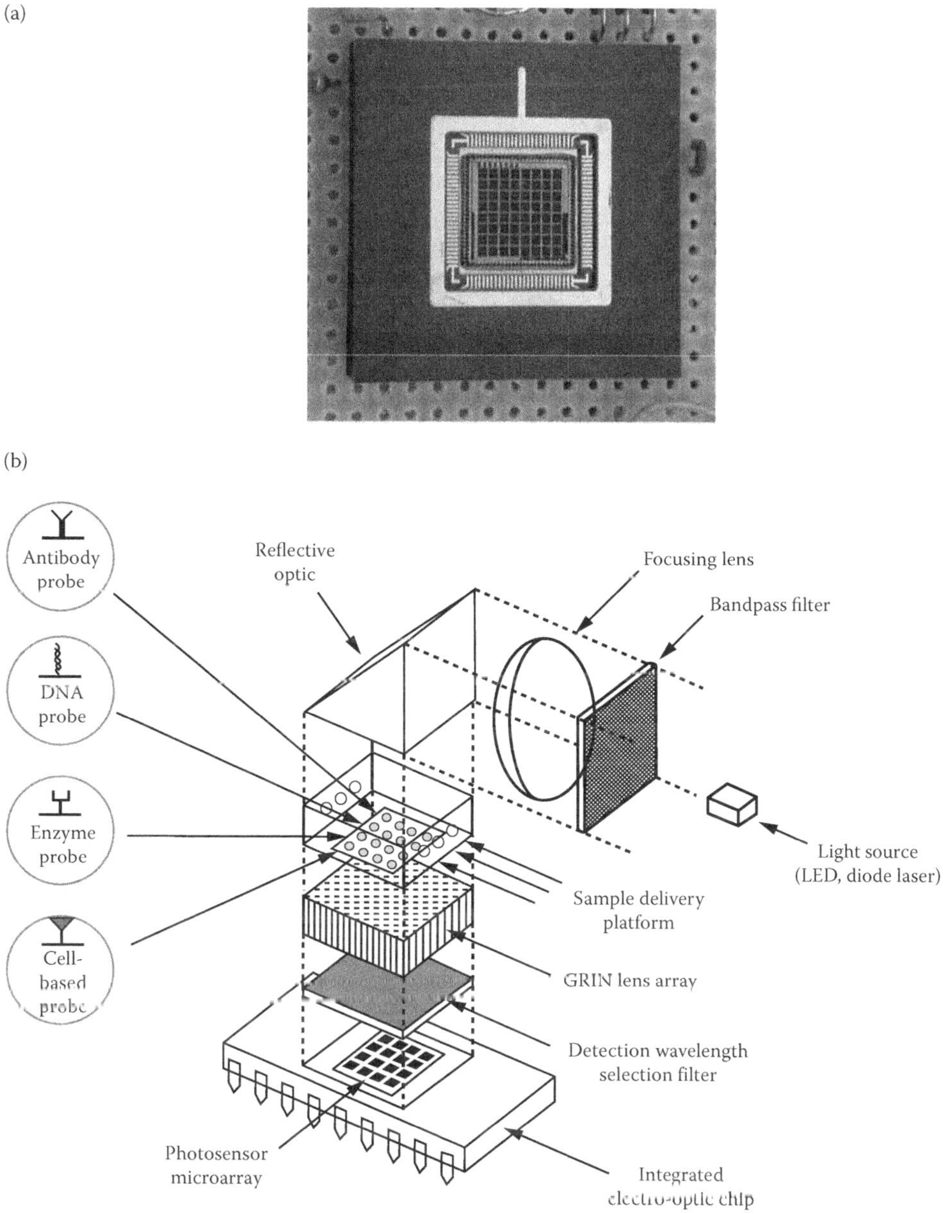

Figure 26.1 (a) Photograph of the 8×8 IC microchip. (b) Schematic diagram of an integrated biochip. Integrated biochips also include an Integrated Circuit (IC) microsensor, which makes these devices very portable and inexpensive. These devices generally have medium-density probe arrays (10–100 probes) and are most appropriate for medical diagnostics at the physician's office or at the point of care in the field. (From Vo-Dinh T.; Griffin, G.D., *Biomedical Photonics Handbook: Biomedical Diagnostics*, Boca Raton, FL, CRC Press, 2014. With permission.)

approach to biological and biomedical research. In the past, researchers investigated one or a few genes at a time. With new, automated, high-throughput, microarray, and biochip technologies, they can now study a medical problem systematically and on a large scale. They can examine all the genes in a genome or all the gene products in a particular tissue, tumor, or organ in the context of the interconnected pathways of a living system. Such knowledge will have a profound impact on the manner by which disorders are diagnosed, treated, and/or prevented and can bring about revolutionary changes in biomedical research and clinical practice [7]. Looking further into the future, the ultimate challenge in biochip research is to realize a truly implantable sensor for reliable, real-time, *in vivo* health monitoring. In order to reach this goal, issues of biocompatibility, remote detection, wireless telemetry, and miniaturization will have to be successfully addressed [7].

26.3.3 DNA probes

Nucleic acids have been widely used as bioreceptors for microarray and biochip systems [16,17]. In DNA biochips, the biorecognition mechanism involves hybridization of DNA or RNA, which are the building blocks of genetics. The microarrays of probes on DNA biochips serve as reaction sites, each reaction site containing single strands of a specific sequence of a DNA fragment. These fragments can either be short oligonucleotide (about 18–24) sequences or longer strands of cDNA. The sequence of any known part of DNA (target) can be amplified by the polymerase chain reaction (PCR) and labeled with an optically detectable compound (e.g., a fluorescent label) inserted during the PCR process. When the targets contain more than one type of sample, each is labeled with a different tag so that they can be detected simultaneously. The complementarity of adenine:thymine (A:T) and cytosine:guanine (C:G) pairing in DNA provides the basis for the specificity of biorecognition in DNA biochips. When unknown fragments of single-strand DNA react (or hybridize) with the probes on the chip, double-strand DNA fragments form only when the target and the probe are complementary according to the base-pairing rule. Finally, the sample is tested for hybridization to the microarray by detecting the presence of the attached labels. Probes based on a synthetic biorecognition element, peptide nucleic acid (PNA), have also been developed [18]. PNA is an artificial oligo amide that is capable of binding very strongly to complementary oligonucleotide sequences.

26.4 *IN VIVO* DIAGNOSTIC APPLICATIONS

Photonic techniques can provide powerful tools for detecting the presence of disease, especially early neoplastic changes. Detection of early neoplastic changes is critical because once carcinoma becomes invasive and metastatic treatment is difficult. At present, excisional biopsy followed by histology is considered to be the "gold standard" for the diagnosis of early cancer. In some cases, cytology rather than excisional biopsy is performed. The use of staining and processing can enhance the contrast and specificity of histopathology and provide high-resolution spatial and morphological information of the cellular and subcellular structures. However, physical removal of tissue specimens is required followed by processing in the laboratory. As a result, these procedures incur a relatively high cost and diagnostic information is not available in real time. More importantly, in the context of detecting early neoplastic changes, both excisional biopsy and cytology can have unacceptable false negative rates often arising from sampling errors due to the limited number of biopsies, which can be practically collected from each location. Furthermore, biopsies cannot be obtained, at all, from certain tissues, e.g., some neurological tissues. Photonic technologies have the potential to perform *in situ* diagnosis, without the need for sample excision and processing, providing diagnostic information in real time. In addition, because removal of tissue is not required for optical diagnostics, a more complete examination of the organ of interest can be achieved than with a finite number of excisional biopsies or cytology. Currently used optical diagnostic technologies can be broadly divided into two categories: (1) spectroscopic diagnostics and (2) optical imaging.

26.4.1 Spectroscopic diagnostics

Spectroscopy is the detection of spectral properties that are related to the molecular composition and/or structure of biochemical species in the tissue of interest. There are several spectroscopic methods that are utilized for optical diagnostics: fluorescence, elastic scattering, Raman (inelastic) scattering, IR absorption, etc. Each of these techniques, which have been studied for the purpose of disease diagnosis with varying degrees of success, will be described in Sections 26.4.1.1 through 26.4.1.3. It should be noted that the application of spectroscopic detection in two dimensions can result in, so-called, spectroscopic imaging.

26.4.1.1 FLUORESCENCE SPECTROSCOPY

Fluorescence tools are important for medical diagnostics [19] and can be grouped into two main categories: (1) methods that detect endogenous fluorophores in tissues, often referred to as autofluorescence, which usually detect variations in the biochemistry of the tissue and, thus, infer the presence or absence of disease based on the changes of the fluorescent chromophores; and (2) methods that detect or use exogenous fluorophores or fluorophore precursors, such as 5-aminolevulinic acid (ALA). These agents are explicitly synthesized so that they target specific tissue types (e.g., dysplasia versus normal) or are activated by functional changes in the tissue. In either case, the tissue is exposed to excitation light at some specific wavelength, typically near ultraviolet or visible, which excites the tissue molecules and results in fluorescence emission. The emission spectrum (emission intensity versus wavelength) is then measured as a function of wavelength. The shape of the emission spectrum is characteristic of the endogenous or exogenous fluorophores present and its intensity is analogous to their concentration.

A number of research groups have investigated laser-induced fluorescence (LIF) as a method to discriminate tumors from normal tissues. However, due to the limited penetration of optical wavelengths into biological tissues, the most common type fluorescence analyses performed *in vivo* are cancer diagnoses of optically accessible tissues. Vo-Dinh and coworkers have developed a LIF diagnostic procedure for *in vivo* detection of gastrointestinal cancer that uses 410 nm laser light from a nitrogen-pumped dye laser passed through a fiber optic probe to excite the tissue. The sensitivity of this method in classifying normal tissue and malignant tumors is 98% [20]. The ability to distinguish between various types of tissues, *in vivo*, based upon multicomponent analysis has also been demonstrated [21]. Richards-Kortum and coworkers have used LIF, employing 337 nm excitation to differentiate in vivo cervical intraepithelial neoplasia, nonneoplastic abnormal, and normal cervical tissues from one another [22]. In a study of lung cancer, it was found that the sensitivity of the autofluorescence bronchoscopy was 86%, which is 50% better than conventional white light bronchoscopy, for the detection of dysplasia and carcinoma *in situ* [23,24]. Like other cancers involving mucosal membranes, oral and laryngeal carcinomas have also been studied by autofluorescence. Fluorescence spectroscopy has been used to differentiate normal tissue from dysplastic or cancerous tissue with a sensitivity of 90% and a specificity of 88% in a training set, and a sensitivity of 100% and a specificity of 98% in a validation set [25]. An alternative approach to conventional fixed excitation fluorescence is the synchronous luminescence method, which involves scanning both excitation and emission wavelengths simultaneously while keeping a constant wavelength interval among them [26]. This method has been developed for multicomponent analysis and has been used to obtain fingerprints of samples with enhanced selectivity in an assay of complex systems.

A major limitation of most of the hardware used in the early studies clinical application of fluorescence spectroscopy was that they could only perform point measurements, leading to inadequate sampling of the tissue or organ under investigation. Fluorescence imaging allows a more global view of the target, although fluorescent images are formed using only selected emission wavelength bands. Emission bands are selected using special optical filters, and then detected by separate cameras to form the final displayed fluorescence image (Figure 26.2) in real time as false color maps [27]. The first clinical demonstration involved imaging of the bronchus to screening for dysplasia and carcinoma in high-risk patients [28]. This led to a commercial system (LIFE-Lung; Xillix Technologies Corp., Richmond, BC, Canada). Using LIFE in combination with white-light endoscopy, the detection of moderate to high-grade bronchial dysplasia was increased by 171%, compared with white light bronchoscopy alone, with only a 22% decrease in specificity [29]. Similar systems are currently evaluated for gastrointestinal endoscopy [30,31].

Exogenous fluorophores are used in many clinical applications. The reason for using such compounds is, often, to provide a contrasting agent, which would make the medical diagnoses easier. The most common of the exogenous fluorophores used for these studies are photosensitizers that are being developed for photodynamic therapy (PDT) treatments, which will be covered in Section 26.5.2. These drugs generally exhibit strong fluorescence properties and preferentially accumulate in malignant tissues. The photosensitizer Photogem,

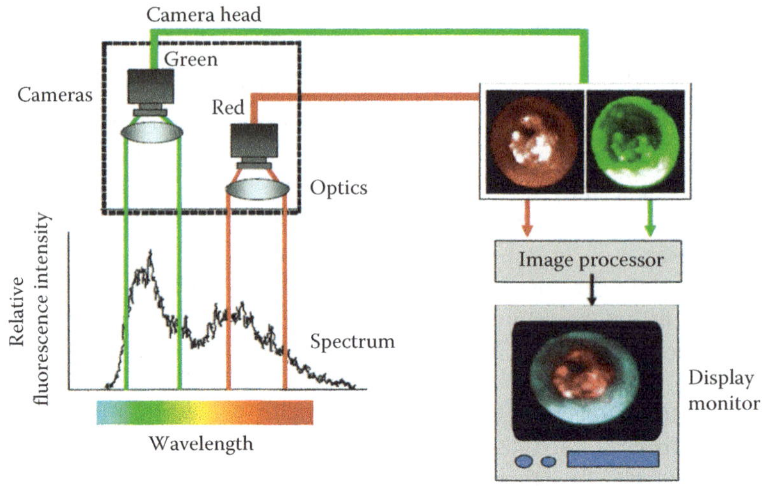

Figure 26.2 Schematic diagram illustrating the individual red and green fluorescence (reflectance) channels that make up the final displayed fluorescence pseudocolored image. (From DaCosta, R.S.; Wilson, B.C.; Marcon, N.E., *Sci. World J.*, 7, 2046–2071, 2007. With permission.)

for lung, larynx, skin, gastric, esophageal, and gynecological cancers [32]; ALA-induced protoporphyrin IX, for tumor demarcation of liver adenocarcinoma and aggressive basal cell carcinoma [33]; Photofrin-enhanced LIF, for Barrett's metaplastic epithelium and esophageal adenocarcinoma [34], and diagnosis of bladder cancer based on the LIF of exogenous fluorophores [35] are some of the studies that have been performed. Another class of compounds, which is beginning to be tested for *in vivo* tumor demarcation, is fluorescently labeled antibodies. Fluorescence analyses revealed that the dye-labeled antibodies localize preferentially in the tumor tissue. Such immune photodiagnoses may prove very useful in the clinical setting for rapid tumor demarcation and surgical guidance in the colon and potentially other organs [36].

26.4.1.2 ELASTIC SCATTERING SPECTROSCOPY

Elastic scattering spectroscopy involves detection of the backscattering of a broadband light source irradiating the tissue [37]. In general, the tissue is illuminated with the excitation light delivered to a specific point location via an optical fiber, and the scattered light is measured from a nearby location. A spectrometer records the backscattered light at various wavelengths and produces a spectrum that dependents on tissue structure as well as chromophore constituents. The physical quantities affecting the measurements are the absorption and scattering properties of the sample and/or the wavelength dependence of these properties. Several characteristics of the tissue structure can be deduced from such a spectrum, e.g., the distribution of nuclear size (inferred from the oscillatory nature of the wavelength dependence of the backscattering spectrum). Nuclear size and its distribution are the important parameters for the diagnosis of malignancies, such as cancer, because they are a critical part of a pathologist's assessment when determining his/her diagnosis during histological examination. Calculations based on Monte Carlo methods have been used to investigate the photon scattering process. This has led to some variations of this type of methodology where the optical transport properties of the tissue can be measured directly. Elastic scattering techniques have been developed for *in vivo* cancer diagnostics [38,39].

26.4.1.3 RAMAN AND IR SPECTROSCOPY

Raman scattering and IR absorption spectroscopies both exhibit spectral variations due to transitions between vibrational energy levels. They are, typically, highly specific because they provide vibrational information directly related to the molecular bonds of the samples. Thus, they can be used for qualitative identification of biological compounds as well as for *in vivo* medical diagnostics [40]. The selection rules and relative intensities of IR and Raman peaks are not similar, so they are often viewed as complementary techniques.

With IR spectroscopy, the ever-present intense absorption bands of water (present in all biological samples), which overlap with most of the other tissue component spectra, hamper possible *in vivo* applications. Most biological molecules are Raman active with fingerprint (i.e., specific) spectral characteristics. Hence, vibrational spectrometry can provide another alternative for diagnosis. For this reason, Raman spectrometry has been investigated for the detection of cancer in many organs [41,42]. Additional applications include the detection of infectious diseases [43,44].

However, the magnitude of Raman scattering is typically small, making either high-illumination intensities or relatively long measurement times necessary in order to obtain good signal-to-noise ratios comparable to fluorescence techniques. One way to overcome this limitation is to take advantage of the phenomenon of surface enhancement [45]. SERS is a variation of Raman spectroscopy that offers significant enhancement of the signal (up to 10^{14} times), thus making detection faster, simpler, and more accurate. The enhancement is a result of the effect of plasmon resonance, i.e., the unison oscillation of electrons on the surface of a metallic nanostructure as a result of incident light of the right, resonant, and frequency. These oscillations produce an enhanced electromagnetic field in the proximity of the surface. If a sample is within a few nanometers from the nanostructure, it will experience this enhanced field and exhibit a stronger Raman signal. The enhancement is such that even single molecules can be detected [46,47].

26.4.2 Imaging

Optical imaging offers a number of advantages over other radiological imaging techniques because it (1) can be performed noninvasively or minimally invasively especially using optical fibers; (2) significantly reduces patient exposure to harmful radiation by using nonionizing radiation, which includes visible and IR light; (3) can easily distinguish soft tissues with contrast provided by absorption and scattering and, in certain cases, can provide unprecedented resolution; (4) can be easily combined with other imaging techniques, such as magnetic resonance imaging (MRI) or X-rays, to provide enhanced information; and (5) can multiplex different wavelengths to interrogate multiple tissue properties. There are various optical techniques by which tissues can be imaged *in vivo* and *in situ*. Each method has a different contrast mechanism and provides different levels of penetration and resolution. The choice depends on the tissue properties and the specific application.

26.4.2.1 VARIOUS TYPES OF MICROSCOPY

Microscopy has been the workhorse of medical imaging for centuries. From the humble beginnings of the single-lens microscope to today's modern devices, it has become an invaluable part of medical, clinical, and scientific work. The use of staining, either with dies or antibodies, provides the necessary contrast to delineate even the most subtle features of disease in tissue biopsies, thus becoming the "gold standard" of diagnosis. Over the years, novel microscopic approaches were developed to exploit additional contrast mechanisms (dark field, phase, fluorescence) without the need for staining thus making microscopy, in some cases, compatible with *in vitro* imaging of live cells.

In dark-field microscopy, contrast is generated in the image by altering the illumination pattern so as to reject the light directly passing through the sample. In phase contrast and differential interference contrast (DIC) microscopy, the illumination is also modified. However, in these cases, complementary optical accessories (e.g., filters or prisms) condition the light before it strikes the specimen and manipulate the light after it has interacted with the specimen. These alterations provide contrast based on the phase difference of light passing through the sample versus through air and result in desirable imaging features such as high resolution and reduced artifacts [48]. Fluorescence microscopy incorporates excitation and emission wavelength-selecting filters to visualize the fluorescence emission from tissue samples. The florescence can originate from either endogenous or exogenous fluorophores. It, thus, provides tissue-specific contrast and even functional characterization because the exogenous fluorophores can be designed to bind to specific cellular moieties.

Confocal microscopy and multiphoton microscopy are two variations of microscopy offering optical sectioning capabilities, thus allowing three-dimensional visualization of tissue samples down to 500–800 μm below the surface. In confocal microscopy, the light from out-of-focus planes is rejected by a pinhole placed at the so-called confocal plane. A laser beam is usually scanned to create the cross-sectional image in reflectance, transmission, or fluorescence mode [49]. In multiphoton microscopy, as the name implies, two or more photons are needed to provide the necessary excitation of the fluorophores in the tissue sample. Because the simultaneous absorption of two or more photons is probabilistically an unlikely event, this effect occurs only at the focus of the laser beam where the photon concentration is higher. Because there is no fluorescence from out of focus planes, a cross-sectional image can be obtained without the need for additional optical elements [50]. In stimulated emission depletion (STED) microscopy, which has recently received much attention due to the Nobel prize in chemistry, an additional donut-shaped beam is used to deplete most of the excited molecules at the periphery of the focus, thus allowing fluorescence emission only from a region even smaller than what would be allowed by the diffraction limit of the focusing lens [51] (Figure 26.3).

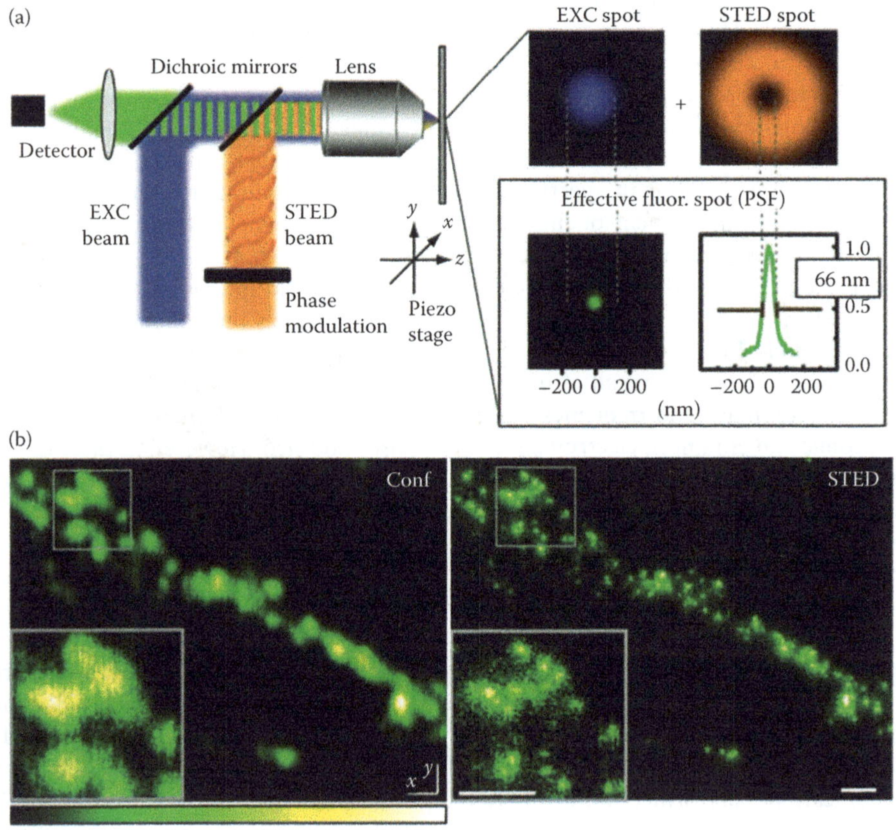

Figure 26.3 (a) Principles of operation. Although the blue excitation (EXC) beam is focused to a diffraction-limited excitation spot, shown in the adjacent panel in blue, the orange STED beam is able to de-excite molecules. The STED beam is phase modulated to form the focal doughnut shown in the top right panel. Superimposition of the two focal spots confines the area in which fluorescence is possible to the doughnut center, yielding the effective fluorescent spot of subdiffraction size shown in green in the lower panel. All spots represent measured data and are drawn to scale. The profile of the green effective fluorescent spot has an Full width at halt maximum (FWHM) of 66 nm as well as a sharp peak. The green spot shows an 11-fold reduction in focal area beyond the excitation diffraction value (compare with blue spot). (b) Comparison of confocal (left) and STED (right) counterpart images of a labeled preparation reveals a marked increase in resolution by STED. Scale bar: 500 nm. (From Willig, K.I.; Rizzoli, S.O.; Westphal, V.; Jahn, R.; Hell, S.W., *Nature*, 440(7086), 935–939, 2006. With permission.)

26.4.2.2 OPTICAL COHERENCE TOMOGRAPHY

Optical coherence tomography (OCT) is a technique that provides high-resolution cross-sectional images of tissue microstructure *in vivo, in situ*, and in real time [52]. It is analogous to ultrasound imaging, but, instead of acoustic waves, it measures the backscattered intensity of light from structures within the tissue. In contrast to ultrasound, because the velocity of light is extremely high, the echo time delay of reflected light cannot be measured directly. Interferometric detection techniques must therefore be used [47]. An image is formed by performing repeated axial measurements at different transverse positions as the optical beam is scanned across the tissue. The resulting data yield a two-dimensional map of backscattering or reflectance from the internal architectural morphology and cellular structure in the tissue. The axial resolution is 1–10 μm, i.e., higher than any clinically available diagnostic modality. Imaging depth is limited to 2–4 mm due to optical attenuation from scattering and absorption. OCT was originally developed and applied to tomographic imaging in ophthalmology to noninvasively provide high-resolution images of the retina. It is now routinely used in clinical ophthalmologic practice [53]. OCT devices can also be constructed with fiber optic probes, which can be incorporated into catheter-based or endoscopic systems.

OCT is a promising imaging technology because it allows real-time and *in situ* visualization of tissue microstructure without the need to excisionally remove and process a specimen as in conventional biopsy and histopathology. It is a diagnostic tool that is complementary to spectroscopic techniques and has great potential for *in situ* microscopic imaging of the cellular attributes of malignancies and precancers. The microstructural images generated rival those of a histopathologist examining a tissue biopsy specimen under a microscope. Although staining techniques and contrast agents have not yet been designed for this technique, the microstructural information and the measurement of the sizes of cellular and subcellular elements *in vivo* can provide unique insights into dysplastic and malignant process and can be linked to therapeutic procedures once a suspicious area is identified. OCT has been applied in vivo to (1) image arterial pathology, where it can differentiate plaque morphology and monitor stent placement; (2) image the gastrointestinal track, where it can detect esophageal dysplasia and intestinal cancers (Figure 26.4); (3) guide surgical breast

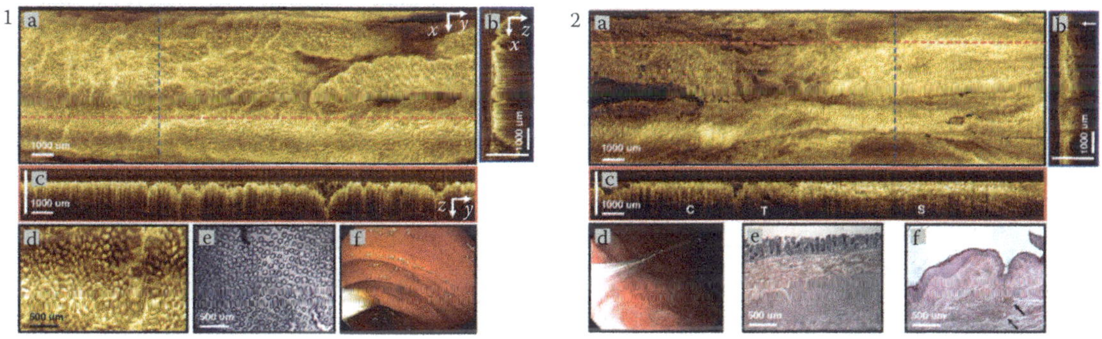

Figure 26.4 (1) 3D-Optical Coherence Tomography (OCT) images of columnar epithelial tissue in the human colon. (a) En face image constructed by axial summation of the entire data set. Dashed lines show locations of cross sections. (b) XZ cross section showing typical columnar structure. (c) YZ cross section. (d) Enlarged view of (a), showing an en face crypt pattern. (e) Representative en face histology of human colon. (f) White light video endoscopy image of region analyzed with 3D-OCT. (2) 3D-OCT images near the dentate line. (a) En face image constructed by axial summation of entire dataset. Dashed lines show locations of cross sections. (b) XZ cross section showing typical squamous structure. (c) YZ cross section showing shift from columnar C to squamous S epithelium over a transition zone T. (d) White light video endoscopy image of region analyzed with 3D-OCT. (e) Representative cross-sectional histology of columnar epithelium. (f) Representative cross-sectional histology of squamous epithelium. Arrows in b and f indicate normal anal vessels. (From Adler, D.C.; Zhou, C.; Tsai, T.H.; Schmitt, J.; Huang, Q.; Mashimo, H.; Fujimoto, J.G., *Opt. Express*, 17(2), 784–796, 2009. With permission.)

tumor excision, where it can identify the tumor margins; and (4) image dental structures, where it can identify the enamel layers and caries, among others [54]. Several preclinical, life science, and material science applications of OCT are also being investigated [55].

26.4.2.3 PHOTOACOUSTIC IMAGING

Photoacoustic tomography, also referred to as optoacoustic tomography, is an emerging biomedical imaging method based on the photoacoustic effect [57]. This effect was reported by Alexander Graham Bell in 1880 and refers to the generation of acoustic waves by substances illuminated by light of varying power. The photoacoustic effect results from the oscillatory thermal expansion within the material as it absorbs the light energy. Because absorption reflects a material's characteristics, physical and chemical information can be gathered by studying the photoacoustic signals [58]. Photoacoustic imaging is a three-dimensional imaging method. Pulsed light illumination is delivered to the tissue where it generates photoacoustic signals from the illuminated volume [59]. The imaging systems, then, records these time-resolved pressure signals around the boundary of the tissue volume. The initial pressure distribution, resulting from optical absorption and subsequent thermal expansion, can be reconstructed. Two-dimensional imaging of an entire volume at once can be demanding in terms of detector elements needed and geometric constrains. Other approaches include focusing the ultrasound detection on a plane through the illuminated volume [60] or sampling one point in the field of view at a time, using spherically focused ultrasound detection or focused light [61]. Applications include cancer imaging [62] and quantitative methods [63] including contrast agents for optoacoustic imaging [64]. A recent development in photoacoustics is multispectral optoacoustic tomography, an imaging approach using multiple excitation wavelengths and, thus, simultaneously interrogating a broad range of biological processes (Figure 26.5) [54]. At the same time,

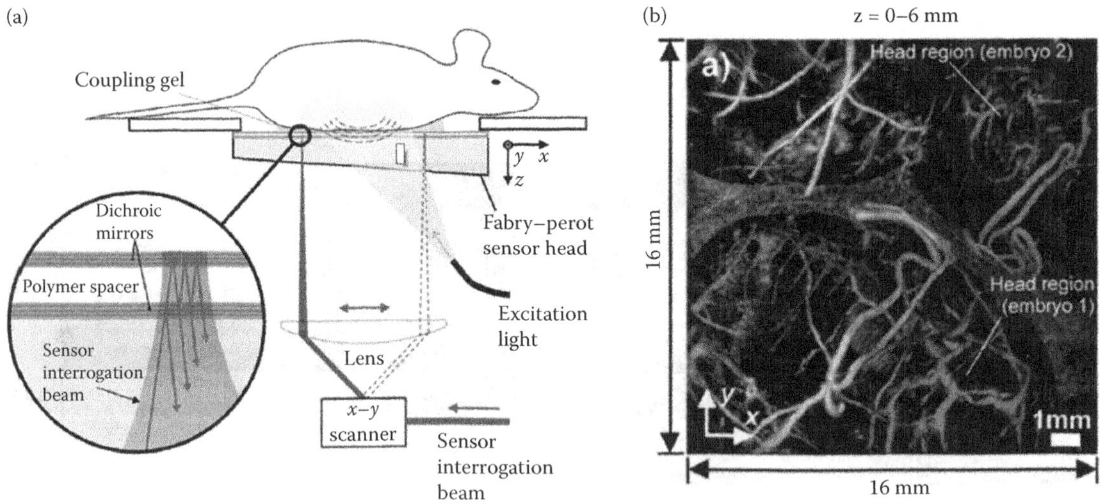

Figure 26.5 (a) Schematic illustrating the operation of the Fabry–Perot interferometer (FPI)-based photoacoustic imaging system. Photoacoustic waves are generated by the absorption of nanosecond optical pulses provided by a wavelength-tunable Optical Parametric Oscillator (OPO) laser and detected by a transparent Fabry–Perot polymer film ultrasound sensor. The sensor comprises a pair of dichroic mirrors separated by a 40-μm-thick polymer spacer, thus forming an FPI. The waves are mapped in two-dimensions by raster-scanning a Continuous Wave (CW)-focused interrogation laser beam across the sensor and recording the acoustically induced modulation of the reflectivity of the FPI at each scanning point. (b) Maximum amplitude projection of the complete three-dimensional image data set (depth: 0–6 mm), showing two embryos (indicated by the arrows). (From Xia, J.; Yao, J.; Wang, L.V., Photoacoustic Tomography: Principles and Advances. *Electromagnetic Waves*, Cambridge, MA, 147, 1, 2014. With permission.)

the development of real-time handheld scanners for clinical use [65] is promoting the translation of photoacoustic imaging to novel applications, such as breast, vascular, and skin imaging.

26.4.2.4 PHOTON MIGRATION AND DIFFUSE OPTICAL IMAGING

Photon migration techniques can be utilized to perform point measurements or imaging in deep tissues (several centimeters) by use of multiply scattered light. Unlike X-rays, direct imaging with visible or NIR light cannot be performed due to multiple scattering. Various techniques and associated instrumentation, both time domain and frequency domain, based on photon migration in tissues, have been investigated to overcome this limitation [66]. In time domain methods, the time delay of the photons coming out from another point on the surface, a few centimeters away from the illuminating source, can be measured by using picosecond laser pulses launched into the tissue via optical fibers. The transit time is the optical path length divided by the speed of light in the tissue medium, and the attenuation of the signal intensity is due to the absorption within the tissue through which the light has migrated. In frequency domain techniques, the incident light intensity is modulated, thus producing a photon number density, which oscillates in space and time. The resultant photon density waves scatter from tissue inhomogeneities (e.g., tumors in a tissue volume) and, if properly recorded, can be used to reconstruct the unknown inhomogeneity distribution into a two-dimensional image [67,68].

Diffuse imaging is still a research tool because it has not been clear, so far, how optical imaging could replace any of the established imaging methods. However, many niche applications exist where optical imaging could have a significant impact as a stand-alone tool or in conjunction with other methods [69]. The most established of these applications, at the moment, is optical topography of babies and infants, where all rival methods have significant safety drawbacks [70]. Diffuse image could also play a role in breast cancer. Although X-ray mammography is a sound tool for screening, it is not well suited for imaging younger women or for repeated imaging for monitoring of response to treatment, which could be performed optically [71].

26.5 *IN VIVO* THERAPEUTIC APPLICATIONS

26.5.1 Laser therapy and surgery

Laser-induced photocoagulation or ablation can be used to alter the tissue shape for surgical or other therapeutic purposes. It is based on the absorption of high-intensity pulses by the targeted tissues causing either protein denaturation or complete evaporation without carbonizing or bleeding. The precise control of the wavelength as well as temporal and power parameters of laser therapeutic techniques can restrict the interaction to specific target areas of tissue. Laser therapy is the current standard of care for the treatment of some retinal diseases such as proliferative diabetic retinopathy, diabetic macular edema, and some types of subretinal neovascularization [72]. Vision correction using photorefractive keratectomy or laser-assisted *in situ* keratomileusis is also based on this effect [73]. In dermatology, careful control of laser parameters permits selective destruction of specific loci in the skin, for example, in tattoo removal, treatment of port-wine stains, and various cosmetic applications (Figure 26.6) [74].

Interstitial laser photocoagulation (ILP) is a laser-based procedure that uses optical fibers (typically 0.2–0.4-mm core diameter) inserted directly into the target tissue (usually through a needle of about 18 gauge) so laser light is delivered to the tissue from the end of the fiber as a point source (bare tip fiber) or emitted from the end section of the fiber (diffuser fiber, where the diffuser section can be up to several centimeters long) [75]. One or more fibers can be used. The energy is absorbed within the surrounding tissues causing local thermal necrosis. The heating is designed to prevent tissue carbonization and vaporization, thus permitting the body's immune system to gradually remove the dead cells over a period of time. For surgical applications, effects ranging from thermal cautery with hemostasis to precision ablation can be achieved. By virtue of the way that they are delivered, optical therapeutic techniques can often reduce the invasiveness of conventional surgical procedures or enable new procedures that are not possible with conventional surgical tools. For example, ILP has recently been approved for treatment of benign prostatic hyperplasia as an outpatient procedure. For treatment of

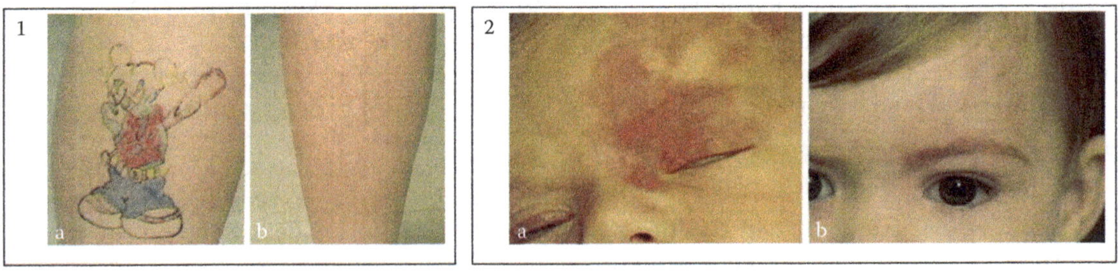

Figure 26.6 (1) Multicolored professional tattoo before (a) and after (b) lesional clearance after nine treatments with quality-switched 755-nm alexandrite laser (blue/black inks) and four treatments with frequency-doubled 532-nm neodymium: yttrium–aluminum–garnet laser (yellow/red inks). (2) Port-wine stain in infant before treatment (a) and resolution after 8585-nm pulsed dye laser treatments (b). (From Tanzi, E.L.; Lupton, J.R.; Alster, T.S., *J. Am. Acad. Dermatol.*, 49(1), 1–34, 2003. With permission.)

tumors of the breast, the procedure can be carried out under local anesthetic and mild sedation. Up to four needles can be placed directly into the tumor under either ultrasound or MRI guidance [76].

26.5.2 Photodynamic therapy

An important application of optical technologies for site-selective therapy is the use of PDT [77]. In PDT, a photoactive compound, with some degree of selective affinity for cancerous tissue, is administered topically, orally, or intravenously. After a period of time (typically 6–48 h), the compound accumulates selectively in areas of malignancy. The molecule is then photoactivated with light of the appropriate wavelength, producing singlet oxygen ($^1O_2$) preferentially in the malignant tissues. Although the exact mechanism of cell death is still under investigation, it has been shown that the presence of $^1O_2$ has a cytotoxic effect on the target cells. The interaction of light at a suitable wavelength with the photosensitizer produces an excited triplet state photosensitizer that gives rise to $^1O_2$ [78]. The highly reactive $^1O_2$ has a short lifetime (<0.04 μs) in the biological milieu and therefore a short radius of action (<0.02 μm). Consequently, $^1O_2$-mediated oxidative damage will occur in the immediate vicinity of the subcellular site of photosensitizer localization. Depending on photosensitizer pharmacokinetics, these sites can be varied and numerous, resulting in a large and complex array of tissue and cellular effects. The relative importance of each has yet to be fully determined. PDT is most beneficial if the light is delivered when the concentration of photosensitizer in the tumor tissue is greater than that of adjacent normal tissue. Development of a noninvasive technique to estimate photosensitizer concentration in tissue, preferably in real time, is desirable. Although the fluorescence of the photosensitizers may be used to localize tumors, it does not provide a quantitative measure of drug concentration.

Originally developed for treatment of various solid cancers, the applications of PDT have been expanded to include treatment of precancerous conditions (e.g., actinic keratoses and high-grade dysplasia in Barrett's esophagus) and noncancerous conditions (e.g., various eye diseases such as age-related macular degeneration) [79]. PDT is also being investigated for applications in several other clinical fields, including skin cancer, bladder cancer, carcinoma of the gastrointestinal tract, and lung cancer [74].

26.5.3 Low-level laser therapy

Low-level laser therapy (LLLT), also known as photobiomodulation, involves exposing cells or tissues to low levels of red and/or NIR light at energy densities below the tissue damage threshold [80]. Although LLLT is now widely used, it remains controversial as a therapy for two main reasons: (1) the underlying biochemical mechanisms remain poorly understood; and (2) a large number of parameters such as the wavelength, fluence, power density, pulse structure, and timing of the applied light are empirically chosen for each treatment [75]. Dosimetry in LLLT is a major challenge due to the considerable level of complexity resulting in, largely, empirical choice of parameters. As far as the LLLT mechanism of action, evidence now suggests that the laser irradiation acts on the mitochondria, increases adenosine triphosphate production, modulates reactive oxygen species, and induces transcription factors. Immune cells, in particular, appear to be strongly affected by LLLT leading to increased infiltration of the illuminated tissues by leukocytes. LLLT also enhances the proliferation, maturation, and motility of fibroblasts, and increases the production of basic fibroblast growth factor [75].

LLLT is used for three main purposes: (1) to promote wound healing, tissue repair, and the prevention of tissue death; (2) to relieve inflammation and edema because of injuries or chronic diseases; and (3) as an analgesic and a treatment for neurological problems. These applications appear in a wide range of clinical settings, ranging from dentistry, to dermatology, to rheumatology, and to physiotherapy. LLLT is also being considered as a viable treatment for serious neurological conditions such as traumatic brain injury, stroke, spinal cord injury, and degenerative central nervous system disease. One of the most commercially successful applications of LLLT is the stimulation of hair regrowth in balding individuals [75].

26.6 FUTURE APPLICATIONS OF PHOTONICS IN MEDICINE

26.6.1 Point-of-care testing

Correct diagnosis of >50% of all diseases relies on a large number of laboratory analyses that, additionally, aid in the monitoring of drug therapy in many other cases. Therefore, laboratory medicine is a vital component of the differential diagnosis [81]. Most laboratory analyses are, currently, offered by centralized facilities. However, in recent years, there has been a trend to a more efficient and decentralized approach, located at the site of primary care or first contact, using so-called point-of-care testing (PoCT). PoCT shortens the laboratory testing turnaround time but it is only useful if the results produced lead to immediate therapeutic decisions. However, evidence of improved patient outcomes is slow to emerge [82]. Although the numbers vary between reports, the total *in vitro* diagnostics market was believed to be worth US$51 billion in 2011, of which approximately US$15 billion was PoCT with a projected annual growth of 4% [83].

A typical classification of PoCT technology splits devices into small handheld ones, including quantitative and qualitative strips, and those that are larger benchtop devices with more complex built-in fluidics, often variants of those used in conventional laboratories [78]. Irrespective of size, all PoCT devices should, as far as possible, be simple to use, use reagents and consumables that are robust in storage and usage, provide results that are concordant with an established laboratory method, and, together with associated reagents and consumables, be safe to use [84].

A myriad of small devices exists for PoCT that range from the, so-called, dipstick to the sophisticated, small cartridge devices used in blood gas analysis. These devices are portable and, typically, used by the patient themselves or by health-care professionals at the bedside, in the clinic, or at the patient's home. However, most of the technologies currently being used for PoCT have not fundamentally changed in the last decades but only improved incrementally through advances in materials, electronics, and computing technologies [79,85]. The quest for better technologies has focused on (1) the continuing efforts to develop the biochip concept, particularly for microbiology and infectious disease applications where serological assays are being replaced by molecular testing; and (2) the development of paper-based analytical devices that are cheaper, can be massed produced, and require only small sample volumes. Comprehensive reviews of recent progress in the development of paper-based PoCT, so-called lab-on-a-stamp, devices can be found in the literature [86]. Examples of other technologies, which look promising for the future, include contact lens glucose sensors [87], tattoo-based sensors [88], and smart holograms [89].

26.6.2 Optical molecular imaging and theranostics

Optically active or activatable contrast agents, accumulating at the lesion location prior to imaging, can be used to visualize a broad range of molecular variations and therefore facilitate the detection of disease at the earliest possible stages [90]. Furthermore, the ability of optical molecular imaging to noninvasively image the spatial and temporal distribution of multiple biomarkers simultaneously has the potential to improve treatment through better selection of targeted therapeutic agents, real-time imaging for *in situ* guidance, and assessment and monitoring without the need for biopsy (Figure 26.7) [85]. Optical molecular imaging systems consist broadly of three components: (1) an optically active or activatable contrast agent targeting a specific biomarker of clinical relevance; (2) a method to safely deliver the contrast agent to the tissue at risk; and (3) an optical imaging system to acquire, process, and interpret the resulting images of the labeled tissue.

A molecular theranostic (therapeutic+diagnostic) system is one that combines the diagnostic information provided by molecular imaging with the delivery of therapy, to patients who are most likely to benefit from the specific agent, as well as monitor the response to that therapy. The main reason for the tremendous excitement regarding theranostics is the promise of improved therapy selection, based on specific molecular features of the disease, greater predictive power to avoid adverse effects, and new ways to objectively monitor therapy response. These properties are fundamental elements of the drive for more personalized medicine [91]. Ideally, the molecular imaging component of a theranostic system would provide crucial diagnostic information regarding the presence and anatomic location of cellular targets for which the therapeutic agent is intended [86]. At present, the most widely used *in vivo* optical molecular imaging modalities are bioluminescence and fluorescence imaging. Because these tools are low cost, are portable, and can be miniaturized, they are capable of expanding access to early detection and improving minimally invasive treatment in a wide variety of urban and rural health-care settings. However, achieving the full potential of theranostics requires coordinated efforts in biomarker discovery and validation, design and delivery of contrast agents, and engineering of optical instrumentation [92,93].

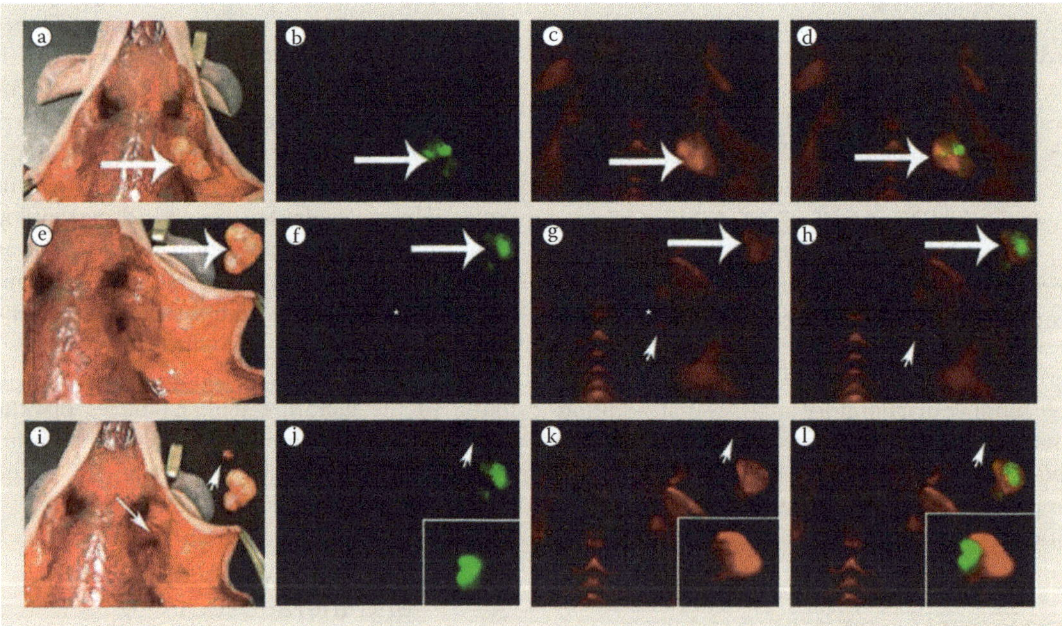

Figure 26.7 Activateable cell-penetrating peptides delineate residual tumor in the surgical margin. (a–d) Activatable cell-penetrating peptides labeled with Cy5 delineate a green fluorescent protein-expressing human melanoma cell line (MDA-MB 435) xenograft (large arrows) in the tumor bed. (e and f) Following excision, the tumor bed appears to be tumor free (*) under white light and Green Fluorescent Protein (GFP) signal. (g and h) However, the Cy5 channel identifies residual fluorescence signal (small arrows) in the surgical margin. (i) Using Cy5 to guide resection, a small piece of residual tumor is identified under the muscle. (j–l) The residual tumor is removed and clear surgical margins confirmed by the green fluorescent protein and Cy5 channels. Insets depict the excised residual tumor magnified and brightened ×5. (From Burtis, C.A.; Ashwood, E.R.; Bruns, D.E., *Tietz Textbook of Clinical Chemistry and Molecular Diagnostics*, St. Louis, Elsevier Health Sciences, 2012. With permission.)

26.6.3 Recommendations of the National Research Council

A recent report by the National Research Council of the National Academies, Washington, DC [94] included the following recommendations regarding optics and photonics as key enabling technologies:

- Develop new instrumentation to allow simultaneous measurement of all immune system cell types in a blood sample (key recommendation).
- Develop new approaches, or dramatic improvements in existing methods and instruments, to increase the rate at which new pharmaceuticals can be safely developed and proved effective (key recommendation).
- Prioritize the development of low-cost diagnostics for extremely drug-resistant and multidrug-resistant tuberculosis, malaria, HIV, and other dangerous pathogens, and low-cost blood serum and tissue analysis technology to potentially save millions of lives per year.
- Prioritize the development of new optical instruments and integrated incubation technology capable of imaging expanding and differentiating cell cultures *in vitro* and *in vivo* to provide important tools for predicting the safety and efficacy of stem cell-derived tissue transplants.
- Prioritize the development of new software methods automating the extracting, quantifying, and highlighting the important features in large, two-, and three-dimensional data sets to optimize the utility of the latest generation of imaging instruments.
- Prioritize the development of the next generation of super-high-throughput sequencing devices, required for lowering the cost of sequencing down to the target cost of $1000 per genome.
- The U.S. government should expand investment in multidisciplinary centers (e.g., at universities with medical and engineering schools) at which critical developments combining medical and engineering discoveries can be efficiently fostered.
- The U.S. government, in cooperation with scientific and medical societies, should facilitate the creation of an information technology infrastructure for sharing large amounts of medical and clinical data (e.g., quantitative imaging and molecular data) and open-source analysis tools.

26.7 CONCLUSIONS

In conclusion, photonic technologies have the ability to provide rapid *in vivo* diagnosis of disease. Sensitivity and specificity in various organ systems will be the deciding factor on which technology is best suited for application. Combination of various optical technologies may be important for improving the sensitivity and specificity. With combined detection modalities, new information can be gathered, thus improving our understanding of the basic pathophysiology of not only cancers but also other, nonmalignant, diseases such as benign structures and inflammatory diseases whose mechanisms are currently poorly understood. Areas of increased research interest include technologies for the noninvasive measurement of the concentration of various drugs and biological species in tissues and advances in biochip technologies for the development of inexpensive, portable, or handheld PoCT devices. In the long term, even if only a few of the promises of optical diagnostic technologies are realized, both the cost savings and the improvement in patient prognosis and quality of life will still be substantial, thus justifying expending more effort and resources in this field.

ACKNOWLEDGMENTS

C. Pitris: Research partially supported by the Research Promotion Foundation of Cyprus and the KIOS Research Center for Intelligent Systems and Networks.

T. Vo-Dinh: Work sponsored by the National Institutes of Health (RO1 CA88787-01) and by the U.S. Department of Energy Office of Biological and Environmental Research, under contract DEAC05-00OR22725 with UT-Battelle, LLC.

REFERENCES

1. Tuchin, V.V., Light–tissue interactions, in Vo-Dinh, T. ed., 2014. *Biomedical Photonics Handbook: Biomedical Diagnostics*. Boca Raton, FL: CRC Press, pp. 123–168.

2. Sanger, F., Nicklen, S. and Coulson, A.R., 1977. DNA sequencing with chain-terminating inhibitors. *Proceedings of the National Academy of Sciences*, 74(12), pp. 5463–5467.
3. Smith, L.M., Sanders, J.Z., Kaiser, R.J., Hughes, P., Dodd, C., Connell, C.R., Heiner, C., Kent, S.B. and Hood, L.E., 1985. Fluorescence detection in automated DNA sequence analysis. *Nature*, 321(6071), pp. 674–679.
4. Slatko, B.E., Kieleczawa, J., Ju, J., Gardner, A.F., Hendrickson, C.L. and Ausubel, F.M., 2011. "First generation" automated DNA sequencing technology. *Current Protocols in Molecular Biology*, 96, pp. 7.2.1–7.2.28.
5. International Human Genome Sequencing Consortium. 2004. Finishing the euchromatic sequence of the human genome. *Nature*, 431(7011), pp. 931–945.
6. Anderson, M.W. and Schrijver, I., 2010. Next generation DNA sequencing and the future of genomic medicine. *Genes*, 1(1), pp. 38–69.
7. Vo-Dinh T. and Griffin, G.D., Biochips and microarrays: Tools for new medicine, in Vo-Dinh, T. ed., 2014. *Biomedical Photonics Handbook: Biomedical Diagnostics*. Boca Raton, FL: CRC Press, pp. 77–124.
8. Vo-Dinh, T., Griffin, G., Stokes, D.L. and Wintenberg, A., 2003. Multi-functional biochip for medical diagnostics and pathogen detection. *Sensors and Actuators B: Chemical*, 90(1), pp. 104–111.
9. Vo-Dinh, T., Allain, L.R. and Stokes, D.L., 2002. Cancer gene detection using surface-enhanced Raman scattering (SERS). *Journal of Raman Spectroscopy*, 33(7), pp. 511–516.
10. Vo-Dinh, T., Isola, N., Alarie, J.P., Landis, D., Griffin, G.D. and Allison, S., 1998. Development of a multiarray biosensor for DNA diagnostics. *Instrumentation Science and Technology*, 26(5), pp. 503–514.
11. Vo-Dinh, T. and Askari, M., 2001. Micro arrays and biochips: Applications and potential in genomics and proteomics. *Current Genomics*, 2(4), pp. 399–415.
12. Walt, D.R. and Franz, D.R., 2000. Peer reviewed: Biological warfare detection. *Analytical Chemistry*, 72(23), pp. 738-A.
13. Hacia, J.G., Edgemon, K., Hunt, N., Collins, F.S., Woski, S.A., Fidanza, J., McGall, G. and Fodor, S.P., 1998. Enhanced high density oligonucleotide array-based sequence analysis using modified nucleoside triphosphates. *Nucleic Acids Research*, 26(21), pp. 4975–4982.
14. Schena, M., Shalon, D., Davis, R.W. and Brown, P.O., 1995. Quantitative monitoring of gene expression patterns with a complementary DNA microarray. *Science*, 270(5235), p. 467.
15. Lipshutz, R.J., Fodor, S.P., Gingeras, T.R. and Lockhart, D.J., 1999. High density synthetic oligonucleotide arrays. *Nature Genetics*, 21, pp. 20–24.
16. Wang, J., Rivas, G., Fernandes, J.R., Paz, J.L.L., Jiang, M. and Waymire, R., 1998. Indicator-free electrochemical DNA hybridization biosensor. *Analytica Chimica Acta*, 375(3), pp. 197–203.
17. Niemeyer, C.M., Boldt, L., Ceyhan, B. and Blohm, D., 1999. DNA-directed immobilization: Efficient, reversible, and site-selective surface binding of proteins by means of covalent DNA-streptavidin conjugates. *Analytical Biochemistry*, 268(1), pp. 54–63.
18. Sawata, S., Kai, E., Ikebukuro, K., Iida, T., Honda, T. and Karube, I., 1999. Application of peptide nucleic acid to the direct detection of deoxyribonucleic acid amplified by polymerase chain reaction. *Biosensors and Bioelectronics*, 14(4), pp. 397–404.
19. Vo-Dinh T. and Cullum B., Fluorescence spectrometry in biomedical diagnostics, in Vo-Dinh, T. ed., 2014. *Biomedical Photonics Handbook: Biomedical Diagnostics*. Boca Raton, FL: CRC Press, pp. 485–542.
20. Vo-Dinh, T., Panjehpour, M. and Overholt, B.F., 1998. Laser-induced fluorescence for esophageal cancer and dysplasia diagnosis. *Annals of the New York Academy of Sciences*, 838(1), pp. 116–122.
21. Sacks, P.G., Savage, H.E., Levine, J., Kolli, V.R., Alfano, R.R. and Schantz, S.P., 1996. Native cellular fluorescence identifies terminal squamous differentiation of normal oral epithelial cells in culture: A potential chemoprevention biomarker. *Cancer Letters*, 104(2), pp. 171–181.

22. Ramanujam, N., Mitchell, M.F., Mahadevan, A., Warren, S., Thomsen, S., Silva, E. and Richards-Kortum, R., 1994. In vivo diagnosis of cervical intraepithelial neoplasia using 337-nm-excited laser-induced fluorescence. *Proceedings of the National Academy of Sciences*, 91(21), pp. 10193–10197.
23. Lam, S., Hung, J.Y., Kennedy, S.M., Leriche, J.C., Vedal, S., Nelems, B., MacAulay, C.E. and Palcic, B., 1992. Detection of dysplasia and carcinoma in situ by ratio fluorometry. *American Review of Respiratory Disease*, 146, pp. 1458–1461.
24. Zellweger, M., Grosjean, P., Goujon, D., Monnier, P., van den Bergh, H. and Wagnieres, G., 2001. In vivo autofluorescence spectroscopy of human bronchial tissue to optimize the detection and imaging of early cancers. *Journal of Biomedical Optics*, 6(1), pp. 41–51.
25. Heintzelman, D.L., Utzinger, U., Fuchs, H., Zuluaga, A., Gossage, K., Gillenwater, A.M., Jacob, R., Kemp, B. and Richards-Kortum, R.R., 2000. Optimal excitation wavelengths for in vivo detection of oral neoplasia using fluorescence spectroscopy. *Photochemistry and Photobiology*, 72(1), pp. 103–113.
26. Vo-Dinh, T., 1982. Synchronous luminescence spectroscopy: Methodology and applicability. *Applied Spectroscopy*, 36(5), pp. 576–581.
27. DaCosta, R.S., Wilson, B.C. and Marcon, N.E., 2007. Fluorescence and spectral imaging. *The Scientific World Journal*, 7, pp. 2046–2071.
28. Harries, M.L., Lam, S., MacAulay, C., Qu, J. and Palcic, B., 1995. Diagnostic imaging of the larynx: Autofluorescence of laryngeal tumours using the helium-cadmium laser. *The Journal of Laryngology and Otology*, 109(02), pp. 108–110.
29. Lam, S., MacAulay, C., leRiche, J.C. and Palcic, B., 2000. Detection and localization of early lung cancer by fluorescence bronchoscopy. *Cancer*, 89(S11), pp. 2468–2473.
30. Haringsma, J., Tytgat, G.N., Yano, H., Iishi, H., Tatsuta, M., Ogihara, T., Watanabe, H., Sato, N., Marcon, N., Wilson, B.C. and Cline, R.W., 2001. Autofluorescence endoscopy: Feasibility of detection of GI neoplasms unapparent to white light endoscopy with an evolving technology. *Gastrointestinal Endoscopy*, 53(6), pp. 642–650.
31. DaCosta, R.S., Wilson, B.C. and Marcon, N.E., 2000. Light-induced fluorescence endoscopy of the gastrointestinal tract. *Gastrointestinal Endoscopy Clinics of North America*, 10(1), pp. 37–69.
32. Chissov, V.I., Sokolov, V.V., Filonenko, E.V., Menenkov, V.D., Zharkova, N.N., Kozlov, D.N., Polivanov, I., Prokhorov, A.M., Pykhov, R.L. and Smirnov, V.V., 1994. Clinical fluorescent diagnosis of tumors using photosensitizer photogem. *Khirurgiia*, 5, pp. 37–41.
33. Heyerdahl, H., Wang, I., Liu, D.L., Berg, R., Andersson-Engels, S., Peng, Q., Moan, J., Svanberg, S. and Svanberg, K., 1997. Pharmacokinetic studies on 5-aminolevulinic acid-induced protoporphyrin IX accumulation in tumours and normal tissues. *Cancer Letters*, 112(2), pp. 225–231.
34. von Holstein, C.S., Nilsson, A.M., Andersson-Engels, S., Willen, R., Walther, B. and Svanberg, K., 1996. Detection of adenocarcinoma in Barrett's oesophagus by means of laser induced fluorescence. *Gut*, 39(5), pp. 711–716.
35. Baert, L., Berg, R., Van Damme, B., D'Hallewin, M.A., Johansson, J., Svanberg, K. and Svanberg, S., 1993. Clinical fluorescence diagnosis of human bladder carcinoma following low dose photofrin injection. *Urology*, 41(4), pp. 322–330.
36. Folli, S., Wagnieres, G., Pelegrin, A., Calmes, J.M., Braichotte, D., Buchegger, F., Chalandon, Y., Hardman, N., Heusser, C.H. and Givel, J.C., 1992. Immunophotodiagnosis of colon carcinomas in patients injected with fluoresceinated chimeric antibodies against carcinoembryonic antigen. *Proceedings of the National Academy of Sciences*, 89(17), pp. 7973–7977.
37. Mourant, J. R. and Bigio, I. J., Elastic scattering spectrometry and diffuse reflectance, in Vo-Dinh, T. ed., 2014. *Biomedical Photonics Handbook: Biomedical Diagnostics*. Boca Raton, FL: CRC Press, pp. 543–564.

38. Gurjar, R.S., Backman, V., Perelman, L.T., Georgakoudi, I., Badizadegan, K., Itzkan, I., Dasari, R.R. and Feld, M.S., 2001. Imaging human epithelial properties with polarized light-scattering spectroscopy. *Nature Medicine*, 7(11), pp. 1245–1248.
39. Mourant, J.R., Bigio, I.J., Boyer, J., Conn, R.L., Johnson, T. and Shimada, T., 1995. Spectroscopic diagnosis of bladder cancer with elastic light scattering. *Lasers in Surgery and Medicine*, 17(4), pp. 350–357.
40. Mahadevan-Jansen, A., Patil, C.A. and Pence, I.J., Raman spectroscopy: From benchtop to bedside, in Vo-Dinh, T. ed., 2014. *Biomedical Photonics Handbook: Biomedical Diagnostics*. Boca Raton, FL: CRC Press, pp. 759–802.
41. Mahadevan-Jansen, A., Mitchell, M.F., Ramanujam, N., Utzinger, U. and Richards-Kortum, R., 1998. Development of a fiber optic probe to measure NIR Raman spectra of cervical tissue in vivo. *Photochemistry and Photobiology*, 68(3), pp. 427–431.
42. Huang, Z., McWilliams, A., Lui, H., McLean, D.I., Lam, S. and Zeng, H., 2003. Near-infrared Raman spectroscopy for optical diagnosis of lung cancer. *International Journal of Cancer*, 107(6), pp. 1047–1052.
43. Jarvis, R.M. and Goodacre, R., 2004. Discrimination of bacteria using surface-enhanced Raman spectroscopy. *Analytical Chemistry*, 76(1), pp. 40–47.
44. Kastanos, E.K., Kyriakides, A., Hadjigeorgiou, K. and Pitris, C., 2010. A novel method for urinary tract infection diagnosis and antibiogram using Raman spectroscopy. *Journal of Raman Spectroscopy*, 41(9), pp. 958–963.
45. Stiles, P.L., Dieringer, J.A., Shah, N.C. and Van Duyne, R.P., 2008. Surface-enhanced Raman spectroscopy. *Annual Review Analytical Chemistry*, 1, pp. 601–626.
46. Kneipp, K., Wang, Y., Kneipp, H., Perelman, L.T., Itzkan, I., Dasari, R.R. and Feld, M.S., 1997. Single molecule detection using surface-enhanced Raman scattering (SERS). *Physical Review Letters*, 78(9), p. 1667.
47. Vo-Dinh, T., Yan, F. and Stokes, D.L., Plasmonics-based nanostructures for surface-enhanced Raman scattering bioanalysis, in Vo-Dinh, T. ed., 2005. *Protein Nanotechnology: Protocols, Instrumentation, and Applications*. Totowa, NJ: Humana Press Inc., pp. 255–283.
48. Frohlich, V.C., 2008. Phase contrast and differential interference contrast (DIC) microscopy. *Journal of Visualized Experiments: JoVE*, 6(17), p. 844.
49. Webb, R.H., 1996. Confocal optical microscopy. *Reports on Progress in Physics*, 59(3), p. 427.
50. König, K., 2000. Multiphoton microscopy in life sciences. *Journal of Microscopy*, 200(2), pp. 83–104.
51. Willig, K.I., Rizzoli, S.O., Westphal, V., Jahn, R. and Hell, S.W., 2006. STED microscopy reveals that synaptotagmin remains clustered after synaptic vesicle exocytosis. *Nature*, 440(7086), pp. 935–939.
52. Chen, Y., Bousie, E., Pitris, C. and Fujimoto, J.G., Optical coherence tomography: Introduction and theory, in Boas, D.A., Pitris, C. and Ramanujam, N. eds., 2011. *Handbook of Biomedical Optics*. Boca Raton, FL: CRC Press, pp. 255–280.
53. Schuman, J.S., Puliafito, C.A., Fujimoto, J.G. and Duker, J.S. eds., 2004. *Optical Coherence Tomography of Ocular Diseases*. Thorofare, NJ: Slack, pp. 1–698.
54. Goldberg, B.D., Suter, M.J., Tearney, G.J. and Bouma, B.E., Optical coherence tomography: Clinical applications, in Boas, D.A., Pitris, C. and Ramanujam, N. eds., 2011. *Handbook of Biomedical Optics*. Boca Raton, FL: CRC Press, pp. 303–318.
55. Skala, M.C., Tao, Y.K, Davis, A.M. and Izatt, J.A., Functional optical coherence tomography in preclinical models, in Boas, D.A., Pitris, C. and Ramanujam, N. eds., 2011. *Handbook of Biomedical Optics*. Boca Raton, FL: CRC Press, pp. 281–302.
56. Adler, D.C., Zhou, C., Tsai, T.H., Schmitt, J., Huang, Q., Mashimo, H. and Fujimoto, J.G., 2009. Three-dimensional endomicroscopy of the human colon using optical coherence tomography. *Optics Express*, 17(2), pp. 784–796.
57. Busse, G. and Rosencwaig, A., 1980. Subsurface imaging with photoacoustics. *Applied Physics Letters*, 36, p. 815.

58. Xia, J., Yao, J. and Wang, L.V., 2014. Photoacoustic tomography: Principles and advances. *Electromagnetic Waves* (Cambridge, MA), 147, p. 1.
59. Taruttis, A. and Ntziachristos, V., 2015. Advances in real-time multispectral optoacoustic imaging and its applications. *Nature Photonics*, 9(4), pp. 219–227.
60. Xia, J., Chatni, M.R., Maslov, K., Guo, Z., Wang, K., Anastasio, M. and Wang, L.V., 2012. Whole-body ring-shaped confocal photoacoustic computed tomography of small animals in vivo. *Journal of Biomedical Optics*, 17(5), pp. 0505061–0505063.
61. Wang, L.V. and Hu, S., 2012. Photoacoustic tomography: In vivo imaging from organelles to organs. *Science*, 335(6075), pp. 1458–1462.
62. Zackrisson, S., van de Ven, S.M.W.Y. and Gambhir, S.S., 2014. Light in and sound out: Emerging translational strategies for photoacoustic imaging. *Cancer Research*, 74(4), pp. 979–1004.
63. Cox, B., Laufer, J.G., Arridge, S.R. and Beard, P.C., 2012. Quantitative spectroscopic photoacoustic imaging: A review. *Journal of Biomedical Optics*, 17(6), pp. 0612021–06120222.
64. Luke, G.P., Yeager, D. and Emelianov, S.Y., 2012. Biomedical applications of photoacoustic imaging with exogenous contrast agents. *Annals of Biomedical Engineering*, 40(2), pp. 422–437.
65. Kim, C., Erpelding, T.N., Maslov, K., Jankovic, L., Akers, W.J., Song, L., Achilefu, S., Margenthaler, J.A., Pashley, M.D. and Wang, L.V., 2010. Handheld array-based photoacoustic probe for guiding needle biopsy of sentinel lymph nodes. *Journal of Biomedical Optics*, 15(4), pp. 046010–046010.
66. Sevick-Muraka, E. M., Near infrared fluorescence imaging and spectrometry in random media and tissues, in Vo-Dinh, T. ed., 2014. *Biomedical Photonics Handbook: Biomedical Diagnostics*. Boca Raton, FL: CRC Press, 587–652.
67. Jacques, S.L., Ramanujam, N., Vishnoi, G., Choe, R. and Chance, B., 2000. Modeling photon transport in transabdominal fetal oximetry. *Journal of Biomedical Optics*, 5(3), pp. 277–282.
68. Norton, S.J. and Vo-Dinh, T., 1998. Diffraction tomographic imaging with photon density waves: An explicit solution. *JOSA A*, 15(10), pp. 2670–2677.
69. Gibson, A. and Dehghani, H., 2009. Diffuse optical imaging. *Philosophical Transactions of the Royal Society of London A: Mathematical, Physical and Engineering Sciences*, 367(1900), pp. 3055–3072.
70. Austin, T., Gibson, A.P., Branco, G., Yusof, R.M., Arridge, S.R., Meek, J.H., Wyatt, J.S., Delpy, D.T. and Hebden, J.C., 2006. Three dimensional optical imaging of blood volume and oxygenation in the neonatal brain. *Neuroimage*, 31(4), pp. 1426–1433.
71. Tromberg, B.J., Pogue, B.W., Paulsen, K.D., Yodh, A.G., Boas, D.A. and Cerussi, A.E., 2008. Assessing the future of diffuse optical imaging technologies for breast cancer management. *Medical physics*, 35(6), pp. 2443–2451.
72. Dorin, G., 2004. Evolution of retinal laser therapy: Minimum intensity photocoagulation (MIP). Can the laser heal the retina without harming it? *Seminars in Ophthalmology*, 19(1–2), pp. 62–68.
73. Shortt, A.J, Allan, B.D.S. and Evans, J.R., 2013. Laser-assisted in-situ keratomileusis (LASIK) versus photorefractive keratectomy (PRK) for myopia. *Cochrane Database of Systematic Reviews 2013*, 1(1), p. CD005135. DOI: 10.1002/14651858.CD005135.pub3.
74. Tanzi, E.L., Lupton, J.R. and Alster, T.S., 2003. Lasers in dermatology: Four decades of progress. *Journal of the American Academy of Dermatology*, 49(1), pp. 1–34.
75. Bown, S.G., 1983. Phototherapy of tumors. *World Journal of Surgery*, 7(6), pp. 700–709.
76. Briggs, G., Lee, A. and Bown, S., Laser treatment of breast tumors, in Vo-Dinh, T. ed., 2003. *Biomedical Photonics Handbook: Biomedical Diagnostics*. Boca Raton, FL: CRC Press pp. 47-1–47-14.
77. Dougherty, T.J., Gomer, C.J., Henderson, B.W., Jori, G., Kessel, D., Korbelik, M., Moan, J. and Peng, Q., 1998. Photodynamic therapy. *Journal of the National Cancer Institute*, 90(12), pp. 889–905.
78. Henderson, B. and Gollnick, S.O., Mechanistic principles of photodynamic therapy, in Vo-Dinh, T. ed., 2014. *Biomedical*

Photonics Handbook: Biomedical Diagnostics. Boca Raton, FL: CRC Press, pp. 3–30.

79. Dougherty, T.J. and Levy, G.J., Photodynamic therapy (PDT) and clinical applications, in Vo-Dinh, T. ed., 2003. *Biomedical Photonics Handbook: Biomedical Diagnostics*. Boca Raton, FL: CRC Press. pp. 38-1–38-16.
80. Chung, H., Dai, T., Sharma, S.K., Huang, Y.Y., Carroll, J.D. and Hamblin, M.R., 2012. The nuts and bolts of low-level laser (light) therapy. *Annals of Biomedical Engineering*, 40(2), pp. 516–533.
81. Luppa, P.B., Müller, C., Schlichtiger, A. and Schlebusch, H., 2011. Point-of-care testing (POCT): Current techniques and future perspectives. *TrAC Trends in Analytical Chemistry*, 30(6), pp. 887–898.
82. Kost, G.J., 1995. Guidelines for point-of-care testing. Improving patient outcomes. *American Journal of Clinical Pathology*, 104(4 Suppl 1), pp. S111–S127.
83. St John, A. and Price, C.P., 2014. Existing and emerging technologies for point-of-care testing. *The Clinical Biochemist Reviews*, 35(3), p. 155.
84. Peeling, R.W., Holmes, K.K, Mabey, D. and Ronald, A., 2006. Rapid tests for sexually transmitted infections (STIs): The way forward. Sexually Transmitted Infections, 82(Suppl V), pp. v1–v6.
85. Burtis, C.A., Ashwood, E.R. and Bruns, D.E., 2012. *Tietz Textbook of Clinical Chemistry and Molecular Diagnostics*. St. Louis: Elsevier Health Sciences.
86. Yetisen, A.K., Akram, M.S. and Lowe, C.R., 2013. Paper-based microfluidic point-of-care diagnostic devices. *Lab on a Chip*, 13(12), pp. 2210–2251.
87. Tweed, K., Google Working on Smart Contact Lens to Monitor Diabetes, http://spectrum.ieee.org/tech-talk/biomedical/devices/google-working-on-smart-contact-lens-to-monitor-diabetes (last accessed 20/12/2016).
88. Bandodkar, A.J., Hung, V.W., Jia, W., Valdés-Ramírez, G., Windmiller, J.R., Martinez, A.G., Ramírez, J., Chan, G., Kerman, K. and Wang, J., 2013. Tattoo-based potentiometric ion-selective sensors for epidermal pH monitoring. *Analyst*, 138(1), pp. 123–128.
89. Yetisen, A.K., 2015. The prospects for holographic sensors, in *Holographic Sensors*. Cham, Switzerland Springer International Publishing. pp. 149–162.
90. Hellebust, A. and Richards-Kortum, R., 2012. Advances in molecular imaging: Targeted optical contrast agents for cancer diagnostics. *Nanomedicine*, 7(3), pp. 429–445.
91. Lee, D.Y. and Li, K.C.P., 2011. Molecular theranostics: A primer for the imaging professional. *American Journal of Roentgenology*, 197(2), pp. 318–324.
92. Hellebust, A. and Richards-Kortum, R., 2012. Advances in molecular imaging: Targeted optical contrast agents for cancer diagnostics. *Nanomedicine*, 7(3), pp. 429–445.
93. Frangioni J.V., 2008. New technologies for human cancer imaging. *Journal Clinical Oncology*, 26(24), pp. 4012–4021.
94. National Research Council, 2012. *Optics and Photonics: Essential Technologies for Our Nation*. National Academies of Science, Division on Engineering and Physical Sciences: National Materials and Manufacturing Board. Washington, DC: National Academies Press. Retrieved October, 23, p. 2014.

27

Breath analysis with mid-infrared diagnostics

VJEKOSLAV KOKORIC, ERHAN TÜTÜNCÜ, FELICIA SEICHTER, ANDREAS WILK, PAULA R. FORTES, IVO M. RAIMUNDO, AND BORIS MIZAIKOFF
University of Ulm

27.1 INTRODUCTION

Exhaled human breath (EB) contains a variety of volatile disease biomarkers, which could either be endogenously produced or administered and subsequently metabolized within the human body. The ancient Greek physicians knew that the specific odor of EB could be associated with certain diseases or health conditions. For example, a sweet smell in the breath may be an indication of diabetes, a fishy smell could be a result of a liver disease, and a urine-like smell could be related to kidney failures. Modern interest in human breath analysis was initiated in 1971, when Pauling et al. (1971) published a method to detect ~250 volatile organic compounds (VOCs) in breath via gas liquid partition chromatography. Over the past few decades, an increasing number of analytical detection methods have been developed for addressing breath constituents, and several biogenic trace gases have been identified as markers for cancer, schizophrenia, diabetes, and other diseases. Currently, the most common breath tests rely on the detection of nitric oxide in EB serving as a diagnostic indicator of airway inflammation and on the quantification of exhaled ^{13}C-labeled carbon dioxide indicative of an infection of the gastric mucosa originating from *Helicobacter pylori*. To date, the breath test for *H. pylori* infection is one of the very few breath tests approved by the U.S. Food and Drug Administration.

The major components in EB are nitrogen (~78%), oxygen (~17%), carbon dioxide (~4%), and water vapor. In addition, several hundred volatile organic species in the concentration range of few tens of parts per billion by volume (ppbv) and below are found in human breath. However, human breath contains only few VOCs at concentrations higher than a few tens of parts per billion. Due to these very low concentrations, enrichment strategies are frequently required for reliable detection/quantification. A variety of preconcentration techniques are applied in breath analysis, including, solid-phase microextraction,

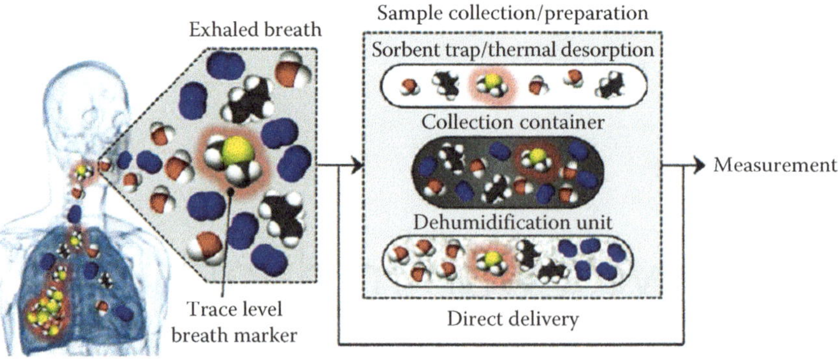

Figure 27.1 Schematic of commonly applied breath sampling and measurement routines. (Reprinted with permission from Konvalina, G., and Haick, H., *Accounts of Chemical Research* 47 (1): 66–76. Copyright 2014 American Chemical Society.)

membrane extraction with a sorbent interface, and sorbent enrichment methods (Ras et al. 2009). Preconcentration of VOCs onto solid sorbents followed by thermal desorption is among the most commonly applied techniques when analyzing breath (Alonso and Sanchez 2013).

One of the major advantageous characteristics of exhaled breath analysis methods is the noninvasive nature of the procedure. Consequently, analyzing exhaled biomarkers enables noninvasive monitoring of patients for disease detection and diagnosis, and/or therapy progress monitoring as well as compliance testing during, for example, clinical drug trials. Ideally, EB analysis is applied as a direct detection scheme or after brief sample preconcentration rather than discontinuous laboratory analysis that requires extensive breath sampling, storage, and sample preparation.

A particular advantage of breath analysis based on optical/spectroscopic techniques is the capability of *in situ* and online measurements, along with a substantial potential for system miniaturization toward compact—ideally handheld—device footprints. Here, the direct analysis of EB in real time—or quasi-real time if a brief preconcentration step is required—is considered an online measurement herein; hence, the clinician is provided with an instantaneous response rather than awaiting the results of conventional laboratory analyses. Nevertheless, in routine clinical diagnostics, for example, using gas chromatography coupled to mass spectrometry (GC-MS), offline analysis remains the most commonly applied breath diagnostic strategy, and EB is usually collected into inert Tedlar bags and electropolished stainless steel canisters, or adsorbed onto specific sorbent materials for subsequent analysis, as schematically shown in Figure 27.1. Thus, collected breath samples are then either preconcentrated or separated into the pattern of VOCs using laboratory-based separation and analysis techniques (i.e., offline breath analysis).

Besides the time offset to obtain diagnostically relevant data, potential problems associated with offline methods include the reproducibility and repeatability of the breath collection and storage routine, and potential changes of the sample composition or contamination during sampling or sample storage. Hence, analyzing/monitoring EB (quasi) online or at least without the requirement of sample storage may overcome these problems next to providing direct feedback to the clinician or physician.

27.2 EQUIPMENT

Infrared spectroscopy is an optical measurement technique that provides direct access to the fundamental vibrational and vibro-rotational signatures of molecular constituents, and is thus ideally suited as a molecularly selective sensing technique specifically for VOC analysis. IR sensing instrumentation is usually based on three main components: (1) broad- or narrowband radiation source, (2) transducer/waveguide/gas cell (or a combination thereof), and (3) radiation detector, as schematically illustrated in Figure 27.2. Radiation propagation in between these components may be guided either in free space using appropriate optics (e.g., mirrors, lenses, etc.) or within optical waveguides (e.g., optical fibers, light pipes, etc.).

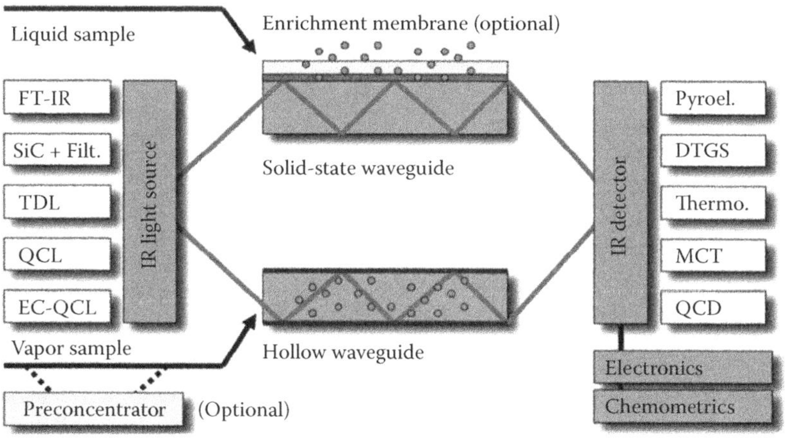

Figure 27.2 Schematic overview of the fundamental measurement principles commonly applied in IR sensing devices. (Mizaikoff, B., *Chem. Soc.* Rev., 42 (22), 8683, 2013. Reproduced by permission of The Royal Society of Chemistry.)

Conventionally, infrared spectrometers—especially Fourier transform infrared (FTIR) spectrometers—equipped with a broadband blackbody thermal emitter (e.g., SiC filament) serve as light source. Thus, the entire MIR spectral range (2.5–20 μm; 500–5000 cm^{-1}) is available for analysis. Alternatively, given their much more compact dimensions, recent technological advancements in terms of emission wavelength regime, tunability, robustness, ease of operation, and commercial availability, infrared laser light sources are increasingly used for advanced IR gas sensing systems:

- Tunable diode lasers (TDLs) covering the spectral range from approximately 0.75 to 2 μm; MIR TDLs (e.g., lead salt laser diodes) have largely been replaced by QCLs (explained next). Optical mixing technique (e.g., difference frequency generation) is commonly employed to extend the useful spectral range.
- Quantum cascade lasers (QCLs), which rely on epitaxially grown superlattices (i.e., a stack of bandstructure-engineered semiconductor layers). The thicknesses of the layers (i.e., the quantum well) largely govern the emission wavelength of the QCL. By integrating a diffraction grating within the device, the laser can be forced to emit at a single frequency (i.e., distributed feedback laser). Control of the device temperature provides tunability in a narrow range (typically <10 cm^{-1}). Operating the QCL in combination with an external cavity (EC-QCL) and an appropriate grating enables tuning across a wavelength band up to several hundreds of wavenumbers.
- Interband cascade lasers (ICLs) use similar semiconductor structures and device concepts as QCLs, yet photons are generated via interband transitions rather than intersubband transitions, which should in principle facilitate lower (electrical) power device operation.

Depending on the targeted analyte and application (e.g., requiring portable systems), several detector types are available for MIR breath diagnostics:

- Bolometers, which register changes in resistance
- Thermopiles relying on the thermoelectric effect
- Pyroelectric detectors using materials such as deuterated (L-alanine doped) triglycine sulfate (DTGS and DLaTGS)
- Photovoltaic or photoconductive semiconductor-based devices, including mercury–cadmium–telluride (MCT)
- Intersubband detectors, which may be operated either in photoconductive mode (quantum well infrared photodetector) or in photovoltaic mode (quantum cascade detector (QCD))

Given the required sensitivities in the gas phase and specifically in breath diagnostic applications, usually cooled photoconductive or intersubband detectors are applied, whereas thermal detectors

Table 27.1 Selected endogenous trace gases/biomarkers found in exhaled breath

Breath constituent	Spectral fingerprint (μm)	Average fraction
Acetone	5.7	0–1 ppm
Ammonia (NH_3)	10.0	0–1 ppm
Carbon monoxide (CO)	4.6	1–10 ppm
Carbonyl sulfide (OCS)	4.9	0–10 ppb
Ethane (C_2H_6)	3.4	0–5 ppb
Isoprene (C_5H_8)	11.1	50–200 ppb
Methane (CH_4)	3.35	0–20 ppm
Nitric oxide (NO)	5.0	10–30 ppb
Nitrous oxide (N_2O)	4.5	0–20 ppb
Pentane (C_5H_{12})	3.4	0–5 ppb

Source: NATO Advanced Research Workshop on Middle Infrared Coherent Sources (MICS), 2007.

such as thermopiles are useful for higher concentration levels.

Most commonly, DTGS and MCT detectors are used. Although the latter provide the highest sensitivity, they have to be operated at cryogenic temperatures for maximum detectivity, which requires liquid nitrogen, thermoelectric or mechanical (e.g., Stirling) cooling; pyroelectric detectors provide the advantage of room temperature operation without additional cooling requirements, albeit at lower detectivities.

In the following text, the sample compartment responsible for reproducible interaction between photons and gas-phase molecules (i.e., gas cell) will be discussed in detail, as light source and detectors are largely the same for all MIR breath diagnostic devices. Smartly structuring the gas cell and/or hollow waveguide structure enables tailoring the sensitivity of MIR breath diagnostics in terms of maximizing the absorption path length and/or the achievable signal-to-noise ratio (SNR).

The required sensitivities are largely governed by the occurrence of volatile biomarkers within the exhaled breath matrix. Table 27.1 gives a brief overview of selected biomarkers and the anticipated concentration range to set the stage for the ultra-trace sensitivities demanded in breath diagnostics.

27.3 SINGLE- AND MULTIPASS ABSORPTION-BASED SENSING METHODS

Direct absorption spectroscopy is probably the most robust optical detection method in gas analysis, and may detect and discriminate a wide variety of molecular species at trace levels. As previously indicated, the basic optical setup comprises a (broadband or narrowband) light source, a sampling cell/transducer, and a photon detector. Usually, the MIR beam is passed through a so-called gas cell, which encompasses the gas phase sample/analyte(s), and provides for a well-defined absorption path length; the reduced radiation transmittance due to analyte absorption is monitored as a function of the wavelength and quantified via the Beer–Lambert law ($A = \log I_0/I$). Due to the usually minute analyte concentrations and/or concentration changes to be monitored within EB (compare Table 27.1), it is essential to maximize (1) the signal intensity (by selecting an appropriate absorption path length) and (2) the signal-to-noise ratio for discriminating small concentration changes against the background noise of the sensor system. Maximizing the analytical signal is usually achieved by propagating the IR beam multiple times within the gas cell, thereby extending the absorption path length. White- or Herriott-type multipass cells have successfully been used in gas phase MIR spectroscopy, extending the effective optical path length up to several tens or even several hundreds of meters using sophisticated mirror arrangements. Yet, it should be noted that such long-path gas cells require substantial volumes of sample gas (i.e., several hundreds of milliliters to several liters), thereby limiting the utility of this strategy to acceptable volumes and sample transient times in case of continuous monitoring applications.

For example, Jouy et al. (2014) have reported the application of a multipass gas cell-based sensing system combining QCLs and QCDs for the analysis of the $^{13}CO_2/^{12}CO_2$ isotope ratios. This isotope ratio (aka, "delta value") is a commonly used marker value in breath monitoring reflecting a variety of pulmonary conditions such as, for example, sepsis (Wilk et al. 2012). A compact sensing system (Figure 27.3) has been developed, comprising a III/V semiconductor QCL lasing at 2310 cm^{-1} combined with an innovative toroidal mirror cell (Figure 27.4), generating an absorption path length of up to 40 m, yet maintaining a device footprint of 13 cm × 30.5 cm.

To refocus the laser beam, the mirror surface of the toroidal cell is appropriately carved at the inside of the metallic cylinder with an inner diameter of 8 cm, thus making it act as a multipass gas cell (cell volume: 40 cm^3). The suitability of this sensor concept was verified by achieving an Allan deviation—for 400 parts per million by volume carbon dioxide—of 2‰ after 1 s acquisition time and 0.2‰ after 600 s.

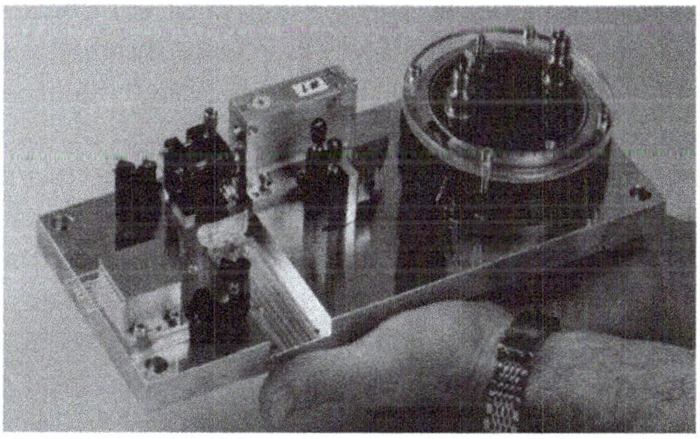

Figure 27.3 Miniaturized gas sensor with QCL, toroidal multipass gas cell, and QCD. , (Jouy, P. et al., *Analyst*, 139 (9), 2039–46,2014. Reproduced by permission of The Royal Society of Chemistry.)

Although the detectivity of QCDs is lower compared to conventionally applied MCT detectors, their high level of integration and potential for assisting compact device footprints provides an attractive strategy toward portable devices.

Further reduction of the size, mass, and volume (i.e., to a few hundreds of microliters or less) of such sensors may be achieved by the application of so-called hollow waveguides (HWGs), which may simultaneously serve as miniaturized gas cells, and as light conduits for propagating MIR photons. Light guiding is established by reflection along the inside walls of a hollow core tube coated, for example, with a metal film (e.g., Ag or Au) and a layer of a dielectric material (e.g., AgI). For applications with reduced availability of sampling gas (e.g., investigating mouse breath, minute gas volumes above plant or animal cells, etc.), HWGs offer considerable advantages compared to conventional multipass gas cells. Recently, Mizaikoff and collaborators (Wilk et al. 2013) developed a new generation of devices, the so-called substrate-integrated HWGs (iHWGs), which pave the way toward ultra-compact MIR gas sensing platforms providing a yet unprecedented level of integration. These waveguides are based on the integration of a waveguide channel structure within a solid substrate such as aluminum (Figure 27.5), brass, plastics, and semiconductors,

(a) (b)

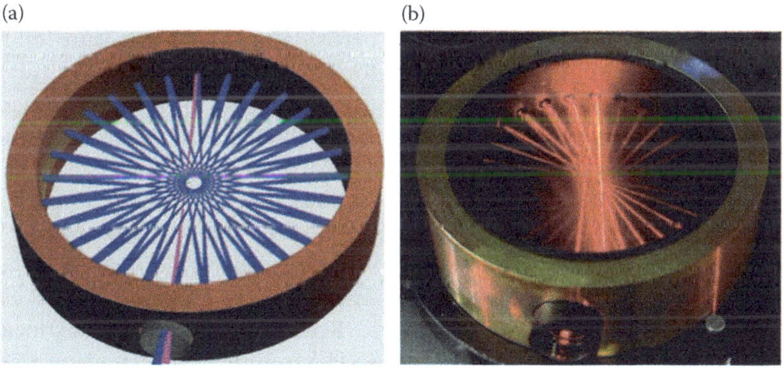

Figure 27.4 (a) Schematic beam propagation within the toroidal multipass gas cell simulated by ray tracing. (b) Multiple reflections demonstrated by a red laser diode reflected within the toroidal gas cell. (Jouy, P. et al., *Analyst*, 139 (9), 2039–46,2014. Reproduced by permission of The Royal Society of Chemistry.)

Figure 27.5 Examples of iHWG structures based on uncoated and Au-coated Al substrates. (Mizaikoff, B., *Chem. Soc. Rev.*, 42 (22), 8683, 2013. Reproduced by permission of The Royal Society of Chemistry.)

which may be conveniently coated with any kind of additional material at the surface (e.g., for enhancing the reflectivity) due to the open channel structure. A remarkable feature of the iHWG is the opportunity to fully integrate complementary sensors (e.g., optical oxygen sensors, temperature and pressure sensors, etc.) within the iHWG, and even to fully integrate a light source and a detector into the same substrate without the requirement of additional coupling optics. Hence, it is anticipated that iHWGs will be the future platform technology facilitating MIR-lab-on-a-chip functionality for establishing ultra-compact vapor- phase sensing systems applicable in EB diagnostics.

Recently, the Mizaikoff team has demonstrated the utility of an iHWG with only 22 cm optical path length (OPL) in combination with a miniaturized preconcentration unit (i.e., same footprint as the iHWG) for the analysis of volatile biomarkers in EB (Perez-Guaita et al. 2014). Isoprene (C_5H_8), an endogenous VOC in human breath, was analyzed with the iHWG coupled to a compact BrukerIR-Cube FTIR spectrometer. A limit of detection (LOD) of 106 ppbv was achieved, thus, covering medium to high concentrations of isoprene in EB. With the developed preconcentration device, an improvement by a factor of 120 was obtained in comparison to direct sensing. Additionally, EB from a human smoker was analyzed using that sensing device—it could determine a concentration of 467 ± 18 ppbv of isoprene in the EB. This functional demonstration combining iHWG technology and miniaturized preconcentrator devices indicates the potential for addressing even low concentrated breath biomarkers that have not been detectable using MIR sensing techniques to date.

27.3.1 Photoacoustic sensing techniques

Another frequently applied sensing technique for trace gas analysis in breath diagnostics is photoacoustic spectroscopy (PAS), which is nowadays frequently performed in combination with QCLs. In contrast to conventional absorption spectroscopy, PAS does not directly measure the absorption of light. Instead, modulated light generates an acoustic wave (i.e., pressure wave) that can be detected using a simple yet sensitive microphone and a phase-sensitive lock-in amplifier. Here, the concentration of the analyte gas is directly proportional to the amplitude of the sound wave. Hence, a calibrated system allows the direct determination of absolute constituent concentrations within gas-phase samples. The sensitivity of the sensor may be improved by optimizing the modulation of the (laser) radiation at a frequency equivalent to the acoustic mode supported by the gas cell (i.e., operating the photoacoustic cell in resonance). Additionally, due to the acoustic detection, which results only from interactions of the photons with the absorber, the background signal is substantially reduced. PAS may be operated with broadband and narrowband light sources such as lasers; hence, using high optical power readily translates into sensitivities at ppbv to parts per trillion by volume (pptv) concentration levels. Some innovative PAS-based concepts are briefly summarized here.

As an interesting biomedical example, human skin damage caused by ultraviolet radiation may be monitored via the ethylene (C_2H_4) concentration in EB. In particular, Harren et al. (1999) have identified ethylene as a suitable biomarker for lipid peroxidation upon exposure of human subjects to solarium-produced UV (ultraviolet) radiation. Combining an infrared CO_2 laser with an intracavity photoacoustic cell, the authors reported on a sensing system that was able to detect concentrations down to approx. 6 pptv of ethylene. As depicted by the profile shown in Figure 27.6, the ethylene concentration in EB starts to increase 2 min after the onset of UV light exposure.

Recently, Wojtas et al. (2014) measured ammonia (NH_3) in human breath using a so-called quartz-enhanced PAS (aka, QEPAS) technique combined with a widely tunable EC-QCL. Ammonia has been established as relevant biomarker for liver disease, stomach ulcers, and duodenal ulcers caused by

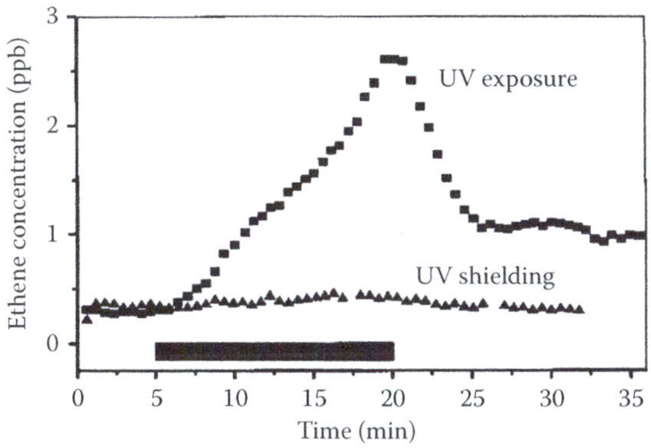

Figure 27.6 Ethylene concentration in EB with and without UV shielding. (Reprinted with permission from Harren, F. J. M. et al., *Appl. Phys. Lett.* 74 (12), 1761. Copyright 1999, American Institute of Physics.)

Helicobacter pylori. Here, a further improvement in PAS sensitivity was achieved by using a resonant quartz tuning fork (QTF) instead of the conventional microphone for detecting the pressure wave. QEPAS sensors comprise a simple design and high stability against environmental acoustic noise. Yet, the major advantage of this technique is the rather small dimension of the tuning fork. Miniaturized sensor systems demanded in handheld diagnostics generally rely on ultra-low sample volumes and accordingly small sensing components. Hence, QEPAS sensors may play an important role in future breath gas diagnostics, as only ~1 mm^3 of sampling volume is required for analysis. The Wojtas group has recently compared conventional PAS with the QEPAS technique (Figure 27.7). In the PAS setup, the laser beam was propagated through a 9 cm path length photoacoustic cell, whereas the QEPAS setup used a QTF detection module of ~1 cm length placed after a spatial filter. The EC-QCL was locked to the peak of the absorption line of ammonia at 10.5 μm. With both sensor systems, ammonia LODs of ~3 ppbv were achieved. Recently, PAS and QEPAS sensing techniques have been proven versatile tools not only in breath diagnostics, but also for environmental, and trace gas sensing in general.

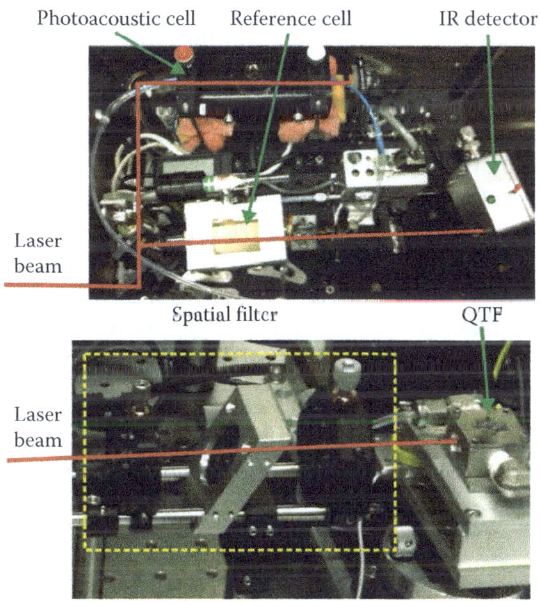

Figure 27.7 PAS and QEPAS setup. (Reprinted with permission from Wojtas, J. et al., *Int. J. Thermophys.*, 35 (12), 2215–25. Copyright 2014, American Institute of Physics.)

27.3.2 Cavity-enhanced sensing techniques

Cavity-enhanced spectroscopic techniques are based on high-finesse optical cavities for increasing the interaction of the period of light with the analyte(s) of interest.

Amongst these techniques, cavity ring-down spectroscopy (CRDS) is probably the most commonly applied sensing technique, and has been established since decades in trace gas analysis. CRDS uses a high-finesse cavity encompassing the gas sample. Basically, a laser beam is trapped in between two highly reflective mirrors establishing a cavity. Once sufficient photons have been injected into the cavity, the source is turned off (e.g., by an acousto-optic modulator). With time, the intensity of the light propagating in between the cavity mirrors decays exponentially, as with each reflection, the intensity is minutely reduced (i.e., leaking out of the cavity). In a CRDS setup, a photodetector located behind the second mirror monitors this decay (aka, "ring-down" of the light intensity). For calculation of the concentration, both "ring-down times" of an empty cavity and a cavity containing an absorbing gas are compared.

Using cavities, extended effective absorption path lengths of up to 10 km have been achieved (NATO Advanced Research Workshop on Middle Infrared Coherent Sources (MICS) et al. 2007).

Cavity-enhanced absorption spectroscopy (CEAS) is considered a variation of the ring-down technique, and thus reveals strong similarities with CRDS. In CEAS, the absorption is calculated from the mirror-transmitted intensities with and without absorption of sample molecule within the cavity. Using broadband sources, CEAS is capable of detecting larger molecules with spectrally broader characteristics.

A portable ultra-sensitive nitric oxide (NO) CEAS sensor (Wojtas et al. 2014) has already been demonstrated (Figure 27.8). The portable case hosts a compact DBF-QCL lasing at 5.26 μm. The laser head is mounted inside a housing suitable for dissipating high heat loads, which is temperature-stabilized by a water-cooled heat changer. Besides, the actual sensor comprises a cavity, a simple gas sample cell, a detection module, and a signal processing unit. At room temperature operating conditions, a LOD of 30 ppbv NO was reported. For asthma patients, this portable sensor presents a comfortable and noninvasive opportunity to monitor the nitric oxide fraction within their EB.

Dahnke et al. (2001) have reported on the real-time analysis of ethane traces in EB by MIR cavity leak-out spectroscopy (CALOS) in the 3 μm spectral region. While CALOS and CRDS are based on the same multipass scheme to enhance the sensitivity, CALOS is a continuous wave (CW) variation of CRDS. Here, relatively low laser powers are required; yet, in turn a higher spectral resolution is provided.

The CALOS sensing system developed by Dahnke et al. was successfully demonstrated for precise time-resolved measurements of the ethane fraction in human breath after smoking. A

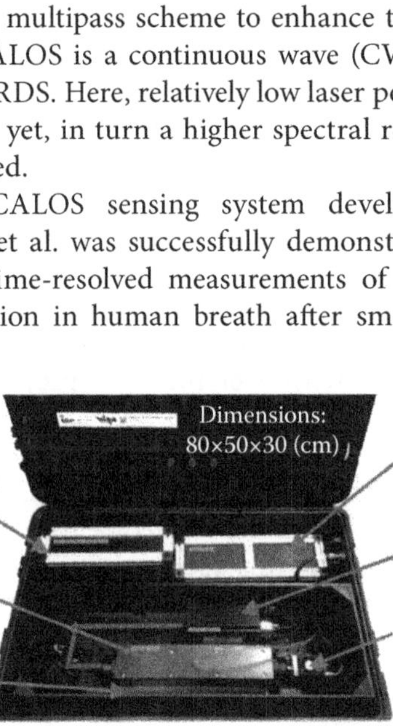

Figure 27.8 Portable NO sensor. (Reprinted with permission from Wojtas, J. et al., *Int. J. Thermophys.*, 35 (12), 2215–25. Copyright 2014, American Institute of Physics.)

CO overtone laser output frequency of 2.6–4.0 μm, and a gas cell operating at 760 torr were used. An LOD of the order of 100 pptv ethane in human breath was achieved during a 5 s integration time, thus rendering sample preconcentration unnecessary. The measurements performed during this study showed a decaying ethane exhalation 1 to 3 h after smoking. Moreover, this technique is capable of ethane detection within single exhalations, thereby, rendering this tool suitable for breath-to-breath monitoring. Like ethylene, ethane is considered a volatile biomarker for the noninvasive investigation of lipid peroxidation.

27.3.3 Chemometrics in breath diagnostics

Initially, the data obtained from most spectroscopic analytical methods were evaluated using univariate strategies simply based on the Beer–Lambert law. However, nowadays—and particularly in the case of complex multi-signal breath matrices and the diagnosis of medical conditions via breath usually involving panels of biomarkers—multivariate data analysis methods (aka, chemometric methods) are essential. Instead of using only a single wavelength (i.e., univariate data evaluation), chemometrics utilizes the entire spectrum or several selected features or feature ranges. The advantages of multivariate calibration and data evaluation techniques compared to less calculation-intensive univariate methods may be summarized as follows:

Interfering constituents such as water (Perez-Guaita 2013), or systematic variation due to changes of the optical setup as well as instrument fluctuations are taken into account or eliminated, thereby improving the sensitivity;

Simultaneous multi-component analysis even with highly overlapping spectral features is enabled;

Using more sophisticated algorithms, robustness (i.e., of the calibration) against unknown constituents and uncalibrated features may be achieved;

Multivariate classification of samples/group allocation determining, for example, different medical health status or disease progression via patient breath samples is attainable using spectral pattern recognition.

Multivariate strategies may be assigned by their designated goal into two general groups: (1) classification methods using pattern recognition, for

Table 27.2 Common multivariate algorithms

Classification method	Quantification method
k-Nearest neighbor	Classical least squares
Discriminant analysis (DA)	Inverse least squares, multiple linear regression
Partial least squares-discriminant analysis	Principal component analysis/regression (PCR)
Soft independent modeling of class analogies	Partial least squares regression
Bayesian classifiers	

Note: Artificial neural networks and support vector machines may be adapted for both data evaluation strategies. Principal component analysis is an explorative data analysis technique that maybe combined with either a classification (i.e., DA) or a regression algorithm (i.e., PCR)

example, assigning/grouping clinical symptoms, and (2) quantification methods for determining the presence and concentration of individual or multiple target analytes/biomarkers.

Some of the most common algorithms are briefly summarized in Table 27.2.

We will only explain PLS (partial least squares) and PCA/PCR (principal component analysis/regression) in detail as they are the most commonly applied multivariate algorithms in literature.

PCA is based on the mathematical concept of singular value decomposition (SVD). Simplified, SVA condenses the entire spectrum into few eigenvectors, which are frequently called principal components (PCs). The spectra of each individual sample can be reconstructed from a set of eigenvectors using a scores matrix (see scheme in Figure 27.9).

The scores can later be used in a regression called (PCR) to calculate the corresponding analyte content (i.e., referring back to the known concentrations of the constituents within the calibration samples). The disadvantage of this method is that a priori knowledge of the constituent concentrations of the calibration samples used to build the calibration model is not immediately utilized during the decomposition (i.e., SVD) step, which may sometimes reduce the predictive power of the PCR when dealing with minimal spectral changes caused by a constituent compared to the general, and potentially high intensity of the spectral features as such.

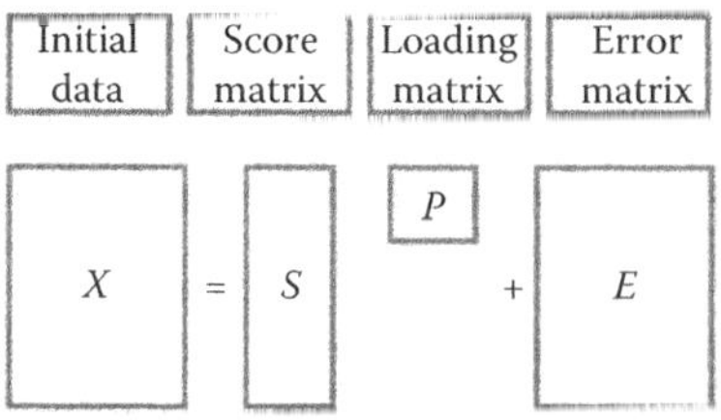

Figure 27.9 Schematic principle of PCA. A data matrix *X* (samples×spectral dimension) is decomposed by SVD into a scores matrix *S* (samples×number of PCs) and a principal components matrix *P* (spectra dimension×number of PCs); residuals of the SVD are contained in the error matrix *E*.

This is improved when using PLS, which readily considers the known constituent concentrations within the calibration samples during the SVD. The eigenvectors in PLS are commonly called latent variables (LV).

As an example, PLS was used to separate the overlapping $^{13}CO_2/^{12}CO_2$ peaks (see Figure 27.10) during isotope ratio studies via an FTIR spectrometer coupled to a hollow waveguide gas cell for analyzing these constituents in exhaled mouse breath samples. The so-called tracer-to-tracee ratio (TTR) after administering ^{13}C labeled glucose in exhaled mouse breath was determined with a precision of +0.02% (Seichter 2013, Multivariat ^{13}C).

Nowadays, most statistical software packages (e.g., MATLAB, SAS, SPSS, R, etc.) offer ready-to-use routines for the most common algorithms. Moreover, standalone software packages (e.g., The Unscrambler, SIMCA, Pirouette, PLS Toolbox for MATLAB/Solo, etc.) are available and may readily be used for evaluating complex IR breath data.

27.4 FUTURE DIRECTIONS

In summary, breath analysis based on MIR sensing techniques shows great potential for noninvasively monitoring physiologically relevant patient conditions. Breath analysis provides information

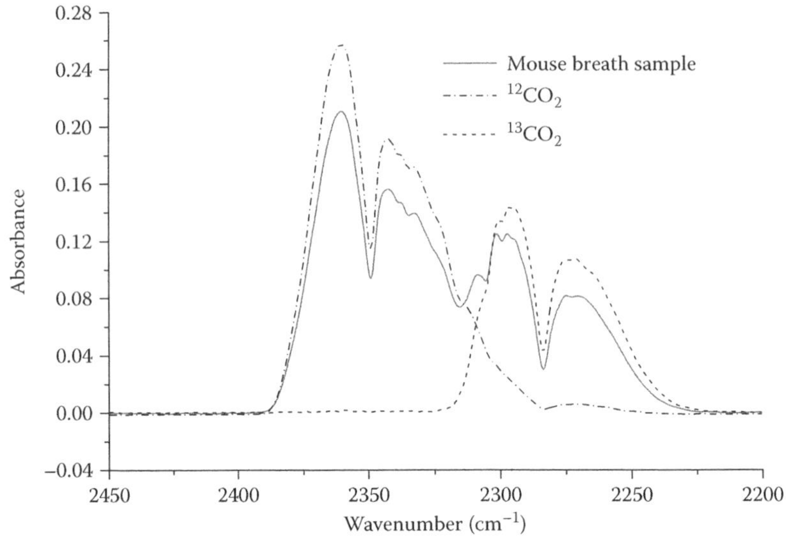

Figure 27.10 Example of overlapping spectral features. IR spectrum of a mouse breath sample (solid red line), $^{12}CO_2$ (dashed–dotted line), and$^{13}CO_2$ (dashed line). (With kind permission from Springer Science+Business Media: *Analytical and Bioanalytical Chemistry*, Multivariate determination of $^{13}CO_2/^{12}CO_2$ ratios in exhaled mouse breath with mid-infrared hollow waveguide gas sensors, 405, 2013, 4945–51, Seichter, F. et al.)

on the health condition based on several hundreds of VOCs present in the exhaled breath matrix at ulra-trace-to-trace concentrations. Although lab-on-a-chip functionality remains to be demonstrated, in particular, iHWG technologies pave the way toward potential on-chip integration and ultra-compact vapor-phase sensing/analyzer systems applicable in EB diagnostics.

Taking advantage of most recent laser technologies—QCLs and ICLs—IR breath diagnostics is gaining increased attention as real-time sensing technology suitable for analyzing and monitoring exhaled trace gases with the required sensitivity, molecular selectivity, and time constant. Hence, breath analyzers for bedside monitoring or even handheld analyzers for physician or patient use may be conceived. However, given the complexity of biomarker panels associated with a wide range of pulmonary diseases and also complex diseases such as various cancers, the reliable detection, diagnosis, and disease or therapy progression monitoring require further substantial research into the disease pathogenesis, the associated breath pattern, and the potential combination of complementary sensing techniques for reliable clinical applications.

REFERENCES

Alonso, M., and J. M. Sanchez. 2013. Analytical challenges in breath analysis and its application to exposure monitoring. *TrAC Trends in Analytical Chemistry* 44: 78–89.

Dahnke, H., D. Kleine, P. Hering, and M. Mürtz. 2001. Real-time monitoring of ethane in human breath using mid-infrared cavity leak-out spectroscopy. *Applied Physics B* 72 (8): 971–75.

Harren, F. J. M., R. Berkelmans, K. Kuiper, S. te Lintel Hekkert, P. Scheepers, R. Dekhuijzen, P. Hollander, and D. H. Parker. 1999. On-line laser photoacoustic detection of ethene in exhaled air as biomarker of ultraviolet radiation damage of the human skin. *Applied Physics Letters* 74 (12): 1761.

Jouy, P., M. Mangold, B. Tuzson, L. Emmenegger, Y.-C. Chang, L. Hvozdara, H. P. Herzig, P. Wägli, A. Homsy, N. F. de Rooij, A. Wirthmueller, D. Hofstetter, H. Looser, and J. Faist. 2014. Mid-infrared spectroscopy for gases and liquids based on quantum cascade technologies. *Analyst* 139 (9): 2039–46.

Konvalina, G., and H. Haick. 2014. Sensors for breath testing: From nanomaterials to comprehensive disease detection. *Accounts of Chemical Research* 47 (1): 66–76. Mizaikoff, B. 2013. Waveguide-enhanced mid-infrared chem/bio sensors. *Chemical Society Reviews* 42 (22): 8683.

NATO Advanced Research Workshop on Middle Infrared Coherent Sources (MICS), M. Ebrahim-Zadeh, I. T. Sorokina, North Atlantic Treaty Organization, and Public Diplomacy Division. 2007. *Mid-Infrared Coherent Sources and Applications*. Dordrecht: Springer.

Pauling, L., A. B. Robinson, R. Teranishi, and P. Cary. 1971. Quantitative analysis of urine vapor and breath by gas-liquid partition chromatography. *Proceedings of the National Academy of Sciences* 68 (10): 2374–76.

Perez-Guaita, D., V. Kokoric, A. Wilk, S. Garrigues, and B. Mizaikoff. 2014. Towards the determination of isoprene in human breath using substrate-integrated hollow waveguide mid-infrared sensors. *Journal of Breath Research* 8 (2): 026003.

Ras, M. R., F. Borrull, and R. M. Marcé. 2009. Sampling and preconcentration techniques for determination of volatile organic compounds in air samples. *TrAC Trends in Analytical Chemistry* 28 (3): 347–61.

Seichter, F., A. Wilk, K. Wörle, S. S. Kim, J. Vogt, U. Wachter, P. Radermacher, and B. Mizaikoff. 2013. Multivariate determination of $^{13}CO_2/^{12}CO_2$ ratios in exhaled mouse breath with mid-infrared hollow waveguide gas sensors. *Analytical and Bioanalytical Chemistry* 405 (14): 4945–51. Wilk, A., J. C. Carter, M. Chrisp, A. M. Manuel, P. Mirkarimi, J. B. Alameda, and B. Mizaikoff. 2013. Substrate-integrated hollow waveguides: A new level of integration in mid-infrared gas sensing. *Analytical Chemistry* 85 (23): 11205–10.

Wilk, A., F. Seichter, S.-S. Kim, E. Tütüncü, B. Mizaikoff, J. A. Vogt, U. Wachter, and P. Radermacher. 2012. Toward the quantification of the $^{13}CO_2/^{12}CO_2$ ratio in exhaled mouse breath with mid-infrared hollow waveguide gas sensors. *Analytical and Bioanalytical Chemistry* 402 (1): 397–404.

Wojtas, J., F. K. Tittel, T. Stacewicz, Z. Bielecki, R. Lewicki, J. Mikolajczyk, M. Nowakowski, D. Szabra, P. Stefanski, and J. Tarka. 2014. Cavity-enhanced absorption spectroscopy and photoacoustic spectroscopy for human breath analysis. *International Journal of Thermophysics* 35 (12): 2215–25.

28

Fiber optic manometry catheters for *in vivo* monitoring of peristalsis in the human gut

JOHN ARKWRIGHT
Flinders University

PHIL DINNING
Flinders University and Flinders Medical Centre

28.1 FIBER OPTIC PRESSURE SENSORS

The human body presents some unique challenges for sensors. The sensing volume is not large, and the point of interest is never more than a few centimeters away from the "outside world"; however, many regions of the body have stubbornly resisted attempts at detailed measurement without simultaneously perturbing the self-same systems and processes under investigation.

The gastrointestinal tract is a case in point. Viewed as a single organ, it stretches up to 8 m in length and spans from mouth to anus in a complex and convoluted fashion. Accessing the regions below the stomach in a minimally invasive fashion requires skill and ingenuity from both medical and engineering communities. Technologies such as SmartPills, Pillcams, magnetic resonance, and scintigraphy (Dinning et al. 2010) can record images, pH, temperature, anatomy, and luminal transit to varying degrees; however, the "pill" technologies only gather information from their immediate vicinity as they traverse the gut, and the latter two technologies only provide snapshots of the anatomy and luminal content. To generate time-varying data from physiologically significant lengths of the gut, it is necessary to resort to intraluminal manometry. Manometry involves placing a catheter with one or more pressure sensors into the lumen of the gut to record pressures and contact forces generated as the gut wall contracts to mix, transport, and extract nutrient from digesta. A variant on manometry has been the introduction of large numbers of sensors spaced at ~10 mm intervals along a catheter to generate a pseudo-continuous image of muscular activity occurring along the gut that greatly assists the interpretation of the recorded data. This approach is referred to as high-resolution manometry (HRM) (Clouse 2001). Although the activity in the oesophagus and the anorectum is now being described in detail using HRM, detailed descriptions of contractile patterns below the stomach have only been superficially investigated.

The unique advantages of optical fiber, including high flexibility, small size, and the ability to define multiple information channels using spectrally separated windows, make this technology ideal for monitoring this region deep within the human gut in a minimally invasive manner (Arkwright et al. 2009; Voigt et al. 2010; Singlehurst et al. 2012).

In this chapter, we present the design and realization of our fiber optic HRM catheter and report the status of an on-going series of clinical trials designed to study motility (the movement of digesta within and through the gastrointestinal tract) using these devices.

28.2 FIBER OPTIC CATHETER DESIGN

The transducers used to form the fiber optic HRM catheters are based on fiber Bragg gratings (FBGs) (Kashyap 1999), which, in their simplest form, can be thought of as localized optical fiber strain gauges. The transduction mechanism has been described in detail elsewhere (Arkwright et al. 2014); in brief, the design consists of a curved FBG rigidly bonded at each end to a stainless steel substrate. The structure is covered with a pressure sensitive diaphragm so that any changes in pressure or contact force distort the FBG sideways, changing its reflected wavelength (see Figure 28.1). Compared to the sideways deflection of a linear fiber, the curved aspect of the fiber significantly increases the sensitivity of the transducer and also allows pressure and vacuum signals to be differentiated. By using an array of spectrally separated FBGs, a large number of discrete, individually addressable transducers can be located along a single length of fiber.

To provide the necessary strength and also enable a 10 mm spacing between each FBG, draw tower gratings (DTGs) from FBGS (FBGS International NV, Geel, Belgium) were used. Each FBG was 3 mm in length with center wavelengths ranging from ~1515 to ~1590 nm spaced at ~2 nm intervals. A solid-state spectrograph with a wavelength range of 1510–1595 nm and four optical inputs was used (FBG-scan 804D; FBGS International NV). In practice, approximately 36 FBGs can be located in a single fiber without risk of wavelengths merging when pressure is applied to a single sensing element. This enables up to 144 sensors to be detected at total acquisition rates of up to 500 Hz, which is more than required for gastrointestinal investigation (we typically use 10 Hz acquisition rates for colonic motility tests). The sensor elements are linear over the range of −50 to +200 mmHg; hence, to map the wavelength change onto variations in pressure, the catheters were calibrated in a pressure vessel at 0 and 100 mmHg prior to each *in vivo* use.

28.3 *IN VIVO* RECORDINGS

Catheters are currently being used in a number of ongoing clinical studies to record muscular activity in the human colon. To date, we have recorded data in healthy volunteers (N = 17) and in patients suffering from slow transit constipation (N = 21), fecal

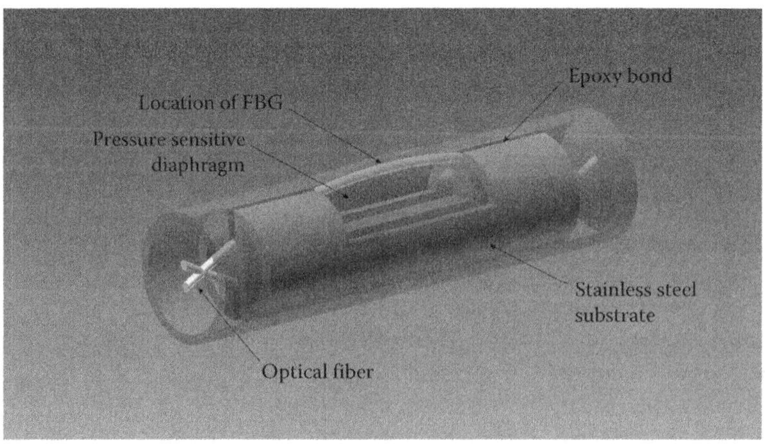

Figure 28.1 FBG pressure transducer.

incontinence (N=14), and irritable bowel syndrome (N = 6). All participants in the studies had given written, informed consent, and the studies were approved by the Human Ethics Committees of the South Eastern Area Health Service, Sydney, and the University of New South Wales (05/122; May 2010), and The Southern Adelaide Health Service/Flinders University Human Research Ethics Committee (419.10; March 2011). On the day prior to the procedure, the bowel was cleared using sodium picosulfate and polyethylene glycol (Pharmatel Fresenius Kabi Pty Ltd., Hornsby, Australia). All subjects were asked to drink only clear fluids overnight.

The manometry catheter was fed into the colon via the rectum using a colonoscope with the subjects under light sedation. The catheter was advanced towards the cecum (close to the junction with the small bowel), and the distal tip of the catheter was clipped to the mucosa of the colon using two endoclips (Resolution Clip; Boston Scientific, Marlborough, MA, USA). This ensured that the catheter remained in place throughout the subsequent recordings. Figure 28.2 shows an X-ray taken of a catheter placed in the colon of a healthy volunteer.

Recording was started within 2 h of placement and continued for 4–24 h depending on the nature of the study.

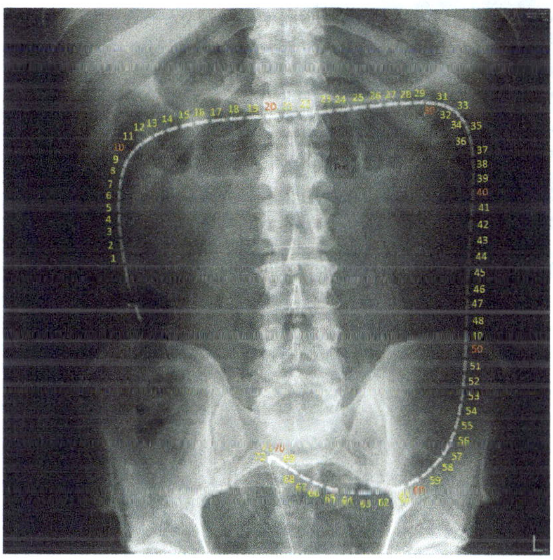

Figure 28.2 X-ray image of a catheter placed in the colon of a healthy volunteer. Every 10th sensor is marked in red to indicate the sensor spacing typically available using traditional technologies.

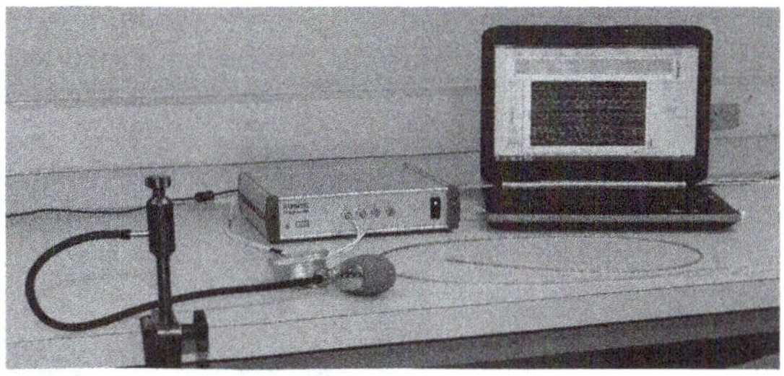

Figure 28.3 Fiber optic manometry system, including catheter, data acquisition unit, calibration tube, and controlling computer.

28.4 DATA ACQUISITION

The data from the catheters was recorded using custom software written in LabVIEW© (*National Instruments*, Austin, TX, USA) and was then analyzed using PlotHRM software written in MATLAB© (The MathWorks, Natick, MA, USA) and JavaTM (Sun Microsystems, Santa Clara, CA, USA) (Dinning et al. 2013). A photograph of the catheter, data acquisition unit, and computer is shown in Figure 28.3. For analysis, the data can be viewed in either a line plot or an intensity plot format. Figure 28.4a shows a typical section of data recorded using the fiber optic catheter but displayed at low spatial resolution to simulate the output from a traditional solid-state manometry catheter (indicated by the sensor numbers shown in red in Figure 28.2). Figure 28.4b shows the same data at high resolution using the full data set recorded at 1 cm spacing. Figure 28.4c shows an interpolated intensity plot of the same data. In Figure 28.4b and c, the direction, extent, and connectivity of the contractions running along the gut can clearly be seen, whereas the low-resolution plot results in considerable ambiguity regarding the extent and even the direction of the contractions.

28.5 CLINICAL INSIGHTS

Using the fiber optic catheter has allowed us to extend the range of HRM over the full length of the colon for the first time. Unlike other regions of the digestive tract, content moves through the colon very slowly, typically taking 24–48 h to traverse its length. This slow progress allows the colon to perform its physiological functions, which includes the absorption of water and electrolytes. Colonic studies using low-resolution manometry

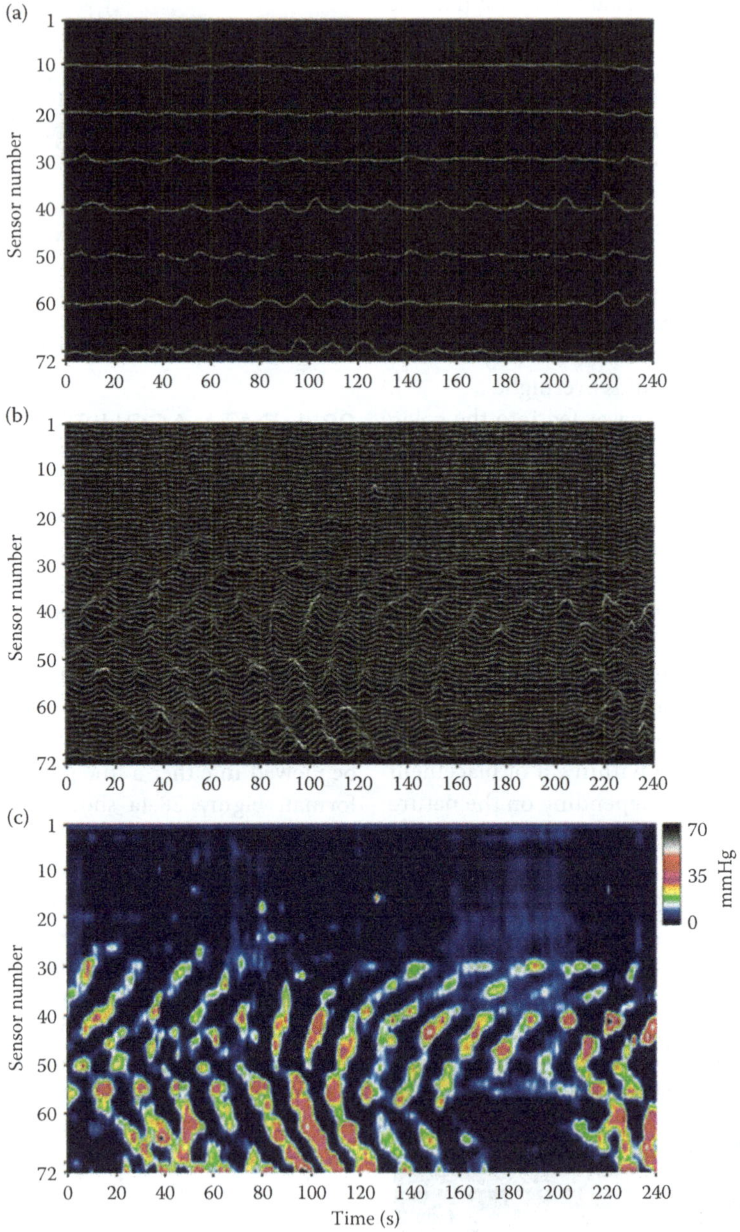

Figure 28.4 Data recorded using the fiber optic manometry catheter shown in Figure 28.2, showing (a) data recorded by every 10th sensor to simulate the output of a traditional colonic catheter; (b) data recorded by every sensor in the array showing the increased levels of information available from HRM; and (c) the same high-resolution data shown on an interpolated intensity plot.

have identified some characteristic propagating motor patterns, the most prominent of which was labeled the high-amplitude propagating sequence (HAPS). These motor patterns are known to be associated with mass movements of colonic content and with defecation, and historically have been used as the best differentiator between health and disease. Other motor patterns that were identified using low-resolution manometry included low-amplitude propagating sequences, a variety of nonpropagating contractions, and occasional episodes of retrograde (oral) propagating pressure waves. However, these descriptions were all based on recordings made by colonic catheters with recording sites spaced at least 7 cm apart (the majority >12 cm), and they have not provided any satisfactory links between symptoms and diagnoses, nor have they improved our understanding of the underlying cause of disorders such as constipation and incontinence.

With the development and use of high-resolution fiber optic manometry catheters, we have reexamined the colonic motor patterns and have been able to demonstrate that the established definitions of motor patterns are very inaccurate (Dinning et al. 2013). Our high-resolution data strongly suggest that conventional sensor spacing (>7 cm) leads to very significant underestimations of the number of propagating sequences that are present in the human colon and that most of these events are, in fact, contractions that propagate over short distances, often in the retrograde direction. These newly quantified retrograde patterns are likely to explain the slow progress of content through the colon and may also play an important role in normal maintenance of fecal and gas continence.

In our latest research, we have been able to characterize all propagating pressure waves (amplitude, gradient, and duration) in the healthy human colon, and then, using discriminant and multivariate analysis, we have been able to statistically separate these propagating motor patterns into two distinct groups. The first group consists of the HAPSs, which appear to be influenced by input from the central nervous system via neural pathways from the spinal column. The second statistically significant set of motor patterns is generated and controlled locally within the muscle of the colon.

The ability to differentiate motor patterns into these two types is significant for two reasons. First, it will allow us to better understand the nature of colonic activity in health and disease, and second, it is likely to provide insight into the root cause of different subtypes of motility disorders.

We have also used our high-resolution catheters to examine the effects of interventions upon colonic function. In patients with fecal incontinence, a novel treatment known as sacral nerve stimulation (SNS) has been shown to dramatically improve the patients' ability to maintain continence of their stool. However, the action of SNS is still a matter of conjecture. In a recent study on SNS, our data have revealed that in patients with fecal incontinence, this stimulation increased the locally controlled retrograde propagating motor patterns in the distal parts of the colon. This is presumed to reduce premature rectal filling, which in turn helps the patient to maintain continence (Patton et al. 2013).

Although we are still in the early stages of clinical investigation using these fiber optic catheters, the association of different motor patterns with health and diseased states is very encouraging and is already being hailed as a paradigm shift in our understanding of colonic motor disorders. The realistic hope now is that we will be able to use this high-resolution data to identify various subtypes of colonic disorders and hence develop more effective targeted therapies for a range of patients suffering from these debilitating disorders.

28.6 CONCLUSIONS

The unique properties of optical fiber lend themselves particularly well to monitoring inside the human body. The small size and extreme flexibility of the fiber enables intubation deep into the complex and convoluted regions of the gastrointestinal tract below the stomach. Using wavelength division multiplexing techniques to simultaneously monitor large numbers of sensing regions along the fiber has allowed us to gather data at an unprecedented spatial resolution that is providing new insights into the nature and cause of colonic motor disorders.

ACKNOWLEDGMENTS

The authors acknowledge the invaluable contributions by Ian Underhill and Neil Blenman (design and fabrication of the catheter), Simon Maunder and Lukasz Wiklendt (data acquisition software and the data analysis software respectively), and David Lubowski, and Vicki Patton for their expert knowledge relating to the use and placement of the catheters.

REFERENCES

Arkwright, J. W., Underhill, I. D., Maunder, S. A., Blenman, N., Szczesniak, M. M., Wiklendt, L., Cook, I. J., Lubowski, D. Z., Dinning, P. G.

(2009). Design of a high-sensor count fibre optic manometry catheter for *in-vivo* colonic diagnostics. *Optics Express* 17:22423–22431.

Arkwright, J. W., Underhill, I. D., Maunder, S. A., Jafari, A., Cartwright, N., Lemckert, C. (2014). Fibre optic pressure sensing arrays for monitoring horizontal and vertical pressures generated by travelling water waves. *IEEE Sensors Journal* 14(8):2739–2742.

Clouse, R. E. (2001). Topographic manometry: An evolving method for motility. *Journal of Pediatric Gastroenterology and Nutrition* 32(Suppl 1):S10–S11.

Dinning, P.G., Arkwright, J.W., Gregersen, H., O'Grady, G., Scott, S.M. (2010). Technical advances in monitoring human motility patterns. *Neurogastroenterology and Motility* 22:366–380.

Dinning, P. G., Wiklendt, L., Gibbins, I., Patton, V., Bampton, P., Lubowski, D.Z., Cook, I. J., Arkwright, J. W. (2013). Low-resolution colonic manometry leads to a gross misinterpretation of the frequency and polarity of propagating sequences: Initial results from fiber-optic high-resolution manometry studies. *Neurogastroenterology and Motility: The Official Journal of the European Gastrointestinal Motility Society* 25(10):e640–e649.

Kashyap, R. (1999). *Fiber Bragg Gratings*. San Diego: Academic Press.

Patton, V., Wiklendt, L., Arkwright, J. W., Lubowski, D. Z., Dinning, P.G. (2013). The effect of sacral nerve stimulation on distal colonic motility in patients with faecal incontinence. *The British Journal of Surgery* 100:959–968.

Singlehurst, D. A., Dennison, C. R., Wild, P.M. (2012). A distributed pressure measurement system comprising multiplexed in-fibre Bragg gratings within a flexible superstructure. *Journal of Lightwave Technology* 30: 123–129.

Voigt, S., Rothhardt, M., Becker, M., Lupke, T., Thieroff, C., Teubner, A., Mehner, J. (2010). Homogeneous catheter for esophagus high-resolution manometry using fiber Bragg gratings. *Proc. SPIE* 7559.

29

Optical coherence tomography in medicine

MICHAEL A. MARCUS

LUMETRICS INC.

29.1 INTRODUCTION

Optical coherence tomography (OCT) is rapidly becoming an established medical imaging technique for capturing high (micrometer)-resolution, two (2D) and three-dimensional (3D) images from optically scattering media such as biological tissue. Today, commercially available optical coherence tomography systems are employed for many diverse applications of diagnostic medicine. It is particularly useful in ophthalmology, where it can be used to obtain detailed images from areas within the retinal layers, for examination of subsurface layers of the skin, for tumor investigation, for examination of the gastrointestinal tract, and for interventional cardiology (where, for example, it can help to diagnose coronary artery disease). It is also a valuable aid for subsurface examination of works of art, becoming an essential tool for the detection of forgeries and to assist art conservation.

OCT is based on low-coherence interferometry (LCI), typically using a near-infrared light source, allowing it to penetrate better into a scattering medium. The data obtained from OCT are, in many ways, similar to those obtained from ultrasound "B-mode" imaging, another method that allows cross-sectional imaging of the tissue microstructure as a function of depth. However, instead of acoustic waves, OCT uses light and performs imaging by measuring the backscattered intensity of light from structures in tissues. In contrast to ultrasound, which utilizes intensity versus echo time delay, coherent (interferometric) detection techniques are employed. Since the wavelength of light is so much shorter, the depth resolution is much better than is possible with ultrasound.

With OCT, an optical beam is directed into the tissue and the light scattered or reflected from the microstructure at different depths is measured using interferometry. The axial resolution is typically 1–10 µm, far better than that from any other clinically available diagnostic modality. Imaging depth is media dependent and it is limited by optical attenuation due to scattering and absorption. In human tissue, penetration depths are on the order of 1–4 mm. The data can yield a 2D map of the structure if a simple line scan over the surface is made, using the backscattering or reflectance from internal architectural morphology and cellular structure in the tissue. This can be extended to give 3D image information by performing repeated axial measurements along many such lines, with the light beam scanned to eventually enter at all the different points over the 2D surface of the tissue.

Today, many OCT instruments employ fiber optic probes to provide more compact and flexible optics. They are used for a variety of applications, including dental imaging and catheter-based or endoscopic systems for gastrointestinal tract and arterial imaging. "Histology-like" images are obtained in real time with no need for excision, providing great advantages for diagnosis of early neoplasia and for surgical guidance. OCT has been applied *in vitro* to image arterial pathologies, where it can differentiate plaque morphology with a resolution superior to that possible with ultrasound. OCT is a promising imaging technology because it can permit real-time and *in situ* visualization of tissue microstructure without the need to excise it and later process a specimen in a laboratory, as would normally be necessary in conventional biopsy and histopathology [1].

OCT was originally developed and applied to tomographic diagnostics in ophthalmology, and the first cross-sectional retinal images were demonstrated in 1991 [2]. The first commercial ophthalmic OCT instrument became available in 1996, and it received the Federal Drug Administration clearance in 2002 [3]. At the time of writing, ophthalmic OCT is presently the largest application area for OCT. The technique has revolutionized the treatment of eye diseases, because it provides detailed images of the retina with resolutions of a few micrometers. It is now used routinely on a global basis to make clinical decisions regarding patients having potentially blinding diseases, such as macular degeneration, diabetic retinopathy, and glaucoma; it is also used during many surgical procedures on the eye.

OCT provides a diagnostic tool that is complementary to spectroscopic techniques and has great potential for *in situ* microscopic imaging of cellular features. It can detect attributes of malignancies and precancers, in a way that rivals that of a histopathologist examining a tissue biopsy specimen under a microscope. Although cell staining techniques have not yet been designed to improve the contrast of this technique, the histopathologist can still observe the orientation of the tissue within a 3D matrix and measure the sizes of cellular and subcellular elements *in vivo*. This information can provide unique information and insights into the dysplastic and malignant process and can be linked to therapeutic procedures, once a suspicious area is identified.

29.2 BASIC ARRANGEMENT OF OCT INSTRUMENTS

The standard method of OCT is to use white light interferometry or LCI, where light from the light source is split into two branches, one interacting with the sample and the other passing through a reference branch. Light signals from the two branches are recombined such that the interference is detected. The optical setup is usually constructed in a Michelson configuration, as described next, and illustrated in Figures 29.1 and Figures 29.3 through 29.5. A Michelson interferometer is composed of an input branch (I), an output branch

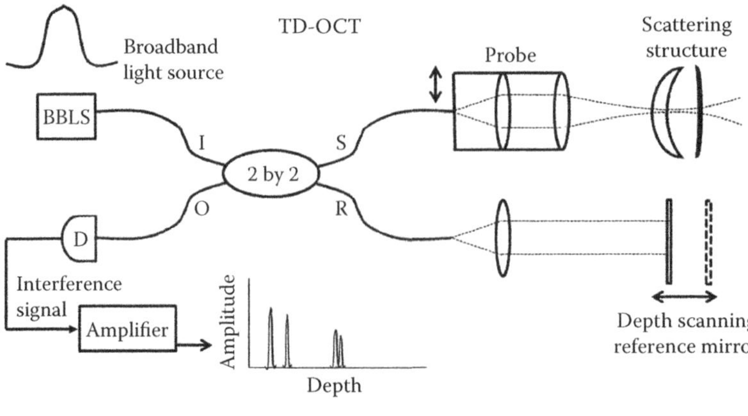

Figure 29.1 Schematic of a TD-OCT instrument.

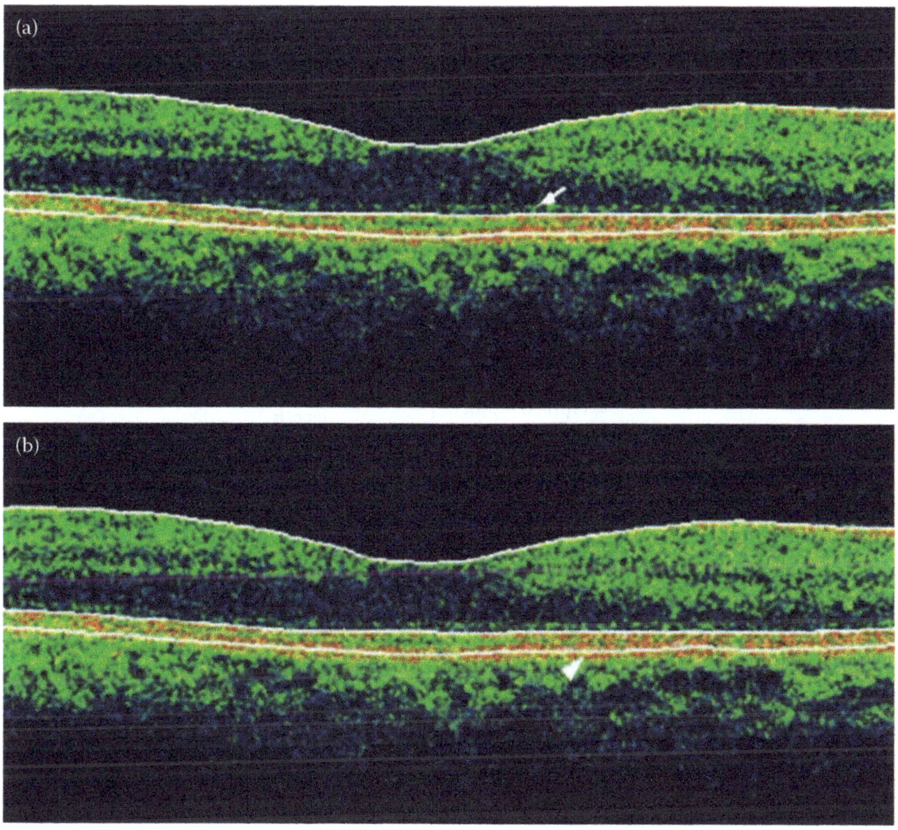

Figure 29.2 False color images of a normal retina using (a) TD-OCT and (b) FD-OCT.

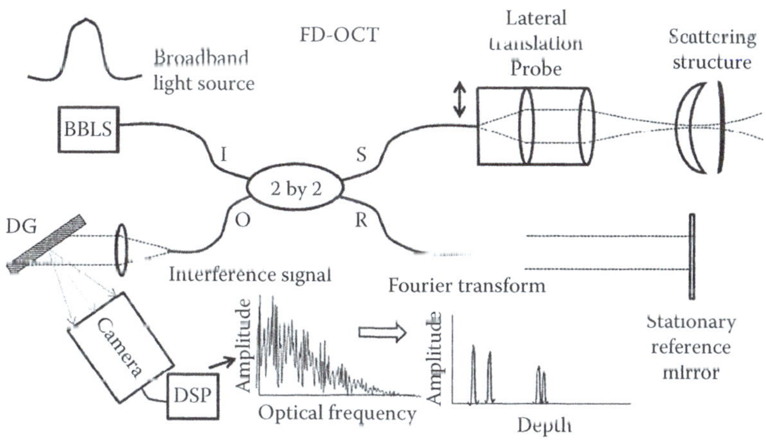

Figure 29.3 Schematic of an FD-OCT instrument.

(O), a sample branch (S), and a reference branch (R). Light from a low-coherence light source traveling along the input branch is split into two beams that travel along the reference (R) and sample branches (S), respectively. Light signals reflected off the sample located in the sample branch (S) and other light signals reflected back from the reference branch (R) are recombined into the output branch (O) of the interferometer, where the light intensity, resulting from coherent interference, is

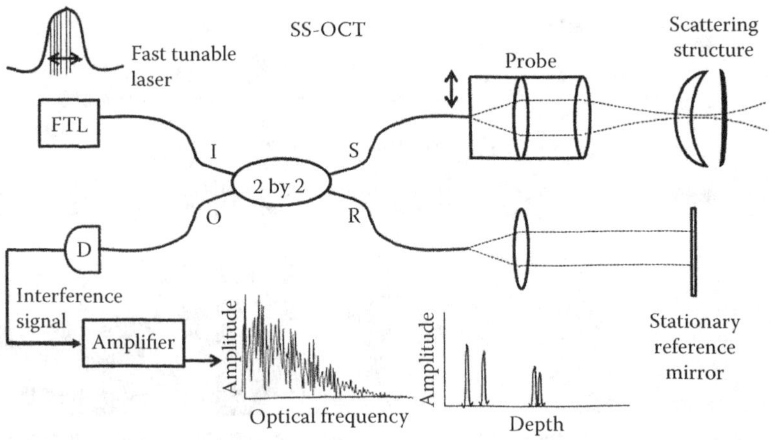

Figure 29.4 Schematic of an SS-OCT instrument.

measured by a detector. Constructive interference with visible fringes only occurs at distances where the optical path lengths of light reflected off the sample and reference branches are the same.

Today, there are a variety of modalities of OCT instruments that are commercially available. These include time-domain OCT (TD-OCT), frequency-domain OCT (FD-OCT), swept-source OCT (SS-OCT), and full-field OCT (FF-OCT), also called *en face* OCT. Examples of these are shown in Figures 29.1 through 29.4, respectively. All of these OCT modalities are based on LCI. Conventional interferometers use lasers with long coherence lengths as their light source, so that interference of light with visible fringes would occur over distances of meters. In OCT, however, visible interference fringes occur only over distances of a few to tens of micrometers, owing to the use of broad-bandwidth light sources having short coherence lengths. Examples of these are superluminescent diodes or lasers with artificially broadened spectra. The latter can be swept-wavelength tunable lasers or other lasers with extremely short pulses (femtosecond lasers), where the spectrum is broadened because of the short pulse length (a very short pulse has a broad Fourier spectrum).

29.3 TIME-DOMAIN OCT

The original OCT instruments were time-domain instruments, and an example time-domain configuration is shown in Figure 29.1. Light from a broadband light source (labeled BBLS), typically a superluminescent light-emitting diode located in the input branch (I) of the Michelson interferometer, is coupled into a single-mode optical fiber. Using a single-mode optical fiber coupler, the light is split into two single-mode optical fiber paths to form a sample branch (S) and a reference branch (R). Light traveling along the sample branch (S) is incident on an optical probe which focuses light on a sample, which we will call as the "scattering structure." Light is reflected and scattered off of each of the optical interfaces of the sample and a portion of this reflected and scattered light is collected by the optical probe and returns via the sample optical fiber. Light traveling along the reference branch (R) is collimated by a lens and is incident on a depth scanning reference mirror. A portion of the light incident on the reflector is reflected back through the collimating lens and returns via the reference optical fiber. The light reflected from the sample and reference mirror is recombined at the 2 by 2 optical coupler and sent through the output optical fiber (O) and into a photodetector (D). This signal is amplified, demodulated, conditioned and analyzed.

In TD-OCT, the path length of the reference arm is increased by arranging the path to be translated longitudinally in time. In LCI, interference (i.e., a series of light and dark bands, or "fringes") only occurs when the optical path length difference between the sample and reference branches lies within the coherence length of the light source. Demodulation of the interference signal produces the fringes having an envelope varying in intensity as the depth of the reference reflector is varied.

The location of the peak of the envelope (point of maximum fringe visibility) corresponds to the path length matching condition and the full width at half maximum of the peak is equal to the coherence length of the BBLS (typically 5–20 μm).

The optical path lengths of the sample and reference branches of the interferometer are closely matched and the reference depth-scanning mirror is scanned with a path-length change sufficient to observe all the desired optical interfaces in the sample. By scanning the depth of the mirror in the reference arm, a reflectivity profile of the sample (scattering structure) is obtained. Areas of the sample that reflect more light will create greater interference than areas that reflect weakly. Any light that has a path difference outside the short coherence length of the source will not noticeably interfere to produce visible fringes. This reflectivity profile as a function of depth is also called an A-scan, and it contains information about the spatial dimensions and location of structures within the item of interest. An example amplified signal obtained for the example scattering structure as a function of depth for a single depth scan of the reference mirror is included in Figure 29.1. The four major peaks shown in the depth scan correspond to the four optical surfaces in the sample scattering structure.

The optical probe in the sample branch of the interferometer can also be scanned laterally, as indicated by the double arrow on the probe in Figure 29.1. A cross-sectional tomographic image (B-scan) may be achieved by laterally combining a series of these axial depth scans (A-scan). Usually, a false color image or a grey scale image is used to represent the intensity of the interfering light as a function of position. In a false color image, white and red colors usually represent high reflectivity regions, whereas black and blue colors represent low reflectivity regions. Examples of false-color OCT images of a retina are shown in Figure 29.2. Figure 29.2a shows a TD-OCT image and Figure 29.2b shows an FD OCT image [4]. The arrow indicates the inner segment/outer segment photoreceptor junction and the arrow head indicates the inner border of the retinal pigment epithelium. The measurement speed of the TD-OCT configuration is limited by the time it takes to scan the entire scan depth of the reference mirror, which is typically 3–10 mm—this limits the measurement repetition rates to a few kilohertz or less.

29.4 FREQUENCY-DOMAIN OCT

In FD-OCT, the interference pattern is acquired using spectrally separated detectors. This can be performed by either dispersing the interfering low-coherence light in a spectrometer (FD-OCT) or using a frequency-swept narrowband light source and a single detector (SS-OCT). Figure 29.3 shows a typical FD-OCT instrument configuration. The only differences from the TD-OCT setup shown in Figure 29.1 is that the reference mirror is now stationary and the detector is replaced with a spectrometer. The output fiber in Figure 29.3 is shown to be coupled to a collimator, which directs near-parallel light onto a diffraction grating (DG). The light from the DG is diffracted to leave at angles that are a function of wavelength, defined by the usual Bragg relationship. The diffracted light is sent to a line scan camera, where the diffraction pattern is processed in a digital signal processing unit and converted to an amplitude versus depth plot.

The raw camera data are in the form of amplitude versus optical frequency, shown as the interference signal in Figure 29.3. By taking the Fourier transform of this interference signal, the result is the amplitude versus depth plot. In the diagram, this is shown to the right of the interference signal, which is similar to that obtained in the time-domain instrument. Since all the data are obtained effectively and simultaneously, much higher measurement rates are obtainable than with TD-OCT. Usually, high-speed linear charge-coupled devices (CCDs) are used and measurement rates up to 50 kHz or higher have been realized in practice. However, the signal-to-noise ratio is typically not as high as with TD-OCT.

29.5 SWEPT-SOURCE OCT

In swept-source OCT (SS-OCT) the low-coherence light source is replaced by a rapidly tuned laser (Figure 29.4). The detector is again a single-point, high-speed photodetector, as in TD-OCT, and the reference reflector is also stationary, as in FD-OCT. As in FD-OCT, the amplitude versus optical frequency plot is first generated for each wavelength scan, and then the Fourier transform is taken to provide amplitude versus depth information. SS-OCT has advantages in both SNR and measurement rates. Measurement rates >200 kHz have been obtained to date.

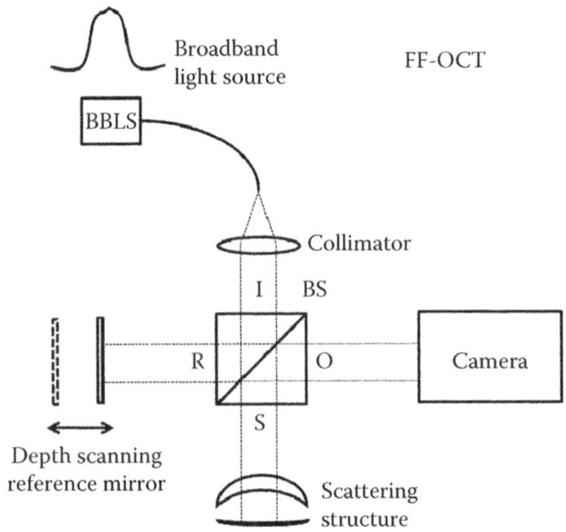

Figure 29.5 Schematic of an FF-OCT instrument.

29.6 FULL-FIELD (*EN FACE*) OCT

In FF-OCT, a 2D charge coupled device (CCD) or complementary metal oxide semiconductor (CMOS) high-speed camera is utilized to produce successive images of the interfering light, with the reference mirror located at a single plane in each frame (also called an *en face* image). During operation, sequential images are obtained at different monotonically increasing or decreasing depths of the reference mirror. This results in a 2D grid of interference images at each pixel of the image. Figure 29.5 shows a typical FF-OCT measurement arrangement. A BBLS is coupled to an optical fiber. The light at the end of the fiber is collimated and set through a beam splitter, where it is split into sample and reference beams. An area of the sample is illuminated all at once, so that lateral scanning is not required. The area of light that is reflected and scattered off the sample is recombined with the area of light that is reflected back from the reference reflector. The recombined interfering light is incident on a camera. Each pixel of the camera forms an interferogram as a function of scan depth. 3D images of the depth scan are obtained similar to that in the TD-OCT case, but now all in one depth scan.

29.7 CONCLUSIONS

The technique of optical coherence tomography, despite being a relatively new diagnostic technique for medical and biomedical applications, has already become established as a vital clinical tool. It is being used routinely to obtain high resolution images of both the anterior segment of the eye and the retina, and in the diagnosis of many eye diseases, including macular degeneration, diabetic retinopathy, glaucoma, and uveitis. It has also begun to be utilized in the area of interventional cardiology, to evaluate obstructions in the gastrointestinal tract, and to evaluate skin cancers. Equipment is still evolving at a dramatic pace, and it is expected that the capability and economic viability of equipment will continue to improve to further future medical diagnostics. OCT also enables subsurface analysis of many other living organisms and many inanimate objects, such as paints, varnishes, and multilayer coatings. It can be used to assess valuable works of art. This highly versatile tool can be used to evaluate almost any material that is semitransparent and contains reflective and light-scattering particles or regions. As a result, OCT is starting to be used in a variety of industrial applications such as nondestructive testing and material thickness measurements.

REFERENCES

1. Bezerra, H.G., Costa, M.A., Guagliumi, G., Rollins, A.M., and Simon, D.I. Intracoronary optical coherence tomography: A comprehensive review, *JACC: Cardiovascular Interventions* 2(11), 1035–1046 (2009).
2. Huang, D., Swanson, E.A., Lin, C.P., Schuman, J.S., Stinson, W.G., Chang, W., Hee, M.R., Flotte, T., Gregory, K., Puliafito, C.A. and Fujimoto, J.G. Optical coherence tomography, *Science* 254, 1178–1181 (1991).
3. Zysk, A.M., Nguyen, F.T., Oldenburg, A.L., Marks, D.L., and Boppart, S.A. Optical coherence tomography: A review of clinical development from bench to bedside, *Journal of Biomedical Optics* 12(5), 051403 (2007).
4. Ko, B.W., Shin, Y.W., Lee, J.M., Song, Y., and Lee, B.R. Comparison of macular thickness measurements between Fourier-domain and time-domain optical coherence tomography in normal eyes and eyes with macular diseases, *Journal of the Korean Ophthalmological Society*, 50(11), 1661–1668 (2009).

PART X

Home and mobile portable equipment applications

This chapter is an overview of applications for home and mobile portable equipment outlined first in the form of a summary table. Please see the last column of the table for where more detailed information on specific applications outlined here can be found.

Table X.1 Summary of applications of optoelectronics for home and entertainment

Application	Technology	Advantages	Disadvantages	Current situation (at the Time of writing)	More reading
Home lighting	Now almost exclusively light-emitting diode (LED) lighting, whether for new or replacement installation	High efficiency, long lifetime, instant turn-on, with no wait to reach full intensity	None	Have developed rapidly to take over all applications. For replacement purposes, new LED bulbs and other lighting units (luminaires) are available in the same form, geometry, electrical base and connectors to match earlier types	See Volume 1, Chapter 10 (LEDs), and this section.
Television (TV)	Solid-state camera to create images, radio, satellite, or fiber network transmission to distribute signal and solid-state technology using liquid crystal, plasma, or LED array to display the images	Image size, quality of display, and power efficiency are improving to give home cinema capability. A host of associated features, traditionally available only on home computers, are becoming common parts of modern TVs. Three-dimensional (3D) TV is available.	None, except perhaps for lack of physical exercise and undesirable distraction of children at times!	The flat TV screens are starting to be being replaced by larger curved displays. A host of associated features, traditionally available only on home computers, are becoming common parts of modern TVs. 3D TV is available. The sci-fi dream of fully realistic holographic type images may eventually arrive!	See Volume II, Part II (Cameras and Display Technologies).

(Continued)

Table X.1 (*Continued*) Summary of applications of optoelectronics for home and entertainment

Application	Technology	Advantages	Disadvantages	Current situation (at the time of writing)	More reading
Recording and playback of TV video signals	Now mainly (non-optical) *digital versatile disks* (DVDs) for purchased recordings or digital flash memories for home recording. High-density optically recorded disks are a potential competitor.	Modern recorders can be equipped with TV network program data, capability for presetting recordings at future times, and parallel recording of one or more programs while watching.	None, as memory capability is expanding at least as fast as the storage capacity needed for higher and higher definition TV signals	Storage capability is currently enough to store of order 1000 programs or TV films. As communications bandwidth to the home increases, programs may eventually all become on-demand in type.	See Volume II, Chapter 13, noting that nonoptical digital flash memory is currently the most common.
Optical remote control units	Handheld coded transmitters, to control electrical equipment via line-of-sight optical link	Appliances can be controlled from the armchair. The main example is control of TV channels and video recorders.	Lack of exercise!	As costs reduce, the use of such controllers is expanding, to control many other appliances, garage (carport) doors, entrance gates, etc.	See Chapter 31 on free-space Optical Communications.

(*Continued*)

Table X.1 (*Continued*) Summary of applications of optoelectronics for home and entertainment

Application	Technology	Advantages	Disadvantages	Current situation (at the time of writing)	More reading
Pocket calculator	Optoelectronic display, usually (because of need for low-power consumption) a liquid crystal display (LCD) type, although LED displays sometimes used. To give better visibility, LED backlighting of LCD display is also used. Often use small solar cell to avoid need for battery.	An optoelectronic display is an essential feature! Solar powering is highly desirable.	Reduced battery life if LED display or backlighting is used. Also, the very small solar cells used will not normally provide enough power for LED displays.	Nearly, every home has one. However, the same functions are available on home personal computers (PCs) mobile phones, and even the TV, thus the calculator is now becoming less necessary.	See Volume II, Part II (displays).
Home PC	The major computing power in the home, utilizing LCD, plasma, or LED screens for display	Has revolutionized many areas, including news updates, financial housekeeping, online purchase of travel, hotels, food, and many electrical appliances. Uses much of the same display technology as the TV. The mouse can also be equipped with optical movement detection.	Often occupies too much time!	The home computer, just like the telephone landline, is, in many homes, being replaced by the portable tablet or i-phone. Later, might be the smart watch (or even medical implants!!??).	See Volume II, Part II (displays).

(*Continued*)

Table X.1 (*Continued*) Summary of applications of optoelectronics for home and entertainment

Application	Technology	Advantages	Disadvantages	Current situation (at the time of writing)	More reading
Home printer, scanner, photocopier, and facsimile	Laser or dot matrix printer, often combined with line-scan document reader to provide photocopier or facsimile function	Combining units in one case saves considerable cost and space.	None	This is a standard add-on unit to most desktop home computers.	See Volume II, Part II (imaging and displays).
Energy	Use of sunlight to provide thermal or electrical energy. Thermal systems simply heat water from absorbed sunlight, so not really involve optoelectronics. Electrical power is provided by large photovoltaic (PV) panels.	It is only necessary to use a tiny fraction of the sunlight energy striking the Earth, to provide all our energy needs. Thermal solar systems use this energy in a more efficient manner, as no energy conversion involved.	Cost and relatively poor efficiency of current PV panels (~20% maximum at the time of writing) No generation at night and very little on cloudy days. Requires a breakthrough to provide low-cost electrical energy storage to give huge boost to viability, but this may come from batteries of future electrically powered cars.	The efficiency of panels is improving slowly, but costs of panels are reducing dramatically. Use in temperate zones still needs to be encouraged by subsidies. Development of low-cost batteries for electric cars is proceeding rapidly and would give huge boost to use of domestic-roof solar panels.	See Volume II, Chapter 16, and Part VIII (Energy Applications) in this volume.

(*Continued*)

Table X.1 (*Continued*) Summary of applications of optoelectronics for home and entertainment

Application	Technology	Advantages	Disadvantages	Current situation (at the time of writing)	More reading
Home security systems	Use of pyroelectric infrared (IR) intruder detectors, or beam interruption systems. These are used to switch on alarms or LED area lights as an intruder deterrent. CCTV video cameras to record events	Systems are relatively cheap, and cameras can provide useful evidence for prosecution.	Intruders, thieves, etc. can wear hoods or masks! Some people dislike the lack of personal privacy.	Are becoming more common in almost every "walk" of life.	See Volume I, Chapter 12 on detectors and Volume I, Chapter 10 on LEDs.
Mobile phones and "smartphones" (e.g., iPhone)	All recent models have a built-in solid-state cameras and LCD display, with LED backlighting. Many also have a LED flash unit.	Provide a mobile telephone and camera in a very compact case. Photos can be transmitted using the phone function. Can replace need for telephone landline. iPhones also provide a portable platform for many of the functions of the home PC. The widespread use of cameras is becoming a useful deterrent to criminals.	One of the most common targets of robbers, particularly muggers! Persistent callers can be a nuisance. Many personal accidents involve "selfie" photographs. Some regard the camera as an invasion of privacy.	The volume of manufacture and miniaturization of these devices has driven down the cost and size of many electronic components. The cost of high-end devices is high, but still remarkably low, if the internal complexity is considered.	See Volume II, Chapter 4 on cameras, Volume II, Chapters 5–9 on displays, and Volume I, Chapter 10 on LEDs.

(*Continued*)

Table X.1 (*Continued*) Summary of applications of optoelectronics for home and entertainment

Application	Technology	Advantages	Disadvantages	Current situation (at the time of writing)	More reading
Tablet computer	Essentially, a portable version of the home PC, with higher processing power and higher definition display than the iPhone. With Wi-Fi or "dongle," online links are available in most places and telephone interface can transmit photos instantly.	As above, but less compact, and needs a linked smartphone or an additional dongle interface to provide the telephony	As above	Often now used instead of home PC	See Volume II, Chapter 4 on cameras, Volume II, Chapters 5–9 on displays, and Chapter 10 on LEDs.
Electronic books (e.g., "kindle")	Just a Wi-Fi or hardwire interface needed to download digital books and display pages on LCD screen, usually with LED backlight. This is a cut-down version of tablet computer, with less processing power and no camera.	Lowers cost of reading	Less easy to read than a real book!	Becoming less common, as the tablet computer contains all the necessary functions	See Volume II, Chapter 4 on cameras, Volume II, Chapters 5–9 on displays, and Chapter 10 on LEDs.

(*Continued*)

Table X.1 (*Continued*) Summary of applications of optoelectronics for home and entertainment

Application	Technology	Advantages	Disadvantages	Current situation (at the time of writing)	More reading
Personal cameras, home video cameras, and "action cameras"	Solid-state cameras, with costs dependent on the size, image quality, and pixel resolution. The viewfinder is usually an LCD display, except in reflex cameras. Video cameras can take and record many images per second. Action cameras are small versions worn on headband or clothing.	No need to buy or develop photographic film! Many hundreds of images can be stored on flash memory. Multiple shots can be made at no extra cost, and the best ones selected later. Digital technology allows easy processing, storage, and transmission of images. Tiny video cameras in vehicles, or worn on headgear or clothing, are becoming a useful record of events and deterrent to criminals.	Action cameras worn on the head have been blamed for increasing vulnerability to accidents.	Cameras, iPhone cameras, and tablet cameras are everywhere!	See Volume II, Chapter 4 on cameras, Volume II, Chapters 5–9 on displays, and Chapter 10 on LEDs.
Multifunction "smart" watch	Essentially like an iPhone worn on the wrist	Depending on users' opinions, it is more available than the iPphone.	Needs to be VERY small in order to be comfortable, but then screen is very small to read.	Strongly hyped as the next must-have gadget	See Volume II, Chapter 4 on cameras, Volume II, Chapters 5–9 on displays, and Chapter 10 on LEDs.

30

Applications for home and mobile portable equipment

JOHN P. DAKIN
University of Southampton

MICHAEL A. MARCUS
Lumetrics Inc.

30.1 OPTOELECTRONICS FOR HOME AND LEISURE ACTIVITIES

Optoelectronic technology is making an ever-increasing impact on home and social life, with the ubiquitous TV set being a feature in almost every dwelling. Far more optoelectronic devices and systems are now being used in the home, and the major application areas are described briefly in this chapter. Naturally, most readers will be familiar with the majority of these, as they will probably be present in their own homes, but we hope that nonspecialists will still be able to gain useful knowledge and insights from the chapter.

30.2 HOME LIGHTING

The development of electrically powered artificial lighting has made a dramatic effect on people's lives, both at work and in the home, offering convenient illumination and removing the fire risks from candles and from oil and gas lamps. A brief summary of the types of lighting units that have been used is presented here, but the earlier Volume I, Chapters 3, 10, 11, and 19 cover the technology in far more detail.

Figure 30.1 shows a selection of various types of home lighting (candle, safety oil lamp, incandescent lamp, compact fluorescent lamp, and LED lamp). All three on the right have approximately the same illumination power (~500 lumens) but have power consumptions of 60, 10, and 7 W, respectively.

The first electrical lights used incandescent carbon filaments, formed from charred organic fabric threads, which were heated by electrical currents (ohmic I^2R heating). Their development led to the need for the first widespread electricity grids, taking electrical power to every home. The lifetime of these first lamps was very poor, but was improved to ~1000 h using tungsten wire to replace the carbon filament. Efficiency and brightness were later improved by incorporating halogen in the bulb to redeposit vaporized metal on the filament, enabling the filament to be run at a higher temperature for a reasonable lifetime.

Most home incandescent bulbs and home lighting units have been replaced in recent years by far more efficient (and longer lived, giving ~8000 h use) compact fluorescent light bulbs, but, because they still have major disadvantages (they contain toxic mercury and take several minutes to reach full brightness after switching on), they are, in turn, being rapidly replaced by LED lighting, as the output power and radiance (spot brightness) improve and their cost reduces. Provided they are not overdriven, LED sources have even better lifetime than the fluorescent sources, so much so that low-radiance types never normally have to be replaced. As already mentioned, these lamps are shown in Figure 30.1, along with a far-less-bright candle and old safety oil lamp for comparison.

In a similar example, from the home of one of the authors, a 750-lumen recessed lighting incandescent BR30 65 W light bulb was replaced, first by a 13 W compact fluorescent unit and later by 8.5 W LED lights. The incandescent BR30 light bulb had an expected lifetime of 2000 h. The compact fluorescent BR30 variety should last about 10,000 h but requires a few minutes of warm up time to reach full illumination, whereas the LED BR30 equivalent turns on to full power in 1 ms, will last 30,000 h, and will be dimmable.

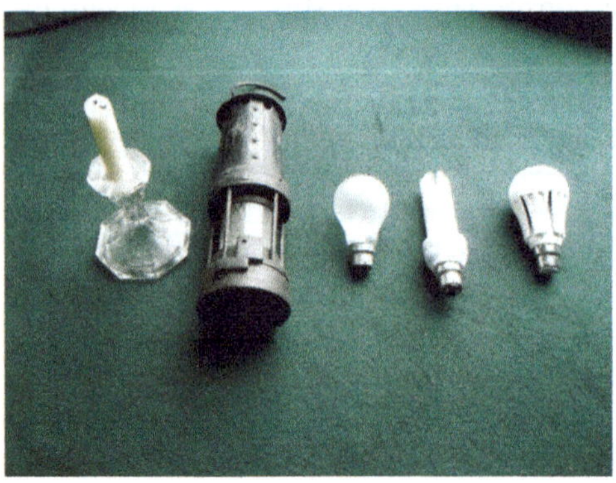

Figure 30.1 Evolution of home lighting.

An interesting side application of home lighting is for entrance path or garden path lighting. Here, a low-cost solar cell is used to charge a battery during the day, and, when dusk is detected (no signal from the solar cell), a small low-brightness LED light is switched on to illuminate the path. The brightness has to be low to conserve charge, but very little is needed to show the line of the path in the dark! Inexpensive photo cells are also being utilized to turn on lamp posts at night and to control outdoor lighting on porches only when people are detected. This is illustrated and discussed in more detail next, when home security is considered.

The compact nature and the relatively low heat output of LED lights mean that they can be employed in far more flexible ways. A common example of this is narrow light strips using linear arrays of LEDs mounted on electrically conductive ribbons, which allow their installation in cylindrical tubular luminaires designed to replace fluorescent tubes, which can now be made of polymer rather than glass. The ribbon is flexible, allowing it to be bent into more complex shapes if desired. Wired arrays of LEDs are also commonly available as decorative lights, for example, for trees.

The LED arrays can also be two dimensional, and large arrays have been developed for the back-lighting of large-screen liquid crystal display (LCD) TVs. They are also available, therefore, as large-area, ultra-low-profile light panels, which can be attached to walls or ceilings to provide a dazzle-free illumination source, which occupies less space and is more easily cleaned than more protruding fittings.

30.3 TELEVISION

John Logie Baird made the first demonstration of televised moving images in 1926, but this was achieved using a very crude electromechanical system to record and display the images. A major advance was made using "cathode ray" tubes

(CRTs), where electronically scanned electron beams were used to scan and interrogate light-sensitive spots on the face of a vidicon camera in order to record images of a photographed scene. A similar electronically scanned electron beam was then used to cause illumination of special phosphor powders in a CRT display unit. Despite the rather bulky, and somewhat hazardous nature of the CRT (they could implode violently if the glass envelope was ruptured and require very high voltages to operate), this led to the widespread use of home TVs in the mid-1930s.

Since these early days, there have been a series of remarkable advances, using many of the camera and display technologies described in the earlier chapters. These advances have led to flat, large-area screens, brighter pictures, and superior power efficiency. Plasma displays are also used in some modern TVs. Newer technology includes ultra high definition (UHD) TVs having 3840- by 2160-pixel resolution and curved screen UHD TVs. The curved screen TVs have a better capability at removing reflections from light sources in the room and other reflective objects. The curved TV also provides the viewer with a more immersive experience, and the screen appears to have more depth to the images. Contrast is also improved over the equivalent sized flat-screen TV, and the viewing angle is increased. Another new technology is the organic light emitting diode (OLED) TV, which has improved contrast and dynamic range over LEDs because blacks are really black. Figure 30.2 shows a pictorial representation of the history of TV over the past nine decades.

Cameras now use solid-state technology, such as charge-coupled devices (CCD) or complementary metal oxide semiconductor (CMOS) (see Chapter 4, Volume II), and displays have virtually all moved to a flat-screen format, using a variety of technologies, such as LED-illuminated LCDs, and brighter, faster responding display screens use arrays of plasma light generating points or large arrays of color LEDs.

The latest advances in the technology are helping to make 3D TV more common and are also directed at producing the flexible or curved screens

Figure 30.2 Evolution of TV. (Courtesy of the dailystar.net http://www.thedailystar.net/the-evolution-of-television-an-infograph-42861.)

discussed earlier to allow a more cinematic home movie experience. The other advances of a non-optical nature are in the electronic hardware and software areas, adding computing power, memory for storing channel information, Wi-Fi and other modems, and incorporating recording and playback technology (see descriptions next). These are blurring the differences between the capabilities of the home PC and the TV to an extent where the only need for separate devices is to have two displays to permit two users to do different tasks.

Regarding the transmission of signals from the broadcast transmitter to the receivers, this used to be done using analog (amplitude modulated vestigial sideband) transmission, but digital transmission of encoded signals is now the norm. This has led to rapid advances in digital recording equipment to store whole TV programs, and to "smart" TVs, where they contain most of the elements of a home computer (see next). The other advance is in cable TV, where large numbers of channels were "piped" along high bandwidth coaxial cables, but are now multiplexed digitally on ultra-high-bandwidth optical cables.

Another important aspect of TV is how the video signals are encoded at the transmitter and decoded in the home TV set. Because of the ubiquitous nature of TV, a logical person might have thought that this would be standardized in the early days, but politics and commercial interests have been unsuccessful at reaching international agreement on standards, even in the present day. In the days of analog TV, extending till the early days of this century, there were still three main decoding systems, each with several perceived advantages and disadvantages. In Europe, with the exception of France, the phase alternating line (PAL) system with 625 horizontal lines in the raster scan was prevalent, whereas in the Americas, the lower resolution National Television System Committee system with ~425 lines was used. In France, Russia, and many former French colonies, particularly in West Africa, the SECAM (*Séquentiel Couleur à Mémoire*) system was used.

Because digital TV and newer high-definition standards, of typically 1000-line resolution or more, have been developed, there are still at least three or four separate standards, depending on geographical location. Fortunately, now complex digital processing chips are commonplace, most commercial TVs available have decoders for different standards already built into them. For example, it is possible to buy a TV in France, where SECAM analog and its newer digital TV standard existed side by side, and expect it to self-adapt in an intelligent manner to operate successfully in European countries where PAL was the analog costandard. Of course, now most countries have adopted digital TV and phased out their old analog transmissions, but this was inevitably a slow process because so many homes still had the older TVs, which were incapable of decoding digital signals.

For further reading on TV decoding and standards, there are of course numerous textbooks, but, because of the huge and dynamic interest in this topic, it is an area where there are also excellent reliable articles on Wikipedia, which cover these in useful detail.

30.4 RECORDING AND PLAYBACK OF VIDEO AND TV SIGNALS

The earliest TV recorders used in studios stored analog video signals on very wide, high-transit-speed magnetic tape, but these have been replaced in turn by DVD recorders (see below) and compact solid-state flash memories, the latter not using any optoelectronics in their construction. Various forms of optical disc and more recently Blu-ray and other storage/playback units have become a very common part of home entertainment equipment. The optical disc was discussed in detail in Chapter 13, Volume II, but a brief review is in order here. This optoelectronic device uses a focused laser to encode patterns in a spinning disk coated with suitable recording materials. This sounds simple to achieve, but in reality, it is an extremely complex optoelectronic system, as it is necessary to control the laser focal position to fractions of a micron, control the spot to follow the recording tracks faithfully, and modulate when used to "burn" coded patterns. The system then has to allow conversion to readout mode to detect reflected intensity changes as the disk spins, again following the tracks, while maintaining correct focusing. That this has been achieved, at a price little more than one can purchase a simple laboratory laser, is a remarkable achievement.

Technology has inevitably moved on in this area, and now most home TV recorders now use hard-disk or solid-state flash memories, so this use of optoelectronics is reducing. However, it might eventually return in the form of Blu-ray recording devices where blue lasers are used to increase the storage density on an optical disk. DVD and Blue-ray disks are of the same size. However, the DVD format uses a red laser at 650 nm and has a 4.7 Gb capacity, whereas the Blu-ray players use a 405 nm laser with a smaller focal spot size and has a 25 Gb capacity. Double-layer Blu-ray disks have a 50 Gb capacity. The DVD player has a 480 horizontal line resolution, whereas the Blu-ray player has a 1080 horizontal line resolution.

30.5 OPTICAL REMOTE CONTROL UNITS

All modern TVs are supplied with a handheld remote control unit, which now nearly always uses line-of-sight optoelectronic signaling. According to Wikipedia, however, the first remote control intended to control a TV remote (perhaps appropriately called "Lazy Bones"!) was developed by Zenith Radio Corporation and was physically connected to the TV by an electrical wire. Modern controls use a broad-beam 940 nm LED transmitter to send pulse-coded signals to a detector/decoder built into the TV set. The brightness is such that accurate pointing is not necessary, and it can often code the TV with diffusely reflected light from a wall.

Such controls are now used for far more home appliances rather than just the TV, being supplied with digital and satellite TV decoder set-top boxes, TV program recorders, children's toys of various kinds, motorized windows and curtains, garage (carport) doors, and automatic gates, to name just a few.

Sadly, for optoelectronic enthusiasts, this is one area where many such remote devices are being replaced by units using new "Bluetooth" technology, which is based on ultra high frequency radio signals.

30.6 OPTICAL FIBER COMMUNICATIONS TO THE HOME

Optical fiber communications dominated the market for long-distance terrestrial communications and also for medium-scale networks for many years, but it is only at the time of writing that these links are now being extensively used to communicate into and out of the home.

This trend is continuing rapidly, as fiber optics enables broadband multimedia communications for all the information-hungry devices being developed for the home. The technologies for this form of communication have already been described in detail in Volume II, Part I, so their operating systems will not be repeated here. The main thing to emphasize is that this trend is revolutionizing the speed of the Internet and other systems. The links to the home now allow, in particular, ultra-fast downloading of video signals and movies, which contain ever more information content as ever-higher definition standards are becoming the standard.

30.7 OTHER OPTICAL COMMUNICATIONS IN HOME EQUIPMENT

Optical fiber cables are starting to be used to interconnect in-home electronics systems. For example, many present-day home theater sound systems are now using optical digital cables for connecting from the TV to the surround sound system. It is anticipated, however, that more use will be made in future of more direct optical line-of-sight modems using LED to optical detector links. Because of the small size of most rooms, and even most of the private gardens, diffuse reflections off ceilings and walls will usually give sufficient signal to give the desired signal-to-noise ratio, even when no direct line of sight is possible. The big advantage of such a technology is that it can reduce the number of trailing cables, with only a short power cable (or rechargeable internal battery) needed to each device. Also, the bandwidth capability is far higher than that of simple cables, and the link can be designed to be free of cross talk and electrical interference.

The home of the future may see a whole variety of "smart" appliances, all interconnected via optical line-of-site or diffusely reflected optical links. If coupled to the home TV, a PC, Wi-Fi, or the Internet, these may then be programmed, either locally or remotely. It is then possible, for example, to switch appliances on or off or even control them in a predetermined fashion, in a far more intelligent manner than possible with simple timers.

30.8 POCKET CALCULATOR

The pocket calculator was once a feature of every home and workplace, but now is becoming less of a necessary feature, as nearly all PCs and mobile phones have the same facility. The calculator must, of course, have a display to check for correct input data and to read the result. It requires very low-power consumption to allow the use of either a small solar cell or a small battery to provide its power. Usually, they use LCDs to achieve this; as being essentially operated by electrostatic forces, they draw hardly any current (the electrical drive to perform the actual calculation requires very little energy, as the time period for which it consumes power is very small).

If they have to be operated in darker conditions, a higher power LED display (or LCD, which back lit by a LED) must be used, as most LCD displays in calculators usually rely on reflected ambient light. This normally then precludes the use of solar power, as the surface area available for the solar cell is very small, and often the ambient light level is a lot lower than that present in direct sunlight.

30.9 HOME PC

The home PC uses essentially the same display technologies as the TV. Just like the TV, early PCs used heavy and bulky CRT display tubes, but now all use flat screens or curved screens of various types. The PC generates the display signals in digital form with a plug-in circuit board card, from signals generated by software in the PC or from image information received via a modem from a communications network or "off-air" radio signal. This is again an area where technology has advanced remarkably. The computing power of the home PC is now many orders of magnitude higher than the early mainframe computers, which used to fill whole rooms of buildings. The technology developed for the larger desktop computer has spun off into smaller units, such as small notebook and iPad units, and microminiature computers built into mobile phones and even wrist watches (see later paragraphs for more details of these).

30.10 HOME PRINTER UNITS

Apart from the obvious display unit, the PC usually has a number of peripherals, which also use optoelectronics. Most home PCs use a printer, which can produce text or images on conventional mat paper or on glossy photographic paper. There are various printing technologies, as discussed in Volume II, Chapter 10, but the most commonly used are currently dot matrix ink projectors (which, as the name implies, produce a number of ink spots in a matrix) or laser printers where a laser fuses a powder to produce the images.

Most home printers also include scanner, photocopy, and facsimile functions. In most printers, the document to be scanned is placed on a flat glass plate, and a flat white cover is placed over the document. An intense lamp illuminates the document, and a line scan head, including mirrors, lenses, filters, and a CCD detector array, is moved slowly over the surface of the document by a belt attached to a stepper motor until the entire document is scanned. Some scanners use three-color filters, one for each primary color to perform color image scanning.

Exciting recent developments have been in the area of 3D printing, where a number of 2D images are built up by stepping the printer in the orthogonal direction to produce a step-by-step 3D image. Many of these are nonoptical in nature, with nozzle-ejected material being built up, layer by layer, as the printer scans, but some use lasers or ultraviolet lamps to fuse or cure the material, which is ejected to produce a solid material.

30.11 ENERGY IN THE HOME

There are two main ways in which solar power is utilized in the home. The first, which does not really use optoelectronics as such, involves the use of highly absorbing panels (or blackened water pipes or heat pipes, within evacuated tubes) to convert the incident solar energy to useful heat and is mainly used to heat water using a fluid (usually water with added antifreeze and anticorrosion agents) to transfer heat to a hot water storage tank in the home. This type of system is highly efficient, using close to 100% of the energy passing through the optical windows of the collector. The only way that electronics may be involved here is when temperature sensors are used to control water circulation pumps, which are activated by an electronic controller when the panels become hot.

The second way of using solar energy is using PV optical-to-electrical energy conversion ("microgeneration"), where the incident energy is collected by semiconducting panels, which produce DC electricity, albeit with currently rather poor efficiency, on the order of 10%–20%, although the technology is advancing very rapidly. This topic was dealt with at far greater length in Volume I, Chapter 12 and

Volume II, Chapter 16, but, for completeness, it is useful to comment here on aspects that are specific to home use.

It must first be admitted that, because of the poor efficiency and rather high cost of panels, this form of solar energy collection has still, at least at the time of writing, only marginal *real* economic benefit, except in countries with near-constant sunshine, where it is already a viable option. However, the situation is changing rapidly, and, for future environmental reasons, solar PV has been actively encouraged in most countries, even in ones with more limited hours of solar incidence. This is achieved by means of generous state subsidies to homeowners or installers, usually arranged in the form of feed-in tariffs, where the state pays for any excess energy generated by the system, which is not used in the home. This excess is fed back into the electricity grid.

Home solar panels are usually connected in serial arrays, giving a high-voltage DC output (typically 30–40 V for each panel, so on the order of 500 V for a 14-panel array). In order to match this to the home electrical system, a high-power "pure sine wave" inverter is needed, which converts high-voltage DC from a series array of solar panels to AC mains power for the home (typically 240 V AC in Europe, 110 V AC in the United States). The output power of this is synchronized with the frequency of the town electricity grid and it can control feed-in of excess power, not used by the home, back into the grid.

An essential safety feature of this arrangement is that the system must shut down instantly if the grid supply to the house is interrupted by a power failure, or repair electricians working inside the house, or even externally, would be in great danger of electrocution!

The feed-in arrangement to export excess electricity not used by the home is a very important economic feature, because energy generation tends to be maximum in the middle of the day, when all but retired people and housewives/househusbands are usually at work, and when electricity demand for the home itself is usually at a minimum. Fortunately, at this time of day, the industry still needs more electricity than PV generation is currently providing. Of course, in future, this situation could change, if nearly every home and factory has panels, and if more farmland factory roofs and derelict city grounds are used for larger-scale arrays!

In some countries, the state subsidizes home solar "microgeneration" by paying a substantial part of the cost of the system cost, or even pays the whole initial cost, and recovers it from the consumer via regular electricity bill payments. These arrangements are called names to appeal to environmentalists, such as "green deals."

It is the high reliability and passive nature of panels, plus the initial generous state subsidies, which enable householders to make good long-time returns, leading to increasing use, even though the environmental benefit is still only marginal except in countries with high levels of daily sunshine. Fortunately, however, as a consequence of this, the rapidly increasing scale of manufacture of panel is driving down costs at a remarkable rate. We would expect that by the time of publication, or soon after, most systems will reach "grid parity" (e.g., the point at which they become the most economic medium-term option for in-home electricity generation), even in countries with only moderate annual solar irradiance.

One remaining disadvantage is that there is only a direct benefit during daylight hours! The eventual hope is that, if suitable economically viable energy storage systems can eventually be developed, solar PV should be able to power the complete energy uses of most homes and factories without any difficulty. The best hopes here are either in new battery or ultracapacitor technology or in novel hybrid energy storage systems using gyroscopes, compressed air, or liquefied gases, in conjunction with motor + alternator energy conversion/reconversion systems. These energy storage developments are already being pursued actively for electrical road vehicles and vehicles with kinetic energy storage and huge investments are being made by all the major automobile and truck manufacturers.

A diagram (Figure 30.3) of the various energy transfer possibilities into and out of the home has kindly been provided by Dr. Jann Binder of ZSW (Zentrum für Sonnenenergie- und Wasserstoff-Forschung Baden-Württemberg or Centre for Solar Energy and Hydrogen Research, Stuttgart, Germany).

In Figures 30.4 and 30.5, some photographs of the roof-mounted solar panels and the power inverter for the obtained standard AC mains voltage from the DC output of panel arrays and controlling inputs to the grid are shown.

As mentioned above, a solar power inverter is used to convert high-voltage DC from a series array

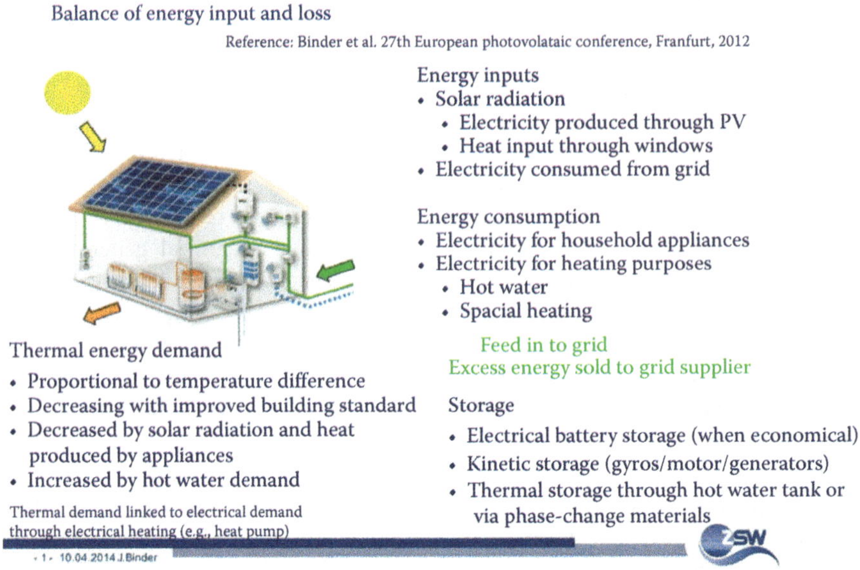

Figure 30.3 Diagram showing balance of energy input and loss for a solar-powered home. (Courtesy of Dr. Jann Binder, ZSW, Stuttgart, Germany)

Figure 30.4 A home with 14-panel solar collector array on the roof. Total power ~3.5 kW in full sunlight.

Figure 30.5 A 4 kW DC-to-AC inverter in attic space of a home.

of solar panels to AC mains power for the home. This synchronizes with the frequency of the town electricity grid and can control feed-in of excess power not used by the home back into the grid.

30.12 HOME SECURITY SYSTEMS

As set out below, there are many commercially available systems available for improving home security. (We shall only discuss these very briefly here, as there is a separate Volume III, Part III on security and surveillance.)

First, darkness is the friend of the criminal intruder, so good lighting systems are the first weapon for improved home security. The most popular method of achieving this, without energy wastage, is to have a light, which is triggered by movement or by the proximity of the intruder. There are now many energy-efficient commercial systems, which achieve this using three key components:

1. An IR detector, usually based on the pyroelectric effect, as these also respond best to a moving (IR) signature. This is used to switch on the illuminator, when people are detected.
2. An illuminator designed to cover the area to be protected. This not only makes the intruder visible but also acts as a strong deterrent to any further intrusion. Most of these now use arrays of high-brightness white light LEDs.
3. A power supply, a detector signal processor, and an electronic switch control unit for the illuminator. This can be powered from the main

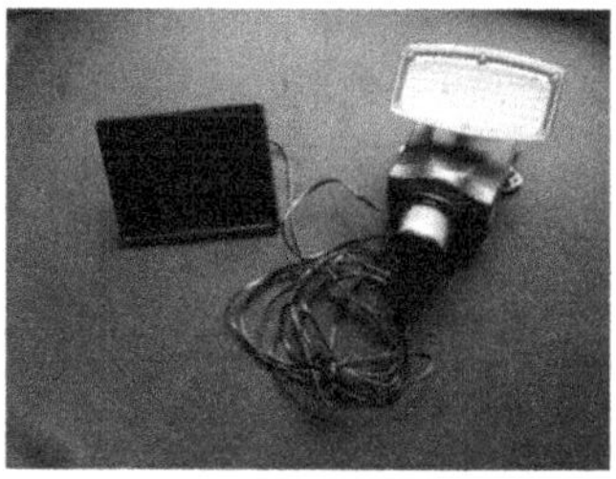

Figure 30.6 A solar-powered home illuminator/security light with pyroelectric detector.

electricity supply, or from a battery, which is kept charged by a solar panel. Very little average power is needed, as the light only comes on when personnel are detected (Figure 30.6).

The pyroelectric detector is also commonly employed within homes to detect the presence of intruders who have actually entered and then to sound an alarm. Of course, although the IR controlled lighting system above is useful to light dark areas for the benefit of the home owner, it can usefully be left active most of the time (particularly if powered from a solar panel). However, the internal intruder detector clearly needs to be switched off while the home is normally occupied to avoid false alarms.

A third major security aid is the CCTV camera. Again, this can be arranged to be only switched on when an intruder is detected, using the same type of IR detector. The camera can either be an IR camera, to allow recording in darkness, or be a conventional low-cost visible CCTV camera, making use of the LED illuminator. Usually, camera systems are mounted on a wall at a height sufficient to be out of easy reach, and often have a flashing LED lamp so they can be seen, thus acting as a psychological deterrent to intruders.

Storage of high-bandwidth video security-related data was originally a big problem, but now, as with the TV recorders discussed earlier, compact digital memories of large capacity (both flash memories and DVD-based ones) can allow many hours of recording. Of course, if the system is only activated intermittently by sensors such as an IR event detector, there is even less need for large-volume data storage. There are then less data to be sorted, if it is wished to try to play back and observe only the key event period from the information in the data store.

30.13 MOBILE PHONES AND OTHER PORTABLE DEVICES

The mobile phone is, as the name implies, a phone that can be carried on the person. It is usually powered by rechargeable lithium batteries and can be rapidly recharged via mains power points, automobile connectors, from standby primary cells, or even using solar panel or portable mechanical generators, driven by physical movement.

It is only a very short time since the mobile phone was first introduced, but now it is so ubiquitous that every child wishes to have one and most in the prosperous nations already do. They are also becoming remarkably common in poorer third world countries too, because the cable infrastructure is so badly developed, so they are the only viable means of rapidly setting up the vital communications links needed for fast developing societies.

Technology here has advanced rapidly and the usual mobile phone is a now a multifunction device, with all variety of "smartphone" features, which the original telephone inventor, A G Bell, would simply marvel at. Whereas the only optoelectronics is the early generation mobile phones was a simple LCD display to show the numbers dialed, current smartphones have multiple software features and high-resolution cameras and LCD color displays to enable functions such as electronic clock, still-photo camera, action videophone, and often a geographical mapping and satellite navigation device. The early cameras in mobile phones had rather low picture resolution, but now, despite their small size, they are capable of taking pictures of quality approaching that of lower-cost, single function, digital cameras, and video cameras.

This latter feature is having as dramatic an influence on activities in our world as did the introduction of TV, as now nothing outside (and often inside) the home is hidden from view. There are now so many widely distributed mobile devices, each capable of filming and recording events, that nothing can be guaranteed to occur in a private unobserved manner. Although this is in some ways deeply disturbing to our human rights of privacy, it also has real advantages of making life very difficult for inconsiderate people, bad highway users, criminals, and even repressive totalitarian regimes to hide their actions.

30.14 TABLET COMPUTER

This portable device, which is essentially a thin and generally compact version of the PC, manages to incorporate a digital camera, a reasonably high-resolution display, high computing power, and many of the features of the smartphone into a device that can easily be carried in a handbag.

30.15 ELECTRONIC BOOKS (E.G., THE KINDLE)

Simpler versions of a tablet computer, without many of the functions, are available as electronic books. The main advantage is that many such books can be downloaded via networks and stored for future reading. However, because this same function can be performed by the more recently introduced tablet, which now costs little more than the e-book, it is likely that the single function book devices will probably soon disappear from use.

30.16 MULTIFUNCTION SMART WATCH

As miniaturization of computer technology continues, apparently unabated, it is likely that many smartphones will eventually take the form of a wristwatch, as they can combine the primarily desired function of electronic clock, with all the other features. Of course, wearing on the wrist makes it easier to access, yet less likely to be accidentally lost. Such devices are already here, but still requiring size reduction to be more aesthetically pleasing and less mechanically intrusive.

30.17 HOME PHOTOGRAPHY, VIDEO PHOTOGRAPHY, AND ACTION CAMERA

The science and engineering behind modern cameras has been dealt with in great detail in Chapter 4, Volumes I and II. Photographic film is now hardly ever used, and the normal camera in personal and professional use is now based on digital solid-state technology. This not only replaces the need for photographic film and its associated development time and cost but greatly eases the ability to record, process, and transmit the pictures to anywhere in the world. Video photography is merely an extension of this, where photographs are taken in rapid succession.

The technology has also greatly reduced the size and weight of typical cameras, taking miniaturization to a level, which would have been unthinkable, even at the end of the last century.

A recent "must-have" item of technology for the adventurous outdoor enthusiast is the compact action camera that can be worn on the body or on the head. This is a compact digital video camera capable of recording, with the aid of semiconductor flash memories, many minutes, or even a few hours of dramatic moving images. For the sports enthusiast or activity vacation tourist, it can provide home movies to record the more dramatic moments. There has been some controversy of the safety aspects of wearing cameras, particularly head cameras, while taking part in dynamic sports such as horse jumping, skiing, and surfing, so clearly care is needed, but otherwise the use of such devices is becoming more and more popular. As discussed earlier, similar wearable cameras are also worn (either visibly, or hidden) as an aid to personal security, for recording of incidents such as abusive incidents or criminal behavior. Police officers now usually have these cameras.

30.18 FUTURE DEVELOPMENTS

Many of the future advancements will, of course, be ongoing developments of the devices mentioned earlier, with a tendency to add ever more internal processing power, as the cost of electronics reduces. However, there are many desirable features, which are just wish lists, others, which are already starting to appear, and others with a real possibility of being developed in future.

Perhaps the greatest desire for visual home entertainment is to have full holographic TV, which would be capable of showing true 3D images in an apparently open space. Such images have been a frequent feature of science fiction films, where the emphasis is on giving a vision of what the future will create. The closest to this so far has been the creation of monochromatic 3D images using laser beams, scattered from a holographic screen. Unfortunately, real 3D systems having sufficient fidelity and a realistic visual effect in full color appear to be still some way off and will probably always require the presence of some form of screen or mist to scatter light.

Another area, which is the subject of extensive R&D, is that of having an automaton, for example, a robotic housewife or househusband to do the chores. This is a dream which is slowly becoming reality for some simple tasks. These are the ones where household appliances move in two dimensions to cover larger areas, such as vacuum cleaners, floor washers, and lawn mowers, as they are battery powered using compact lithium cells and can be programmed to cover all the floor areas without too much overlap of areas already covered and, with the aid of sensors, such as video cameras, optical proximity detectors, or acoustic transponders, avoid obstacles and detect the edges of the area to be covered (e.g., house wall or lawn edge).

For the robots to work in the complex 3D volume of a house and handle delicate objects such as glass, move light papers and flower arrangements, and so on., they require several other levels of capability and artificial intelligence, but progress in this area is very rapid, so it cannot be discounted. Walking and stair climbing robots are already here, so it is just the visual appreciation and handling capabilities, which still need to be fully developed, apart from the cost aspects!

Perhaps the aspect that will most change our homes in the near to mid-future will be the level of centralized monitoring and control that is being envisaged for the "smart" home. The basic concept is to have all the electronics-based (or electronics-compatible) devices in the home connected to a central control system, such as a super PC, in such a way that all can be monitored, controlled, and preprogrammed to perform their desired functions at any time of day or night. This way, a householder could, just as one example, preprogram breakfast to be cooked and ready soon after a PC-controlled wake-up call has raised them from slumber. Optoelectronics would find its place in cameras to allow the activity to be monitored and the communications links could be achieved by line-of-sight or diffusely scattered remote control units, such as those used for TV. If the householder is at work or on vacation, the house could still be programmed and observed (via cameras in the home and local displays on his/her tablet PC) from afar, using a combination of Wi-Fi and Internet links, in this case taking advantage of the huge bandwidth available over optical fiber cables to the home.

Clearly, the extent to which technology will enter our homes will, as always, be limited not by only technical limits but also by affordability and the willingness to add complexity to our lives. History so far has suggested that the last two factors of cost and complexity are sometimes not as powerful as one might think, and many people have a pressing desire not just to have the latest gadgets, despite the expense and extensive learning curve needed to operate them, but to be seen to have them!

A CRT TV of a few decades ago (note the very deep profile, compared to modern TVs).

A more recent flat TV design, circa 2004.

Curved screen UHD TV.

Curved TV, with liquid crystal display, from circa 2008.

BIBLIOGRAPHY

http://en.wikipedia.org/wiki/Electric_light.

History of Television on Wikipedia http://en.wikipedia.org/wiki/History_of_television.

The History of Television, 1942 to 2000, Albert Abramson, McFarland, 2003, ISBN 0786412208, 9780786412204.

Videodisc and Optical Memory Systems, Jordan Isailovic Prentice Hall, 1984, ISBN 10: 0139420533 ISBN 13: 9780139420535.

PART XI

Free space optical communications

In earlier chapters (Part 1, Volume II), the subject of communication over optical fiber waveguides, one of the principle modes of use of optical telecommunications, was covered in extensive technical detail. Therefore, it is not discussed further in this chapter, allowing us to concentrate more on the growing use and future potential of free-space optical links.

Table XI.1 shows a number of different applications for free-space communications. Outdoor free-space communications is commonly referred to as free-space optics (FSO). Links operating over ranges from tens of meters (e.g., between buildings) up to several hundred thousand kilometers through space fall into this category. In indoor environments, optical wireless systems can conveniently use free-space transmission to provide wireless connectivity without electrical interference with other devices. There has also been a limited amount of work on free-space transmission between and inside racks of electronic equipment inside data centers.

This chapter provides a brief introduction to FSO and rack-to-rack communications, and then focuses on indoor optical wireless communications.

Table XI.1 Summary of technology and applications for free-space optical line-of-sight communications

Application	Technology	Advantages	Disadvantages	Current situation (at the time of writing)	More reading
Ultra-long-distance communications Range > 1000 km	Modulated laser beams, expanded with lens to give a wide-diameter beam to reduce diffraction	Because of shorter wavelength, (less diffraction spread), beam collimation is much better than possible with microwaves or radio waves.	Limited to line-of-sight applications. Can be disrupted by bad weather conditions, except when used in space applications	Used for communications between satellites and to and from satellites to the earth. Has also been used for attempts to communicate with alien planets [search for extraterrestrial intelligence (SETI)]	This chapter
Long-distance (outdoor) terrestrial communications. 1000 km > range >10 km	Modulated lasers, with beam expansion if necessary, to reduce beam divergence	Beam collimation is much better than possible with microwaves or radio waves.	Limited to line-of-sight applications. Can be disrupted by bad weather conditions, such as fog, mist, and even thermoclynes	Not used very much, because it can be disrupted by bad weather conditions	This chapter
Medium-distance (outdoor) terrestrial communications 10 km > range > 100 m	Modulated and directed narrow-beam lasers, or wider-angle broadcast systems using LEDs or lasers	Very useful for high-bandwidth transmission over moderate distances, for example, building to building	Limited to line-of-sight or diffusely reflected applications. Can be disrupted by bad weather conditions	Systems under development for communication to personal handsets and to moving vehicles	This chapter
Medium- and short-distance (indoor) communications (less than 100m)	Mainly wide-angle broadcast systems using LEDs or lasers. Will often make use of diffuse reflections from walls and ceilings	Very useful and convenient for high-bandwidth transmission without need for wires .Great potential for links between cabinets of equipment. Very flexible and can use wavelength division multiplexing (WDM) Being indoor, it is not affected by weather.	Limited to line-of-sight or diffusely reflected applications	Established technology for remote control of home appliances, and children toys but otherwise not used a lot at time of writing. Wider use likely to expand as technology improves	This chapter (see also Chapter 30, Volume III for home use)

(*Continued*)

Table XI.1 (*Continued*) Summary of technology and applications for free-space optical line-of-sight communications

Application	Technology	Advantages	Disadvantages	Current situation (at the time of writing)	More reading
Intra-cabinet and intra-device communications	Wide-angle broadcast systems or directed-beam systems using LEDs or lasers	Very high bandwidth possible, particularly with WDM Adds flexibility to communications within electronics equipment, without disruption of, or interference to, electrical connections No earth loop problems	No real disadvantages	Not used a lot at the time of writing, but potential is huge, now that component costs are low Use likely to expand dramatically, as technology improves	This chapter
Inter-board and inter-chip communications	Wide-angle broadcast systems or directed-beam systems using LEDs or lasers	Very high bandwidth possible, particularly with WDM Adds flexibility to communications within electronics equipment, without disruption of, or interference to, electrical connections No earth loop problems and directed beams can cross over in space without crosstalk.	Most systems still being developed	Not used a lot at the time of writing, but potential is huge, now that component costs are low Use likely to expand dramatically, as technology improves	This chapter
Intra-chip communications	Directed-beam systems using LEDs or lasers	Very high bandwidth possible, particularly with WDM Adds flexibility to communications within electronics equipment, without disruption of, or interference to, electrical connections No earth loop problems	Most systems still being developed	Not used a lot at the time of writing, but potential is huge now that component costs are low Use likely to expand dramatically, as technology improves	This chapter

31

Optical communications through free space

DOMINIC O'BRIEN
University of Oxford

31.1 INTRODUCTION TO FREE-SPACE OPTICAL COMMUNICATIONS

Optical communication via line of sight has been a key feature of the natural world. The earliest humans used signal beacons, with a human acting as an optical receiver. Later, more sophisticated semaphore systems using the same principle were deployed and heliographs were also developed. Perhaps the earliest FSO link with an electronic receiver was the photophone developed by Bell [1]. This used a microphone to modulate a mirror, which in turn modulated a transmitted beam of light. An electronic receiver used a selenium photocell to create an electrical signal that modulated the microphone.

FSO is now extensively researched, with a number of companies selling links, over a wide range of different ranges and data rates. Figure 31.1 shows a typical FSO system. A laser or light-emitting diode (LED) source is used as a transmitter, and optics is used to control the emitted light, creating a transmitted beam with known divergence. This beam propagates to a receiver, where collecting optics focuses it onto a photodetector. A transimpedance or other amplifier is used to amplify the photocurrent output from the detector and create an

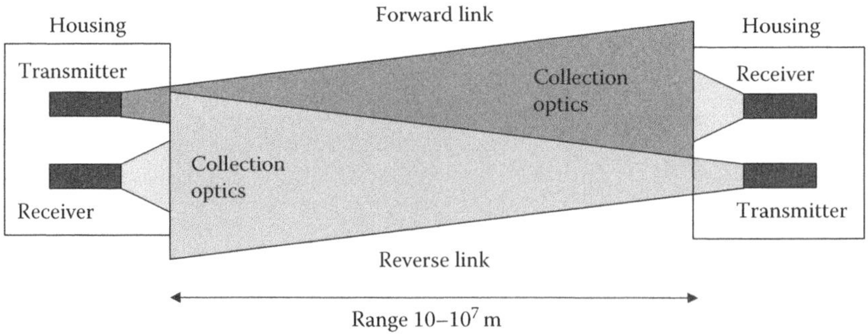

Figure 31.1 FSO system.

electrical signal, from which data is recovered. Links are normally bidirectional and are created by using a pair of identical transmitters and receivers.

At the transmitter, the diameter of the transmission aperture is usually set by eye safety considerations, in order to ensure that the optical power density at the aperture is below allowable limits. The beam divergence depends on application and range. For links between buildings, flexing of the structures causes terminal movement, and for short-range links (typically hundreds of meters), using a divergence of several degrees removes the need for any active alignment of the transmitter and receiver to compensate for this. Tracking is required (due to terminal movement) for narrow divergence links, but the lower path loss created by the low divergence allows these to operate over longer ranges, or have an enhanced link margin.

There are a number of commercially available Gbit/s class links (for instance, [2,3]). These operate in either the near-infrared (IR) (<1000 nm) wavelength region, taking advantage of low-cost optoelectronic devices, or near 1500 nm, taking advantage of optical fiber telecommunications components. The major impairments to the operation of such links are fog (see, for instance, [4]) and atmospheric turbulence [5]. Modest gains in link performance in fog can be obtained at longer wavelengths, but these are not sufficient to overcome the attenuation of hundreds of decibels per kilometer that fog can create.

The market for such links is greatest where visibility is good, or link lengths are short (<1000 m or so). Many applications are in data communications, for example, to create redundant links or implement campus networks. More recently, the low latency of these links compared with an indirect fiber route has found application in communications for high-frequency trading on financial markets.

A number of different approaches to implementing the terminals have also been taken. Links using light from an optical fiber, which is then collimated, propagated through free space, and coupled back into a receiving fiber, have also been reported. In this case, the free-space link is transparent to the particular data format and communications wavelength. Tbit/s capacity links can be achieved using this approach [6]. However, adaptive optical systems are required for longer links in order to ensure that any distortion of the optical wavefront can be corrected to allow efficient coupling of light into the optical fiber at the receiver [7].

A different variant is to use a modulated retroreflector (MRR)-based terminal. In this case, light from the transmitter propagates to the receiver and is returned to it by the action of a retroreflector. A receiver colocated with the transmitter decodes the returned signal. The data is imposed on the beam of light by the use of a modulator in the beam path at the retroreflector. A number of links have been demonstrated using this principle [8]. The advantage is that the MRR is lightweight, and no source of light is required at the MRR terminal. In addition, the retroreflector removes the need for the MRR terminal to perform tracking. This allows a low mass, low power terminal which makes it suitable for placement on airborne mobile terminals such as unmanned aerial vehicles (UAVs) [9].

A number of longer link experiments have been performed, from aircraft to ground, aircraft to satellite, satellite to satellite [10], and the moon to the earth [11]. The European Data Relay Satellite

(EDRS) network incorporates long-range optical links from low earth orbit (LEO) satellites to geostationary earth orbit (GEO) satellites as part of its space data relay network [12].

31.2 FREE-SPACE LINKS INSIDE ELECTRONIC EQUIPMENT

Data centers and warehouse-scale computing environments use optical fibers to connect racks together, and are beginning to deploy rack-level optical communications. This is because the speed of the electrical interconnect required between boards requires complex electronic transmission protocols, high-speed impedance matched lines with consequent high-power requirements due to terminations, and complex expensive multilayer printed circuit boards.

Board-to-board free-space interconnection was widely researched several decades ago. The major challenge is to create free-space links that can align themselves to the tolerances created by the electronic racks in which the transmitter and receiver boards are placed and to create an optical interconnect structure that is compatible with the tolerances and fabrication techniques of electronics.

Broadly, three approaches have been followed to achieve this:

1. Build alignment-free links to match the tolerances available. Such short-range links between boards have been demonstrated. An array of sources and a detector array are used, and the particular source–detector pair that creates a link is selected after the boards are positioned [13]. Other approaches to reduce misalignment using link redundancy have also been reported [14].
2. Build precise, independently stable optical bus structures, with precision connectors to allow alignment of boards into these. This is the approach taken in [15,16], and has the advantage that high-density interconnection can be achieved, albeit at the cost of bulky, expensive components.
3. Use adaptive alignment techniques to compensate for misalignments. This is the approach taken in [17], where a crossbar architecture is used to allow reconfigurable interconnection between a transmitter array and a receiver array. Figure 31.2 shows a schematic of this system. Light from an array

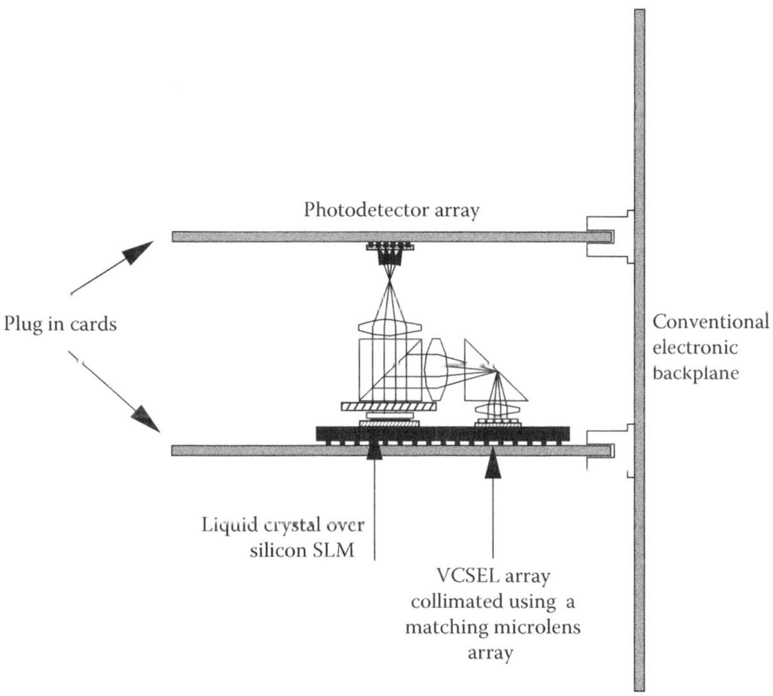

Figure 31.2 Adaptive board-to-board optical interconnect.

of lasers is steered to an array of detectors using a spatial light modulator (SLM). This offers a great degree of flexibility in the interconnection but requires complex costly components. A similar approach, using a different type of liquid crystal beamsteering, is reported in [18].

Work on free-space approaches has been superseded by the use of optical fibers or waveguides. One approach that incorporates a wave-guiding layer in a conventional printed circuit board and uses custom connector interfaces to launch and detect light is reported in [19]. This has the advantage that it is compatible with existing electronic rack-based systems and printed circuit board manufacture. There are a number of methods used to create the waveguide layer, but typically multimode waveguides are used. This approach is gaining widespread acceptance and is likely to be incorporated into backplanes of storage networks and other applications with high bandwidth requirements in the near future.

31.3 INDOOR OPTICAL WIRELESS COMMUNICATIONS

Radiofrequency (RF) wireless systems have been immensely successful, and users now see Wi-Fi and mobile data services as a utility that will always be available. These systems use carrier frequencies in the low gigahertz region of the electromagnetic spectrum, which offers an excellent coverage, due to the diffraction and reflection from the built environment. The key challenge for RF systems is that the available spectrum is limited and not able to support the data rates that will be required in the future. High-frequency RF approaches can provide this capacity, but coverage is more challenging, as the RF carrier propagates in a similar manner as a beam of light.

Optical wireless systems have access to hundreds of terahertz of spectrum, with much of it accessible using low-cost sources and detectors (falling within the sensitivity region of silicon). This is unlicensed and regulated only by the need to consider eye safety. The availability of almost unlimited spectrum is the major attraction of optical wireless systems. The key challenge for optical wireless systems is the link margin that is available, which is much lower than equivalent RF systems, due to the limitations of the receivers used and the non-coherent optical detection.

In this section, we present an introduction to optical wireless systems, covering both the IR and visible regions of the spectrum (where optical wireless communications is known as visible light communications (VLC) or Li-Fi).

31.3.1 Basic configurations

Figure 31.3 shows a schematic of a typical optical wireless system. In each case, an optical source illuminates the room. This illumination (which could be in the IR or visible region of the optical spectrum) is modulated to transmit information. The illumination propagates to the receiver, either via some intermediate reflection from surfaces within the room or via a line of sight. The receiver collects this radiation and converts the optical signal to an electrical signal from which the information can be recovered.

Optical wireless systems can be classified according to whether they use line of sight or diffuse paths. Figure 31.3a shows a diffuse configuration, where good coverage is obtained by allowing light to reflect off intermediate surfaces and create a large number of paths from the transmitter to the receiver. This configuration is robust to blocking of any particular path but has high loss and the different path lengths lead to dispersion. (This is described in more detail in later Section 31.6.4.2). Controlling the beam spread of the optical source restricts paths to line of sight and the configurations in Figure 31.3b through d show various degrees of restriction. Narrow beam configurations allow lower path loss, which allows higher data rate systems, due to the increased power density at the receiver.

31.4 HISTORY

Early work in optical wireless communications focused on diffuse IR systems [20], using LEDs and low-speed photodetectors. In [21], a 50 Mbit/s diffuse system was demonstrated. Architectures suitable for high-speed communications have also been developed, including imaging diversity [22,23], angle diversity [24], and quasi-diffuse architectures, with recent predictions of multi-gigabit systems [25]. More recent systems

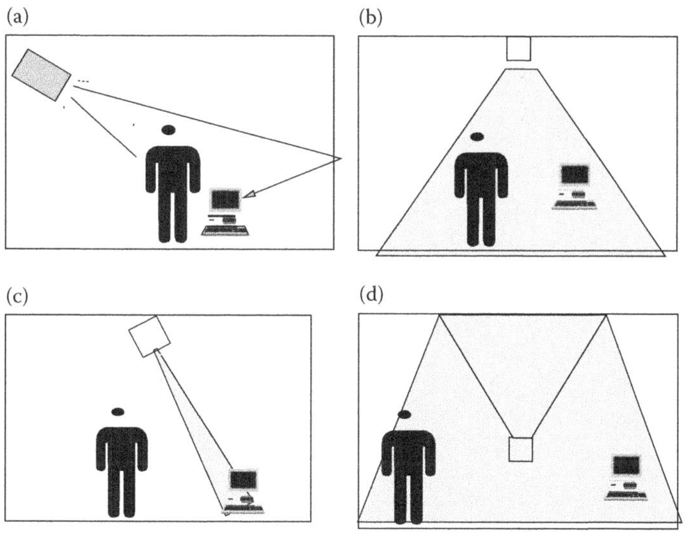

Figure 31.3 Typical room showing optical wireless configurations: (a) Diffuse; (b) wide line of sight; (c) narrow line of sight. (d) Quasi-diffuse.

demonstrations include [26] where Gbit/s communications was demonstrated.

31.4.1 Visible light communications

VLC has been enabled by the adoption of solid-state lighting using LEDs. Lighting LEDs can be modulated at rapid rates compared with fluorescent or incandescent alternatives, offering the potential to combine illumination with communications. VLC originated in Japan, with much early work undertaken by members of the Visible Light Communications Consortium (VLCC) [27–30]. In the past decade, research in this area has grown rapidly, including the Smart Lighting Engineering Research Center in the United States [31], groups in Europe and Asia, and a number of start-up companies [32].

31.5 SYSTEM DESCRIPTION

31.5.1 Transmitter

VLC typically uses white light LEDs as both an illumination and a communications source. White light is generated using either a mixture of light from red, green, and blue (RGB) emitters packaged together or a blue gallium nitride (GaN) emitter that is combined with a yellow inorganic phosphor. The RGB approach has the advantage that the color can be tuned to that desired by weighting the emission of each component appropriately. Each emitter can also be modulated separately, offering the potential for data transmission using wavelength division multiplexing (WDM), or a data symbol constellation using color as a degree of freedom [33,34]. Most commercially available luminaires used for general lighting purposes use a GaN LED that illuminates a phosphor coating applied to the LED chip. This phosphor absorbs a proportion of the blue light and reemits a broad spectrum yellow color. The combination of the blue light that passes through the coating with this yellow light results in a white emission. Figure 31.4 shows a typical emission spectrum, with a narrow blue emission peak from the GaN emitter and a broad yellow emission peak from the phosphor. Typically, the modulation bandwidth of the white emitter is several megahertz, due to the slow phosphor response [35]. Using an optical filter that blocks the yellow component at the receiver allows only the blue channel to be used for information transmission, which increases the bandwidth to 10 MHz or so for a typical commercial device [35].

A number of approaches have been used to improve the bandwidth of the emitter. In [36], an array of LEDs is used, each with their peak optical output frequency tuned to a different point, using an analog equalization technique. This created a white light emitter with a bandwidth of 25 MHz.

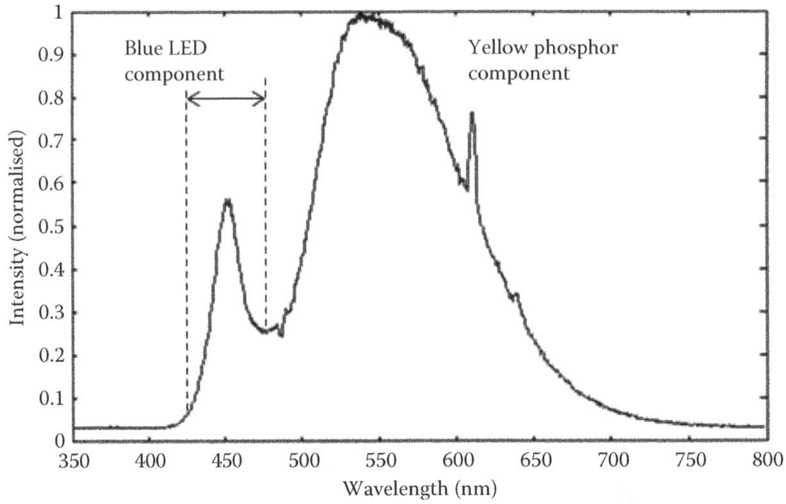

Figure 31.4 Emission spectrum of GaN/phosphor emitter.

Other circuit-based pre-emphasis techniques have been demonstrated, with a 45 MHz bandwidth emitter reported in [37]. The GaN emitter bandwidth can also be increased by reducing the size of the emitter area, and GaN micro-LEDs have been demonstrated with electrical–optical bandwidths of 400 MHz [38].

Some recent work has focused on combining micro-LEDs with blue-to-yellow color converters that have high bandwidth. A "fast-white" LED that uses a GaN micro-LED combined with an organic color converter has recently been reported [39]. In this case, the converter has a bandwidth of several hundred megahertz, which is much greater than that of the micro-LED.

IR systems can use either LEDs or lasers as sources. LEDs provide low-bandwidth, low-cost sources. Lasers can be used to provide high-bandwidth sources that are more efficient than LEDs, at the cost of greater complexity. Eye safety is of concern for emitters in the IR. For regions between visible wavelengths and 1400 nm, the eye can still focus radiation onto the retina, and there is a retinal thermal hazard [40]. This is a function of the emitter power, the position of the eye relative to the emission, and the apparent angular subtense of the emitter (which governs the size of the focused spot on the retina). Hazard evaluation is complex, but for a simple point source, the allowed accessible emission limit (AEL) might be <1 mW at 800 nm. At wavelengths longer than 1400 nm, water absorption stops radiation reaching the retina, and the hazard is that of corneal damage. This leads to a much greater AEL. (As an example, the AEL is 10 mW at 1500 nm.)

High-emission powers that are safe can be achieved in the IR region by using a diffuser to increase the apparent emitting area. Some systems use Lambertian diffusers such as ground glass or opal. More recently, large area holographic diffusers with controlled beam angle have been developed for the displays industry. These provide a low-cost alternative, and in [26], a 180 mW laser-based class 1 eye-safe emitter that uses this approach is reported.

Achieving high data rates using IR systems requires relatively narrow beams in order to provide the power density required at the receiver. Multiple beam, or tracking, systems are therefore required to achieve a wide field of view coverage at these rates. Both angle diversity [24] and imaging diversity [22] architectures can be used to achieve this. Figure 31.5 shows both approaches. In the angle diversity case, sources with a limited field of view are combined to increase the overall value. Imaging receivers achieve the same result by using an array of sources combined with a lens system. In both cases, the transmitter and the receiver both have matched fields of view.

VLC systems typically use multiple sources to illuminate an indoor coverage space, so there is the potential to send different data on each LED or separate lighting fixture. Such optical multiple input multiple output (MIMO) systems are an area of growing interest as they allow increased data

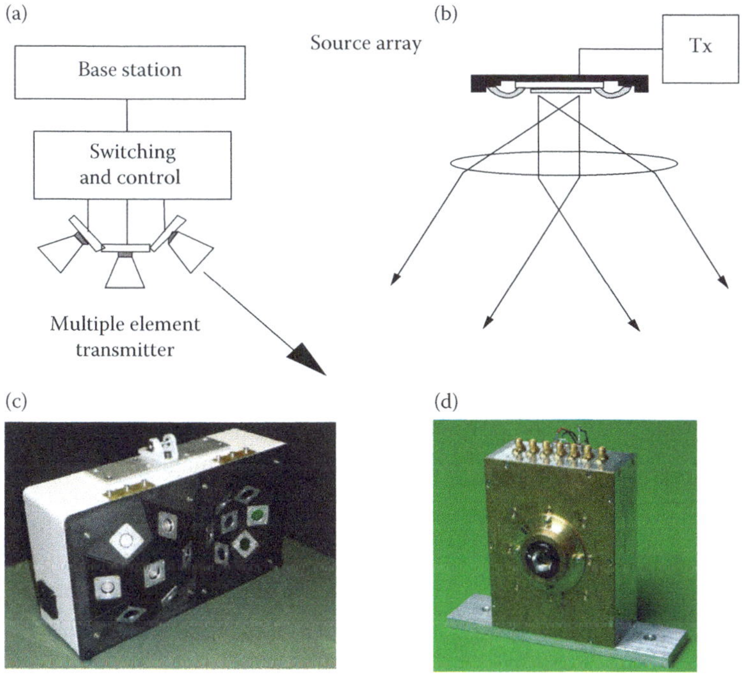

Figure 31.5 Multielement transmitters of different types: (a) Angle diversity transmitter schematic; (b) imaging diversity transmitter schematic; (c) angle diversity terminal showing the receiver (right) and the transmitter (left); (d) imaging diversity transmitter. Note that (c) shows a terminal incorporating both the transmitter and the receiver.

rate compared with a single information stream (see [41] for an early example).

31.5.2 Receiver

Figure 31.6 shows a typical optical receiver. Light passes through an optical filter, and is then focused using either a concentrator or lens system. The resulting output beam then illuminates a photodetector. This converts the optical signal to an electrical output, which is then amplified and passed to a demodulation stage. Each of these elements is described in the following sections.

31.5.1.1 OPTICAL FILTER

White light VLC systems typically use a filter to remove the yellow light (which has low modulation bandwidth) emitted from the phosphor, whereas in IR systems, a bandpass filter is used to only allow the data transmission wavelengths to be detected, thus removing unwanted ambient light. Optical filters use material absorption to remove unwanted optical wavelengths, and these typically have similar performance for a wide range of illumination angles. Interference filters have an excellent rejection of unwanted wavelengths with lower insertion loss, but at the cost of sensitivity to the illumination angle. For a receiver with a wide field of view,

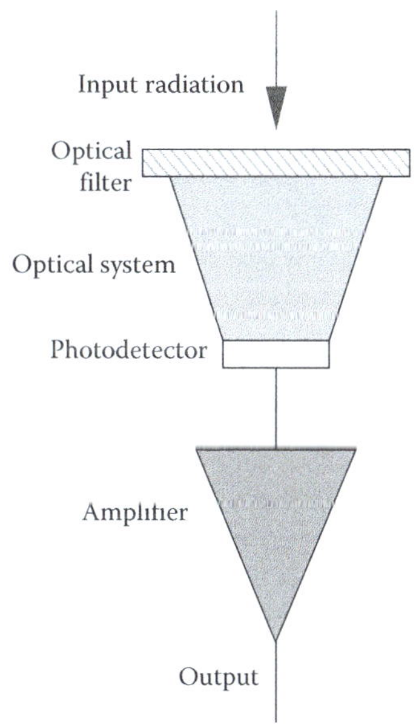

Figure 31.6 Typical optical receiver.

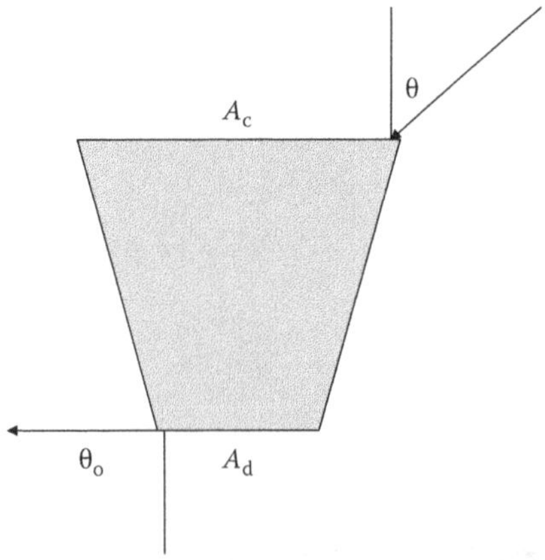

Figure 31.7 Optical concentrator.

this can have a substantial effect, which must be considered in receiver design.

31.5.1.2 CONCENTRATING OPTICS

An ideal receiver would accept light over a wide field of view, have a large collection area, and concentrate it to a small photodetector. Nonimaging optical concentrators (for a recent example, see [42]) provide the most efficient method of achieving this. The performance of all optical systems is constrained by the conservation of *etendue*, and nonimaging concentrators offer performance that can theoretically achieve that predicted by this constraint. Figure 31.7 shows a typical concentrator, with field of view θ and collection area A_c. Conservation of *etendue* leads to

$$\frac{A_c}{A_d} \le \frac{n^2}{\sin^2\theta}$$

where A_d is the detector area and n is the concentrator refractive index.

31.5.1.3 PHOTODETECTOR/PREAMPLIFIER

The combination of photodetector and preamplifier is critical in determining the sensitivity, and hence the overall performance of the communications link. Both positive–intrinsic–negative (PIN) photodiodes and avalanche photodiode types are used. The detector area is a key factor, as increasing the area brings a proportionate increase in collected power. However, this increased area can lead to high capacitance, which limits the bandwidth, and hence the system data rate. An ideal detector would have low capacitance per unit area, and the ideal preamplifier would be tolerant to high input capacitance. There has been some work on optimizing these components [43–45], but substantial gains are still thought to be available.

31.5.3 Multielement receivers

The capacitance and optical gain constraints of single channel receivers can be circumvented by using multiple receivers, using similar approaches as that used in the transmitter. Figure 31.8 shows angle and imaging diversity examples. Angle diversity uses single-channel receivers with a limited field of view (and thus high gain), which are arranged to cover a large range of angles. In imaging diversity receivers, a planar detector array is combined with optics to allow small individual detectors (and therefore high speed), yet have large overall field of view. These receivers can be used for single input single output systems where only one data stream is received. In this case, some form of receiver selection is required, so the best (in some sense) received signal is recovered. For MIMO applications, processing to separate multiple data streams [41] or to ascertain the location of active transmitters [46] has also been developed.

31.5.4 Channel modeling and characterization techniques

The performance of the system is highly dependent on that of the channel, which can be characterized by a measure of path loss and temporal dispersion. Path loss is a function of the transmitter and receiver geometry and relative location in the case of line-of-sight systems. For fully diffuse systems, the environment also determines path loss. There is no dispersion in a line-of-sight system, but where there are multiple paths, incoherent interference between the multiple copies of the data signal leads to intersymbol interference.

31.5.4.1 LINE OF SIGHT

Estimation of the path loss of a line-of-sight link can be achieved by ray tracing if models of the transmitter and receiver structures are used or by analytical techniques. Typically, a higher order Lambertian model of the emitter output beam

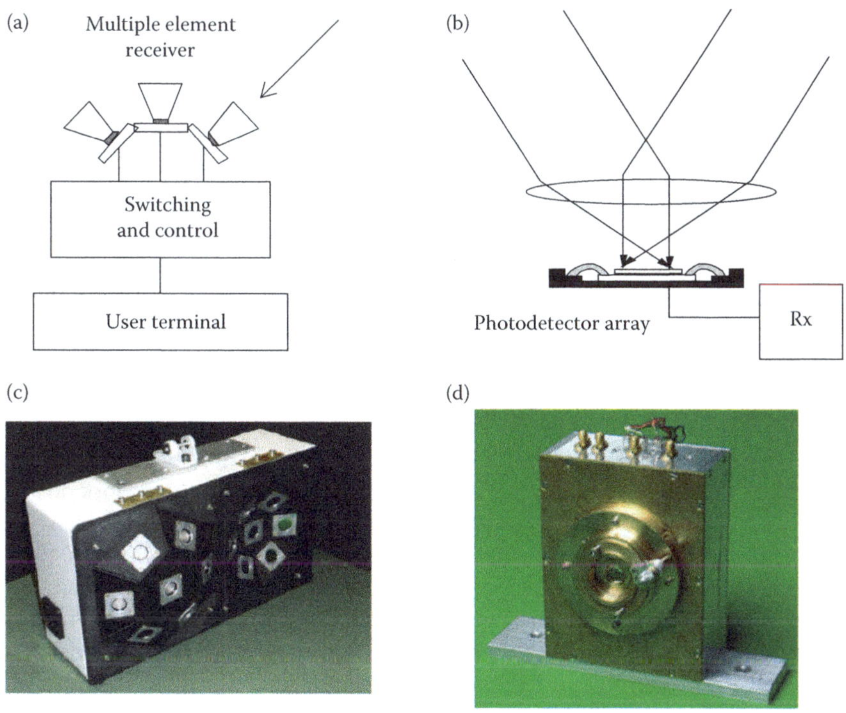

Figure 31.8 Multielement receivers of different types: (a) Angle diversity receiver schematic; (b) imaging diversity receiver schematic; (c) angle diversity terminal showing the receiver (right) and the transmitter (left); (d) imaging diversity receiver. Note that (c) shows a terminal that incorporates both the transmitter and the receiver.

profile is assumed, and this allows the intensity at the receiver to be determined. Knowledge of the receiver optical system, typically, the concentrator gain, and the photodetector area then allows link loss to be calculated. A full calculation is shown in [26]. The path loss increases rapidly with an increasing field of view in line-of-sight links, and for IR systems, the margins available mean that high-speed links (Gbit/s and greater) require more than one transmitter and receiver to achieve good coverage.

31.5.4.2 DIFFUSE

Diffuse channel models based on ray tracing (see, for instance, [47]) and integrating sphere [48] have been developed. In a fully diffuse environment, all surfaces are evenly illuminated, and typical losses are in the range of 65–80 dB cm^2 Sr^{-1} [48,49]. Bandwidths as low as 10 MHz have been measured for a fully diffuse environment [48], but for Rician-like channels with a dominant path, there may be no channel limitation due to dispersion (such "transparency" is noted in [48] and measured in [50]).

The importance of channel modeling and characterization depends on the particular application and wavelength regime used. For VLC, the power levels are set by illumination constraints, and for a well-illuminated space, a level of 400 lux is required. Using a communications system with a typical PIN receiver, this leads to very high received signal-to-noise ratio (SNR > 40dB) and no detailed channel modeling is required to determine path loss.The bandwidth of the VLC channel is typically in the hundreds of megahertz [51], and as this is far greater than the bandwidth available from the LED transmitters, the channel dispersion is not significant and therefore does not require detailed modeling.

For the IR region, the power levels used are typically much lower, and path loss is an important factor that requires detailed investigation.

31.5.5 Modulation

Optical wireless systems typically use intensity modulation combined with direct detection, and both pulse-based modulation, and linear modulation schemes can be used over such a channel. On–off keying offers simple transmission and detection circuitry and high spectral efficiency in channels with modest SNR. It has been used extensively, including in the fastest system demonstrations [26]. There has also been extensive investigation of pulse position modulation (PPM) and variants [52,53].

The most used scheme for VLC is orthogonal frequency division multiplexing (OFDM) (see, for instance, [54,55]). There are many different variants of this scheme, but all use Hermitian symmetry to create real-valued data for transmission. The VLC channel has very high SNR at low frequencies, leading to usable SNRs much beyond the 3dB bandwidth of the channel (including transmitter and receiver). OFDM allows each carrier to be optimized to the SNR available, and therefore efficiently uses the resources available. In addition, it can easily combat any dispersion that there may be. There has been extensive research in how to compensate for the nonlinearities in LED transmitters, accommodate dimming, and provide multiple access. This has led to several high-rate data transmission demonstrations, including [56,57].

31.6 DEMONSTRATIONS, COMMERCIAL PRODUCTS, AND FUTURE PROSPECTS

31.6.1 IR

IR remote control is ubiquitous, and diffuse IR finds limited use in some shelf-labeling systems (see, for example, [58]). There has been recent interest in very high-speed short-range links [59]. Links operating at 1500 nm, using telecommunications components, have also been demonstrated. In [60], a WDM link operating at ~50 Gbit/s is demonstrated. Fiber-to-fiber links using orbital angular momentum representations have achieved 100 Tbit/s data rates [61]. In [62], a fiber-to-fiber link with a field of view of 60° and a rate of ~100 Gbits/s is reported.

31.6.2 Visible light communications

VLC was initiated by members of the VLCC. A wide range of demonstrations have been reported, including information broadcast from luminaires, display screens, music broadcast, and long-range point-to-point communications. Short-range high-speed links have also been used to connect peripherals [63]. A full-scale demonstration of information broadcasting within a room was reported as part of a European Community-funded project [64]. Recently, a number of companies have begun to commercialize this technology (see, for instance, [32,65–67]). The use of cellphone cameras as software-defined communications receivers has been demonstrated by a number of different groups (see, for instance, [68]). A number of approaches to mitigate the effects of the slow camera frame rate and the order in which pixels are interrogated have been developed, and a special interest group has started work on a revision to the Institute of Electrical and Electronics Engineers standard on VLC (IEEE 802.15.7 [69]).

Positioning and location-based information systems are an area of growing interest, as systems relying on the Global Positioning System do not operate reliably indoors. The use of luminaires as positioning beacons is being pursued by both a number of academic groups (see [70] for an early example) and commercial organizations [71,72].

31.6.3 Challenges and future prospects

Figure 31.9 shows a possible future wireless system for an indoor environment. RF wireless is ubiquitous and can provide reliable broad area coverage. VLC provides an excellent downlink capacity and can support a very high user density; a provision of 1–10 Gbit/s per user is likely to be achievable using this approach. For rates higher than this point-to-point IR links using telecommunications components is an attractive approach, as the challenge of interfacing the wired and wireless network is much reduced compared with a distributed VLC infrastructure.

In all cases, it is the cooperation between different wireless networks that will allow the user to obtain the data rates they require. Widespread adoption of this approach will require research at the network, transceiver, and individual component level, together with a more widespread adoption of LED lighting in the case of VLC and "fiber to the desk" in the case of IR-based systems.

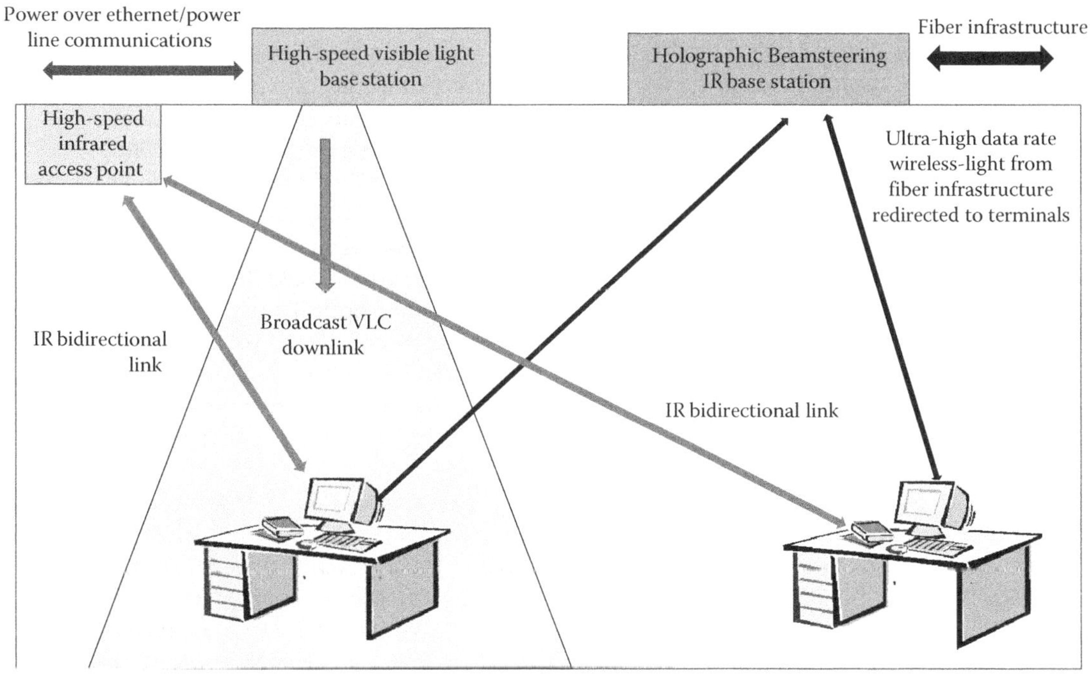

Figure 31.9 Future cooperative wireless landscape.

ACKNOWLEDGMENTS

The author expresses his thanks to the members of the Optical Wireless Communications Group at Oxford and our current and past collaborators for their contributions to the work described in this chapter. Special thanks are due to Dr. Sujan Rajbhandari, who read and commented on a draft.

REFERENCES

1. Photophone. (Accessed 2015). Wikipedia entry. https://en.wikipedia.org/wiki/Photophone.
2. Cablefree. (Accessed 2015). www.cablefree.net.
3. Lightpointe. (Accessed 2015). www.lightpointe.com.
4. M. Gebhart, E. Leitgeb, S. Sheikh Muhammad, B. Flecker, C. Chlestil, M. Al Naboulsi, F. de Fornel, and H. Sizun, Measurement of light attenuation in dense fog conditions for FSO applications,SPIE, pp. 58910K-1–58910K-12, 2005.
5. L. C. Andrews, and R. L. Phillips, *Laser Propagation through Random Media.* Bellingham, WA: SPIE, 2005.
6. G. Parca, A. Shahpari, V. Carrozzo, G. M. Tosi Beleffi, and A. L. J. Teixeira, Optical wireless transmission at 1.6-Tbit/s (16×100 Gbit/s) for next-generation convergent urban infrastructures, *Optical Engineering*, vol. 52, pp. 116102–116102, 2013.
7. M. J. Northcott, A. McClaren, J. E. Graves, J. Phillips, D. Driver, D. Abelson, D. W. Young, J. E. Sluz, J. C. Juarez, M. B. Airola, R. M. Sova, H. Hurt, and J. Foshee, Long distance laser communications demonstration, SPIE, pp. 65780S-1–65780S-8, 2007.
8. W. S. Rabinovich, C. I. Moore, H. R. Burris, J. L. Murphy, M. R. Suite, R. Mahon, M. S. Ferraro, P. G. Goetz, L. M. Thomas, C. Font, G. C. Gilbreath, B. Xu, S. Binari, K. Hacker, S. Reese, W. T. Freeman, S. Frawley, E. Saint-Georges, S. Uecke, and J. Sender, Free space optical

communications research at the U.S. Naval Research Laboratory, SPIE, vol. 7587, pp. 758702-1–758702-15, 2010.
9. C. Quintana, G. Erry, A. Gomez, Y. Thueux, G. Faulkner, and D. C. O'Brien, Novel non-mechanical fine tracking module for retroreflective free space optics, in *SPIE Security and Defence Europe*, SPIE, Amsterdam, 2014.
10. Z. Sodnik, B. Furch, and H. Lutz, Optical intersatellite communication, *IEEE Journal of Selected Topics in Quantum Electronics*, vol. 16, pp. 1051–1057, 2010.
11. D. M. Boroson, B. S. Robinson, D. V. Murphy, D. A. Burianek, F. Khatri, J. M. Kovalik, Z. Sodnik, and D. M. Cornwell, Overview and results of the lunar laser communication demonstration, SPIE, pp. 89710S-1–89710S-11, 2014.
12. F. Heine, G. Mühlnikel, H. Zech, D. Tröndle, S. Seel, M. Motzigemba, R. Meyer, S. Philipp-May, and E. Benzi, LCT for the European data relay system: In orbit commissioning of the Alphasat and Sentinel 1A LCTs, SPIE, vol. 9354, pp. 93540G-1–93540G-6, 2015.
13. D. J. Goodwill, D. Kabal, and P. Palacharla, Free space optical interconnect at 1.25 Gb/s/channel using adaptive alignment, in *Optical Fiber Communication Conference, 1999, and the International Conference on Integrated Optics and Optical Fiber Communication. OFC/IOOC '99. Technical Digest*, vol. 2, pp. 259–261, 1999.
14. G. C. Boisset, B. Robertson, and H. S. Hinton, Design and construction of an active alignment demonstrator for a free-space optical interconnect, *IEEE Photonics Technology Letters*, vol. 7, pp. 676–678, 1995.
15. D. V. Plant, B. Robertson, H. S. Hinton, M. H. Ayliffe, G. C. Boisset, D. J. Goodwill, D. Kabal, R. Iyer, Y. S. Liu, D. R. Rolston, H. Venditti, T. H. Szmanski, W. M. Robertson, and M. R. Taghizadeh, Optical, optomechanical, and optoelectronic design and operational testing of a multi-stage optical backplane demonstration system, in *Proceedings of the Third International Conference on Massively Parallel Processing Using Optical Interconnections*, 1996, pp. 306–312, 1996.
16. D. V. Plant, M. B. Venditti, E. Laprise, J. Faucher, K. Razavi, M. Chateauneuf, A. G. Kirk, and J. S. Ahearn, 256-channel bidirectional optical interconnect using VCSELs and photodiodes on CMOS, *Journal-of-Lightwave-Technology*, vol. 19, pp. 1093–1103, 2001.
17. T. D. Wilkinson, C. Henderson, D. Gil-Leyva, B. Robertson, D. O'Brien, and G. Faulkner, Adaptive optical interconnect using an FLC SLM, *Ferroelectrics*, vol. 312, pp. 81–85, 2004.
18. K. Hirabayashi, T. Yamamoto, S. Matsuo, and S. Hino, Board-to-board free-space optical interconnections passing through boards for a bookshelf-assembled terabit-per-second-class ATM switch, *Applied-Optics*, vol. 37, pp. 2985–2995, 1998.
19. N. Bamiedakis, C. Jian, P. Westbergh, J. S. Gustavsson, A. Larsson, R. V. Penty, and I. H. White, 40 Gb/s Data transmission over a 1-m-Long multimode polymer spiral wave-guide for board-level optical interconnects, *Journal of Lightwave Technology*, vol. 33, pp. 882–888, 2015.
20. F. R. Gfeller, and U. Bapst, Wireless in-house data communication via diffuse infrared radiation, *Proceedings of the IEEE*, vol. 67, pp. 1474–1486, 1979.
21. G. W. Marsh, and J. M. Kahn, 50-Mb/s diffuse infrared free-space link using on-off keying with decision-feedback equalization, *IEEE Photonics Technology Letters*, vol. 6, pp. 1268–1270, 1994.
22. G. Yun, and M. Kavehrad, Spot-diffusing and fly-eye receivers for indoor infrared wireless communications, in *Proceedings of IEEE International Conference on Selected Topics in Wireless Communications*, 1992, pp. 262–265, 1992.
23. A. P. Tang, J. M. Kahn, and K.-P. Ho, Wireless infrared communication links using multi-beam transmittersand imaging receivers, in *IEEE International Conference on Communications*, Dallas, pp. 180–186, 1996.
24. J. B. Carruthers and J. M. Kahn, Angle diversity for nondirected wireless infrared communication, in *Communications, 1998. ICC 98. Conference Record. 1998 IEEE International Conference on* 1998, vol. 3, pp. 1665–1670, 1998.

25. M. T. Alresheedi, and J. M. H. Elmirghani, Performance evaluation of 5 Gbit/s and 10 Gbit/s mobile optical wirelesssystems employing beam angle and power adaptation with diversity receivers, *IEEE Journal on Selected Areas in Communications*, vol. 29, pp. 1328–1340, 2011.
26. D. C. O'Brien, R. Turnbull, H. Le Minh, G. Faulkner, O. Bouchet, P. Porcon, M. El Tabach, E. Gueutier, M. Wolf, L. Grobe, and J. H. Li, High-speed optical wirelessdemonstrators: Conclusions and future directions, *Journal of Lightwave Technology*, vol. 30, pp. 2181–2187, 2012.
27. vlcc. (2008). *Visible Light Communications Consortium*. www.vlcc.net.
28. M. Akanegawa, Y. Tanaka, and M. Nakagawa, Basic study on traffic information system using LED traffic lights, *IEEE Transactions on Intelligent Transportation Systems*, vol. 2, pp. 197–203, 2001.
29. S. Kitano, S. Haruyama, and M. Nakagawa, LED road illumination communications system, in *2003 IEEE 58th Vehicular Technology Conference. VTC 2003 Fall. Orlando, FL. 6 9 Oct. 2003.*, IEEE, Piscataway, NJ, USA, 2003.
30. T. Komine and M. Nakagawa, Fundamental analysis for visible-light communication system using LED lights, *IEEE Transactions on Consumer Electronics*, vol. 50, pp. 100 107, 2004.
31. Smartlighting. (Accessed 2015). *Smartlighting Engineering Research Center.* www.smartlighting.org.
32. PureLifi. (Accessed 2015). www.purelifi.com.
33. R. J. Drost and B. M. Sadler, Constellation design for color-shift keying using billiards algorithms, in *GLOBECOM Workshops (GC Wkshps), 2010 IEEE*, pp. 980–984, 2010.
34. R. Singh, T. O'Farrell, and J. David, An enhanced colour shift keying modulation scheme for high speed wireless visible light communications, *Journal of Lightwave Technology*, vol. 32, no 14, pp. 2582–2592, 2014.
35. H. Le Minh, D. O'Brien, G. Faulkner, L. B. Zeng, K. Lee, D. Jung, Y. Oh, and E. T. Won, 100-Mb/s NRZ visible light communications using a postequalized white LED, *IEEE Photonics Technology Letters*, vol. 21, pp. 1063–1065, 2009.
36. H. Le-Minh, D. O'Brien, G. Faulkner, L. Zeng, K. Lee, D. Jung, and Y. Oh, High-speed visible light communications using multiple-resonant equalization, *IEEE Photonics Technology Letters*, vol. 20, pp. 1243–5, 2008.
37. H. Le-Minh, D. O'Brien, G. Faulkner, L. Zeng, L. Kyungwoo, J. Daekwang, and O. YunJe, 80 Mbit/s visible light communications using pre-equalized white LED, in *Optical Communication, 2008. ECOC 2008. 34th European Conference on* 2008, pp. 1–2, 2008.
38. Z. Shuailong, S. Watson, J. J. D. McKendry, D. Massoubre, A. Cogman, G. Erdan, R. K. Henderson, A. E. Kelly, and M. D. Dawson, 1.5 Gbit/s multi-channel visible light communications using CMOS-controlled GaN-based LEDs, *Journal of Lightwave Technology*, vol. 31, pp. 1211–1216, 2013.
39. H. Chun, P. Manousiadis, S. Rajbhandari, D. A. Vithanage, G. Faulkner, D. Tsonev, J. J. D. McKendry, S. Videv, E. Xie, E. Gu, M. D. Dawson, H. Haas, G. A. Turnbull, I. D. W. Samuel, and D. O'Brien, Visible light communication using a blue GaN μLED and fluorescent polymer colour converter, *IEEE Photonics Technology Letters*, [Under reivew].
40. *Safety of laser products-Part 1: Equipment classification and requirements*, BSI, 2007.
41. K. Dambul, D. C. O'Brien, and G. Faulkner, Indoor optical wireless MIMO System with an Imaging Receiver, *IEEE Photonics Technology Letters*, vol. 23, pp. 97–99, 2011.
42. H. Al Hajjar, B. Fracasso, and F. Lamarque, Mini optical concentrator design for indoor high bit rate optical wireless communications, in *Optical Wireless Communications (IWOW), 2013 2nd International Workshop on*, 2013, pp. 147–151.
43. J. Zeng, V. Joyner, L. Jun, D. Shengling, and H. Zhaoran, A 5Gb/s 7-channel current-mode imaging receiver front-end for free-space optical MIMO, in *Circuits and Systems, 2009. MWSCAS '09. 52nd IEEE International Midwest Symposium on* 2009, pp. 148–151, 2009.
44. Z. Yiling and V. Joyner, An analog front-end receiver with desensitization to input capacitance for free space optical

communication, in *Communication Systems, Networks and Digital Signal Processing, 2008. CNSDSP 2008. 6th International Symposium on* 2008, pp. 183– 186, 2008.
45. D. O'Brien, G. Faulkner, K. Jim, D. J. Edwards, E. B. Zyambo, P. Stavrinou, G. Parry, J. Bellon, M. J. Sibley, R. J. Samsudin, D. M. Holburn, V. A. Lalithambika, V. M. Joyner, and R. J. Mears, Experimental characterization of integrated optical wireless components, *IEEE Photonics Technology Letters*, vol. 18, pp. 977–979, 2006.
46. W. O. Popoola and H. Haas, Demonstration of the merit and limitation of generalised space shift keying for indoor visible light communications, *Journal of Lightwave Technology*, vol. 32, pp. 1960–1965, 2014.
47. O. Gonzalez, C. Militello, S. Rodriguez, R. Prez-Jimenez, and A. Ayala, Error estimation of the impulse response on diffuse wireless infrared indoor channels using a Monte Carlo ray-tracing algorithm, *IEE Proceedings-Optoelectronics*, vol. 149, pp. 222–227, 2002.
48. V. Jungnickel, V. Pohl, S. Nonnig, and C. von Helmolt, A physical model of the wireless infrared communication channel, *IEEE Journal on Selected Areas in Communications*, vol. 20, pp. 631–640, 2002.
49. J. M. Kahn, W. J. Krause, and J. B. Carruthers, Experimental characterization of non-directed indoor infrared channels, *IEEE Transactions on Communications*, vol. 43, pp. 1613–1623, 1995.
50. D. P. Manage, G. E. Faulkner, D. C. O'Brien, and D. J. Edwards, A novel system for the imaging of optical multipaths, in *Proceedings of the SPIE The International Society for Optical Engineering*, 2004, pp. 47–54.
51. J. Grubor, S. Randel, K.-D. Langer, and J. W. Walewski, Broadband information broadcasting using LED-based interior lighting, *Journal of Lightwave Technology*, vol. 26, pp. 3883–3892, 2008.
52. M. D. Audeh, and J. M. Kahn, Performance evaluation of L-pulse-position modulation on non-directed indoor infrared channels, in *Communications, 1994. ICC '94, SUPERCOMM/ICC '94, Conference Record, 'Serving Humanity Through Communications.' IEEE International Conference on* 1994, vol. 2, pp. 660–664, 1994.
53. H. Chan, K. L. Sterckx, J. M. H. Elmirghani, and R. A. Cryan, Performance of optical wireless OOK and PPM systems under the constraints of ambient noise and multipath dispersion, *IEEE Communications Magazine*, vol. 36, pp. 83–87, 1998.
54. M. Z. Afgani, H. Haas, H. Elgala, and D. A. K. D. Knipp, Visible light communication using OFDM, in *Testbeds and Research Infrastructures for the Development of Networks and Communities, 2006. TRIDENTCOM 2006. 2nd International Conference on*, 2006, p. 6.
55. J. Armstrong, OFDM for Optical Communications, *Journal of Lightwave Technology*, vol. 27, pp. 189–204, 2009.
56. K.-D. Langer, D. S. Jonas Hilt, F. Lassak, F. Hartlieb, C. Kottke, L. Grobe, V. Jungnickel, and A. Paraskevopoulos, Rate-adaptive visible light communication at 500Mb/s arrives at plug and play, in *SPIE*, 2013. SPIE Newsroom. DOI: 10.1117/2.1201311.005196
57. A. M. Khalid, G. Cossu, R. Corsini, P. Choudhury, and E. Ciaramella, 1-Gb/s transmission over a phosphorescent white LED by using rate-adaptive discrete multitone modulation, *IEEE Photonics Journal*, vol. 4, pp. 1465–1473, 2012.
58. P. s. labelling. (Accessed 2015). www.pricer.com.
59. M. Faulwasser, F. Deicke, and T. Schneider, 10 Gbit/s bidirectional optical wireless communication module for docking devices, in *Globecom Workshops (GC Wkshps), 2014*, 2014, pp. 512–517.
60. W. Ke, A. Nirmalathas, C. Lim, and E. Skafidas, 4×12.5 Gb/s WDM optical wirelesscommunication system for indoor applications, *Journal of Lightwave Technology*, vol. 29, pp. 1988–1996, 2011.
61. H. Hao, X. Guodong, Y. Yan, N. Ahmed, R. Yongxiong, Y. Yang, D. Rogawski, M. Tur, B. Erkmen, K. Birnbaum, S. Dolinar, M. Lavery, M. J. Padgett, and A. E. Willner,

100 Tbit/s free-space data link using orbital angular momentum mode division multiplexing combined with wavelength division multiplexing, in *Optical Fiber Communication Conference and Exposition and the National Fiber Optic Engineers Conference (OFC/NFOEC), 2013*, 2013, pp. 1–3, 2013.
62. A. Gomez, S. Kai, C. Quintana, M. Sato, G. Faulkner, B. C. Thomsen, and D. O'Brien, Beyond 100-Gb/s indoor wide field-of-view optical wirelesscommunications,*IEEE Photonics Technology Letters*, vol. 27, pp. 367–370, 2015.
63. S. Hongseok, S. B. Park, D. K. Jung, Y. M. Lee, S. Seoksu, and P. Jinwoo, VLC transceiver design for short-range wireless data interfaces, in *ICT Convergence (ICTC), 2011 International Conference on*, 2011, pp. 689–690.
64. Omega.(Accessed 2015). *Omega Home Area Network Project.* http://cordis.europa.eu/project/rcn/85271_en.html.
65. Nakagawa. (Accessed 2015). *Nakagawa Laboratories.* http://www.naka-lab.jp/index_e.html.
66. Oledcomm. (Accessed 2015) www.oledcomm.com.
67. Luciom. (Accessed 2015).www.luciom.com.
68. R. D. Roberts, Undersampled frequency shift ON-OFF keying (UFSOOK) for camera communications (CamCom), in *Wireless and Optical Communication Conference (WOCC), 2013 22nd*, pp. 645–648, 2013.
69. R. D. Roberts, S. Rajagopal, and L. Sang-Kyu, IEEE 802.15.7 physical layer summary, in *GLOBECOM Workshops (GC Wkshps), 2011 IEEE*, 2011, pp. 772–776.
70. M. Yoshino, S. Haruyama, and M. Nakagawa, High-accuracy positioning system using visible LED lights and image sensor, in *Radio and Wireless Symposium, 2008 IEEE*, 2008, pp. 439–442.
71. Bytelight. (Accessed 2015) http://hydrel.acuitybrands.com/sitecore/content/acuity-brandscorporate/home/solutions/services/bytelight-services-indoor-positioning.
72. Casio. (Accessed 2015) http://picalico.casio.com/en/.

Index

B

C

D

#0006 - 230917 - C464 - 254/178/25 [27] - CB - 9781138102262